SPACETIME AND GEOMETRY

Spacetime and Geometry

An Introduction to General Relativity

Sean Carroll

Pearson

Original Edition, Entitled Spacetime and Geometry: An Introduction to General Relativity, 1st Edition, by Carroll, Sean, published by Pearson Education, Limited.

ISBN 978-93-325-7165-5

First Impression, 2016

Published by Pearson India Education Services Pvt.Ltd,CIN:U72200TN2005PTC057128.
Formerly known as TutorVista Global Pvt Ltd, licensees of Pearson Education in South Asia

Head Office: 7th Floor, knowledge Boulevard, A-8(A) Sector-62, Noida (U.P) 201309, India

Registered Office: 4th floor, Software Block, Elnet Software City, TS 140 Block 2 & 9, Rajiv Gandhi Salai, Taramani, Chennai - 600 113, Tamil Nadu, Fax: 080-30461003, Phone: 080-30461060, www.pearson.co.in Email: companysecretary.india@pearson.com

Digitally printed in India by Saurabh Printers Pvt. Ltd. in the year of 2017.

"For if each Star is little more a mathematical Point, located upon the Hemisphere of Heaven by Right Ascension and Declination, then all the Stars, taken together, tho' innumerable, must like any other set of points, in turn represent some single gigantick Equation, to the mind of God as straightforward as, say, the Equation of a Sphere,—to us unreadable, incalculable. A lonely, uncompensated, perhaps even impossible Task,—yet some of us must ever be seeking, I suppose."

—Thomas Pynchon, *Mason & Dixon*

Preface

General relativity is the most beautiful physical theory ever invented. It describes one of the most pervasive features of the world we experience—gravitation—in terms of an elegant mathematical structure—the differential geometry of curved spacetime—leading to unambiguous predictions that have received spectacular experimental confirmation. Consequences of general relativity, from the big bang to black holes, often get young people first interested in physics, and it is an unalloyed joy to finally reach the point in one's studies where these phenomena may be understood at a rigorous quantitative level. If you are contemplating reading this book, that point is here.

In recent decades, general relativity (GR) has become an integral and indispensable part of modern physics. For a long time after it was proposed by Einstein in 1916, GR was counted as a shining achievement that lay somewhat outside the mainstream of interesting research. Increasingly, however, contemporary students in a variety of specialties are finding it necessary to study Einstein's theory. In addition to being an active research area in its own right, GR is part of the standard syllabus for anyone interested in astrophysics, cosmology, string theory, and even particle physics. This is not to slight the more pragmatic uses of GR, including the workings of the Global Positioning System (GPS) satellite network.

There is no shortage of books on GR, and many of them are excellent. Indeed, approximately thirty years ago witnessed the appearance of no fewer than three books in the subject, each of which has become a classic in its own right: those by Weinberg (1972), Misner, Thorne, and Wheeler (1973), and Hawking and Ellis (1975). Each of these books is suffused with a strongly-held point of view advocated by the authors. This has led to a love-hate relationship between these works and their readers; in each case, it takes little effort to find students who will declare them to be the best textbook ever written, or other students who find them completely unpalatable. For the individuals in question, these judgments may very well be correct; there are many different ways to approach this subject.

The present book has a single purpose: to provide a clear introduction to general relativity, suitable for graduate students or advanced undergraduates. I have attempted to include enough material so that almost any one-semester introductory course on GR can find the appropriate subjects covered in the text, but not too much more than that. In particular, I have tried to resist the temptation to write a comprehensive reference book. The only goal of this book is to teach you GR.

An intentional effort has been made to prefer the conventional over the idiosyncratic. If I can be accused of any particular ideological bias, it would be a tendency to think of general relativity as a field theory, a point of view that helps one to appreciate the connections among GR, particle physics, and string theory. At the same time, there are a number of exciting astrophysical applications of GR (black holes, gravitational lensing, the production and detection of gravitational waves, the early universe, the late universe, the cosmological constant), and I have endeavored to include at least enough background discussion of these issues to prepare students to tackle the current literature.

The primary question facing any introductory treatment of general relativity is the level of mathematical rigor at which to operate. There is no uniquely proper solution, as different students will respond with different levels of understanding and enthusiasm to different approaches. Recognizing this, I have tried to provide something for everyone. I have not shied away from detailed formalism, but have also attempted to include concrete examples and informal discussion of the concepts under consideration. Much of the most mathematical material has been relegated to the Appendices. Some of the material in the Appendices is actually an integral part of the course (for example, the discussion of conformal diagrams), but an individual reader or instructor can decide just when it is appropriate to delve into them; signposts are included in the body of the text.

Surprisingly, there are very few formal prerequisites for learning general relativity; most of the material is developed as we go along. Certainly no prior exposure to Riemannian geometry is assumed, nor would it necessarily be helpful. It would be nice to have already studied some special relativity; although a discussion is included in Chapter 1, its purpose is more to review the basics and and introduce some notation, rather than to provide a self-contained introduction. Beyond that, some exposure to electromagnetism, Lagrangian mechanics, and linear algebra might be useful, but the essentials are included here.

The structure of the book should be clear. The first chapter is a review of special relativity and basic tensor algebra, including a brief discussion of classical field theory. The next two chapters introduce manifolds and curvature in some detail; some motivational physics is included, but building a mathematical framework is the primary goal. General relativity proper is introduced in Chapter 4, along with some discussion of alternative theories. The next four chapters discuss the three major applications of GR: black holes (two chapters), perturbation theory and gravitational waves, and cosmology. Each of these subjects has witnessed an explosion of research in recent years, so the discussions here will be necessarily introductory, but I have tried to emphasize issues of relevance to current work. These three applications can be covered in any order, although there are interdependencies highlighted in the text. Discussions of experimental tests are sprinkled through these chapters. Chapter 9 is a brief introduction to quantum field theory in curved spacetime; this is not a necessary part of a first look at GR, but has become increasingly important to work in quantum gravity and cosmology, and therefore deserves some mention. On the other hand, a few topics are scandalously neglected; the initial-value problem and cosmological perturbation theory come to mind, but there are others. Fortunately there is no shortage of other resources. The Appendices serve various purposes: There are discussions of technical points that were avoided in the body of the book, crucial concepts that could have been put in various places, and extra topics that are useful but outside the main development.

Since the goal of the book is pedagogy rather than originality, I have often leaned heavily on other books (listed in the bibliography) when their expositions seemed perfectly sensible to me. When this leaning was especially heavy, I have indicated it in the text itself. It will be clear that a primary resource was the book by Wald (1984), which has become a standard reference in the field; readers of this book will hopefully be well-prepared to jump into the more advanced sections of Wald's book.

This book grew out of a set of lecture notes that were prepared when I taught a course on GR at MIT. These notes are available on the web for free, and will continue to be so; they will be linked to the website listed below. Perhaps a little over half of the material here is contained in the notes, although the advantages of owning the book (several copies, even) should go without saying.

Countless people have contributed greatly both to my own understanding of general relativity and to this book in particular—too many to acknowledge with any hope of completeness. Some people, however, deserve special mention. Ted Pyne learned the subject along with me, taught me a great deal, and collaborated with me the first time we taught a GR course, as a seminar in the astronomy department at Harvard; parts of this book are based on our mutual notes. Nick Warner taught the course at MIT from which I first learned GR, and his lectures were certainly a very heavy influence on what appears here. Neil Cornish was kind enough to provide a wealth of exercises, many of which have been included at the end of each chapter. And among the many people who have read parts of the manuscript and offered suggestions, Sanaz Arkani-Hamed was kind enough to go through the entire thing in great detail.

I would also like to thank everyone who either commented in person or by email on different parts of the book; these include Tigran Aivazian, Teodora Beloreshka, Ed Bertschinger, Patrick Brady, Peter Brown, Jennifer Chen, Michele Ferraz Figueiró, Eanna Flanagan, Jacques Fric, Ygor Geurts, Marco Godina, Monica Guica, Jim Hartle, Tamás Hauer, Daniel Holz, Ted Jacobson, Akash Kansagra, Chuck Keeton, Arthur Kosowsky, Eugene Lim, Jorma Louko, Robert A. McNees, Hayri Mutluay, Simon Ross, Itai Seggev, Robert Wald, and Barton Zwiebach. Apologies are due to anyone I may have neglected to mention. And along the way I was fortunate to be the recipient of wisdom and perspective from numerous people, including Shadi Bartsch, George Field, Deryn Fogg, Ilana Harrus, Gretchen Helfrich, Mari Ruti, Maria Spiropulu, Mark Trodden, and of course my family. (This wisdom often came in the form, "What were you thinking?") Finally, I would like to thank the students in my GR classes, on whom the strategies deployed here were first tested, and express my gratitude to my students and collaborators, for excusing my book-related absences when I should have been doing research.

My friends who have written textbooks themselves tell me that the first printing of a book will sometimes contain mistakes. In the unlikely event that this happens here, there will be a list of errata kept at the website for the book:

`http://spacetimeandgeometry.net/`

The website will also contain other relevant links of interest to readers.

During the time I was working on this book, I was supported by the National Science Foundation, the Department of Energy, the Alfred P. Sloan Foundation, and the David and Lucile Packard Foundation.

Sean Carroll
Chicago, Illinois
June 2003

Contents

CHAPTER 1

Special Relativity and Flat Spacetime

1.1 ■ PRELUDE

General relativity (GR) is Einstein's theory of space, time, and gravitation. At heart it is a very simple subject (compared, for example, to anything involving quantum mechanics). The essential idea is perfectly straightforward: while most forces of nature are represented by fields defined on spacetime (such as the electromagnetic field, or the short-range fields characteristic of subnuclear forces), gravity is inherent in spacetime itself. In particular, what we experience as "gravity" is a manifestation of the *curvature* of spacetime.

Our task, then, is clear. We need to understand spacetime, we need to understand curvature, and we need to understand how curvature becomes gravity. Roughly, the first two chapters of this book are devoted to an exploration of spacetime, the third is about curvature, and the fourth explains the relationship between curvature and gravity, before we get into applications of the theory. However, let's indulge ourselves with a short preview of what is to come, which will perhaps motivate the initial steps of our journey.

GR is a theory of gravity, so we can begin by remembering our previous theory of gravity, that of Newton. There are two basic elements: an equation for the gravitational field as influenced by matter, and an equation for the response of matter to this field. The conventional Newtonian statement of these rules is in terms of forces between particles; the force between two objects of masses M and m separated by a vector $\mathbf{r} = r\mathbf{e}_{(r)}$ is the famous inverse-square law,

$$\mathbf{F} = -\frac{GMm}{r^2}\mathbf{e}_{(r)}, \tag{1.1}$$

and this force acts on a particle of mass m to give it an acceleration according to Newton's second law,

$$\mathbf{F} = m\mathbf{a}. \tag{1.2}$$

Equivalently, we could use the language of the gravitational potential Φ; the potential is related to the mass density ρ by Poisson's equation,

$$\nabla^2\Phi = 4\pi G\rho, \tag{1.3}$$

and the acceleration is given by the gradient of the potential,

$$\mathbf{a} = -\nabla\Phi. \tag{1.4}$$

Either (1.1) and (1.2), or (1.3) and (1.4), serve to define Newtonian gravity. To define GR, we need to replace each of them by statements about the curvature of spacetime.

The hard part is the equation governing the response of spacetime curvature to the presence of matter and energy. We will eventually find what we want in the form of Einstein's equation,

$$R_{\mu\nu} - \frac{1}{2}Rg_{\mu\nu} = 8\pi G T_{\mu\nu}. \tag{1.5}$$

This looks more forbidding than it should, largely because of those Greek subscripts. In fact this is simply an equation between 4×4 matrices, and the subscripts label elements of each matrix. The expression on the left-hand side is a measure of the curvature of spacetime, while the right-hand side measures the energy and momentum of matter, so this equation relates energy to curvature, as promised. But we will defer until later a detailed understanding of the inner workings of Einstein's equation.

The response of matter to spacetime curvature is somewhat easier to grasp: Free particles move along paths of "shortest possible distance," or geodesics. In other words, particles try their best to move on straight lines, but in a curved spacetime there might not be any straight lines (in the sense we are familiar with from Euclidean geometry), so they do the next best thing. Their parameterized paths $x^\mu(\lambda)$ obey the geodesic equation:

$$\frac{d^2x^\mu}{d\lambda^2} + \Gamma^\mu_{\rho\sigma}\frac{dx^\rho}{d\lambda}\frac{dx^\sigma}{d\lambda} = 0. \tag{1.6}$$

At this point you aren't expected to understand (1.6) any more than (1.5); but soon enough it will all make sense.

As we will discuss later, the universal nature of geodesic motion is an extremely profound feature of GR. This universality is the origin of our claim that gravity is not actually a "force," but a feature of spacetime. A charged particle in an electric field feels an acceleration, which deflects it from straight-line motion; in contrast, a particle in a gravitational field moves along a path that is the closest thing there is to a straight line. Such particles do not feel acceleration; they are freely falling. Once we become more familiar with the spirit of GR, it will make perfect sense to think of a ball flying through the air as being more truly "unaccelerated" than one sitting on a table; the one sitting on a table is being deflected away from the geodesic it would like to be on (which is why we feel a force on our feet as we stand on Earth).

The basic concept underlying our description of spacetime curvature will be that of the metric tensor, typically denoted by $g_{\mu\nu}$. The metric encodes the geometry of a space by expressing deviations from Pythagoras's theorem, $(\Delta l)^2 = (\Delta x)^2 + (\Delta y)^2$ (where Δl is the distance between two points defined on a Cartesian grid with coordinate separations Δx and Δy). This familiar formula is valid only in conventional Euclidean geometry, where it is implicitly assumed that space is flat. In the presence of curvature our deeply ingrained notions of geometry will begin to fail, and we can characterize the amount of curvature by keeping track of how Pythagoras's relation is altered. This information is contained in the metric tensor. From the metric we will derive the Riemann curvature tensor, used to define Einstein's equation, and also the geodesic equation. Setting up this mathematical apparatus is the subject of the next several chapters.

Despite the need to introduce a certain amount of formalism to discuss curvature in a quantitative way, the essential notion of GR ("gravity is the curvature of spacetime") is quite simple. So why does GR have, at least in some benighted circles, a reputation for difficulty or even abstruseness? Because the elegant truths of Einstein's theory are obscured by the accumulation of certain pre-relativity notions which, although very useful, must first be discarded in order to appreciate the world according to GR. Specifically, we live in a world in which spacetime curvature is very small, and particles are for the most part moving quite slowly compared to the speed of light. Consequently, the mechanics of Galileo and Newton comes very naturally to us, even though it is only an approximation to the deeper story.

So we will set about learning the deeper story by gradually stripping away the layers of useful but misleading Newtonian intuition. The first step, which is the subject of this chapter, will be to explore special relativity (SR), the theory of spacetime in the absence of gravity (curvature). Hopefully this is mostly review, as it will proceed somewhat rapidly. The point will be both to recall what SR is all about, and to introduce tensors and related concepts that will be crucial later on, without the extra complications of curvature on top of everything else. Therefore, for this chapter we will always be working in flat spacetime, and furthermore we will only use inertial (Cartesian-like) coordinates. Needless to say it is possible to do SR in any coordinate system you like, but it turns out that introducing the necessary tools for doing so would take us halfway to curved spaces anyway, so we will put that off for a while.

1.2 ■ SPACE AND TIME, SEPARATELY AND TOGETHER

A purely cold-blooded approach to GR would reverse the order of Chapter 2 (Manifolds) and Chapter 1 (Special Relativity and Flat Spacetime). A *manifold* is the kind of mathematical structure used to describe spacetime, while *special relativity* is a model that invokes a particular kind of spacetime (one with no curvature, and hence no gravity). However, if you are reading this book you presumably have at least some familiarity with special relativity (SR), while you may not know anything about manifolds. So our first step will be to explore the relatively familiar territory of SR, taking advantage of this opportunity to introduce concepts and notation that will be crucial to later developments.

Special relativity is a theory of the structure of spacetime, the background on which particles and fields evolve. SR serves as a replacement for Newtonian mechanics, which also is a theory of the structure of spacetime. In either case, we can distinguish between this basic structure and the various dynamical laws govern-

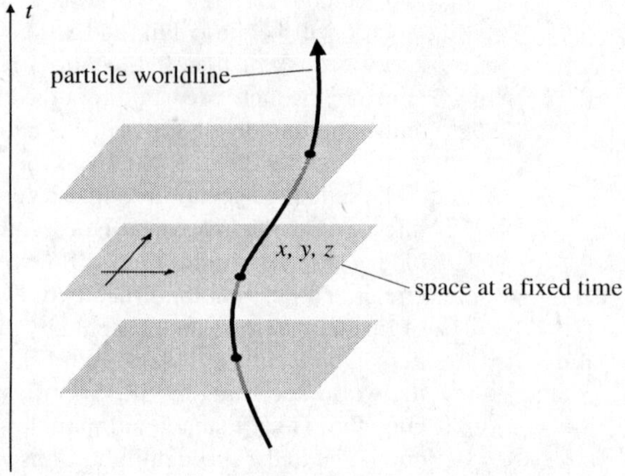

FIGURE 1.1 In Newtonian spacetime there is an absolute slicing into distinct copies of space at different moments in time. Particle worldlines are constrained to move forward in time, but can travel through space at any velocity; there is universal agreement on the question of whether two events at different points in space occur at the same moment of time.

ing specific systems: Newtonian gravity is an example of a dynamical system set within the context of Newtonian mechanics, while Maxwell's electromagnetism is a dynamical system operating within the context of special relativity.

Spacetime is a four-dimensional set, with elements labeled by three dimensions of space and one of time. (We'll do a more rigorous job with the definitions in the next chapter.) An individual point in spacetime is called an **event**. The path of a particle is a curve through spacetime, a parameterized one-dimensional set of events, called the **worldline**. Such a description applies equally to SR and Newtonian mechanics. In either case, it seems clear that "time" is treated somewhat differently than "space"; in particular, particles always travel forward in time, whereas they are free to move back and forth in space.

There is an important difference, however, between the set of allowed paths that particles can take in SR and those in Newton's theory. In Newtonian mechanics, there is a basic division of spacetime into well-defined slices of "all of space at a fixed moment in time." The notion of *simultaneity*, when two events occur at the same time, is unambiguously defined. Trajectories of particles will move ever forward in time, but are otherwise unconstrained; in particular, there is no limit on the relative velocity of two such particles.

In SR the situation is dramatically altered: in particular, *there is no well-defined notion of two separated events occurring "at the same time."* That is not to say that spacetime is completely structureless. Rather, at any event we can define a **light cone**, which is the locus of paths through spacetime that could conceivably be taken by light rays passing through this event. The absolute division, in Newtonian

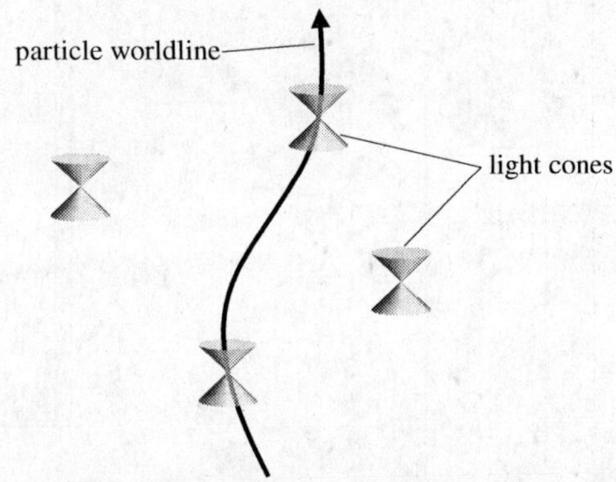

FIGURE 1.2 In special relativity there is no absolute notion of "all of space at one moment in time." Instead, there is a rule that particles always travel at less than or equal to the speed of light. We can therefore define light cones at every event, which locally describe the set of allowed trajectories. For two events that are outside each others' light cones, there is no universal notion of which event occurred earlier in time.

mechanics, of spacetime into unique slices of space parameterized by time, is replaced by a rule that says that physical particles cannot travel faster than light, and consequently move along paths that always remain inside these light cones.

The absence of a preferred time-slicing in SR is at the heart of why the notion of spacetime is more fundamental in this context than in Newtonian mechanics. Of course we can choose specific coordinate systems in spacetime, and once we do, it makes sense to speak of separated events occurring at the same value of the time coordinate in this particular system; but there will also be other possible coordinates, related to the first by "rotating" space and time into each other. This phenomenon is a natural generalization of rotations in Euclidean geometry, to which we now turn.

Consider a garden-variety two-dimensional plane. It is typically convenient to label the points on such a plane by introducing coordinates, for example by defining orthogonal x and y axes and projecting each point onto these axes in the usual way. However, it is clear that most of the interesting geometrical facts about the plane are independent of our choice of coordinates; there aren't any preferred directions. As a simple example, we can consider the distance between two points, given by

$$(\Delta s)^2 = (\Delta x)^2 + (\Delta y)^2. \tag{1.7}$$

In a different Cartesian coordinate system, defined by x' and y' axes that are rotated with respect to the originals, the formula for the distance is unaltered:

$$(\Delta s)^2 = (\Delta x')^2 + (\Delta y')^2. \tag{1.8}$$

We therefore say that the distance is invariant under such changes of coordinates.

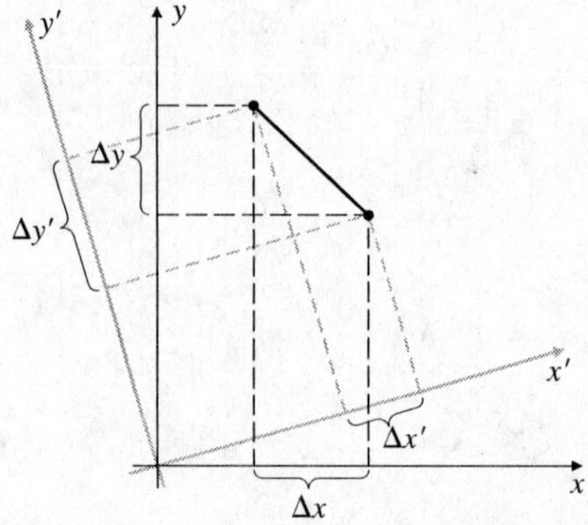

FIGURE 1.3 Two-dimensional Euclidean space, with two different coordinate systems. Notions such as "the distance between two points" are independent of the coordinate system chosen.

This is why it is useful to think of the plane as an intrinsically two-dimensional space, rather than as two fundamentally distinct one-dimensional spaces brought arbitrarily together: Although we use two distinct numbers to label each point, the numbers are not the essence of the geometry, since we can rotate axes into each other while leaving distances unchanged. In Newtonian physics this is not the case with space and time; there is no useful notion of rotating space and time into each other. Rather, the notion of "all of space at a single moment in time" has a meaning independent of coordinates.

SR is a different story. Let us consider coordinates (t, x, y, z) on spacetime, set up in the following way. The spatial coordinates (x, y, z) comprise a standard Cartesian system, constructed for example by welding together rigid rods that meet at right angles. The rods must be moving freely, unaccelerated. The time coordinate is defined by a set of clocks, which are not moving with respect to the spatial coordinates. (Since this is a thought experiment, we can imagine that the rods are infinitely long and there is one clock at every point in space.) The clocks are synchronized in the following sense. Imagine that we send a beam of light from point 1 in space to point 2, in a straight line at a constant velocity c, and then immediately back to 1 (at velocity $-c$). Then the time on the coordinate clock when the light beam reaches point 2, which we label t_2, should be halfway between the time on the coordinate clock when the beam left point 1 (t_1) and the time on that same clock when it returned (t'_1):

$$t_2 = \frac{1}{2}(t'_1 + t_1). \tag{1.9}$$

The coordinate system thus constructed is an **inertial frame**, or simply "inertial coordinates." These coordinates are the natural generalization to spacetime of Cartesian (orthonormal) coordinates in space. (The reason behind the careful

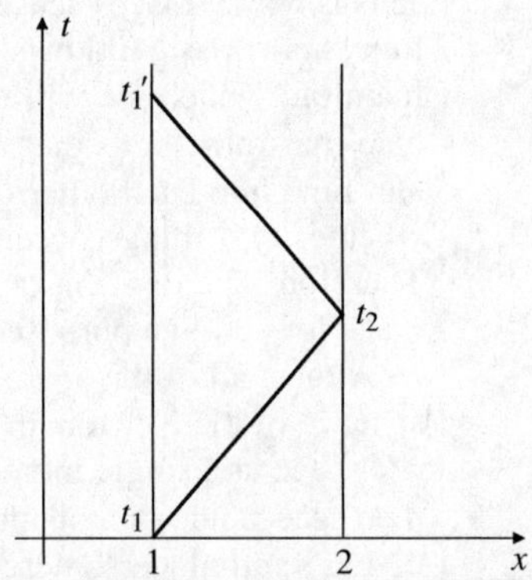

FIGURE 1.4 Synchronizing clocks in an inertial coordinate system. The clocks are synchronized if the time t_2 is halfway between t_1 and t_1' when we bounce a beam of light from point 1 to point 2 and back.

construction is so that we only make comparisons *locally*; never, for example, comparing two far-away clocks to each other at the same time. This kind of care will be even more necessary once we go to general relativity, where there will not be any way to construct inertial coordinates throughout spacetime.)

We can construct any number of inertial frames via this procedure, differing from the first one by an offset in initial position and time, angle, and (constant) velocity. In a Newtonian world, the new coordinates (t', x', y', z') would have the feature that $t' = t +$ constant, independent of spatial coordinates. That is, there is an absolute notion of "two events occurring simultaneously, that is, at the same time." But in SR this isn't true; in general the three-dimensional "spaces" defined by $t =$ constant will differ from those defined by $t' =$ constant.

However, we have not descended completely into chaos. Consider, without any motivation for the moment, what we will call the **spacetime interval** between two events:

$$\boxed{(\Delta s)^2 = -(c\Delta t)^2 + (\Delta x)^2 + (\Delta y)^2 + (\Delta z)^2.} \tag{1.10}$$

(Notice that it can be positive, negative, or zero even for two nonidentical points.) Here, c is some fixed conversion factor between space and time, that is, a fixed velocity. As an empirical matter, it turns out that electromagnetic waves propagate in vacuum at this velocity c, which we therefore refer to as "the speed of light." The important thing, however, is not that photons happen to travel at that speed, but that there exists a c such that *the spacetime interval is invariant under changes of inertial coordinates*. In other words, if we set up a new inertial frame (t', x', y', z'), the interval will be of the same form:

$$(\Delta s)^2 = -(c\Delta t')^2 + (\Delta x')^2 + (\Delta y')^2 + (\Delta z')^2. \tag{1.11}$$

This is why it makes sense to think of SR as a theory of four-dimensional spacetime, known as **Minkowski space**. (This is a special case of a four-dimensional manifold, which we will deal with in detail later.) As we shall see, the coordinate transformations that we have implicitly defined do, in a sense, rotate space and time into each other. There is no absolute notion of "simultaneous events"; whether two things occur at the same time depends on the coordinates used. Therefore, the division of Minkowski space into space and time is a choice we make for our own purposes, not something intrinsic to the situation.

Almost all of the "paradoxes" associated with SR result from a stubborn persistence of the Newtonian notions of a unique time coordinate and the existence of "space at a single moment in time." By thinking in terms of spacetime rather than space and time together, these paradoxes tend to disappear.

Let's introduce some convenient notation. Coordinates on spacetime will be denoted by letters with Greek superscript indices running from 0 to 3, with 0 generally denoting the time coordinate. Thus,

$$x^\mu : \quad \begin{aligned} x^0 &= ct \\ x^1 &= x \\ x^2 &= y \\ x^3 &= z. \end{aligned} \tag{1.12}$$

(Don't start thinking of the superscripts as exponents.) Furthermore, for the sake of simplicity we will choose units in which

$$c = 1; \tag{1.13}$$

we will therefore leave out factors of c in all subsequent formulae. Empirically we know that c is 3×10^8 meters per second; thus, we are working in units where 1 second equals 3×10^8 meters. Sometimes it will be useful to refer to the space and time components of x^μ separately, so we will use Latin superscripts to stand for the space components alone:

$$x^i : \quad \begin{aligned} x^1 &= x \\ x^2 &= y \\ x^3 &= z. \end{aligned} \tag{1.14}$$

It is also convenient to write the spacetime interval in a more compact form. We therefore introduce a 4×4 matrix, the **metric**, which we write using two lower indices:

$$\eta_{\mu\nu} = \begin{pmatrix} -1 & 0 & 0 & 0 \\ 0 & 1 & 0 & 0 \\ 0 & 0 & 1 & 0 \\ 0 & 0 & 0 & 1 \end{pmatrix}. \tag{1.15}$$

(Some references, especially field theory books, define the metric with the opposite sign, so be careful.) We then have the nice formula

$$(\Delta s)^2 = \eta_{\mu\nu} \Delta x^\mu \Delta x^\nu. \tag{1.16}$$

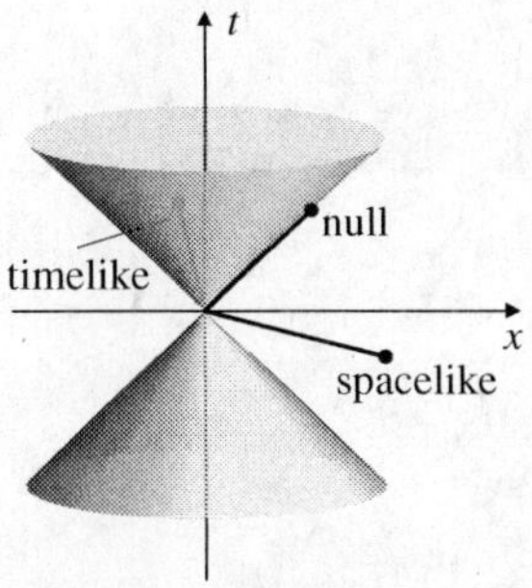

FIGURE 1.5 A light cone, portrayed on a spacetime diagram. Points that are spacelike-, null-, and timelike-separated from the origin are indicated.

This formula introduces the **summation convention**, in which indices appearing both as superscripts and subscripts are summed over. We call such labels **dummy indices**; it is important to remember that they are summed over all possible values, rather than taking any specific one. (It will always turn out to be the case that dummy indices occur strictly in pairs, with one "upstairs" and one "downstairs." More on this later.) The content of (1.16) is therefore exactly the same as (1.10).

An extremely useful tool is the **spacetime diagram**, so let's consider Minkowski space from this point of view. We can begin by portraying the initial t and x axes at right angles, and suppressing the y and z axes. ("Right angles" as drawn on a spacetime diagram don't necessarily imply "orthogonal in spacetime," although that turns out to be true for the t and x axes in this case.) It is enlightening to consider the paths corresponding to travel at the speed $c = 1$, given by $x = \pm t$. A set of points that are all connected to a single event by straight lines moving at the speed of light is the **light cone**, since if we imagine including one more spatial coordinate, the two diagonal lines get completed into a cone. Light cones are naturally divided into future and past; the set of all points inside the future and past light cones of a point p are called **timelike separated** from p, while those outside the light cones are **spacelike separated** and those on the cones are **lightlike** or **null separated** from p. Referring back to (1.10), we see that the interval between timelike separated points is negative, between spacelike separated points is positive, and between null separated points is zero. (The interval is defined to be $(\Delta s)^2$, not the square root of this quantity.)

The fact that the interval is negative for a timelike line (on which a slower-than-light particle will actually move) is annoying, so we define the **proper time** τ to satisfy

$$(\Delta\tau)^2 = -(\Delta s)^2 = -\eta_{\mu\nu}\Delta x^\mu \Delta x^\nu . \tag{1.17}$$

A crucial feature of the spacetime interval is that *the proper time between two events measures the time elapsed as seen by an observer moving on a straight path between the events*. This is easily seen in the very special case that the two events have the same spatial coordinates, and are only separated in time; this corresponds to the observer traveling between the events being at rest in the coordinate system used. Then $(\Delta\tau)^2 = -\eta_{\mu\nu}\Delta x^\mu \Delta x^\nu = (\Delta t)^2$, so $\Delta\tau = \Delta t$, and of course we defined t as the time measured by a clock located at a fixed spatial position. But the spacetime interval is invariant under changes of inertial frame; the proper time (1.17) between two fixed events will be the same when evaluated in an inertial frame where the observer is moving as it is in the frame where the observer is at rest.

A crucial fact is that, for more general trajectories, the proper time and coordinate time are different (although the proper time is always that measured by the clock carried by an observer along the trajectory). Consider two trajectories between events A and C, one a straight line passing through a halfway point marked B, and another traveled by an observer moving away from A at a constant velocity $v = dx/dt$ to a point B' and then back at a constant velocity $-v$ to intersect at

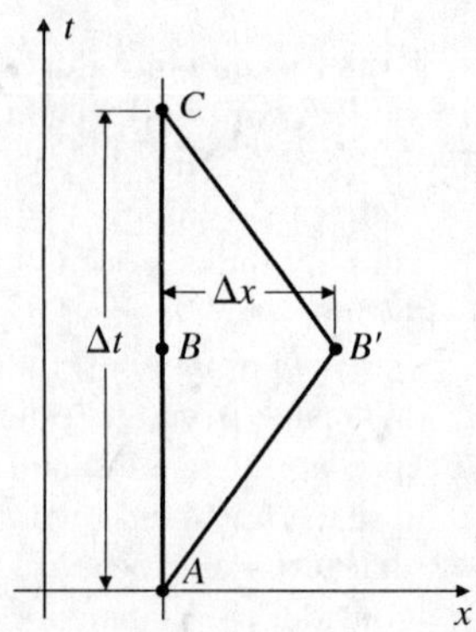

FIGURE 1.6 The twin paradox. A traveler on the straight path through spacetime ABC will age more than someone on the nonstraight path $AB'C$. Since proper time is a measure of distance traveled through spacetime, this should come as no surprise. (The only surprise might be that the straight path is the one of *maximum* proper time; this can be traced to the minus sign for the timelike component of the metric.)

the event C. Choose inertial coordinates such that the straight trajectory describes a motionless particle, with event A located at coordinates $(t, x) = (0, 0)$ and C located at $(\Delta t, 0)$. The two paths then describe an isosceles triangle in spacetime; B has coordinates $(\frac{1}{2}\Delta t, 0)$ and B' has coordinates $(\frac{1}{2}\Delta t, \Delta x)$, with $\Delta x = \frac{1}{2}v\Delta t$. Clearly, $\Delta\tau_{AB} = \frac{1}{2}\Delta t$, but

$$\begin{aligned}\Delta\tau_{AB'} &= \sqrt{(\tfrac{1}{2}\Delta t)^2 - (\Delta x)^2} \\ &= \tfrac{1}{2}\sqrt{1 - v^2}\,\Delta t. \end{aligned} \tag{1.18}$$

It should be obvious that $\Delta\tau_{BC} = \Delta\tau_{AB}$ and $\Delta\tau_{B'C} = \Delta\tau_{AB'}$. Thus, the observer on the straight-line trip from event A to C experiences an elapsed time of $\Delta\tau_{ABC} = \Delta t$, whereas the one who traveled out and returned experiences

$$\Delta\tau_{AB'C} = \sqrt{1 - v^2}\,\Delta t < \Delta t. \tag{1.19}$$

Even though the two observers begin and end at the same points in spacetime, they have aged different amounts. This is the famous "twin paradox," the unfortunate scene of all sorts of misunderstandings and tortured explanations. The truth is straightforward: a nonstraight path in spacetime has a different interval than a straight path, just as a nonstraight path in space has a different length than a straight one. This isn't as trivial as it sounds, of course; the profound insight is the way in which "elapsed time along a worldline" is related to the interval traversed through spacetime. In a Newtonian world, the coordinate t represents a universal flow of time throughout all of spacetime; in relativity, t is just a convenient coordinate, and the elapsed time depends on the path along which you travel. An important distinction is that the nonstraight path has a *shorter* proper time. In space, the shortest distance between two points is a straight line; in spacetime, the longest proper time between two events is a straight trajectory.

Not all trajectories are nice enough to be constructed from pieces of straight lines. In more general circumstances it is useful to introduce the infinitesimal interval, or **line element**:

$$\boxed{ds^2 = \eta_{\mu\nu} dx^\mu dx^\nu,} \tag{1.20}$$

for infinitesimal coordinate displacements dx^μ. (We are being quite informal here, but we'll make amends later on.) From this definition it is tempting to take the square root and integrate along a path to obtain a finite interval, but it is somewhat unclear what $\int \sqrt{\eta_{\mu\nu} dx^\mu dx^\nu}$ is supposed to mean. Instead we consider a path through spacetime as a parameterized curve, $x^\mu(\lambda)$. Note that, unlike conventional practice in Newtonian mechanics, the parameter λ is not necessarily identified with the time coordinate. We can then calculate the derivatives $dx^\mu/d\lambda$, and write the path length along a spacelike curve (one whose infinitesimal intervals are spacelike) as

$$\Delta s = \int \sqrt{\eta_{\mu\nu} \frac{dx^\mu}{d\lambda} \frac{dx^\nu}{d\lambda}}\, d\lambda, \tag{1.21}$$

where the integral is taken over the path. For timelike paths we use the proper time

$$\Delta \tau = \int \sqrt{-\eta_{\mu\nu} \frac{dx^\mu}{d\lambda} \frac{dx^\nu}{d\lambda}}\, d\lambda, \tag{1.22}$$

which will be positive. (For null paths the interval is simply zero.) Of course we may consider paths that are timelike in some places and spacelike in others, but fortunately it is seldom necessary since the paths of physical particles never change their character (massive particles move on timelike paths, massless particles move on null paths). Once again, $\Delta\tau$ really is the time measured by an observer moving along the trajectory.

The notion of *acceleration* in special relativity has a bad reputation, for no good reason. Of course we were careful, in setting up inertial coordinates, to make sure that particles at rest in such coordinates are unaccelerated. However, once we've set up such coordinates, we are free to consider any sort of trajectories for physical particles, whether accelerated or not. In particular, there is no truth to the rumor that SR is unable to deal with accelerated trajectories, and general relativity must be invoked. General relativity becomes relevant in the presence of gravity, when spacetime becomes curved. Any processes in flat spacetime are described within the context of special relativity; in particular, expressions such as (1.22) are perfectly general.

1.3 ■ LORENTZ TRANSFORMATIONS

We can now consider coordinate transformations in spacetime at a somewhat more abstract level than before. We are interested in a formal description of how to relate the various inertial frames constructed via the procedure outlined above; that is, coordinate systems that leave the interval (1.16) invariant. One simple variety are the **translations**, which merely shift the coordinates (in space or time):

$$x^\mu \to x^{\mu'} = \delta^{\mu'}_\mu (x^\mu + a^\mu), \tag{1.23}$$

where a^μ is a set of four fixed numbers and $\delta^{\mu'}_\mu$ is the four-dimensional version of the traditional Kronecker delta symbol:

$$\delta^{\mu'}_\mu = \begin{cases} 1 & \text{when } \mu' = \mu, \\ 0 & \text{when } \mu' \neq \mu. \end{cases} \tag{1.24}$$

Notice that we put the prime on the index, not on the x. The reason for this should become more clear once we start dealing with vectors and tensors; the notation serves to remind us that the geometrical object is the same, but its components are resolved with respect to a different coordinate system. Translations leave the differences Δx^μ unchanged, so it is not remarkable that the interval is unchanged. The other relevant transformations include spatial **rotations** and offsets by a constant velocity vector, or **boosts**; these are linear transformations, described by multiplying x^μ by a (spacetime-independent) matrix:

$$x^{\mu'} = \Lambda^{\mu'}{}_\nu x^\nu, \tag{1.25}$$

or, in more conventional matrix notation,

$$x' = \Lambda x. \tag{1.26}$$

(We will generally use indices, rather than matrix notation, but right now we have an interest in relating our discussion to certain other familiar notions usually described by matrices.) These transformations do not leave the differences Δx^μ unchanged, but multiply them also by the matrix Λ. What kind of matrices will leave the interval invariant? Sticking with the matrix notation, what we would like is

$$\begin{aligned}(\Delta s)^2 = (\Delta x)^{\mathrm{T}} \eta (\Delta x) &= (\Delta x')^{\mathrm{T}} \eta (\Delta x') \\ &= (\Delta x)^{\mathrm{T}} \Lambda^{\mathrm{T}} \eta \Lambda (\Delta x),\end{aligned} \tag{1.27}$$

and therefore

$$\eta = \Lambda^{\mathrm{T}} \eta \Lambda, \tag{1.28}$$

or

$$\eta_{\rho\sigma} = \Lambda^{\mu'}{}_{\rho}\eta_{\mu'\nu'}\Lambda^{\nu'}{}_{\sigma} = \Lambda^{\mu'}{}_{\rho}\Lambda^{\nu'}{}_{\sigma}\eta_{\mu'\nu'}. \tag{1.29}$$

(In matrix notation the order matters, while in index notation it is irrelevant.) We want to find the matrices $\Lambda^{\mu'}{}_{\nu}$ such that the components of the matrix $\eta_{\mu'\nu'}$ are the same as those of $\eta_{\rho\sigma}$; that is what it means for the interval to be invariant under these transformations.

The matrices that satisfy (1.28) are known as the **Lorentz transformations**; the set of them forms a group under matrix multiplication, known as the **Lorentz group**. There is a close analogy between this group and SO(3), the rotation group in three-dimensional space. The rotation group can be thought of as 3×3 matrices R that satisfy $R^{\mathrm{T}}R = \mathbf{1}$, where $\mathbf{1}$ is the 3×3 identity matrix. Such matrices are called *orthogonal*, and the 3×3 ones form the group O(3). This includes not only rotations but also reversals of orientation of the spatial axes (parity transformations). Sometimes we choose to exclude parity transformations by also demanding that the matrices have unit determinant, $|R| = 1$; such matrices are called *special*, and the resulting group is SO(3). The orthogonality condition can be made to look more like (1.28) if we write it as

$$\mathbf{1} = R^{\mathrm{T}}\mathbf{1}R. \tag{1.30}$$

So the difference between the rotation group O(3) and the Lorentz group is the replacement of $\mathbf{1}$, a 3×3 diagonal matrix with all entries equal to $+1$, by η, a 4×4 diagonal matrix with one entry equal to -1 and the rest equal to $+1$. The Lorentz group is therefore often referred to as O(3,1). It includes not only boosts and rotations, but discrete reversals of the time direction as well as parity transformations. As before we can demand that $|\Lambda| = 1$, leaving the "proper Lorentz group" SO(3,1). However, this does not leave us with what we really want, which is the set of continuous Lorentz transformations (those connected smoothly to the identity), since a combination of a time reversal and a parity reversal would have unit determinant. From the $(\rho, \sigma) = (0, 0)$ component of (1.29) we can easily show that $|\Lambda^{0'}{}_{0}| \geq 1$, with negative values corresponding to time reversals. We can therefore demand at last that $\Lambda^{0'}{}_{0} \geq 1$ (in addition to $|\Lambda| = 1$), leaving the "proper orthochronous" or "restricted" Lorentz group. Sometimes this is denoted by something like $\mathrm{SO}(3,1)^{\uparrow}$, but usually we will not bother to make this distinction explicitly. Note that the 3×3 identity matrix is simply the metric for ordinary flat space. Such a metric, in which all of the eigenvalues are positive, is called **Euclidean**, while those such as (1.15), which feature a single minus sign, are called **Lorentzian**.

It is straightforward to write down explicit expressions for simple Lorentz transformations. A familiar rotation in the x-y plane is:

$$\Lambda^{\mu'}{}_{\nu} = \begin{pmatrix} 1 & 0 & 0 & 0 \\ 0 & \cos\theta & \sin\theta & 0 \\ 0 & -\sin\theta & \cos\theta & 0 \\ 0 & 0 & 0 & 1 \end{pmatrix}. \tag{1.31}$$

The rotation angle θ is a periodic variable with period 2π. The boosts may be thought of as "rotations between space and time directions." An example is given by a boost in the x-direction:

$$\Lambda^{\mu'}{}_{\nu} = \begin{pmatrix} \cosh\phi & -\sinh\phi & 0 & 0 \\ -\sinh\phi & \cosh\phi & 0 & 0 \\ 0 & 0 & 1 & 0 \\ 0 & 0 & 0 & 1 \end{pmatrix}. \tag{1.32}$$

The boost parameter ϕ, unlike the rotation angle, is defined from $-\infty$ to ∞. A general transformation can be obtained by multiplying the individual transformations; the explicit expression for this six-parameter matrix (three boosts, three rotations) is not pretty, or sufficiently useful to bother writing down. In general Lorentz transformations will not commute, so the Lorentz group is nonabelian. The set of both translations and Lorentz transformations is a ten-parameter non-abelian group, the **Poincaré group**.

You should not be surprised to learn that the boosts correspond to changing coordinates by moving to a frame that travels at a constant velocity, but let's see it more explicitly. (Don't confuse "boosting" with "accelerating." The difference between boosting to a different reference frame and accelerating an object is the same as the difference between rotating to a different coordinate system and setting an object spinning.) For the transformation given by (1.32), the transformed coordinates t' and x' will be given by

$$\begin{aligned} t' &= t\cosh\phi - x\sinh\phi \\ x' &= -t\sinh\phi + x\cosh\phi. \end{aligned} \tag{1.33}$$

From this we see that the point defined by $x' = 0$ is moving; it has a velocity

$$v = \frac{x}{t} = \frac{\sinh\phi}{\cosh\phi} = \tanh\phi. \tag{1.34}$$

To translate into more pedestrian notation, we can replace $\phi = \tanh^{-1} v$ to obtain

$$\begin{aligned} t' &= \gamma(t - vx) \\ x' &= \gamma(x - vt), \end{aligned} \tag{1.35}$$

where $\gamma = 1/\sqrt{1 - v^2}$. So indeed, our abstract approach has recovered the conventional expressions for Lorentz transformations. Applying these formulae leads to time dilation, length contraction, and so forth.

It's illuminating to consider Lorentz transformations in the context of spacetime diagrams. According to (1.33), under a boost in the x-t plane the x' axis ($t' = 0$) is given by $t = x\tanh\phi$, while the t' axis ($x' = 0$) is given by $t = x/\tanh\phi$. We therefore see that the space and time axes are rotated into each other, although they scissor together instead of remaining orthogonal in the traditional Euclidean sense. (As we shall see, the axes do in fact remain orthogonal in

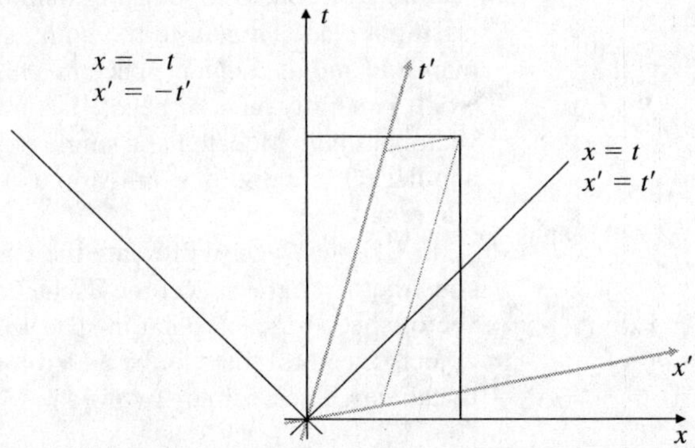

FIGURE 1.7 A Lorentz transformation relates the $\{t', x'\}$ coordinates to the $\{t, x\}$ coordinates. Note that light cones are unchanged.

the Lorentzian sense; that's the implication of the metric remaining invariant under boosts.) This should come as no surprise, since if spacetime behaved just like a four-dimensional version of space the world would be a very different place. We see quite vividly the distinction between this situation and the Newtonian world; in SR, it is impossible to say (in a coordinate-independent way) whether a point that is spacelike separated from p is in the future of p, the past of p, or "at the same time."

Note also that the paths defined by $x' = \pm t'$ are precisely the same as those defined by $x = \pm t$; these trajectories are left invariant under boosts along the x-axis. Of course we know that light travels at this speed; we have therefore found that the speed of light is the same in any inertial frame.

1.4 ■ VECTORS

To probe the structure of Minkowski space in more detail, it is necessary to introduce the concepts of vectors and tensors. We will start with vectors, which should be familiar. Of course, in spacetime vectors are four-dimensional, and are often referred to as **four-vectors**. This turns out to make quite a bit of difference—for example, there is no such thing as a cross product between two four-vectors.

Beyond the simple fact of dimensionality, the most important thing to emphasize is that each vector is located at a given point in spacetime. You may be used to thinking of vectors as stretching from one point to another in space, and even of "free" vectors that you can slide carelessly from point to point. These are not useful concepts outside the context of flat spaces; once we introduce curvature, we lose the ability to draw preferred curves from one point to another, or to move vectors uniquely around a manifold. Rather, to each point p in spacetime we associate the set of all possible vectors located at that point; this set is known as the **tangent space** at p, or T_p. The name is inspired by thinking of the set of vectors attached to a point on a simple curved two-dimensional space as comprising a plane tangent to the point. (This picture relies on an embedding of the manifold and the tangent space in a higher-dimensional external space, which we won't generally have or need.) Inspiration aside, it is important to think of these vectors as being located at a single point, rather than stretching from one point to another (although this won't stop us from drawing them as arrows on spacetime diagrams).

In Chapter 2 we will relate the tangent space at each point to things we can construct from the spacetime itself. For right now, just think of T_p as an abstract vector space for each point in spacetime. A **(real) vector space** is a collection of objects (vectors) that can be added together and multiplied by real numbers in a linear way. Thus, for any two vectors V and W and real numbers a and b, we have

$$(a + b)(V + W) = aV + bV + aW + bW. \tag{1.36}$$

Every vector space has an origin, that is, a zero vector that functions as an identity element under vector addition. In many vector spaces there are additional operations such as taking an inner (dot) product, but this is extra structure over and above the elementary concept of a vector space.

A vector is a perfectly well-defined geometric object, as is a **vector field**, defined as a set of vectors with exactly one at each point in spacetime. [The set of all the tangent spaces of an n-dimensional manifold M can be assembled into a $2n$-dimensional manifold called the **tangent bundle**, $T(M)$. It is a specific example of a "fiber bundle," which is endowed with some extra mathematical structure; we won't need the details for our present purposes.] Nevertheless it is often useful to decompose vectors into components with respect to some set of basis vectors. A **basis** is any set of vectors which both spans the vector space (any vector is a linear combination of basis vectors) and is linearly independent (no vector in the basis

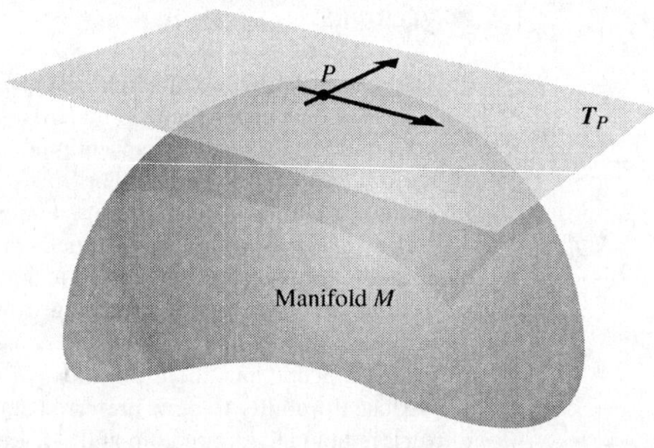

FIGURE 1.8 A suggestive drawing of the tangent space T_p, the space of all vectors at the point p.

is a linear combination of other basis vectors). For any given vector space, there will be an infinite number of possible bases we could choose, but each basis will consist of the same number of vectors, known as the **dimension** of the space. (For a tangent space associated with a point in Minkowski space, the dimension is, of course, four.)

Let us imagine that at each tangent space we set up a basis of four vectors $\hat{e}_{(\mu)}$, with $\mu \in \{0, 1, 2, 3\}$ as usual. In fact let us say that each basis is "adapted to the coordinates x^μ"—that is, the basis vector $\hat{e}_{(1)}$ is what we would normally think of pointing along the x-axis. It is by no means necessary that we choose a basis adapted to any coordinate system at all, although it is often convenient. (As before, we really could be more precise here, but later on we will repeat the discussion at an excruciating level of precision, so some sloppiness now is forgivable.) Then any abstract vector A can be written as a linear combination of basis vectors:

$$A = A^\mu \hat{e}_{(\mu)}. \tag{1.37}$$

The coefficients A^μ are the **components** of the vector A. More often than not we will forget the basis entirely and refer somewhat loosely to "the vector A^μ," but keep in mind that this is shorthand. The real vector is an abstract geometrical entity, while the components are just the coefficients of the basis vectors in some convenient basis. (Since we will usually suppress the explicit basis vectors, the indices usually will label components of vectors and tensors. This is why there are parentheses around the indices on the basis vectors, to remind us that this is a collection of vectors, not components of a single vector.)

A standard example of a vector in spacetime is the tangent vector to a curve. A parameterized curve or path through spacetime is specified by the coordinates as a function of the parameter, for example, $x^\mu(\lambda)$. The tangent vector $V(\lambda)$ has components

$$V^\mu = \frac{dx^\mu}{d\lambda}. \tag{1.38}$$

The entire vector is $V = V^\mu \hat{e}_{(\mu)}$. Under a Lorentz transformation the coordinates x^μ change according to (1.25), while the parameterization λ is unaltered; we can therefore deduce that the components of the tangent vector must change as

$$\boxed{V^\mu \rightarrow V^{\mu'} = \Lambda^{\mu'}{}_{\nu} V^\nu.} \tag{1.39}$$

However, the vector V itself (as opposed to its components in some coordinate system) is invariant under Lorentz transformations. We can use this fact to derive the transformation properties of the basis vectors. Let us refer to the set of basis vectors in the transformed coordinate system as $\hat{e}_{(\nu')}$. Since the vector is invariant, we have

$$V = V^\mu \hat{e}_{(\mu)} = V^{\nu'} \hat{e}_{(\nu')} = \Lambda^{\nu'}{}_{\mu} V^\mu \hat{e}_{(\nu')}. \tag{1.40}$$

But this relation must hold no matter what the numerical values of the components V^μ are. We can therefore say

$$\hat{e}_{(\mu)} = \Lambda^{\nu'}{}_{\mu}\hat{e}_{(\nu')}. \tag{1.41}$$

To get the new basis $\hat{e}_{(\nu')}$ in terms of the old one $\hat{e}_{(\mu)}$, we should multiply by the inverse of the Lorentz transformation $\Lambda^{\nu'}{}_{\mu}$. But the inverse of a Lorentz transformation from the unprimed to the primed coordinates is also a Lorentz transformation, this time from the primed to the unprimed systems. We will therefore introduce a somewhat subtle notation, by using the same symbol for both matrices, just with primed and unprimed indices switched. That is, the Lorentz transformation specified by $\Lambda^{\mu'}{}_{\nu}$ has an inverse transformation written as $\Lambda^{\rho}{}_{\sigma'}$. Operationally this implies

$$\Lambda^{\mu}{}_{\nu'}\Lambda^{\nu'}{}_{\rho} = \delta^{\mu}_{\rho}, \qquad \Lambda^{\sigma'}{}_{\lambda}\Lambda^{\lambda}{}_{\tau'} = \delta^{\sigma'}_{\tau'}. \tag{1.42}$$

From (1.41) we then obtain the transformation rule for basis vectors:

$$\hat{e}_{(\nu')} = \Lambda^{\mu}{}_{\nu'}\hat{e}_{(\mu)}. \tag{1.43}$$

Therefore the set of basis vectors transforms via the inverse Lorentz transformation of the coordinates or vector components.

Let's pause a moment to take all this in. We introduced coordinates labeled by upper indices, which transformed in a certain way under Lorentz transformations. We then considered vector components that also were written with upper indices, which made sense since they transformed in the same way as the coordinate functions. (In a fixed coordinate system, each of the four coordinates x^μ can be thought of as a function on spacetime, as can each of the four components of a vector field.) The basis vectors associated with the coordinate system transformed via the inverse matrix, and were labeled by a lower index. This notation ensured that the invariant object constructed by summing over the components and basis vectors was left unchanged by the transformation, just as we would wish. It's probably not giving too much away to say that this will continue to be the case for tensors, which may have multiple indices.

1.5 ■ DUAL VECTORS (ONE-FORMS)

Once we have set up a vector space, we can define another associated vector space (of equal dimension) known as the **dual vector space**. The dual space is usually denoted by an asterisk, so that the dual space to the tangent space T_p, called the **cotangent space**, is denoted T^*_p. The dual space is the space of all linear maps from the original vector space to the real numbers; in math lingo, if $\omega \in T^*_p$ is a dual vector, then it acts as a map such that

$$\omega(aV + bW) = a\omega(V) + b\omega(W) \in \mathbf{R}, \tag{1.44}$$

where V, W are vectors and a, b are real numbers. The nice thing about these maps is that they form a vector space themselves; thus, if ω and η are dual vectors, we have

$$(a\omega + b\eta)(V) = a\omega(V) + b\eta(V). \tag{1.45}$$

To make this construction somewhat more concrete, we can introduce a set of basis dual vectors $\hat{\theta}^{(\nu)}$ by demanding

$$\hat{\theta}^{(\nu)}(\hat{e}_{(\mu)}) = \delta^{\nu}_{\mu}. \tag{1.46}$$

Then every dual vector can be written in terms of its components, which we label with lower indices:

$$\omega = \omega_\mu \hat{\theta}^{(\mu)}. \tag{1.47}$$

Usually, we will simply write ω_μ, in perfect analogy with vectors, to stand for the entire dual vector. In fact, you will sometimes see elements of T_p (what we have called vectors) referred to as **contravariant vectors**, and elements of T_p^* (what we have called dual vectors) referred to as **covariant vectors**, although in this day and age these terms sound a little dated. If you just refer to ordinary vectors as vectors with upper indices and dual vectors as vectors with lower indices, nobody should be offended. Another name for dual vectors is **one-forms**, a somewhat mysterious designation that will become clearer in Chapter 2.

The component notation leads to a simple way of writing the action of a dual vector on a vector:

$$\begin{aligned}
\omega(V) &= \omega_\mu \hat{\theta}^{(\mu)}(V^\nu \hat{e}_{(\nu)}) \\
&= \omega_\mu V^\nu \hat{\theta}^{(\mu)}(\hat{e}_{(\nu)}) \\
&= \omega_\mu V^\nu \delta^\mu_\nu \\
&= \omega_\mu V^\mu \in \mathbf{R}.
\end{aligned} \tag{1.48}$$

This is why it is rarely necessary to write the basis vectors and dual vectors explicitly; the components do all of the work. The form of (1.48) also suggests that we can think of vectors as linear maps on dual vectors, by defining

$$V(\omega) \equiv \omega(V) = \omega_\mu V^\mu. \tag{1.49}$$

Therefore, the dual space to the dual vector space is the original vector space itself.

Of course in spacetime we will be interested not in a single vector space, but in fields of vectors and dual vectors. [The set of all cotangent spaces over M can be combined into the **cotangent bundle**, $T^*(M)$.] In that case the action of a dual vector field on a vector field is not a single number, but a **scalar** (function) on spacetime. A scalar is a quantity without indices, which is unchanged under

Lorentz transformations; it is a coordinate-independent map from spacetime to the real numbers.

We can use the same arguments that we earlier used for vectors (that geometrical objects are independent of coordinates, even if their components are not) to derive the transformation properties of dual vectors. The answers are, for the components,

$$\boxed{\omega_{\mu'} = \Lambda^{\nu}{}_{\mu'}\omega_{\nu},} \tag{1.50}$$

and for basis dual vectors,

$$\hat{\theta}^{(\rho')} = \Lambda^{\rho'}{}_{\sigma}\hat{\theta}^{(\sigma)}. \tag{1.51}$$

This is just what we would expect from index placement; the components of a dual vector transform under the inverse transformation of those of a vector. Note that this ensures that the scalar (1.48) is invariant under Lorentz transformations, just as it should be.

In spacetime the simplest example of a dual vector is the **gradient** of a scalar function, the set of partial derivatives with respect to the spacetime coordinates, which we denote by a lowercase d:

$$\mathrm{d}\phi = \frac{\partial\phi}{\partial x^{\mu}}\hat{\theta}^{(\mu)}. \tag{1.52}$$

The conventional chain rule used to transform partial derivatives amounts in this case to the transformation rule of components of dual vectors:

$$\begin{aligned}\frac{\partial\phi}{\partial x^{\mu'}} &= \frac{\partial x^{\mu}}{\partial x^{\mu'}}\frac{\partial\phi}{\partial x^{\mu}}\\ &= \Lambda^{\mu}{}_{\mu'}\frac{\partial\phi}{\partial x^{\mu}},\end{aligned} \tag{1.53}$$

where we have used (1.25) to relate the Lorentz transformation to the coordinates. The fact that the gradient is a dual vector leads to the following shorthand notations for partial derivatives:

$$\frac{\partial\phi}{\partial x^{\mu}} = \partial_{\mu}\phi = \phi_{,\mu}. \tag{1.54}$$

So, x^{μ} has an upper index, but when it is in the denominator of a derivative it implies a lower index on the resulting object. In this book we will generally use ∂_{μ} rather than the comma notation. Note that the gradient does in fact act in a natural way on the example we gave above of a vector, the tangent vector to a curve. The result is an ordinary derivative of the function along the curve:

$$\partial_{\mu}\phi\frac{\partial x^{\mu}}{\partial\lambda} = \frac{d\phi}{d\lambda}. \tag{1.55}$$

1.6 ■ TENSORS

A straightforward generalization of vectors and dual vectors is the notion of a **tensor**. Just as a dual vector is a linear map from vectors to $\mathbf{R}$, a tensor T of type (or rank) (k, l) is a multilinear map from a collection of dual vectors and vectors to $\mathbf{R}$:

$$T : \underset{(k \text{ times})}{T_p^* \times \cdots \times T_p^*} \times \underset{(l \text{ times})}{T_p \times \cdots \times T_p} \to \mathbf{R}. \tag{1.56}$$

Here, "$\times$" denotes the Cartesian product, so that for example $T_p \times T_p$ is the space of ordered pairs of vectors. Multilinearity means that the tensor acts linearly in each of its arguments; for instance, for a tensor of type $(1, 1)$, we have

$$\begin{aligned} T(a\omega + b\eta, cV + dW) = acT(\omega, V) \\ + adT(\omega, W) + bcT(\eta, V) + bdT(\eta, W). \end{aligned} \tag{1.57}$$

From this point of view, a scalar is a type $(0, 0)$ tensor, a vector is a type $(1, 0)$ tensor, and a dual vector is a type $(0, 1)$ tensor.

The space of all tensors of a fixed type (k, l) forms a vector space; they can be added together and multiplied by real numbers. To construct a basis for this space, we need to define a new operation known as the **tensor product**, denoted by $\otimes$. If T is a (k, l) tensor and S is an (m, n) tensor, we define a $(k+m, l+n)$ tensor $T \otimes S$ by

$$\begin{aligned} T \otimes S(\omega^{(1)}, \ldots, \omega^{(k)}, \ldots, \omega^{(k+m)}, V^{(1)}, \ldots, V^{(l)}, \ldots, V^{(l+n)}) \\ = T(\omega^{(1)}, \ldots, \omega^{(k)}, V^{(1)}, \ldots, V^{(l)}) \\ \times S(\omega^{(k+1)}, \ldots, \omega^{(k+m)}, V^{(l+1)}, \ldots, V^{(l+n)}). \end{aligned} \tag{1.58}$$

Note that the $\omega^{(i)}$ and $V^{(i)}$ are distinct dual vectors and vectors, not components thereof. In other words, first act T on the appropriate set of dual vectors and vectors, and then act S on the remainder, and then multiply the answers. Note that, in general, tensor products do not commute: $T \otimes S \neq S \otimes T$.

It is now straightforward to construct a basis for the space of all (k, l) tensors, by taking tensor products of basis vectors and dual vectors; this basis will consist of all tensors of the form

$$\hat{e}_{(\mu_1)} \otimes \cdots \otimes \hat{e}_{(\mu_k)} \otimes \hat{\theta}^{(\nu_1)} \otimes \cdots \otimes \hat{\theta}^{(\nu_l)}. \tag{1.59}$$

In a four-dimensional spacetime there will be 4^{k+l} basis tensors in all. In component notation we then write our arbitrary tensor as

$$T = T^{\mu_1 \cdots \mu_k}{}_{\nu_1 \cdots \nu_l} \hat{e}_{(\mu_1)} \otimes \cdots \otimes \hat{e}_{(\mu_k)} \otimes \hat{\theta}^{(\nu_1)} \otimes \cdots \otimes \hat{\theta}^{(\nu_l)}. \tag{1.60}$$

Alternatively, we could define the components by acting the tensor on basis vectors and dual vectors:

$$T^{\mu_1\cdots\mu_k}{}_{\nu_1\cdots\nu_l} = T(\hat{\theta}^{(\mu_1)}, \ldots, \hat{\theta}^{(\mu_k)}, \hat{e}_{(\nu_1)}, \ldots, \hat{e}_{(\nu_l)}). \tag{1.61}$$

You can check for yourself, using (1.46) and so forth, that these equations all hang together properly.

As with vectors, we will usually take the shortcut of denoting the tensor T by its components $T^{\mu_1\cdots\mu_k}{}_{\nu_1\cdots\nu_l}$. The action of the tensors on a set of vectors and dual vectors follows the pattern established in (1.48):

$$T(\omega^{(1)}, \ldots, \omega^{(k)}, V^{(1)}, \ldots, V^{(l)}) = T^{\mu_1\cdots\mu_k}{}_{\nu_1\cdots\nu_l}\omega^{(1)}_{\mu_1}\cdots\omega^{(k)}_{\mu_k}V^{(1)\nu_1}\cdots V^{(l)\nu_l}. \tag{1.62}$$

A (k, l) tensor thus has k upper indices and l lower indices. The order of the indices is obviously important, since the tensor need not act in the same way on its various arguments.

Finally, the transformation of tensor components under Lorentz transformations can be derived by applying what we already know about the transformation of basis vectors and dual vectors. The answer is just what you would expect from index placement,

$$T^{\mu'_1\cdots\mu'_k}{}_{\nu'_1\cdots\nu'_l} = \Lambda^{\mu'_1}{}_{\mu_1}\cdots\Lambda^{\mu'_k}{}_{\mu_k}\Lambda^{\nu_1}{}_{\nu'_1}\cdots\Lambda^{\nu_l}{}_{\nu'_l}T^{\mu_1\cdots\mu_k}{}_{\nu_1\cdots\nu_l}. \tag{1.63}$$

Thus, each upper index gets transformed like a vector, and each lower index gets transformed like a dual vector.

Although we have defined tensors as linear maps from sets of vectors and dual vectors to $\mathbf{R}$, there is nothing that forces us to act on a full collection of arguments. Thus, a (1, 1) tensor also acts as a map from vectors to vectors:

$$T^{\mu}{}_{\nu} : V^{\nu} \to T^{\mu}{}_{\nu}V^{\nu}. \tag{1.64}$$

You can check for yourself that $T^{\mu}{}_{\nu}V^{\nu}$ is a vector (that is, obeys the vector transformation law). Similarly, we can act one tensor on (all or part of) another tensor to obtain a third tensor. For example,

$$U^{\mu}{}_{\nu} = T^{\mu\rho}{}_{\sigma}S^{\sigma}{}_{\rho\nu} \tag{1.65}$$

is a perfectly good (1, 1) tensor.

You may be concerned that this introduction to tensors has been somewhat too brief, given the esoteric nature of the material. In fact, the notion of tensors does not require a great deal of effort to master; it's just a matter of keeping the indices straight, and the rules for manipulating them are very natural. Indeed, a number of books like to *define* tensors as collections of numbers transforming according to (1.63). While this is operationally useful, it tends to obscure the deeper meaning of tensors as geometrical entities with a life independent of any chosen coordinate system. There is, however, one subtlety that we have glossed over. The notions of dual vectors and tensors and bases and linear maps belong to the realm of linear algebra, and are appropriate whenever we have an abstract vector space at hand. In the case of interest to us we have not just a vector space, but a vector space at each point in spacetime. More often than not we are interested in tensor fields, which can be thought of as tensor-valued functions on spacetime. Fortunately, none of the manipulations we defined above really care whether we are dealing with a single vector space or a collection of vector spaces, one for each event. We will be able to get away with simply calling things functions of x^μ when appropriate. However, you should keep straight the logical independence of the notions we have introduced and their specific application to spacetime and relativity.

In spacetime, we have already seen some examples of tensors without calling them that. The most familiar example of a $(0, 2)$ tensor is the metric, $\eta_{\mu\nu}$. The action of the metric on two vectors is so useful that it gets its own name, the **inner product** (or scalar product, or dot product):

$$\eta(V, W) = \eta_{\mu\nu} V^\mu W^\nu = V \cdot W. \tag{1.66}$$

Just as with the conventional Euclidean dot product, we will refer to two vectors whose inner product vanishes as **orthogonal**. Since the inner product is a scalar, it is left invariant under Lorentz transformations; therefore, the basis vectors of any Cartesian inertial frame, which are chosen to be orthogonal by definition, are still orthogonal after a Lorentz transformation (despite the "scissoring together" we noticed earlier). The **norm** of a vector is defined to be inner product of the vector with itself; unlike in Euclidean space, this number is not positive definite:

$$\text{if } \eta_{\mu\nu} V^\mu V^\nu \text{ is } \begin{cases} < 0, & V^\mu \text{ is timelike} \\ = 0, & V^\mu \text{ is lightlike or null} \\ > 0, & V^\mu \text{ is spacelike.} \end{cases}$$

(A vector can have zero norm without being the zero vector.) You will notice that the terminology is the same as that which we used earlier to classify the relationship between two points in spacetime; it's no accident, of course, and we will go into more detail later.

Another tensor is the Kronecker delta δ^μ_ρ, of type $(1, 1)$. Thought of as a map from vectors to vectors (or one-forms to one-forms), the Kronecker delta is simply the identity map. We follow the example of many other references in placing the upper and lower indices in the same column for this unique tensor; purists might write $\delta^\mu{}_\rho$ or $\delta_\rho{}^\mu$, but these would be numerically identical, and we shouldn't get in trouble being careless in this one instance.

Related to the Kronecker delta and the metric is the **inverse metric** $\eta^{\mu\nu}$, a type $(2, 0)$ tensor defined (unsurprisingly) as the "inverse" of the metric:

$$\eta^{\mu\nu}\eta_{\nu\rho} = \eta_{\rho\nu}\eta^{\nu\mu} = \delta^\mu_\rho. \tag{1.67}$$

(It's the inverse metric since, when multiplied by the metric, it yields the identity map.) In fact, as you can check, the inverse metric has exactly the same components as the metric itself. This is only true in flat space in Cartesian coordinates, and will fail to hold in more general situations. There is also the **Levi–Civita symbol**, a (0, 4) tensor:

$$\tilde{\epsilon}_{\mu\nu\rho\sigma} = \begin{cases} +1 & \text{if } \mu\nu\rho\sigma \text{ is an even permutation of } 0123 \\ -1 & \text{if } \mu\nu\rho\sigma \text{ is an odd permutation of } 0123 \\ 0 & \text{otherwise.} \end{cases} \tag{1.68}$$

Here, a "permutation of 0123" is an ordering of the numbers 0, 1, 2, 3, which can be obtained by starting with 0123 and exchanging two of the digits; an even permutation is obtained by an even number of such exchanges, and an odd permutation is obtained by an odd number. Thus, for example, $\tilde{\epsilon}_{0321} = -1$. (The tilde on $\tilde{\epsilon}_{\mu\nu\rho\sigma}$, and referring to it as a symbol rather than simply a tensor, derive from the fact that this object is actually not a tensor in more general geometries or coordinates; instead, it is something called a "tensor density." It is straightforward enough to define a related object that is a tensor, which we will denote by $\epsilon_{\mu\nu\rho\sigma}$ and call the "Levi–Civita tensor." See Chapter 2 for a discussion.)

A remarkable property of the above tensors—the metric, the inverse metric, the Kronecker delta, and the Levi–Civita symbol—is that, even though they all transform according to the tensor transformation law (1.63), their components remain unchanged in *any* inertial coordinate system in flat spacetime. In some sense this makes them nongeneric examples of tensors, since most tensors do not have this property. In fact, these are the *only* tensors with this property, although we won't prove it. The Kronecker delta is even more unusual, in that it has exactly the same components in any coordinate system in any spacetime. This makes sense from the definition of a tensor as a linear map; the Kronecker tensor can be thought of as the identity map from vectors to vectors (or from dual vectors to dual vectors), which clearly must have the same components regardless of coordinate system. Meanwhile, the metric and its inverse characterize the structure of spacetime, while the Levi–Civita symbol is secretly not a true tensor at all. We shall therefore have to treat these objects more carefully when we drop our assumption of flat spacetime.

A more typical example of a tensor is the **electromagnetic field strength tensor**. We all know that the electromagnetic fields are made up of the electric field vector E_i and the magnetic field vector B_i. (Remember that we use Latin indices for spacelike components 1, 2, 3.) Actually these are only "vectors" under rotations in space, not under the full Lorentz group. In fact they are components of a (0, 2) tensor $F_{\mu\nu}$, defined by

$$F_{\mu\nu} = \begin{pmatrix} 0 & -E_1 & -E_2 & -E_3 \\ E_1 & 0 & B_3 & -B_2 \\ E_2 & -B_3 & 0 & B_1 \\ E_3 & B_2 & -B_1 & 0 \end{pmatrix} = -F_{\nu\mu}. \tag{1.69}$$

From this point of view it is easy to transform the electromagnetic fields in one reference frame to those in another, by application of (1.63). The unifying power of the tensor formalism is evident: rather than a collection of two vectors whose relationship and transformation properties are rather mysterious, we have a single tensor field to describe all of electromagnetism. (On the other hand, don't get carried away; sometimes it's more convenient to work in a single coordinate system using the electric and magnetic field vectors.)

1.7 ■ MANIPULATING TENSORS

With these examples in hand we can now be a little more systematic about some properties of tensors. First consider the operation of **contraction**, which turns a (k, l) tensor into a $(k-1, l-1)$ tensor. Contraction proceeds by summing over one upper and one lower index:

$$S^{\mu\rho}{}_{\sigma} = T^{\mu\nu\rho}{}_{\sigma\nu}. \tag{1.70}$$

You can check that the result is a well-defined tensor. It is only permissible to contract an upper index with a lower index (as opposed to two indices of the same type); otherwise the result would *not* be a well-defined tensor. (By well-defined tensor we mean either "transforming according to the tensor transformation law," or "defining a unique multilinear map from a set of vectors and dual vectors to the real numbers"; take your pick.) Note also that the order of the indices matters, so that you can get different tensors by contracting in different ways; thus,

$$T^{\mu\nu\rho}{}_{\sigma\nu} \neq T^{\mu\rho\nu}{}_{\sigma\nu} \tag{1.71}$$

in general.

The metric and inverse metric can be used to **raise and lower indices** on tensors. That is, given a tensor $T^{\alpha\beta}{}_{\gamma\delta}$, we can use the metric to define new tensors, which we choose to denote by the same letter T:

$$\begin{aligned} T^{\alpha\beta\mu}{}_{\delta} &= \eta^{\mu\gamma} T^{\alpha\beta}{}_{\gamma\delta}, \\ T_{\mu}{}^{\beta}{}_{\gamma\delta} &= \eta_{\mu\alpha} T^{\alpha\beta}{}_{\gamma\delta}, \\ T_{\mu\nu}{}^{\rho\sigma} &= \eta_{\mu\alpha}\eta_{\nu\beta}\eta^{\rho\gamma}\eta^{\sigma\delta} T^{\alpha\beta}{}_{\gamma\delta}, \end{aligned} \tag{1.72}$$

and so forth. Notice that raising and lowering does not change the position of an index relative to other indices, and also that free indices (which are *not* summed over) must be the same on both sides of an equation, while dummy indices (which *are* summed over) only appear on one side. As an example, we can turn vectors and dual vectors into each other by raising and lowering indices:

$$\begin{aligned} V_{\mu} &= \eta_{\mu\nu} V^{\nu} \\ \omega^{\mu} &= \eta^{\mu\nu} \omega_{\nu}. \end{aligned} \tag{1.73}$$

Because the metric and inverse metric are truly inverses of each other, we are free to raise and lower simultaneously a pair of indices being contracted over:

$$A^{\lambda} B_{\lambda} = \eta^{\lambda\rho} A_{\rho} \eta_{\lambda\sigma} B^{\sigma} = \delta^{\rho}_{\sigma} A_{\rho} B^{\sigma} = A_{\sigma} B^{\sigma}. \tag{1.74}$$

The ability to raise and lower indices with a metric explains why the gradient in three-dimensional flat Euclidean space is usually thought of as an ordinary vector, even though we have seen that it arises as a dual vector; in Euclidean space (where the metric is diagonal with all entries +1) a dual vector is turned into a vector with precisely the same components when we raise its index. You may then wonder why we have belabored the distinction at all. One simple reason, of course, is that in a Lorentzian spacetime the components are not equal:

$$\omega^\mu = (-\omega_0, \omega_1, \omega_2, \omega_3). \tag{1.75}$$

In a curved spacetime, where the form of the metric is generally more complicated, the difference is rather more dramatic. But there is a deeper reason, namely that tensors generally have a "natural" definition independent of the metric. Even though we will always have a metric available, it is helpful to be aware of the logical status of each mathematical object we introduce. The gradient, with its action on vectors, is perfectly well-defined regardless of any metric, whereas the "gradient with upper indices" is not. (As an example, we will eventually want to take variations of functionals with respect to the metric, and will therefore have to know exactly how the functional depends on the metric, something that is easily obscured by the index notation.)

Continuing our compilation of tensor jargon, we refer to a tensor as **symmetric** in any of its indices if it is unchanged under exchange of those indices. Thus, if

$$S_{\mu\nu\rho} = S_{\nu\mu\rho}, \tag{1.76}$$

we say that $S_{\mu\nu\rho}$ is symmetric in its first two indices, while if

$$S_{\mu\nu\rho} = S_{\mu\rho\nu} = S_{\rho\mu\nu} = S_{\nu\mu\rho} = S_{\nu\rho\mu} = S_{\rho\nu\mu}, \tag{1.77}$$

we say that $S_{\mu\nu\rho}$ is symmetric in all three of its indices. Similarly, a tensor is **antisymmetric** (or skew-symmetric) in any of its indices if it changes sign when those indices are exchanged; thus,

$$A_{\mu\nu\rho} = -A_{\rho\nu\mu} \tag{1.78}$$

means that $A_{\mu\nu\rho}$ is antisymmetric in its first and third indices (or just "antisymmetric in μ and ρ"). If a tensor is (anti-) symmetric in all of its indices, we refer to it as simply (anti-) symmetric (sometimes with the redundant modifier "completely"). As examples, the metric $\eta_{\mu\nu}$ and the inverse metric $\eta^{\mu\nu}$ are symmetric, while the Levi–Civita symbol $\tilde{\epsilon}_{\mu\nu\rho\sigma}$ and the electromagnetic field strength tensor $F_{\mu\nu}$ are antisymmetric. (Check for yourself that if you raise or lower a set of indices that are symmetric or antisymmetric, they remain that way.) Notice that it makes no sense to exchange upper and lower indices with each other, so don't succumb to the temptation to think of the Kronecker delta δ^α_β as symmetric. On the other hand, the fact that lowering an index on δ^α_β gives a symmetric tensor (in fact, the metric) means that the order of indices doesn't really matter, which is why we don't keep track of index placement for this one tensor.

Given any tensor, we can **symmetrize** (or antisymmetrize) any number of its upper or lower indices. To symmetrize, we take the sum of all permutations of the relevant indices and divide by the number of terms:

$$T_{(\mu_1\mu_2\cdots\mu_n)\rho}{}^{\sigma} = \frac{1}{n!}\left(T_{\mu_1\mu_2\cdots\mu_n\rho}{}^{\sigma} + \text{sum over permutations of indices } \mu_1\cdots\mu_n\right), \tag{1.79}$$

while antisymmetrization comes from the alternating sum:

$$T_{[\mu_1\mu_2\cdots\mu_n]\rho}{}^{\sigma} = \frac{1}{n!}\,(T_{\mu_1\mu_2\cdots\mu_n\rho}{}^{\sigma} + \text{alternating sum over permutations of indices } \mu_1\cdots\mu_n). \tag{1.80}$$

By "alternating sum" we mean that permutations that are the result of an odd number of exchanges are given a minus sign, thus:

$$T_{[\mu\nu\rho]\sigma} = \tfrac{1}{6}\left(T_{\mu\nu\rho\sigma} - T_{\mu\rho\nu\sigma} + T_{\rho\mu\nu\sigma} - T_{\nu\mu\rho\sigma} + T_{\nu\rho\mu\sigma} - T_{\rho\nu\mu\sigma}\right). \tag{1.81}$$

Notice that round/square brackets denote symmetrization/antisymmetrization. Furthermore, we may sometimes want to (anti-) symmetrize indices that are not next to each other, in which case we use vertical bars to denote indices not included in the sum:

$$T_{(\mu|\nu|\rho)} = \tfrac{1}{2}\left(T_{\mu\nu\rho} + T_{\rho\nu\mu}\right). \tag{1.82}$$

If we are contracting over a pair of upper indices that are symmetric on one tensor, only the symmetric part of the lower indices will contribute; thus,

$$X^{(\mu\nu)}Y_{\mu\nu} = X^{(\mu\nu)}Y_{(\mu\nu)}, \tag{1.83}$$

regardless of the symmetry properties of $Y_{\mu\nu}$. (Analogous statements hold for antisymmetric indices, or if it's the lower indices that are symmetric to start with.) For any *two* indices, we can decompose a tensor into symmetric and antisymmetric parts,

$$T_{\mu\nu\rho\sigma} = T_{(\mu\nu)\rho\sigma} + T_{[\mu\nu]\rho\sigma}, \tag{1.84}$$

but this will not in general hold for three or more indices,

$$T_{\mu\nu\rho\sigma} \neq T_{(\mu\nu\rho)\sigma} + T_{[\mu\nu\rho]\sigma}, \tag{1.85}$$

because there are parts with mixed symmetry that are not specified by either the symmetric or antisymmetric pieces. Finally, some people use a convention in which the factor of $1/n!$ is omitted. The one used here is a good one, since, for example, a symmetric tensor satisfies

$$S_{\mu_1\cdots\mu_n} = S_{(\mu_1\cdots\mu_n)}, \tag{1.86}$$

and likewise for antisymmetric tensors.

For a $(1, 1)$ tensor $X^\mu{}_\nu$, the **trace** is a scalar, often denoted by leaving off the indices, which is simply the contraction:

$$X = X^\lambda{}_\lambda. \tag{1.87}$$

If we think of $X^\mu{}_\nu$ as a matrix, this is just the sum of the diagonal components, so it makes sense. However, we will also use trace in the context of a $(0, 2)$ tensor $Y_{\mu\nu}$, in which case it means that we should first raise an index ($Y^\mu{}_\nu = g^{\mu\lambda}Y_{\lambda\nu}$) and then contract:

$$Y = Y^\lambda{}_\lambda = \eta^{\mu\nu}Y_{\mu\nu}. \tag{1.88}$$

(It must be this way, since we cannot sum over two lower indices.) Although this is the sum of the diagonal components of $Y^\mu{}_\nu$, it is certainly *not* the sum of the diagonal components of $Y_{\mu\nu}$; we had to raise an index, which in general will change the numerical value of the components. For example, you might guess that the trace of the metric is $-1 + 1 + 1 + 1 = 2$, but it's not:

$$\eta^{\mu\nu}\eta_{\mu\nu} = \delta^\mu_\mu = 4. \tag{1.89}$$

(In n dimensions, $\delta^\mu_\mu = n$.) There is no reason to denote this trace by g (or δ), since it will always be the same number, even after we make the transition to curved spaces where the metric components are more complicated. Note that antisymmetric $(0, 2)$ tensors are always traceless.

We have been careful so far to distinguish clearly between things that are always true (on a manifold with arbitrary metric) and things that are only true in Minkowski space in inertial coordinates. One of the most important distinctions arises with **partial derivatives**. If we are working in flat spacetime with inertial coordinates, then the partial derivative of a (k, l) tensor is a $(k, l + 1)$ tensor; that is,

$$T_\alpha{}^\mu{}_\nu = \partial_\alpha R^\mu{}_\nu \tag{1.90}$$

transforms properly under Lorentz transformations. However, this will no longer be true in more general spacetimes, and we will have to define a covariant derivative to take the place of the partial derivative. Nevertheless, we can still use the fact that partial derivatives give us tensors in this special case, as long as we keep our wits about us. [The one exception to this warning is the partial derivative of a scalar, $\partial_\alpha\phi$, which is a perfectly good tensor (the gradient) in any spacetime.] Of course, if we fix a particular coordinate system, the partial derivative is a perfectly good operator, which we will use all the time; its failure is only that it doesn't transform in the same way as the tensors we will be using (or equivalently, that the map it defines is not coordinate-independent). One of the most useful properties of partial derivatives is that they commute,

$$\partial_\mu \partial_\nu (\cdots) = \partial_\nu \partial_\mu (\cdots), \tag{1.91}$$

no matter what kind of object is being differentiated.

1.8 ■ MAXWELL'S EQUATIONS

We have now accumulated enough tensor know-how to illustrate some of these concepts using actual physics. Specifically, we will examine **Maxwell's equations** of electrodynamics. In 19th-century notation, these are

$$\begin{aligned} \nabla \times \mathbf{B} - \partial_t \mathbf{E} &= \mathbf{J} \\ \nabla \cdot \mathbf{E} &= \rho \\ \nabla \times \mathbf{E} + \partial_t \mathbf{B} &= 0 \\ \nabla \cdot \mathbf{B} &= 0. \end{aligned} \tag{1.92}$$

Here, **E** and **B** are the electric and magnetic field 3-vectors, **J** is the current, ρ is the charge density, and $\nabla\times$ and $\nabla\cdot$ are the conventional curl and divergence. These equations are invariant under Lorentz transformations, of course; that's how the whole business got started. But they don't look obviously invariant; our tensor notation can fix that. Let's begin by writing these equations in component notation,

$$\begin{aligned} \tilde{\epsilon}^{ijk} \partial_j B_k - \partial_0 E^i &= J^i \\ \partial_i E^i &= J^0 \\ \tilde{\epsilon}^{ijk} \partial_j E_k + \partial_0 B^i &= 0 \\ \partial_i B^i &= 0. \end{aligned} \tag{1.93}$$

In these expressions, spatial indices have been raised and lowered with abandon, without any attempt to keep straight where the metric appears, because δ_{ij} is the metric on flat 3-space, with δ^{ij} its inverse (they are equal as matrices). We can therefore raise and lower indices at will, since the components don't change. Meanwhile, the three-dimensional Levi–Civita symbol $\tilde{\epsilon}^{ijk}$ is defined just as the four-dimensional one, although with one fewer index (normalized so that $\tilde{\epsilon}^{123} = \tilde{\epsilon}_{123} = 1$). We have replaced the charge density by J^0; this is legitimate because the density and current together form the **current 4-vector**, $J^\mu = (\rho, J^x, J^y, J^z)$.

From (1.93), and the definition (1.69) of the field strength tensor $F_{\mu\nu}$, it is easy to get a completely tensorial 20th-century version of Maxwell's equations. Begin by noting that we can express the field strength with upper indices as

$$F^{0i} = E^i$$
$$F^{ij} = \tilde{\epsilon}^{ijk} B_k. \tag{1.94}$$

To check this, note for example that $F^{01} = \eta^{00}\eta^{11}F_{01}$ and $F^{12} = \tilde{\epsilon}^{123}B_3$. Then the first two equations in (1.93) become

$$\partial_j F^{ij} - \partial_0 F^{0i} = J^i$$
$$\partial_i F^{0i} = J^0. \tag{1.95}$$

Using the antisymmetry of $F^{\mu\nu}$, we see that these may be combined into the single tensor equation

$$\partial_\mu F^{\nu\mu} = J^\nu. \tag{1.96}$$

A similar line of reasoning, which is left as an exercise, reveals that the third and fourth equations in (1.93) can be written

$$\partial_{[\mu} F_{\nu\lambda]} = 0. \tag{1.97}$$

It's simple to verify that the antisymmetry of $F_{\mu\nu}$ implies that (1.97) can be equivalently expressed as

$$\partial_\mu F_{\nu\lambda} + \partial_\nu F_{\lambda\mu} + \partial_\lambda F_{\mu\nu} = 0. \tag{1.98}$$

The four traditional Maxwell equations are thus replaced by two, vividly demonstrating the economy of tensor notation. More importantly, however, both sides of equations (1.96) and (1.97) manifestly transform as tensors; therefore, if they are true in one inertial frame, they must be true in any Lorentz-transformed frame. This is why tensors are so useful in relativity—we often want to express relationships without recourse to any reference frame, and the quantities on each side of an equation must transform in the same way under changes of coordinates. As a matter of jargon, we will sometimes refer to quantities written in terms of tensors as **covariant** (which has nothing to do with "covariant" as opposed to "contravariant"). Thus, we say that (1.96) and (1.97) together serve as the covariant form of Maxwell's equations, while (1.92) or (1.93) are noncovariant.

1.9 ■ ENERGY AND MOMENTUM

We've now gone over essentially everything there is to know about the care and feeding of tensors. In the next chapter we will look more carefully at the rigorous definitions of manifolds and tensors, but the basic mechanics have been pretty well covered. Before jumping to more abstract mathematics, let's review how physics works in Minkowski spacetime.

Start with the worldline of a single particle. This is specified by a map $\mathbf{R} \to M$, where M is the manifold representing spacetime; we usually think of the path as a parameterized curve $x^\mu(\lambda)$. As mentioned earlier, the tangent vector to this path is $dx^\mu/d\lambda$ (note that it depends on the parameterization). An object of primary interest is the norm of the tangent vector, which serves to characterize the path; if the tangent vector is timelike/null/spacelike at some parameter value λ, we say that the path is timelike/null/spacelike at that point. This explains why the same words are used to classify vectors in the tangent space and intervals between two points—because a straight line connecting, say, two timelike separated points will itself be timelike at every point along the path.

Nevertheless, be aware of the sleight of hand being pulled here. The metric, as a $(0, 2)$ tensor, is a machine that acts on two vectors (or two copies of the same vector) to produce a number. It is therefore very natural to classify tangent vectors according to the sign of their norm. But the interval between two points isn't something quite so natural; it depends on a specific choice of path (a "straight line") that connects the points, and this choice in turn depends on the fact that spacetime is flat (which allows a unique choice of straight line between the points).

Let's move from the consideration of paths in general to the paths of massive particles (which will always be timelike). Since the proper time is measured by a clock traveling on a timelike worldline, it is convenient to use τ as the parameter along the path. That is, we use (1.22) to compute $\tau(\lambda)$, which (if λ is a good parameter in the first place) we can invert to obtain $\lambda(\tau)$, after which we can think of the path as $x^\mu(\tau)$. The tangent vector in this parameterization is known as the **four-velocity**, U^μ:

$$\boxed{U^\mu = \frac{dx^\mu}{d\tau}.} \tag{1.99}$$

Since $d\tau^2 = -\eta_{\mu\nu}dx^\mu dx^\nu$, the four-velocity is automatically normalized:

$$\eta_{\mu\nu}U^\mu U^\nu = -1. \tag{1.100}$$

This absolute normalization is a reflection of the fact that the four-velocity is not a velocity through space, which can of course take on different magnitudes, but a "velocity through spacetime," through which one always travels at the same rate. The norm of the four-velocity will always be negative, since we are only defining it for timelike trajectories. You could define an analogous vector for spacelike paths as well; for null paths the proper time vanishes, so τ can't be used as a parameter, and you have to be more careful. In the rest frame of a particle, its four-velocity has components $U^\mu = (1, 0, 0, 0)$.

A related vector is the **momentum four-vector**, defined by

$$\boxed{p^\mu = mU^\mu,} \tag{1.101}$$

where m is the mass of the particle. The mass is a fixed quantity independent of inertial frame, what you may be used to thinking of as the "rest mass." It turns out to be much more convenient to take this as the mass once and for all, rather than thinking of mass as depending on velocity. The **energy** of a particle is simply $E = p^0$, the timelike component of its momentum vector. Since it's only one component of a four-vector, it is not invariant under Lorentz transformations; that's to be expected, however, since the energy of a particle at rest is not the same as that of the same particle in motion. In the particle's rest frame we have $p^0 = m$; recalling that we have set $c = 1$, we see that we have found the equation that made Einstein a celebrity, $E = mc^2$. (The field equation of general relativity is actually more fundamental than this one, but $R_{\mu\nu} - \frac{1}{2}Rg_{\mu\nu} = 8\pi GT_{\mu\nu}$ doesn't elicit the visceral reaction that you get from $E = mc^2$.) In a moving frame we can find the components of p^μ by performing a Lorentz transformation; for a particle moving with three-velocity $v = dx/dt$ along the x axis we have

$$p^\mu = (\gamma m, v\gamma m, 0, 0), \tag{1.102}$$

where $\gamma = 1/\sqrt{1-v^2}$. For small v, this gives $p^0 = m + \frac{1}{2}mv^2$ (what we usually think of as rest energy plus kinetic energy) and $p^1 = mv$ (what we usually think of as Newtonian momentum). Outside this approximation, we can simply write

$$p_\mu p^\mu = -m^2, \tag{1.103}$$

or

$$E = \sqrt{m^2 + \mathbf{p}^2}, \tag{1.104}$$

where $\mathbf{p}^2 = \delta_{ij}p^i p^j$.

The centerpiece of pre-relativity physics is Newton's Second Law, or $\mathbf{f} = m\mathbf{a} = d\mathbf{p}/dt$. An analogous equation should hold in SR, and the requirement that it be tensorial leads us directly to introduce a force four-vector f^μ satisfying

$$f^\mu = m\frac{d^2}{d\tau^2}x^\mu(\tau) = \frac{d}{d\tau}p^\mu(\tau). \tag{1.105}$$

The simplest example of a force in Newtonian physics is the force due to gravity. In relativity, however, gravity is not described by a force, but rather by the curvature of spacetime itself. Instead, let us consider electromagnetism. The three-dimensional Lorentz force is given by $\mathbf{f} = q(\mathbf{E} + \mathbf{v} \times \mathbf{B})$, where q is the charge on the particle. We would like a tensorial generalization of this equation. There turns out to be a unique answer:

$$f^\mu = -qU^\lambda F_\lambda{}^\mu. \tag{1.106}$$

You can check for yourself that this reduces to the Newtonian version in the limit of small velocities. Notice how the requirement that the equation be tensorial, which is one way of guaranteeing Lorentz invariance, severely restricts the possible expressions we can get. This is an example of a very general phenomenon, in which a small number of an apparently endless variety of possible physical laws are picked out by the demands of symmetry.

Although p^μ provides a complete description of the energy and momentum of an individual particle, we often need to deal with extended systems comprised of huge numbers of particles. Rather than specify the individual momentum vectors of each particle, we instead describe the system as a **fluid**—a continuum characterized by macroscopic quantities such as density, pressure, entropy, viscosity, and so on. Although such a fluid may be composed of many individual particles with different four-velocities, the fluid itself has an overall four-velocity field. Just think of everyday fluids like air or water, where it makes sense to define a velocity for each individual fluid element even though nearby molecules may have appreciable relative velocities.

A single momentum four-vector field is insufficient to describe the energy and momentum of a fluid; we must go further and define the **energy-momentum tensor** (sometimes called the stress-energy tensor), $T^{\mu\nu}$. This symmetric (2, 0) tensor tells us all we need to know about the energy-like aspects of a system: energy density, pressure, stress, and so forth. A general definition of $T^{\mu\nu}$ is "the flux of four-momentum p^μ across a surface of constant x^ν." In fact, this definition is not going to be incredibly useful; in Chapter 4 we will define the energy-momentum tensor in terms of a functional derivative of the action with respect to the metric, which will be a more algorithmic procedure for finding an explicit expression for $T^{\mu\nu}$. But the definition here does afford some physical insight. Consider an infinitesimal element of the fluid in its rest frame, where there are no bulk motions. Then T^{00}, the "flux of p^0 (energy) in the x^0 (time) direction," is simply the rest-frame **energy density** ρ. Similarly, in this frame, $T^{0i} = T^{i0}$ is the momentum density. The spatial components T^{ij} are the momentum flux, or the *stress*; they represent the forces between neighboring infinitesimal elements of the fluid. Off-diagonal terms in T^{ij} represent shearing terms, such as those due to viscosity. A diagonal term such as T^{11} gives the x-component of the force being exerted (per unit area) by a fluid element in the x-direction; this is what we think of as the x-component of the **pressure,** p_x (don't confuse it with the momentum). The pressure has three components, given in the fluid rest frame (in inertial coordinates) by

$$p_i = T^{ii}. \tag{1.107}$$

There is no sum over i.

To make this more concrete, let's start with the simple example of **dust**. (Cosmologists tend to use "matter" as a synonym for dust.) Dust may be defined in flat spacetime as a collection of particles at rest with respect to each other. The four-velocity field $U^\mu(x)$ is clearly going to be the constant four-velocity of the individual particles. Indeed, its components will be the same at each point. Define the **number-flux four-vector** to be

$$N^\mu = nU^\mu, \tag{1.108}$$

where n is the number density of the particles as measured in their rest frame. (This doesn't sound coordinate-invariant, but it is; in any frame, the number density that would be measured if you were in the rest frame is a fixed quantity.) Then N^0 is the number density of particles as measured in any other frame, while N^i is the flux of particles in the x^i direction. Let's now imagine that each of the particles has the same mass m. Then in the rest frame the energy density of the dust is given by

$$\rho = mn. \tag{1.109}$$

By definition, the energy density completely specifies the dust. But ρ only measures the energy density in the rest frame; what about other frames? We notice that both n and m are 0-components of four-vectors in their rest frame; specifically, $N^\mu = (n, 0, 0, 0)$ and $p^\mu = (m, 0, 0, 0)$. Therefore ρ is the $\mu = 0$, $\nu = 0$ component of the tensor $p \otimes N$ as measured in its rest frame. We are therefore led to define the energy-momentum tensor for dust:

$$T^{\mu\nu}_{\text{dust}} = p^\mu N^\nu = mnU^\mu U^\nu = \rho U^\mu U^\nu, \tag{1.110}$$

where ρ is defined as the energy density in the rest frame. (Typically you don't just guess energy-momentum tensors by such a procedure, you derive them from equations of motion or an action principle.) Note that the pressure of the dust in any direction is zero; this should not be surprising, since pressure arises from the random motions of particles within the fluid, and we have defined dust to be free of such motions.

Dust is not sufficiently general to describe most of the interesting fluids that appear in general relativity; we only need a slight generalization, however, to arrive at the concept of a **perfect fluid**. A perfect fluid is one that can be completely specified by two quantities, the rest-frame energy density ρ, and an isotropic rest-frame pressure p. The single parameter p serves to specify the pressure in every direction. A consequence of isotropy is that $T^{\mu\nu}$ is diagonal in its rest frame—there is no net flux of any component of momentum in an orthogonal direction. Furthermore, the nonzero spacelike components must all be equal, $T^{11} = T^{22} = T^{33}$. The only two independent numbers are therefore the energy density $\rho = T^{00}$ and the pressure $p = T^{ii}$; we don't need a subscript on p, since the pressure is equal in every direction. The energy-momentum tensor of a perfect fluid therefore takes the following form in its rest frame:

$$T^{\mu\nu} = \begin{pmatrix} \rho & 0 & 0 & 0 \\ 0 & p & 0 & 0 \\ 0 & 0 & p & 0 \\ 0 & 0 & 0 & p \end{pmatrix}. \tag{1.111}$$

(Remember that we are in flat spacetime; this will change when curvature is introduced.) We would like, of course, a formula that is good in any frame. For dust we had $T^{\mu\nu} = \rho U^\mu U^\nu$, so we might begin by guessing $(\rho + p)U^\mu U^\nu$, which gives

$$\begin{pmatrix} \rho + p & 0 & 0 & 0 \\ 0 & 0 & 0 & 0 \\ 0 & 0 & 0 & 0 \\ 0 & 0 & 0 & 0 \end{pmatrix}. \tag{1.112}$$

This is not a very clever guess, to be honest. But by subtracting this guess from our desired answer, we see that what we need to add is

$$\begin{pmatrix} -p & 0 & 0 & 0 \\ 0 & p & 0 & 0 \\ 0 & 0 & p & 0 \\ 0 & 0 & 0 & p \end{pmatrix}. \tag{1.113}$$

Fortunately, this has an obvious covariant generalization, namely $p\eta^{\mu\nu}$. Thus, the general form of the energy-momentum tensor for a perfect fluid is

$$\boxed{T^{\mu\nu} = (\rho + p)U^\mu U^\nu + p\eta^{\mu\nu}.} \tag{1.114}$$

It may seem that the procedure used to arrive at this formula was somewhat arbitrary, but we can have complete confidence in the result. Given that (1.111) should be the form of $T^{\mu\nu}$ in the rest frame, and that (1.114) is a perfectly tensorial expression that reduces to (1.111) in the rest frame, we know that (1.114) must be the right expression in any frame.

The concept of a perfect fluid is general enough to describe a wide variety of physical forms of matter. To determine the evolution of such a fluid, we specify an equation of state relating the pressure to the energy density, $p = p(\rho)$. Dust is a special case for which $p = 0$, while an isotropic gas of photons has $p = \frac{1}{3}\rho$. A more exotic example is vacuum energy, for which the energy-momentum tensor is proportional to the metric, $T^{\mu\nu} = -\rho_{\rm vac}\eta^{\mu\nu}$. By comparing to (1.114) we find that vacuum energy is a kind of perfect fluid for which $p_{\rm vac} = -\rho_{\rm vac}$. The notion of an energy density in vacuum is completely pointless in special relativity, since in nongravitational physics the absolute value of the energy doesn't matter, only the difference in energy between two states. In general relativity, however, all energy couples to gravity, so the possibility of a nonzero vacuum energy will become an important consideration, which we will discuss more fully in Chapter 4.

Besides being symmetric, $T^{\mu\nu}$ has the even more important property of being *conserved*. In this context, conservation is expressed as the vanishing of the "divergence":

$$\boxed{\partial_\mu T^{\mu\nu} = 0.} \tag{1.115}$$

This expression is a set of four equations, one for each value of ν. The equation with $\nu = 0$ corresponds to conservation of energy, while $\partial_\mu T^{\mu k} = 0$ expresses conservation of the kth component of the momentum. Let's apply this equation to a perfect fluid, for which we have

$$\partial_\mu T^{\mu\nu} = \partial_\mu(\rho + p)U^\mu U^\nu + (\rho + p)(U^\nu \partial_\mu U^\mu + U^\mu \partial_\mu U^\nu) + \partial^\nu p. \quad (1.116)$$

To analyze what this equation means, it is helpful to consider separately what happens when we project it into pieces along and orthogonal to the four-velocity field U^μ. We first note that the normalization $U_\nu U^\nu = -1$ implies the useful identity

$$U_\nu \partial_\mu U^\nu = \tfrac{1}{2}\partial_\mu(U_\nu U^\nu) = 0. \quad (1.117)$$

To project (1.116) along the four-velocity, simply contract it into U_ν:

$$U_\nu \partial_\mu T^{\mu\nu} = -\partial_\mu(\rho U^\mu) - p\partial_\mu U^\mu. \quad (1.118)$$

Setting this to zero gives the relativistic equation of energy conservation for a perfect fluid. It will look more familiar in the nonrelativistic limit, in which

$$U^\mu = (1, v^i), \qquad |v^i| \ll 1, \qquad p \ll \rho. \quad (1.119)$$

The last condition makes sense, because pressure comes from the random motions of the individual particles, and in this limit these motions (as well as the bulk motion described by U^μ) are taken to be small. So in ordinary nonrelativistic language, (1.118) becomes

$$\partial_t \rho + \nabla \cdot (\rho \mathbf{v}) = 0, \quad (1.120)$$

the continuity equation for the energy density. We next consider the part of (1.116) that is orthogonal to the four-velocity. To project a vector orthogonal to U^μ, we multiply it by the projection tensor

$$P^\sigma{}_\nu = \delta^\sigma_\nu + U^\sigma U_\nu. \quad (1.121)$$

To convince yourself this does the trick, check that if we have a vector $V^\mu_\parallel$, parallel to U^μ, and another vector $W^\mu_\perp$, perpendicular to U^μ, the projection tensor will annihilate the parallel vector and preserve the orthogonal one:

$$\begin{aligned} P^\sigma{}_\nu V^\nu_\parallel &= 0 \\ P^\sigma{}_\nu W^\nu_\perp &= W^\sigma_\perp. \end{aligned} \quad (1.122)$$

Applied to $\partial_\mu T^{\mu\nu}$, we obtain

$$P^\sigma{}_\nu \partial_\mu T^{\mu\nu} = (\rho + p)U^\mu \partial_\mu U^\sigma + \partial^\sigma p + U^\sigma U^\mu \partial_\mu p. \quad (1.123)$$

In the nonrelativistic limit given by (1.119), setting the spatial components of this expression equal to zero yields

$$\rho\,[\partial_t \mathbf{v} + (\mathbf{v}\cdot\nabla)\mathbf{v}] + \nabla p + \mathbf{v}(\partial_t p + \mathbf{v}\cdot\nabla p) = 0. \quad (1.124)$$

But notice that the last set of terms involve derivatives of p times the three-velocity $\mathbf{v}$, assumed to be small; these will therefore be negligible compared to the ∇p term, and can be neglected. We are left with

$$\rho\,[\partial_t \mathbf{v} + (\mathbf{v}\cdot\nabla)\mathbf{v}] = -\nabla p, \tag{1.125}$$

which is the Euler equation familiar from fluid mechanics.

1.10 ■ CLASSICAL FIELD THEORY

When we make the transition from special relativity to general relativity, the metric $\eta_{\mu\nu}$ will be promoted to a dynamical tensor field, $g_{\mu\nu}(x)$. GR is thus a particular example of a classical field theory; we can build up some feeling for how such theories work by considering classical fields defined on flat spacetime. (We say classical field theory in contrast with quantum field theory, which is quite a different story; we will discuss it briefly in Chapter 9, but it is outside our main area of interest here.)

Let's begin with the familiar example of the classical mechanics of a single particle in one dimension with coordinate $q(t)$. We can derive the equations of motion for such a particle by using the "principle of least action": we search for critical points (as a function of the trajectory) of an **action** S, written as

$$S = \int dt\, L(q, \dot{q}), \tag{1.126}$$

where the function $L(q, \dot{q})$ is the **Lagrangian**. The Lagrangian in point-particle mechanics is typically of the form

$$L = K - V, \tag{1.127}$$

where K is the kinetic energy and V the potential energy. Following the calculus-of-variations procedure, which is described in any advanced textbook on classical mechanics, we show that critical points of the action [trajectories $q(t)$ for which S remains stationary under small variations] are those that satisfy the **Euler–Lagrange equations**,

$$\frac{\partial L}{\partial q} - \frac{d}{dt}\left(\frac{\partial L}{\partial(\dot{q})}\right) = 0. \tag{1.128}$$

For example, $L = \frac{1}{2}\dot{q}^2 - V(q)$ leads to

$$\ddot{q} = -\frac{dV}{dq}. \tag{1.129}$$

Field theory is a similar story, except that we replace the single coordinate $q(t)$ by a set of spacetime-dependent **fields**, $\Phi^i(x^\mu)$, and the action S becomes a *functional* of these fields. A functional is simply a function of an infinite number of variables, such as the values of a field in some region of spacetime. Functionals are often expressed as integrals. Each Φ^i is a function on spacetime (at least in some coordinate system), and i is an index labeling our individual fields. For example, in electromagnetism (as we will see below) the fields are the four components of a one-form called the "vector potential," A_μ:

$$\Phi^i = \{A_0, A_1, A_2, A_3\}. \tag{1.130}$$

We're being very lowbrow here, in thinking of a one-form field as four different functions rather than a single tensor object. This point of view makes sense so long as we stick to a fixed coordinate system, and it will make our calculations more straightforward.

In field theory, the Lagrangian can be expressed as an integral over space of a **Lagrange density** $\mathcal{L}$, which is a function of the fields Φ^i and their spacetime derivatives $\partial_\mu \Phi^i$:

$$L = \int d^3x\, \mathcal{L}(\Phi^i, \partial_\mu \Phi^i). \tag{1.131}$$

So the action is

$$S = \int dt\, L = \int d^4x\, \mathcal{L}(\Phi^i, \partial_\mu \Phi^i). \tag{1.132}$$

The Lagrange density is a Lorentz scalar. We typically just say "Lagrangian" when we mean "Lagrange density." It will most often be convenient to define a field theory by specifying the Lagrange density, from which all of the equations of motion can be readily derived.

We will use "natural units," in which not only $c = 1$ but also $\hbar = k = 1$, where $\hbar = h/2\pi$, h is Planck's constant, and k is Boltzmann's constant. The objection might be raised that we shouldn't involve $\hbar$ in a purely classical discussion; but all we are doing here is choosing units, not determining physics. (The relevance of $\hbar$ would appear if we were to quantize our field theory and obtain particles, but we won't get that far right now.) In natural units we have

$$[\text{energy}] = [\text{mass}] = [(\text{length})^{-1}] = [(\text{time})^{-1}]. \tag{1.133}$$

We will most often use energy or mass as our fundamental unit. Since the action is an integral of L (with units of energy) over time, it is dimensionless:

$$[S] = [E][T] = M^0. \tag{1.134}$$

The volume element has units

$$[d^4x] = M^{-4}, \tag{1.135}$$

so to get a dimensionless action we require that the Lagrange density have units

$$[\mathcal{L}] = M^4. \tag{1.136}$$

The Euler–Lagrange equations come from requiring that the action be unchanged under small variations of the fields,

$$\Phi^i \to \Phi^i + \delta\Phi^i, \tag{1.137}$$

$$\partial_\mu \Phi^i \to \partial_\mu \Phi^i + \delta(\partial_\mu \Phi^i) = \partial_\mu \Phi^i + \partial_\mu(\delta\Phi^i). \tag{1.138}$$

The expression for the variation in $\partial_\mu \Phi^i$ is simply the derivative of the variation of Φ^i. Since $\delta\Phi^i$ is assumed to be small, we may Taylor-expand the Lagrangian under this variation:

$$\begin{aligned}\mathcal{L}(\Phi^i, \partial_\mu \Phi^i) &\to \mathcal{L}(\Phi^i + \delta\Phi^i, \partial_\mu \Phi^i + \partial_\mu \delta\Phi^i)\\ &= \mathcal{L}(\Phi^i, \partial_\mu \Phi^i) + \frac{\partial \mathcal{L}}{\partial \Phi^i}\delta\Phi^i + \frac{\partial \mathcal{L}}{\partial(\partial_\mu \Phi^i)}\partial_\mu(\delta\Phi^i).\end{aligned} \tag{1.139}$$

Correspondingly, the action goes to $S \to S + \delta S$, with

$$\delta S = \int d^4x \left[\frac{\partial \mathcal{L}}{\partial \Phi^i}\delta\Phi^i + \frac{\partial \mathcal{L}}{\partial(\partial_\mu \Phi^i)}\partial_\mu(\delta\Phi^i)\right]. \tag{1.140}$$

We would like to factor out $\delta\Phi^i$ from the integrand, by integrating the second term by parts:

$$\begin{aligned}\int d^4x \frac{\partial \mathcal{L}}{\partial(\partial_\mu \Phi^i)}\partial_\mu(\delta\Phi^i) = &-\int d^4x\, \partial_\mu \left(\frac{\partial \mathcal{L}}{\partial(\partial_\mu \Phi^i)}\right)\delta\Phi^i\\ &+\int d^4x\, \partial_\mu \left(\frac{\partial \mathcal{L}}{\partial(\partial_\mu \Phi^i)}\delta\Phi^i\right).\end{aligned} \tag{1.141}$$

The final term is a total derivative—the integral of something of the form $\partial_\mu V^\mu$—that can be converted to a surface term by Stokes's theorem (the four-dimensional version, that is; see Appendix E for a discussion). Since we are considering variational problems, we can choose to consider variations that vanish at the boundary (along with their derivatives). It is therefore traditional in such contexts to integrate by parts with complete impunity, always ignoring the boundary contributions. (Sometimes this is not okay, as in instanton calculations in Yang–Mills theory.)

We are therefore left with

$$\delta S = \int d^4x \left[\frac{\partial \mathcal{L}}{\partial \Phi^i} - \partial_\mu \left(\frac{\partial \mathcal{L}}{\partial(\partial_\mu \Phi^i)}\right)\right]\delta\Phi^i. \tag{1.142}$$

The functional derivative $\delta S/\delta\Phi^i$ of a functional S with respect to a function Φ^i is defined to satisfy

$$\delta S = \int d^4x \frac{\delta S}{\delta \Phi^i}\delta\Phi^i, \tag{1.143}$$

when such an expression is valid. We can therefore express the notion that S is at a critical point by saying that the functional derivative vanishes. The final equations of motion for our field theory are thus:

$$\boxed{\frac{\delta S}{\delta \Phi^i} = \frac{\partial \mathcal{L}}{\partial \Phi^i} - \partial_\mu \left(\frac{\partial \mathcal{L}}{\partial(\partial_\mu \Phi^i)} \right) = 0.} \tag{1.144}$$

These are known as the Euler–Lagrange equations for a field theory in flat spacetime.

The simplest example of a field is a real scalar field:

$$\phi(x^\mu) : (\text{spacetime}) \rightarrow \mathbf{R}. \tag{1.145}$$

Slightly more complicated examples would include complex scalar fields, or maps from spacetime to any vector space or even any manifold (sometimes called "nonlinear sigma models"). Upon quantization, excitations of the field are observable as particles. Scalar fields give rise to spinless particles, while vector fields and other tensors give rise to higher-spin particles. If the field were complex instead of real, it would have two degrees of freedom rather than just one, which would be interpreted as a particle and a distinct antiparticle. Real fields are their own antiparticles. An example of a real scalar field would be the neutral π-meson.

So let's consider the classical mechanics of a single real scalar field. It will have an energy density that is a local function of spacetime, and includes various contributions:

$$\begin{array}{rc} \text{kinetic energy :} & \frac{1}{2}\dot{\phi}^2 \\ \text{gradient energy :} & \frac{1}{2}(\nabla\phi)^2 \\ \text{potential energy :} & V(\phi). \end{array} \tag{1.146}$$

Actually, although the potential is a Lorentz-invariant function, the kinetic and gradient energies are not by themselves Lorentz-invariant; but we can combine them into a manifestly Lorentz-invariant form:

$$-\tfrac{1}{2}\eta^{\mu\nu}(\partial_\mu\phi)(\partial_\nu\phi) = \tfrac{1}{2}\dot{\phi}^2 - \tfrac{1}{2}(\nabla\phi)^2. \tag{1.147}$$

[The combination $\eta^{\mu\nu}(\partial_\mu\phi)(\partial_\nu\phi)$ is often abbreviated as $(\partial\phi)^2$.] So a reasonable choice of Lagrangian for our single real scalar field, analogous to $L = K - V$ in the point-particle case, would be

$$\mathcal{L} = -\tfrac{1}{2}\eta^{\mu\nu}(\partial_\mu\phi)(\partial_\nu\phi) - V(\phi). \tag{1.148}$$

This generalizes "kinetic minus potential energy" to "kinetic minus gradient minus potential energy density." Note that since $[\mathcal{L}] = M^4$, we must have $[V] = M^4$. Also, since $[\partial_\mu] = [\partial/\partial x^\mu] = M^1$, we have

$$[\phi] = M^1. \tag{1.149}$$

For the Lagrangian (1.148) we have

$$\frac{\partial \mathcal{L}}{\partial \phi} = -\frac{dV}{d\phi}, \quad \frac{\partial \mathcal{L}}{\partial(\partial_\mu \phi)} = -\eta^{\mu\nu}\partial_\nu \phi. \tag{1.150}$$

The second of these equations is a little tricky, so let's go through it slowly. When differentiating the Lagrangian, the trick is to make sure that the index placement is "compatible" (so that if you have a lower index on the thing being differentiated with respect to, you should have only lower indices when the same kind of object appears in the thing being differentiated), and also that the indices are strictly different. The first of these is already satisfied in our example, since we are differentiating a function of $\partial_\mu \phi$ with respect to $\partial_\mu \phi$. Later on, we will need to be more careful. To fulfill the second, we simply relabel dummy indices:

$$\eta^{\mu\nu}(\partial_\mu \phi)(\partial_\nu \phi) = \eta^{\rho\sigma}(\partial_\rho \phi)(\partial_\sigma \phi). \tag{1.151}$$

Then we can use the general rule, for any object with one index such as V_μ, that

$$\frac{\partial V_\alpha}{\partial V_\beta} = \delta_\alpha^\beta \tag{1.152}$$

because each component of V_α is treated as a distinct variable. So we have

$$\begin{aligned}\frac{\partial}{\partial(\partial_\mu \phi)}\left[\eta^{\rho\sigma}(\partial_\rho \phi)(\partial_\sigma \phi)\right] &= \eta^{\rho\sigma}\left[\delta_\rho^\mu(\partial_\sigma \phi) + (\partial_\rho \phi)\delta_\sigma^\mu\right] \\ &= \eta^{\mu\sigma}(\partial_\sigma \phi) + \eta^{\rho\mu}(\partial_\rho \phi) = 2\eta^{\mu\nu}\partial_\nu \phi. \end{aligned} \tag{1.153}$$

This leads to the second expression in (1.150).

Putting (1.150) into (1.144) leads to the equation of motion

$$\Box \phi - \frac{dV}{d\phi} = 0, \tag{1.154}$$

where $\Box \equiv \eta^{\mu\nu}\partial_\mu \partial_\nu$ is known as the **d'Alembertian**. Note that our metric sign convention $(-+++)$ comes into this equation; with the alternative $(+---)$ convention the sign would have been switched. In flat spacetime (1.154) is equivalent to

$$\ddot{\phi} - \nabla^2 \phi + \frac{dV}{d\phi} = 0. \tag{1.155}$$

A popular choice for the potential V is that of a simple harmonic oscillator, $V(\phi) = \frac{1}{2}m^2\phi^2$. The parameter m is called the mass of the field, and you should notice that the units work out correctly. You may be wondering how a field can have mass. When we quantize the field we find that momentum eigenstates are collections of particles, each with mass m. At the classical level, we think of "mass" as simply a convenient characterization of the field dynamics. Then our equation of motion is

$$\Box\phi - m^2\phi = 0, \tag{1.156}$$

the famous **Klein–Gordon equation**. This is a linear differential equation, so the sum of two solutions is a solution; a complete set of solutions (in the form of plane waves) is easy to find, as you can check for yourself.

A slightly more elaborate example of a field theory is provided by electromagnetism. We mentioned that the relevant field is the **vector potential** A_μ; the timelike component A_0 can be identified with the electrostatic potential Φ, and the spacelike components with the traditional vector potential $\mathbf{A}$ (in terms of which the magnetic field is given by $\mathbf{B} = \nabla \times \mathbf{A}$). The field strength tensor, with components given by (1.69), is related to the vector potential by

$$F_{\mu\nu} = \partial_\mu A_\nu - \partial_\nu A_\mu. \tag{1.157}$$

From this definition we see that the field strength tensor has the important property of **gauge invariance**: when we perform a **gauge transformation** on the vector potential,

$$A_\mu \to A_\mu + \partial_\mu\lambda(x), \tag{1.158}$$

the field strength tensor is left unchanged:

$$F_{\mu\nu} \to F_{\mu\nu} + \partial_\mu\partial_\nu\lambda - \partial_\nu\partial_\mu\lambda = F_{\mu\nu}. \tag{1.159}$$

The last equality follows from the fact that partial derivatives commute, $\partial_\mu\partial_\nu = \partial_\nu\partial_\mu$. Gauge invariance is a symmetry that is fundamental to our understanding of electromagnetism, and all observable quantities must be gauge-invariant. Thus, while the dynamical field of the theory (with respect to which we vary the action to derive equations of motion) is A_μ, physical quantities will generally be expressed in terms of $F_{\mu\nu}$.

We already know that the dynamical equations of electromagnetism are Maxwell's equations, (1.96) and (1.97). Given the definition of the field strength tensor in terms of the vector potential, (1.97) is actually automatic:

$$\partial_{[\mu}F_{\nu\sigma]} = \partial_{[\mu}\partial_\nu A_{\sigma]} - \partial_{[\mu}\partial_\sigma A_{\nu]} = 0, \tag{1.160}$$

again because partial derivatives commute. On the other hand, (1.96) is equivalent to Euler–Lagrange equations of the form

$$\frac{\partial\mathcal{L}}{\partial A_\nu} - \partial_\mu\left(\frac{\partial\mathcal{L}}{\partial(\partial_\mu A_\nu)}\right) = 0, \tag{1.161}$$

if we presciently choose the Lagrangian to be

$$\mathcal{L} = -\tfrac{1}{4}F_{\mu\nu}F^{\mu\nu} + A_\mu J^\mu. \tag{1.162}$$

For this choice, the first term in the Euler–Lagrange equation is straightforward:

$$\frac{\partial \mathcal{L}}{\partial A_\nu} = \delta^\nu_\mu J^\mu = J^\nu. \tag{1.163}$$

The second term is trickier. First we write $F_{\mu\nu}F^{\mu\nu}$ as

$$F_{\mu\nu}F^{\mu\nu} = F_{\alpha\beta}F^{\alpha\beta} = \eta^{\alpha\rho}\eta^{\beta\sigma}F_{\alpha\beta}F_{\rho\sigma}. \tag{1.164}$$

We want to work with lower indices on $F_{\mu\nu}$, since we are differentiating with respect to $\partial_\mu A_\nu$, which has lower indices. Likewise we change the dummy indices on $F_{\mu\nu}F^{\mu\nu}$, since we want to have different indices on the thing being differentiated and the thing we are differentiating with respect to. Once you get familiar with this stuff it will become second nature and you won't need nearly so many steps. This lets us write

$$\frac{\partial(F_{\alpha\beta}F^{\alpha\beta})}{\partial(\partial_\mu A_\nu)} = \eta^{\alpha\rho}\eta^{\beta\sigma}\left[\left(\frac{\partial F_{\alpha\beta}}{\partial(\partial_\mu A_\nu)}\right)F_{\rho\sigma} + F_{\alpha\beta}\left(\frac{\partial F_{\rho\sigma}}{\partial(\partial_\mu A_\nu)}\right)\right]. \tag{1.165}$$

Then, since $F_{\alpha\beta} = \partial_\alpha A_\beta - \partial_\beta A_\alpha$, we have

$$\frac{\partial F_{\alpha\beta}}{\partial(\partial_\mu A_\nu)} = \delta^\mu_\alpha\delta^\nu_\beta - \delta^\mu_\beta\delta^\nu_\alpha. \tag{1.166}$$

Combining (1.166) with (1.165) yields

$$\begin{aligned}\frac{\partial(F_{\alpha\beta}F^{\alpha\beta})}{\partial(\partial_\mu A_\nu)} &= \eta^{\alpha\rho}\eta^{\beta\sigma}\left[(\delta^\mu_\alpha\delta^\nu_\beta - \delta^\mu_\beta\delta^\nu_\alpha)F_{\rho\sigma} + (\delta^\mu_\rho\delta^\nu_\sigma - \delta^\mu_\sigma\delta^\nu_\rho)F_{\alpha\beta}\right] \\ &= (\eta^{\mu\rho}\eta^{\nu\sigma} - \eta^{\nu\rho}\eta^{\mu\sigma})F_{\rho\sigma} + (\eta^{\alpha\mu}\eta^{\beta\nu} - \eta^{\alpha\nu}\eta^{\beta\mu})F_{\alpha\beta} \\ &= F^{\mu\nu} - F^{\nu\mu} + F^{\mu\nu} - F^{\nu\mu} \\ &= 4F^{\mu\nu},\end{aligned} \tag{1.167}$$

so

$$\frac{\partial \mathcal{L}}{\partial(\partial_\mu A_\nu)} = -F^{\mu\nu}. \tag{1.168}$$

Then sticking (1.163) and (1.168) into (1.161) yields precisely (1.96):

$$\partial_\mu F^{\nu\mu} = J^\nu. \tag{1.169}$$

Note that we switched the order of the indices on $F^{\mu\nu}$ in order to save ourselves from an unpleasant minus sign.

You may wonder what the purpose of introducing a Lagrangian formulation is, if we were able to invent the equations of motion before we ever knew the Lagrangian (as Maxwell did for his equations). There are a number of reasons, starting with the basic simplicity of positing a single scalar function of spacetime, the Lagrange density, rather than a number of (perhaps tensor-valued) equations of motion. Another reason is the ease with which symmetries are implemented; demanding that the action be invariant under a symmetry ensures that the dynamics respects the symmetry as well. Finally, as we will see in Chapter 4, the action leads via a direct procedure (involving varying with respect to the metric itself) to a unique energy-momentum tensor. Applying this procedure to (1.148) leads straight to the energy-momentum tensor for a scalar field theory,

$$T^{\mu\nu}_{\text{scalar}} = \eta^{\mu\lambda}\eta^{\nu\sigma}\partial_\lambda\phi\partial_\sigma\phi - \eta^{\mu\nu}\left[\tfrac{1}{2}\eta^{\lambda\sigma}\partial_\lambda\phi\partial_\sigma\phi + V(\phi)\right]. \tag{1.170}$$

Similarly, from (1.162) we can derive the energy-momentum tensor for electromagnetism,

$$T^{\mu\nu}_{\text{EM}} = F^{\mu\lambda}F^{\nu}{}_{\lambda} - \tfrac{1}{4}\eta^{\mu\nu}F^{\lambda\sigma}F_{\lambda\sigma}. \tag{1.171}$$

Using the appropriate equations of motion, you can show that these energy-momentum tensors are conserved, $\partial_\mu T^{\mu\nu} = 0$ (and will be asked to do so in the Exercises).

The two examples we have considered—scalar field theory and electromagnetism—are paradigms for much of our current understanding of nature. The Standard Model of particle physics consists of three types of fields: gauge fields, Higgs fields, and fermions. The gauge fields describe the "forces" of nature, including the strong and weak nuclear forces in addition to electromagnetism. The gauge fields giving rise to the nuclear forces are described by one-form potentials, just as in electromagnetism; the difference is that they are matrix-valued rather than ordinary one-forms, and the symmetry groups corresponding to gauge transformations are therefore noncommutative (nonabelian) symmetries. The Higgs fields are scalar fields much as we have described, although they are also matrix-valued. The fermions include leptons (such as electrons and neutrinos) and quarks, and are not described by any of the tensor fields we have discussed here, but rather by a different kind of field called a **spinor**. We won't get around to discussing spinors in this book, but they play a crucial role in particle physics and their coupling to gravity is interesting and subtle. Upon quantization, these fields give rise to particles of different spins; gauge fields are spin-1, scalar fields are spin-0, and the Standard Model fermions are spin-$\frac{1}{2}$.

Before concluding this chapter, let's ask an embarassingly simple question: Why should we consider one classical field theory rather than some other one? More concretely, let's say that we have discovered some particle in nature, and we know what kind of field we want to use to describe it; how should we pick the Lagrangian for this field? For example, when we wrote down our scalar-field Lagrangian (1.148), why didn't we include a term of the form

$$\mathcal{L}' = \lambda\phi^2\eta^{\mu\nu}(\partial_\mu\phi)(\partial_\nu\phi), \tag{1.172}$$

where λ is a coupling constant? Ultimately, of course, we work by trial and error and try to fit the data given to us by experiment. In classical field theory, there's not much more we could do; generally we would start with a simple Lagrangian, and perhaps make it more complicated if the first try failed to agree with the data. But quantum field theory actually provides some simple guidelines, and since we use classical field theory as an approximation to some underlying quantum theory, it makes sense to take advantage of these principles. To make a long story short, quantum field theory allows "virtual" processes at arbitrarily high energies to contribute to what we observe at low energies. Fortunately, the effect of these processes can be summarized in a low-energy **effective field theory**. In the effective theory, which is what we actually observe, the result of high-energy processes is simply to "renormalize" the coupling constants of our theory. Consider an arbitrary coupling constant, which we can express as a parameter μ (with dimensions of mass) raised to some power, $\lambda = \mu^q$ (unless λ is dimensionless, in which case the discussion becomes more subtle). Very roughly speaking, *the effect of high-energy processes will be to make μ very large*. Slightly more specifically, μ will be pushed up to a scale at which new physics kicks in, whatever that may be. Therefore, potential higher-order terms we might think of adding to a Lagrangian are suppressed, because they are multiplied by coupling constants that are very small. For (1.172), for example, we must have $\lambda = \mu^{-2}$, so λ will be tiny (because μ will be big). Only the lowest-order terms we can put in our Lagrangian will come with dimensionless couplings (or ones with units of mass to a positive power), so we only need bother with those at low energies. This feature of field theory allows for a dramatic simplification in considering all of the models we might want to examine.

As mentioned at the beginning of this section, general relativity itself is a classical field theory, in which the dynamical field is the metric tensor. It is nevertheless fair to think of GR as somehow different; for the most part other classical field theories rely on the existence of a pre-existing spacetime geometry, whereas in GR the geometry is determined by the equations of motion. (There are exceptions to this idea, called topological field theories, in which the metric makes no appearance.) Our task in the next few chapters is to explore the nature of curved geometries as characterized by the spacetime metric, before moving in Chapter 4 to putting these notions to work in constructing a theory of gravitation.

1.11 ■ EXERCISES

1. Consider an inertial frame S with coordinates $x^\mu = (t, x, y, z)$, and a frame S' with coordinates $x^{\mu'}$ related to S by a boost with velocity parameter v along the y-axis. Imagine we have a wall at rest in S', lying along the line $x' = -y'$. From the point of view of S, what is the relationship between the incident angle of a ball hitting the wall (traveling in the x-y plane) and the reflected angle? What about the velocity before and after?

2. Imagine that space (not spacetime) is actually a finite box, or in more sophisticated terms, a three-torus, of size L. By this we mean that there is a coordinate system $x^\mu = (t, x, y, z)$ such that every point with coordinates (t, x, y, z) is *identified* with every point with coordinates $(t, x + L, y, z)$, $(t, x, y + L, z)$, and $(t, x, y, z + L)$. Note that the time coordinate is the same. Now consider two observers; observer A is at rest in this coordinate system (constant spatial coordinates), while observer B moves in the x-direction with constant velocity v. A and B begin at the same event, and while A remains still, B moves once around the universe and comes back to intersect the worldline of A without ever having to accelerate (since the universe is periodic). What are the relative proper times experienced in this interval by A and B? Is this consistent with your understanding of Lorentz invariance?

3. Three events, A, B, C, are seen by observer $\mathcal{O}$ to occur in the order ABC. Another observer, $\bar{\mathcal{O}}$, sees the events to occur in the order CBA. Is it possible that a third observer sees the events in the order ACB? Support your conclusion by drawing a spacetime diagram.

4. Projection effects can trick you into thinking that an astrophysical object is moving "superluminally." Consider a quasar that ejects gas with speed v at an angle θ with respect to the line-of-sight of the observer. Projected onto the sky, the gas appears to travel perpendicular to the line of sight with angular speed $v_{\rm app}/D$, where D is the distance to the quasar and $v_{\rm app}$ is the apparent speed. Derive an expression for $v_{\rm app}$ in terms of v and θ. Show that, for appropriate values of v and θ, $v_{\rm app}$ can be greater than 1.

5. Particle physicists are so used to setting $c = 1$ that they measure mass in units of energy. In particular, they tend to use electron volts (1 eV $= 1.6 \times 10^{-12}$ erg $= 1.8 \times 10^{-33}$ g), or, more commonly, keV, MeV, and GeV (10^3 eV, 10^6 eV, and 10^9 eV, respectively). The muon has been measured to have a mass of 0.106 GeV and a rest frame lifetime of 2.19×10^{-6} seconds. Imagine that such a muon is moving in the circular storage ring of a particle accelerator, 1 kilometer in diameter, such that the muon's total energy is 1000 GeV. How long would it appear to live from the experimenter's point of view? How many radians would it travel around the ring?

6. In Euclidean three-space, let p be the point with coordinates $(x, y, z) = (1, 0, -1)$. Consider the following curves that pass through p:

$$x^i(\lambda) = (\lambda, (\lambda - 1)^2, -\lambda)$$
$$x^i(\mu) = (\cos\mu, \sin\mu, \mu - 1)$$
$$x^i(\sigma) = (\sigma^2, \sigma^3 + \sigma^2, \sigma).$$

(a) Calculate the components of the tangent vectors to these curves at p in the coordinate basis $\{\partial_x, \partial_y, \partial_z\}$.

(b) Let $f = x^2 + y^2 - yz$. Calculate $df/d\lambda$, $df/d\mu$ and $df/d\sigma$.

7. Imagine we have a tensor $X^{\mu\nu}$ and a vector V^μ, with components

$$X^{\mu\nu} = \begin{pmatrix} 2 & 0 & 1 & -1 \\ -1 & 0 & 3 & 2 \\ -1 & 1 & 0 & 0 \\ -2 & 1 & 1 & -2 \end{pmatrix}, \qquad V^\mu = (-1, 2, 0, -2).$$

Find the components of:

(a) $X^{\mu}{}_{\nu}$

(b) $X_{\mu}{}^{\nu}$

(c) $X^{(\mu\nu)}$

(d) $X_{[\mu\nu]}$

(e) $X^{\lambda}{}_{\lambda}$

(f) $V^{\mu}V_{\mu}$

(g) $V_{\mu}X^{\mu\nu}$

8. If $\partial_{\nu}T^{\mu\nu} = Q^{\mu}$, what physically does the spatial vector Q^{i} represent? Use the dust energy momentum tensor to make your case.

9. For a system of discrete point particles the energy-momentum tensor takes the form

$$T_{\mu\nu} = \sum_{a} \frac{p_{\mu}^{(a)} p_{\nu}^{(a)}}{p^{0(a)}} \delta^{(3)}(\mathbf{x} - \mathbf{x}^{(a)}), \tag{1.173}$$

where the index a labels the different particles. Show that, for a dense collection of particles with isotropically distributed velocities, we can smooth over the individual particle worldlines to obtain the perfect-fluid energy-momentum tensor (1.114).

10. Using the tensor transformation law applied to $F_{\mu\nu}$, show how the electric and magnetic field 3-vectors **E** and **B** transform under

(a) a rotation about the y-axis,

(b) a boost along the z-axis.

11. Verify that (1.98) is indeed equivalent to (1.97), and that they are both equivalent to the last two equations in (1.93).

12. Consider the two field theories we explicitly discussed, Maxwell's electromagnetism (let $J^{\mu} = 0$) and the scalar field theory defined by (1.148).

(a) Express the components of the energy-momentum tensors of each theory in three-vector notation, using the divergence, gradient, curl, electric, and magnetic fields, and an overdot to denote time derivatives.

(b) Using the equations of motion, verify (in any notation you like) that the energy-momentum tensors are conserved.

13. Consider adding to the Lagrangian for electromagnetism an additional term of the form $\mathcal{L}' = \tilde{\epsilon}_{\mu\nu\rho\sigma} F^{\mu\nu} F^{\rho\sigma}$.

(a) Express $\mathcal{L}'$ in terms of **E** and **B**.

(b) Show that including $\mathcal{L}'$ does not affect Maxwell's equations. Can you think of a deep reason for this?

CHAPTER 2

Manifolds

2.1 ■ GRAVITY AS GEOMETRY

Gravity is special. In the context of general relativity, we ascribe this specialness to the fact that the dynamical field giving rise to gravitation is the metric tensor describing the curvature of spacetime itself, rather than some additional field propagating through spacetime; this was Einstein's profound insight. The physical principle that led him to this idea was the *universality* of the gravitational interaction, as formalized by the **Principle of Equivalence**. Let's see how this physical principle leads us to the mathematical strategy of describing gravity as the geometry of a curved manifold.

The Principle of Equivalence comes in a variety of forms, the first of which is the **Weak Equivalence Principle**, or WEP. The WEP states that the inertial mass and gravitational mass of any object are equal. To see what this means, think about Newtonian mechanics. The Second Law relates the force exerted on an object to the acceleration it undergoes, setting them proportional to each other with the constant of proportionality being the inertial mass m_i:

$$\mathbf{F} = m_i \mathbf{a}. \tag{2.1}$$

The inertial mass clearly has a universal character, related to the resistance you feel when you try to push on the object; it takes the same value no matter what kind of force is being exerted. We also have Newton's law of gravitation, which can be thought of as stating that the gravitational force exerted on an object is proportional to the gradient of a scalar field Φ, known as the gravitational potential. The constant of proportionality in this case is called the gravitational mass m_g:

$$\mathbf{F}_g = -m_g \nabla\Phi. \tag{2.2}$$

On the face of it, m_g has a very different character than m_i; it is a quantity specific to the gravitational force. If you like, m_g/m_i can be thought of as the "gravitational charge" of the body. Nevertheless, Galileo long ago showed (apocryphally by dropping weights off of the Leaning Tower of Pisa, actually by rolling balls down inclined planes) that the response of matter to gravitation is universal—every object falls at the same rate in a gravitational field, independent of the composition of the object. In Newtonian mechanics this translates into the WEP, which is simply

$$m_i = m_g \tag{2.3}$$

for any object. An immediate consequence is that the behavior of freely-falling test particles is universal, independent of their mass (or any other qualities they may have); in fact, we have

$$\mathbf{a} = -\nabla\Phi. \tag{2.4}$$

Experimentally, the independence of the acceleration due to gravity on the composition of the falling object has been verified to extremely high precision by the Eötvös experiment and its modern successors.

This suggests an equivalent formulation of the WEP: there exists a preferred class of trajectories through spacetime, known as *inertial* (or "freely-falling") trajectories, on which unaccelerated particles travel—where unaccelerated means "subject only to gravity." Clearly this is not true for other forces, such as electromagnetism. In the presence of an electric field, particles with opposite charges will move on quite different trajectories. Every particle, on the other hand, has an identical gravitational charge.

The universality of gravitation, as implied by the WEP, can be stated in another, more popular, form. Imagine that we consider a physicist in a tightly sealed box, unable to observe the outside world, who is doing experiments involving the motion of test particles, for example to measure the local gravitational field. Of course she would obtain different answers if the box were sitting on the moon or on Jupiter than she would on Earth. But the answers would also be different if the box were accelerating at a constant rate; this would change the acceleration of the freely-falling particles with respect to the box. The WEP implies that there is no way to disentangle the effects of a gravitational field from those of being in a uniformly accelerating frame, simply by observing the behavior of freely-falling particles. This follows from the universality of gravitation; in electrodynamics, in contrast, it would be possible to distinguish between uniform acceleration and an electromagnetic field, by observing the behavior of particles with different charges. But with gravity it is impossible, since the "charge" is necessarily proportional to the (inertial) mass.

To be careful, we should limit our claims about the impossibility of distinguishing gravity from uniform acceleration by restricting our attention to "small enough regions of spacetime." If the sealed box were sufficiently big, the gravitational field would change from place to place in an observable way, while the effect of acceleration would always be in the same direction. In a rocket ship or elevator, the particles would always fall straight down. In a very big box in a gravitational field, however, the particles would move toward the center of the Earth, for example, which would be a different direction for widely separated experiments. The WEP can therefore be stated as follows: *The motion of freely-falling particles are the same in a gravitational field and a uniformly accelerated frame, in small enough regions of spacetime.* In larger regions of spacetime there will be inhomogeneities in the gravitational field, which will lead to tidal forces, which can be detected.

After the advent of special relativity, the concept of mass lost some of its uniqueness, as it became clear that mass was simply a manifestation of energy and momentum (as we have seen in Chapter 1). It was therefore natural for Einstein to think about generalizing the WEP to something more inclusive. His idea was simply that there should be no way whatsoever for the physicist in the box to distinguish between uniform acceleration and an external gravitational field, no matter what experiments she did (not only by dropping test particles). This reasonable extrapolation became what is now known as the **Einstein Equivalence Principle**, or EEP: *In small enough regions of spacetime, the laws of physics reduce to those of special relativity; it is impossible to detect the existence of a gravitational field by means of local experiments.*

In fact, it is hard to imagine theories that respect the WEP but violate the EEP. Consider a hydrogen atom, a bound state of a proton and an electron. Its mass is actually less than the sum of the masses of the proton and electron considered individually, because there is a negative binding energy—you have to put energy into the atom to separate the proton and electron. According to the WEP, the gravitational mass of the hydrogen atom is therefore less than the sum of the masses of its constituents; the gravitational field couples to electromagnetism (which holds the atom together) in exactly the right way to make the gravitational mass come out right. This means that not only must gravity couple to rest mass universally, but also to all forms of energy and momentum—which is practically the claim of the EEP. It is possible to come up with counterexamples, however; for example, we could imagine a theory of gravity in which freely falling particles began to rotate as they moved through a gravitational field. Then they could fall along the same paths as they would in an accelerated frame (thereby satisfying the WEP), but you could nevertheless detect the existence of the gravitational field (in violation of the EEP). Such theories seem contrived, but there is no law of nature that forbids them.

Sometimes a distinction is drawn between "gravitational laws of physics" and "nongravitational laws of physics," and the EEP is defined to apply only to the latter. Then the Strong Equivalence Principle (SEP) is defined to include all of the laws of physics, gravitational and otherwise. A theory that violated the SEP but not the EEP would be one in which the *gravitational* binding energy did not contribute equally to the inertial and gravitational mass of a body; thus, for example, test particles with appreciable self-gravity (to the extent that such a concept makes sense) could fall along different trajectories than lighter particles.

It is the EEP that implies (or at least suggests) that we should attribute the action of gravity to the curvature of spacetime. Remember that in special relativity a prominent role is played by inertial frames—while it is not possible to single out some frame of reference as uniquely "at rest," it is possible to single out a family of frames that are "unaccelerated" (inertial). The acceleration of a charged particle in an electromagnetic field is therefore uniquely defined with respect to these frames. The EEP, on the other hand, implies that gravity is inescapable—there is no such thing as a "gravitationally neutral object" with respect to which we can measure the acceleration due to gravity. It follows that the acceleration due to gravity is not something that can be reliably defined, and therefore is of little use.

Instead, it makes more sense to *define* "unaccelerated" as "freely falling," and that is what we shall do. From here we are led to the idea that gravity is not a "force"—a force is something that leads to acceleration, and our definition of zero acceleration is "moving freely in the presence of whatever gravitational field happens to be around."

This seemingly innocuous step has profound implications for the nature of spacetime. In SR, we have a procedure for starting at some point and constructing an inertial frame that stretches throughout spacetime, by joining together rigid rods and attaching clocks to them. But, again due to inhomogeneities in the gravitational field, this is no longer possible. If we start in some freely-falling state and build a large structure out of rigid rods, at some distance away freely-falling objects will look like they are accelerating with respect to this reference frame, as shown in Figure 2.1. The solution is to retain the notion of inertial frames, but to discard the hope that they can be uniquely extended throughout space and time. Instead we can define **locally inertial frames**, those that follow the motion of individual freely falling particles in small enough regions of spacetime. (Every time we say "small enough regions," purists should imagine a limiting procedure in which we take the appropriate spacetime volume to zero.) This is the best we can do, but it forces us to give up a good deal. For example, we can no longer speak with confidence about the relative velocity of far-away objects, since the inertial reference frames appropriate to those objects are completely different from those appropriate to us.

Our job as physicists is to construct mathematical models of the world, and then test the predictions of such models against observations and experiments. Following the implications of the universality of gravitation has led us to give up on the idea of expressing gravity as a force propagating through spacetime,

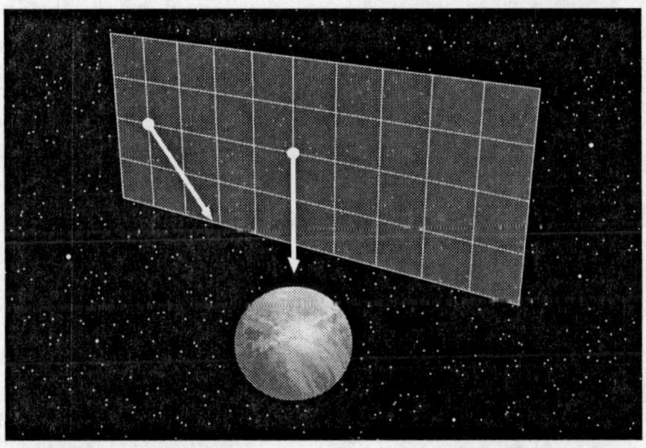

FIGURE 2.1 Failure of global frames. Since every particle feels the influence of gravity, we define "unaccelerating" as "freely falling." As a consequence, it becomes impossible to define globally inertial coordinate systems by the procedure outlined in Chapter 1, since particles initially at rest will begin to move with respect to such a frame.

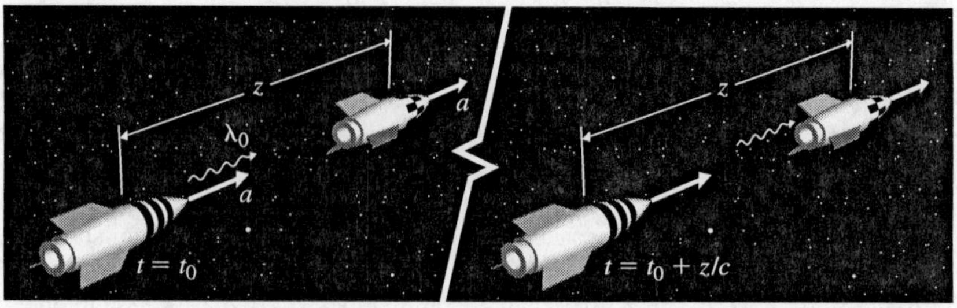

FIGURE 2.2 The Doppler shift as measured by two rockets separated by a distance z, each feeling an acceleration a.

and indeed to give up on the idea of global reference frames stretching throughout spacetime. We therefore need to invoke a mathematical framework in which physical theories can be consistent with these conclusions. The solution will be to imagine that spacetime has a curved geometry, and that gravitation is a manifestation of this curvature. The appropriate mathematical structure used to describe curvature is that of a *differentiable manifold*: essentially, a kind of set that looks locally like flat space, but might have a very different global geometry. (Remember that the EEP can be stated as "the laws of physics reduce to those of special relativity in small regions of spacetime," which matches well with the mathematical notion of a set that locally resembles flat space.)

We cannot prove that gravity should be thought of as the curvature of spacetime; instead we can propose the idea, derive its consequences, and see if the result is a reasonable fit to our experience of the world. Let's set about doing just that.

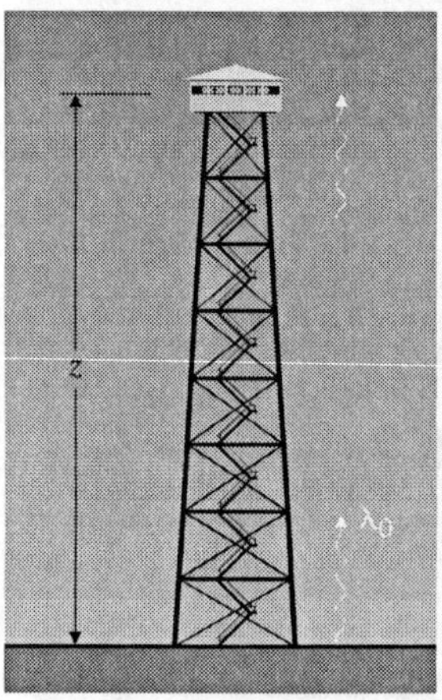

FIGURE 2.3 Gravitational redshift on the surface of the Earth, as measured by observers at different elevations.

Consider one of the celebrated predictions of the EEP, the gravitational redshift. Imagine two rockets, a distance z apart, each moving with some constant acceleration a in a region far away from any gravitational fields, as shown in Figure 2.2. At time t_0 the trailing rocket emits a photon of wavelength λ_0. The rockets remain a constant distance apart, so the photon reaches the leading rocket after a time $\Delta t = z/c$ in our background reference frame. (We assume $\Delta v/c$ is small, so we only work to first order.) In this time the rockets will have picked up an additional velocity $\Delta v = a\Delta t = az/c$. Therefore, the photon reaching the leading rocket will be redshifted by the conventional Doppler effect, by an amount

$$\frac{\Delta\lambda}{\lambda_0} = \frac{\Delta v}{c} = \frac{az}{c^2}. \tag{2.5}$$

According to the EEP, the same thing should happen in a uniform gravitational field. So we imagine a tower of height z sitting on the surface of a planet, with a_g the strength of the gravitational field (what Newton would have called the "acceleration due to gravity"), as portrayed in Figure 2.3. We imagine that observers in the rocket at the top of the tower are able to detect photons emitted from the ground, but are otherwise unable to look outside and see that they are sitting on a tower. In other words, they have no way of distinguishing this situation from that of the accelerating rockets. Therefore, the EEP allows us to conclude immediately that a photon emitted from the ground with wavelength λ_0 will be redshifted by an amount

$$\frac{\Delta\lambda}{\lambda_0} = \frac{a_g z}{c^2}. \tag{2.6}$$

This is the famous gravitational redshift. Notice that it is a direct consequence of the EEP; the details of general relativity were not required.

The formula for the redshift is more often stated in terms of the Newtonian potential Φ, where $\mathbf{a}_g = \nabla\Phi$. (The sign is changed with respect to the usual convention, since we are thinking of $\mathbf{a}_g$ as the acceleration of the reference frame, not of a particle with respect to this reference frame.) A nonconstant gradient of Φ is like a time-varying acceleration, and the equivalent net velocity is given by integrating over the time between emission and absorption of the photon. We then have

$$\begin{aligned}\frac{\Delta\lambda}{\lambda_0} &= \frac{1}{c}\int \nabla\Phi\, dt \\ &= \frac{1}{c^2}\int \partial_z\Phi\, dz \\ &= \Delta\Phi,\end{aligned} \tag{2.7}$$

where $\Delta\Phi$ is the total change in the gravitational potential, and we have once again set $c = 1$. This simple formula for the gravitational redshift continues to be true in more general circumstances. Of course, by using the Newtonian potential at all, we are restricting our domain of validity to weak gravitational fields.

From the EEP we have argued in favor of a gravitational redshift; we may now use this phenomenon to provide further support for the idea that we should think of spacetime as curved. Consider the same experimental setup that we had before, now portrayed on the spacetime diagram in Figure 2.4. A physicist on the ground emits a beam of light with wavelength λ_0 from a height z_0, which travels to the top of the tower at height z_1. The time between when the beginning of any single wavelength of the light is emitted and the end of that same wavelength is emitted is $\Delta t_0 = \lambda_0/c$, and the same time interval for the absorption is $\Delta t_1 = \lambda_1/c$, where time is measured by clocks located at the respective elevations. Since we imagine that the gravitational field is static, the paths through spacetime followed by the leading and trailing edge of the single wave must be precisely congruent. (They are represented by generic curved paths, since we do not pretend that we know just what the paths will be.) Simple geometry seems to imply that the times Δt_0 and Δt_1 must be the same. But of course they are not; the gravitational redshift implies that the elevated experimenters observe fewer wavelengths per second, so that $\Delta t_1 > \Delta t_0$. We can interpret this roughly as "the clock on the tower appears to run more quickly." What went wrong? Simple geometry—the spacetime through which the photons traveled was curved.

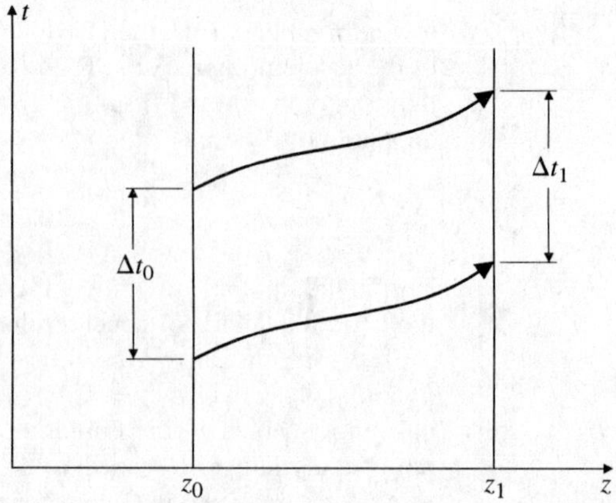

FIGURE 2.4 Spacetime diagram of the gravitational-redshift experiment portrayed in Figure 2.3. Spacetime paths beginning at different moments are congruent, but the time intervals as measured on the ground and on the tower are different, signaling a breakdown of Euclidean geometry.

We therefore would like to describe spacetime as a kind of mathematical structure that looks locally like Minkowski space, but may possess nontrivial curvature over extended regions. The kind of object that encompasses this notion is that of a manifold. In this chapter we will confine ourselves to understanding the concept of manifolds and the structures we may define on them, leaving the precise characterization of curvature for the next chapter.

2.2 ■ WHAT IS A MANIFOLD?

Manifolds (or differentiable manifolds) are one of the most fundamental concepts in mathematics and physics. We are all used to the properties of n-dimensional Euclidean space, $\mathbf{R}^n$, the set of n-tuples $(x^1, \ldots, x^n)$, often equipped with a flat positive-definite metric with components δ_{ij}. Mathematicians have worked for many years to develop the theory of analysis in $\mathbf{R}^n$—differentiation, integration, properties of functions, and so on. But clearly there are other spaces (spheres, for example) which we intuitively think of as "curved" or perhaps topologically complicated, on which we would like to perform analogous operations.

To address this problem we invent the notion of a manifold, which corresponds to a space that may be curved and have a complicated topology, but in local regions looks just like $\mathbf{R}^n$. Here by "looks like" we do not mean that the metric is the same, but only that more primitive notions like functions and coordinates work in a similar way. The entire manifold is constructed by smoothly sewing together these local regions. A crucial point is that the dimensionality n of the Euclidean spaces being used must be the same in every patch of the manifold; we then say that the manifold is of dimension n. With this approach we can analyze functions on such a space by converting them (locally) to functions in a Euclidean space. Examples of manifolds include:

- $\mathbf{R}^n$ itself, including the line ($\mathbf{R}$), the plane ($\mathbf{R}^2$), and so on. This should be obvious, since $\mathbf{R}^n$ looks like $\mathbf{R}^n$ not only locally but globally.
- The n-sphere, S^n. This can be defined as the locus of all points some fixed distance from the origin in $\mathbf{R}^{n+1}$. The circle is of course S^1, and the two-sphere S^2 is one of the most useful examples of a manifold. (The zero-sphere S^0, if you think about it, consists of two points. We say that S^0 is a disconnected zero-dimensional manifold.) It's worth emphasizing that the definition of S^n in terms of an embedding in $\mathbf{R}^{n+1}$ is simply a convenient shortcut; all of the manifolds we will discuss may be defined in their own right, without recourse to higher-dimensional flat spaces.
- The n-torus T^n results from taking an n-dimensional cube and identifying opposite sides. The two-torus T^2 is a square with opposite sides identified, as shown in Figure 2.5. The surface of a doughnut is a familiar example.
- A Riemann surface of genus g is essentially a two-torus with g holes instead of just one, as shown in Figure 2.6. S^2 may be thought of as a Riemann surface of genus zero. In technical terms (not really relevant to our present dis-

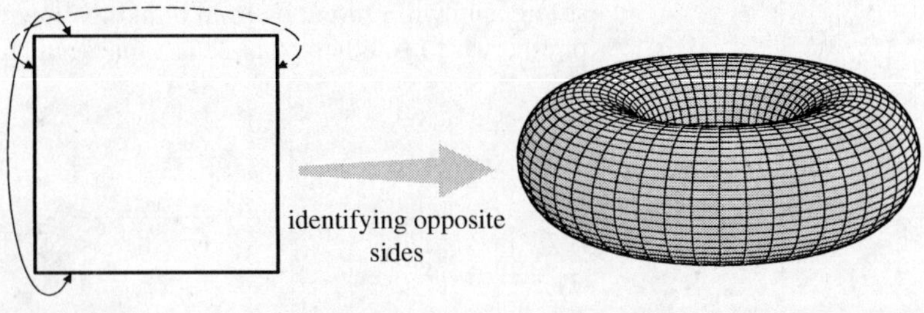

FIGURE 2.5 The torus, T^2, constructed by identifying opposite sides of a square.

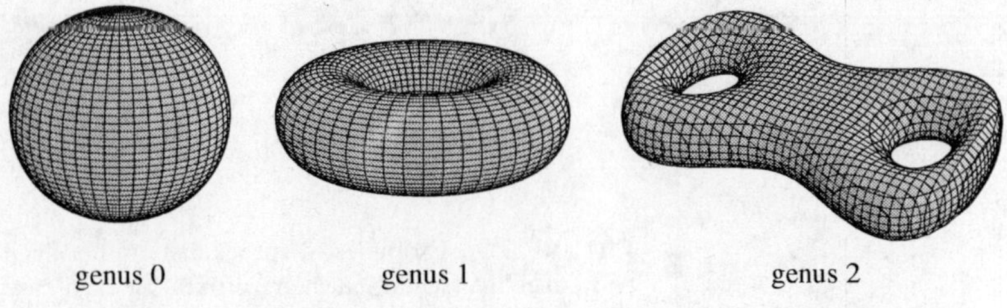

FIGURE 2.6 Riemann surfaces of different genera (plural of "genus").

cussion), every "compact orientable boundaryless" two-dimensional manifold is a Riemann surface of some genus.

- More abstractly, a set of continuous transformations such as rotations in $\mathbf{R}^n$ forms a manifold. Lie groups are manifolds that also have a group structure. So for example SO(2), the set of rotations in two dimensions, is the same manifold as S^1 (although in general group manifolds will be more complicated than spheres).

- The direct product of two manifolds is a manifold. That is, given manifolds M and M' of dimension n and n', we can construct a manifold $M \times M'$, of dimension $n + n'$, consisting of ordered pairs (p, p') with $p \in M$ and $p' \in M'$.

With all of these examples, the notion of a manifold may seem vacuous: what *isn't* a manifold? Plenty of things are not manifolds, because somewhere they do not look locally like $\mathbf{R}^n$. Examples include a one-dimensional line running into a two-dimensional plane, and two cones stuck together at their vertices, as portrayed in Figure 2.7. More subtle examples are shown in Figure 2.8. Consider for example a single (two-dimensional) cone. There is clearly a sense in which the cone looks locally like $\mathbf{R}^2$; at the same time, there is just as clearly something singular about the vertex of the cone. This is where the word "differentiable" in "differentiable manifold" begins to play a role; as we will see when we develop the formal definition, the cone is perfectly smooth as a manifold, even though the curvature is not smooth at its vertex. (Other types of singularities are more severe, and will prevent us from thinking of certain spaces as manifolds, smooth or otherwise.) Another example is a line segment (with endpoints included). This

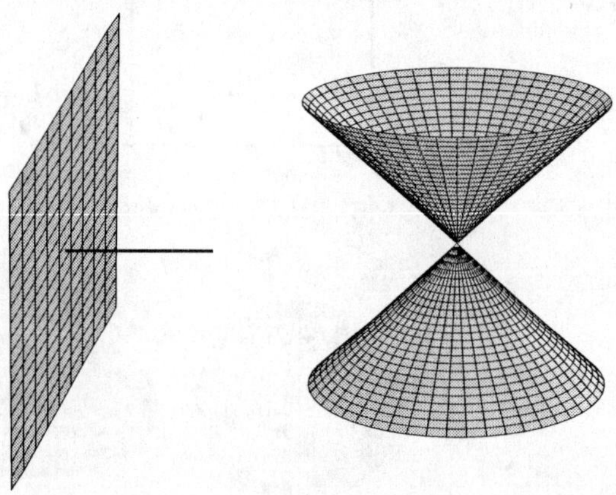

FIGURE 2.7 Examples of spaces that are not manifolds: a line ending on a plane, and two cones intersecting at their vertices. In each case there is a point that does not look locally like a Euclidean space of fixed dimension.

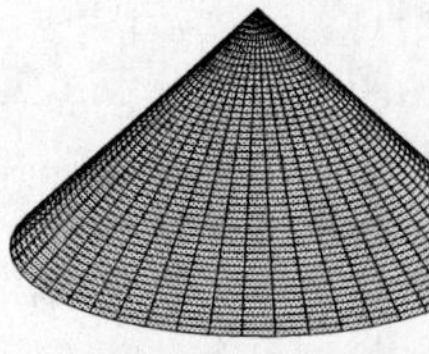

FIGURE 2.8 Subtle examples. The single cone is a smooth manifold, even though the curvature is singular at its vertex. A line segment is not a manifold, but may be described by the more general notion of "manifold with boundary."

certainly will not fit under the definition of manifolds we will develop, due to the endpoints. Nevertheless, we can extend the definition to include "manifolds with boundary," of which the line segment is a paradigmatic example. A brief discussion of manifolds with boundary is in Appendix D.

These subtle cases should convince you of the need for a rigorous definition, which we now begin to construct; our discussion follows that of Wald (1984). The informal idea of a manifold is that of a space consisting of patches that look locally like $\mathbf{R}^n$, and are smoothly sewn together. We therefore need to formalize the notions of "looking locally like $\mathbf{R}^n$" and "smoothly sewn together." We require a number of preliminary definitions, most of which are fairly clear, but it's nice to be complete. The most elementary notion is that of a **map** between two sets. (We assume you know what a set is, or think you do; we won't need to be too precise.) Given two sets M and N, a map $\phi : M \to N$ is a relationship that assigns, to each element of M, exactly one element of N. A map is therefore just a simple generalization of a function. Given two maps $\phi : A \to B$ and $\psi : B \to C$, we define the **composition** $\psi \circ \phi : A \to C$ by the operation $(\psi \circ \phi)(a) = \psi(\phi(a))$, as in Figure 2.9. So $a \in A$, $\phi(a) \in B$, and thus $(\psi \circ \phi)(a) \in C$. The order in which the maps are written makes sense, since the one on the right acts first.

A map ϕ is called **one-to-one** (or injective) if each element of N has at most one element of M mapped into it, and **onto** (or surjective) if each element of N has at least one element of M mapped into it. (If you think about it, better names for "one-to-one" would be "one-from-one" or for that matter "two-to-two.") Consider functions $\phi : \mathbf{R} \to \mathbf{R}$. Then $\phi(x) = e^x$ is one-to-one, but not onto; $\phi(x) = x^3 - x$ is onto, but not one-to-one; $\phi(x) = x^3$ is both; and $\phi(x) = x^2$ is neither, as in Figure 2.10.

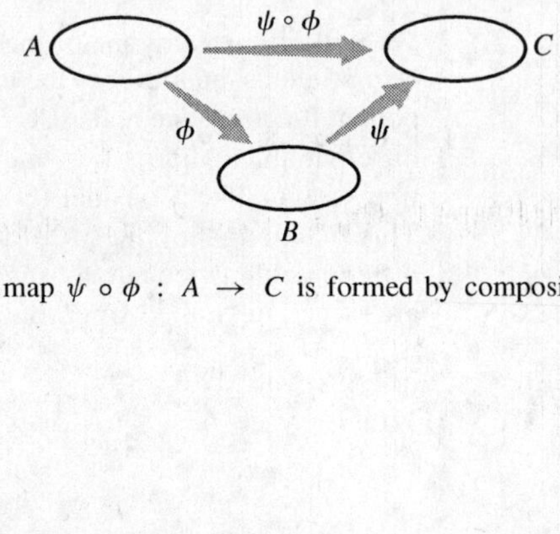

FIGURE 2.9 The map $\psi \circ \phi : A \to C$ is formed by composing $\phi : A \to B$ and $\psi : B \to C$.

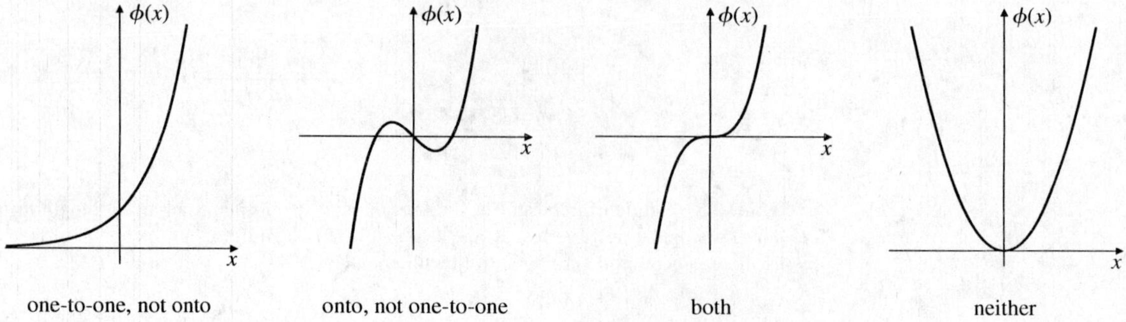

FIGURE 2.10 Types of maps.

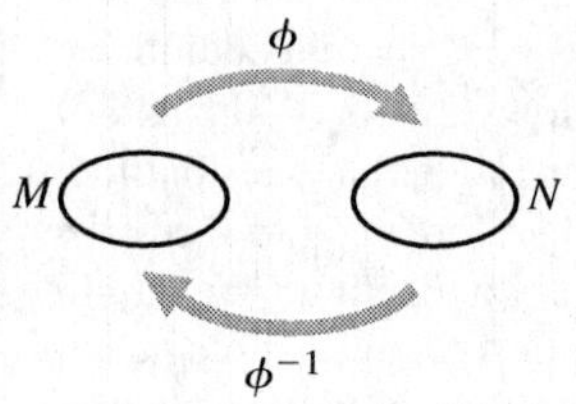

FIGURE 2.11 A map and its inverse.

The set M is known as the **domain** of the map ϕ, and the set of points in N that M gets mapped into is called the **image** of ϕ. For any subset $U \subset N$, the set of elements of M that get mapped to U is called the **preimage** of U under ϕ, or $\phi^{-1}(U)$. A map that is both one-to-one and onto is known as **invertible** (or bijective). In this case we can define the **inverse map** $\phi^{-1} : N \to M$ by $(\phi^{-1} \circ \phi)(a) = a$, as in Figure 2.11. Note that the same symbol ϕ^{-1} is used for both the preimage and the inverse map, even though the former is always defined and the latter is only defined in some special cases.

The notion of **continuity** of a map is actually a very subtle one, the precise formulation of which we won't need. Instead we will assume you understand the concepts of continuity and differentiability as applied to ordinary functions, maps $\phi : \mathbf{R} \to \mathbf{R}$. It will then be useful to extend these notions to maps between more general Euclidean spaces, $\phi : \mathbf{R}^m \to \mathbf{R}^n$. A map from $\mathbf{R}^m$ to $\mathbf{R}^n$ takes an m-tuple $(x^1, x^2, \ldots, x^m)$ to an n-tuple $(y^1, y^2, \ldots, y^n)$, and can therefore be thought of as a collection of n functions ϕ^i of m variables:

$$\begin{aligned} y^1 &= \phi^1(x^1, x^2, \ldots, x^m) \\ y^2 &= \phi^2(x^1, x^2, \ldots, x^m) \\ &\vdots \\ y^n &= \phi^n(x^1, x^2, \ldots, x^m). \end{aligned} \tag{2.8}$$

We will refer to any one of these functions as C^p if its pth derivative exists and is continuous, and refer to the entire map $\phi : \mathbf{R}^m \to \mathbf{R}^n$ as C^p if each of its component functions are at least C^p. Thus a C^0 map is continuous but not necessarily differentiable, while a C^∞ map is continuous and can be differentiated as many times as you like. Consider for example the function of one variable $\phi(x) = |x^3|$. This function is infinitely differentiable everywhere except at $x = 0$, where it is differentiable twice but not three times; we therefore say that it is C^2. C^∞ maps are sometimes called **smooth**.

We will call two sets M and N **diffeomorphic** if there exists a C^∞ map $\phi : M \to N$ with a C^∞ inverse $\phi^{-1} : N \to M$; the map ϕ is then called a diffeomorphism. This is the best notion we have that two spaces are "the same" as manifolds. For example, when we said that SO(2) was the same manifold as S^1, we meant they were diffeomorphic. See Appendix B for more discussion.

These basic definitions may have been familiar to you, even if only vaguely remembered. We will now put them to use in the rigorous definition of a manifold. Unfortunately, a somewhat baroque procedure is required to formalize this relatively intuitive notion. We will first have to define the notion of an open set, on which we can put coordinate systems, and then sew the open sets together in an appropriate way.

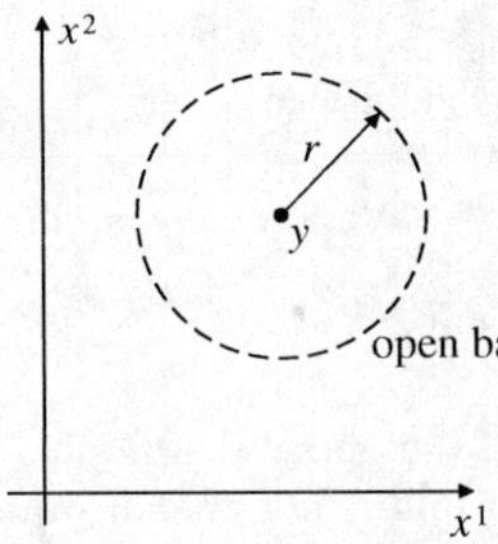

FIGURE 2.12 An open ball defined in $\mathbf{R}^n$.

We start with the notion of an **open ball**, which is the set of all points x in $\mathbf{R}^n$ such that $|x - y| < r$ for some fixed $y \in \mathbf{R}^n$ and $r \in \mathbf{R}$, where $|x - y| = [\sum_i (x^i - y^i)^2]^{1/2}$. Note that this is a strict inequality—the open ball is the interior of an n-sphere of radius r centered at y, as shown in Figure 2.12. An **open set** in $\mathbf{R}^n$ is a set constructed from an arbitrary (maybe infinite) union of open balls. In other words, $V \subset \mathbf{R}^n$ is open if, for any $y \in V$, there is an open ball centered at y that is completely inside V. Roughly speaking, an open set is the interior of some $(n-1)$-dimensional closed surface (or the union of several such interiors).

A **chart** or **coordinate system** consists of a subset U of a set M, along with a one-to-one map $\phi : U \to \mathbf{R}^n$, such that the image $\phi(U)$ is open in $\mathbf{R}^n$, as in Figure 2.13. (Any map is onto its image, so the map $\phi : U \to \phi(U)$ is invertible if it is one-to-one.) We then can say that U is an open set in M. A C^∞ **atlas** is an indexed collection of charts $\{(U_\alpha, \phi_\alpha)\}$ that satisfies two conditions:

1. The union of the U_α is equal to M; that is, the U_α cover M.
2. The charts are smoothly sewn together. More precisely, if two charts overlap, $U_\alpha \cap U_\beta \neq \emptyset$, then the map $(\phi_\alpha \circ \phi_\beta^{-1})$ takes points in $\phi_\beta(U_\alpha \cap U_\beta) \subset \mathbf{R}^n$ *onto* an open set $\phi_\alpha(U_\alpha \cap U_\beta) \subset \mathbf{R}^n$, and all of these maps must be C^∞ where they are defined. This should be clearer from Figure 2.14, adapted from Wald (1984).

So a chart is what we normally think of as a coordinate system on some open set, and an atlas is a system of charts that are smoothly related on their overlaps.

At long last, then: a C^∞ n-dimensional **manifold** (or n-manifold for short) is simply a set M along with a maximal atlas, one that contains every possible

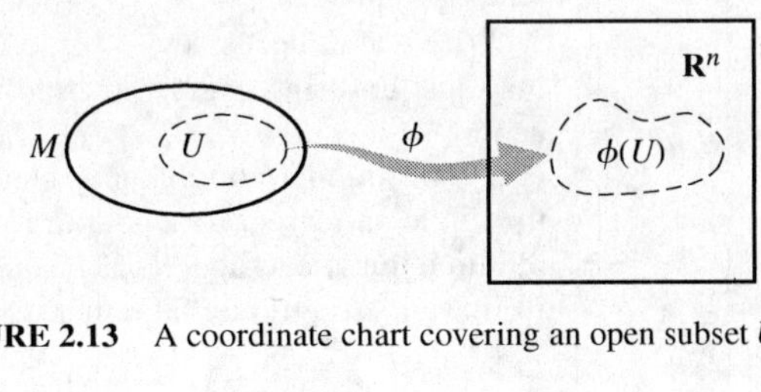

FIGURE 2.13 A coordinate chart covering an open subset U of M.

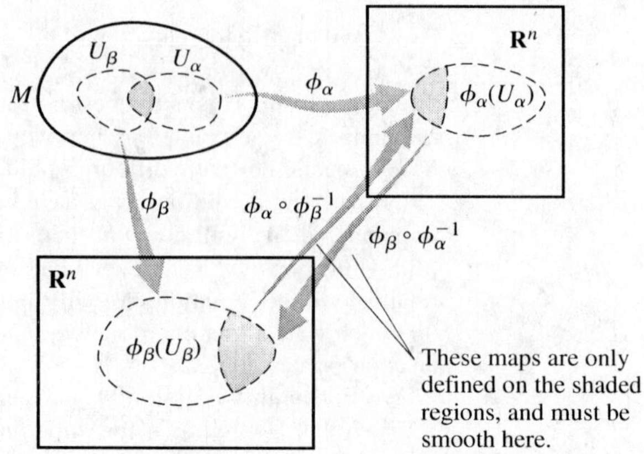

FIGURE 2.14 Overlapping coordinate charts.

compatible chart. We can also replace C^∞ by C^p in all the above definitions. For our purposes the degree of differentiability of a manifold is not crucial; we will always assume that any manifold is as differentiable as necessary for the application under consideration. The requirement that the atlas be maximal is so that two equivalent spaces equipped with different atlases don't count as different manifolds. This definition captures in formal terms our notion of a set that looks locally like $\mathbf{R}^n$. Of course we will rarely have to make use of the full power of the definition, but precision is its own reward.

One nice thing about our definition is that it does not rely on an embedding of the manifold in some higher-dimensional Euclidean space. In fact, any n-dimensional manifold can be embedded in $\mathbf{R}^{2n}$ (Whitney's embedding theorem), and sometimes we will make use of this fact, such as in our definition of the sphere above. (A Klein bottle is an example of a 2-manifold that cannot be embedded in $\mathbf{R}^3$, although it can be embedded in $\mathbf{R}^4$.) But it is important to recognize that the manifold has an individual existence independent of any embedding. It is not necessary to believe, for example, that four-dimensional spacetime is stuck in some larger space. On the other hand, it might be; we really don't know. Recent advances in string theory have led to the suggestion that our visible universe is actually a "brane" (generalization of "membrane") inside a higher-dimensional space. But as far as classical GR is concerned, the four-dimensional view is perfectly adequate.

Why was it necessary to be so finicky about charts and their overlaps, rather than just covering every manifold with a single chart? Because most manifolds cannot be covered with just one chart. Consider the simplest example, S^1. There is a conventional coordinate system, $\theta : S^1 \to \mathbf{R}$, where $\theta = 0$ at the top of the circle and wraps around to 2π. However, in the definition of a chart we have required that the image $\theta(S^1)$ be open in $\mathbf{R}$. If we include either $\theta = 0$ or $\theta = 2\pi$, we have a closed interval rather than an open one; if we exclude both points, we

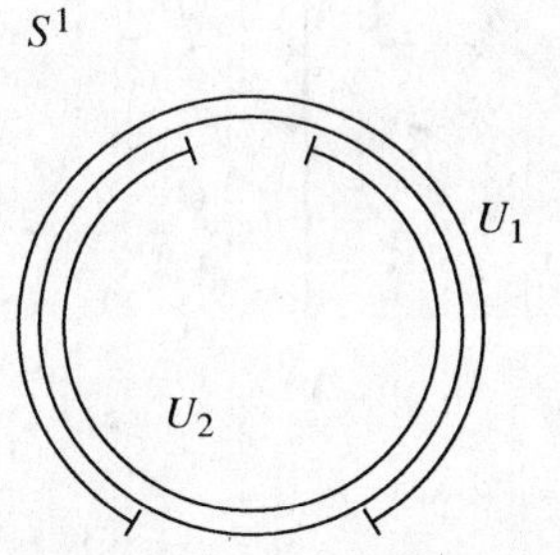

FIGURE 2.15 Two coordinate charts, which together cover S^1.

haven't covered the whole circle. So we need at least two charts, as shown in Figure 2.15.

A somewhat more complicated example is provided by S^2, where once again no single chart will cover the manifold. A Mercator projection, traditionally used for world maps, misses both the North and South poles (as well as the International Date Line, which involves the same problem with θ that we found for S^1.) Let's take S^2 to be the set of points in $\mathbf{R}^3$ defined by $(x^1)^2 + (x^2)^2 + (x^3)^2 = 1$. We can construct a chart from an open set U_1, defined to be the sphere minus the north pole, via stereographic projection, illustrated in Figure 2.16. Thus, we draw a straight line from the north pole to the plane defined by $x^3 = -1$, and assign to the point on S^2 intercepted by the line the Cartesian coordinates (y^1, y^2) of the appropriate point on the plane. Explicitly, the map is given by

$$\phi_1(x^1, x^2, x^3) \equiv (y^1, y^2) = \left(\frac{2x^1}{1-x^3}, \frac{2x^2}{1-x^3}\right). \tag{2.9}$$

Check this for yourself. Another chart (U_2, ϕ_2) is obtained by projecting from the south pole to the plane defined by $x^3 = +1$. The resulting coordinates cover the sphere minus the south pole, and are given by

$$\phi_2(x^1, x^2, x^3) \equiv (z^1, z^2) = \left(\frac{2x^1}{1+x^3}, \frac{2x^2}{1+x^3}\right). \tag{2.10}$$

Together, these two charts cover the entire manifold, and they overlap in the region $-1 < x^3 < +1$. Another thing you can check is that the composition $\phi_2 \circ \phi_1^{-1}$ is given by

$$z^i = \frac{4y^i}{[(y^1)^2 + (y^2)^2]}, \tag{2.11}$$

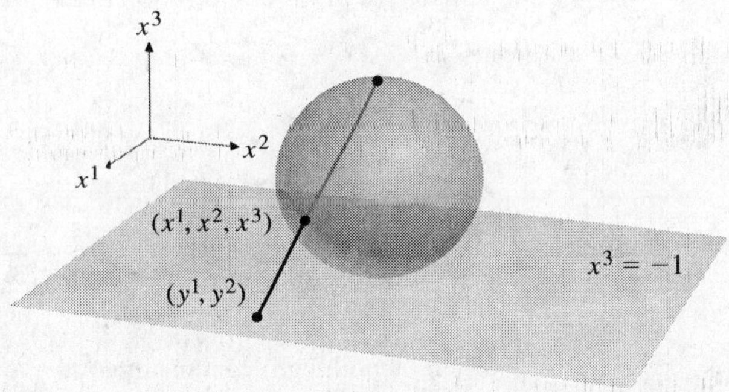

FIGURE 2.16 Defining a stereographic coordinate chart on S^2 by projecting from the north pole down to a plane tangent to the south pole. Such a chart covers all of the sphere except for the north pole itself.

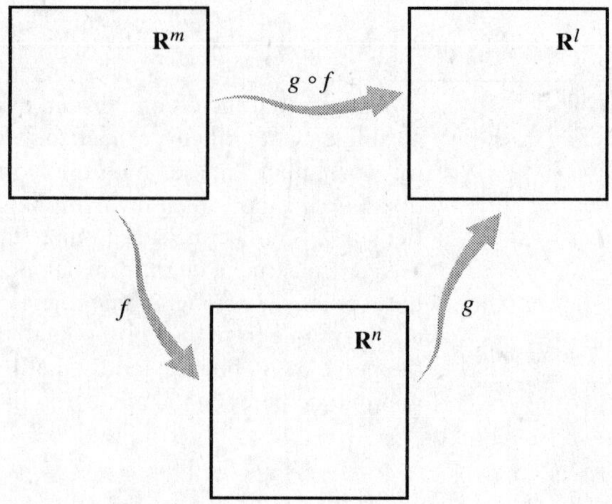

FIGURE 2.17 The chain rule relates the partial derivatives of $g \circ f$ to those of g and f.

and is C^∞ in the region of overlap. As long as we restrict our attention to this region, (2.11) is just what we normally think of as a change of coordinates.

We therefore see the necessity of charts and atlases: Many manifolds cannot be covered with a single coordinate system. Nevertheless, it is often convenient to work with a single chart, and just keep track of the set of points that aren't included.

One piece of conventional calculus that we will need later is the **chain rule**. Let us imagine that we have maps $f : \mathbf{R}^m \to \mathbf{R}^n$ and $g : \mathbf{R}^n \to \mathbf{R}^l$, and therefore the composition $(g \circ f) : \mathbf{R}^m \to \mathbf{R}^l$, as shown in Figure 2.17. We can label points in each space in terms of components: x^a on $\mathbf{R}^m$, y^b on $\mathbf{R}^n$, and z^c on $\mathbf{R}^l$, where the indices range over the appropriate values. The chain rule relates the partial derivatives of the composition to the partial derivatives of the individual maps:

$$\frac{\partial}{\partial x^a}(g \circ f)^c = \sum_b \frac{\partial f^b}{\partial x^a}\frac{\partial g^c}{\partial y^b}. \tag{2.12}$$

This is usually abbreviated to

$$\frac{\partial}{\partial x^a} = \sum_b \frac{\partial y^b}{\partial x^a}\frac{\partial}{\partial y^b}. \tag{2.13}$$

There is nothing illegal or immoral about using this shorthand form of the chain rule, but you should be able to visualize the maps that underlie the construction. Recall that when $m = n$, the determinant of the matrix $\partial y^b / \partial x^a$ is called the **Jacobian** of the map, and the map is invertible whenever the Jacobian is nonzero.

2.3 ■ VECTORS AGAIN

Having constructed this groundwork, we can now proceed to introduce various kinds of structure on manifolds. We begin with vectors and tangent spaces. In our discussion of special relativity we were intentionally vague about the definition of vectors and their relationship to the spacetime. One point we stressed was the notion of a tangent space—the set of all vectors at a single point in spacetime. The reason for this emphasis was to remove from your minds the idea that a vector stretches from one point on the manifold to another, but instead is just an object associated with a single point. What is temporarily lost by adopting this view is a way to make sense of statements like "the vector points in the x direction"—if the tangent space is merely an abstract vector space associated with each point, it's hard to know what this should mean. Now it's time to fix the problem.

Let's imagine that we wanted to construct the tangent space at a point p in a manifold M, using only things that are intrinsic to M (no embeddings in higher-dimensional spaces). A first guess might be to use our intuitive knowledge that there are objects called "tangent vectors to curves," which belong in the tangent space. We might therefore consider the set of all parameterized curves through p—that is, the space of all (nondegenerate) maps $\gamma : \mathbf{R} \to M$, such that p is in the image of γ. The temptation is to define the tangent space as simply the space of all tangent vectors to these curves at the point p. But this is obviously cheating; the tangent space T_p is supposed to be the space of vectors at p, and before we have defined this we don't have an independent notion of what "the tangent vector to a curve" is supposed to mean. In some coordinate system x^μ any curve through p defines an element of $\mathbf{R}^n$ specified by the n real numbers $dx^\mu/d\lambda$ (where λ is the parameter along the curve), but this map is clearly coordinate-dependent, which is not what we want.

Nevertheless we are on the right track, we just have to make things independent of coordinates. To this end we define $\mathcal{F}$ to be the space of all smooth functions on M (that is, C^∞ maps $f : M \to \mathbf{R}$). Then we notice that each curve through p defines an operator on this space, the directional derivative, which maps $f \to df/d\lambda$ (at p). We will make the following claim: *the tangent space T_p can be identified with the space of directional derivative operators along curves through p*. To establish this idea we must demonstrate two things: first, that the space of directional derivatives is a vector space, and second that it is the vector space we want (it has the same dimensionality as M, yields a natural idea of a vector pointing along a certain direction, and so on).

The first claim, that directional derivatives form a vector space, seems straightforward enough. Imagine two operators $d/d\lambda$ and $d/d\eta$ representing derivatives along two curves $x^\mu(\lambda)$ and $x^\mu(\eta)$ through p. There is no problem adding these and scaling by real numbers, to obtain a new operator $a(d/d\lambda) + b(d/d\eta)$. It is not immediately obvious, however, that the space closes; in other words, that the resulting operator is itself a derivative operator. A good derivative operator is one that acts linearly on functions, and obeys the conventional Leibniz (product) rule on products of functions. Our new operator is manifestly linear, so we need to

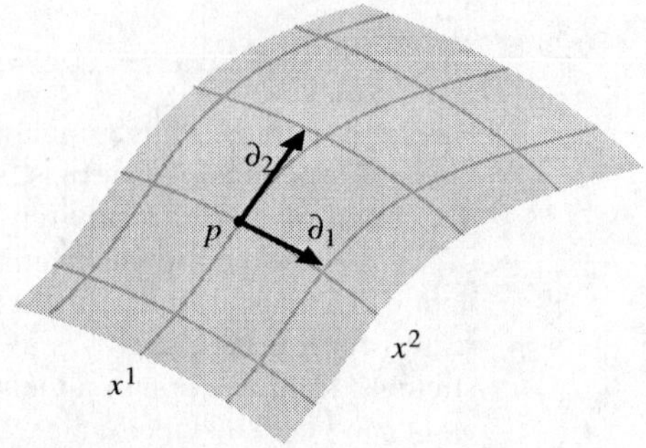

FIGURE 2.18 Partial derivatives define directional derivatives along curves that keep all of the other coordinates constant.

verify that it obeys the Leibniz rule. We have

$$\left(a\frac{d}{d\lambda}+b\frac{d}{d\eta}\right)(fg) = af\frac{dg}{d\lambda}+ag\frac{df}{d\lambda}+bf\frac{dg}{d\eta}+bg\frac{df}{d\eta}$$
$$= \left(a\frac{df}{d\lambda}+b\frac{df}{d\eta}\right)g+\left(a\frac{dg}{d\lambda}+b\frac{dg}{d\eta}\right)f. \quad (2.14)$$

As we had hoped, the product rule is satisfied, and the set of directional derivatives is therefore a vector space.

Is it the vector space that we would like to identify with the tangent space? The easiest way to become convinced is to find a basis for the space. Consider again a coordinate chart with coordinates x^μ. Then there is an obvious set of n directional derivatives at p, namely the partial derivatives ∂_μ at p, as shown in Figure 2.18. Note that this is really the *definition* of the partial derivative with respect to x^μ: the directional derivative along a curve defined by x^ν = constant for all $\nu \neq \mu$, parameterized by x^μ itself. We are now going to claim that the partial derivative operators $\{\partial_\mu\}$ at p form a basis for the tangent space T_p. (It follows immediately that T_p is n-dimensional, since that is the number of basis vectors.) To see this we will show that any directional derivative can be decomposed into a sum of real numbers times partial derivatives. This will just be the familiar expression for the components of a tangent vector, but it's nice to see it from the big-machinery approach. Consider an n-manifold M, a coordinate chart $\phi : M \to \mathbf{R}^n$, a curve $\gamma : \mathbf{R} \to M$, and a function $f : M \to \mathbf{R}$. This leads to the tangle of maps shown in Figure 2.19. If λ is the parameter along γ, we want to express the vector/operator $d/d\lambda$ in terms of the partials ∂_μ. Using the chain rule (2.12), we have

$$\frac{d}{d\lambda}f = \frac{d}{d\lambda}(f\circ\gamma)$$
$$= \frac{d}{d\lambda}[(f\circ\phi^{-1})\circ(\phi\circ\gamma)]$$

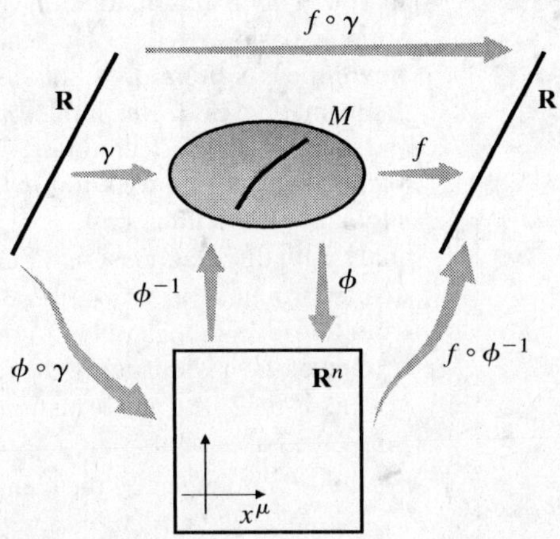

FIGURE 2.19 Decomposing the tangent vector to a curve $\gamma : \mathbf{R} \to M$ in terms of partial derivatives with respect to coordinates on M.

$$\begin{aligned} &= \frac{d(\phi \circ \gamma)^\mu}{d\lambda} \frac{\partial (f \circ \phi^{-1})}{\partial x^\mu} \\ &= \frac{dx^\mu}{d\lambda} \partial_\mu f. \end{aligned} \tag{2.15}$$

The first line simply takes the informal expression on the left-hand side and rewrites it as an honest derivative of the function $(f \circ \gamma) : \mathbf{R} \to \mathbf{R}$. The second line just comes from the definition of the inverse map ϕ^{-1} (and associativity of the operation of composition). The third line is the formal chain rule (2.12), and the last line is a return to the informal notation of the start. Since the function f was arbitrary, we have

$$\frac{d}{d\lambda} = \frac{dx^\mu}{d\lambda} \partial_\mu. \tag{2.16}$$

Thus, the partials $\{\partial_\mu\}$ do indeed represent a good basis for the vector space of directional derivatives, which we can therefore safely identify with the tangent space.

Of course, the vector represented by $d/d\lambda$ is one we already know; it's the tangent vector to the curve with parameter λ. Thus (2.16) can be thought of as a restatement of equation (1.38), where we claimed that the components of the tangent vector were simply $dx^\mu/d\lambda$. The only difference is that we are working on an arbitrary manifold, and we have specified our basis vectors to be $\hat{e}_{(\mu)} = \partial_\mu$.

This particular basis $(\hat{e}_{(\mu)} = \partial_\mu)$ is known as a **coordinate basis** for T_p; it is the formalization of the notion of setting up the basis vectors to point along the coordinate axes. There is no reason why we are limited to coordinate bases when we consider tangent vectors. For example, the coordinate basis vectors are typically not normalized to unity, nor orthogonal to each other, as we shall see shortly. This is not a situation we can define away; on a curved manifold, a coordinate basis will *never* be orthonormal throughout a neighborhood of any point where the curvature does not vanish. Of course we can define noncoordinate orthonormal bases, for example by giving their components in a coordinate basis, and sometimes this technique is useful. However, coordinate bases are very simple and natural, and we will use them almost exclusively throughout the book; for a look at orthonormal bases, see Appendix J. (It is standard in the study of vector analysis in three-dimensional Euclidean space to choose orthonormal bases rather than coordinate bases; you should therefore be careful when applying formulae from GR texts to the study of non-Cartesian coordinates in flat space.)

One of the advantages of the rather abstract point of view we have taken toward vectors is that the transformation law under changes of coordinates is immediate. Since the basis vectors are $\hat{e}_{(\mu)} = \partial_\mu$, the basis vectors in some new coordinate system $x^{\mu'}$ are given by the chain rule (2.13) as

$$\partial_{\mu'} = \frac{\partial x^\mu}{\partial x^{\mu'}} \partial_\mu . \tag{2.17}$$

We can get the transformation law for vector components by the same technique used in flat space, demanding that the vector $V = V^\mu \partial_\mu$ be unchanged by a change of basis. We have

$$\begin{aligned} V^\mu \partial_\mu &= V^{\mu'} \partial_{\mu'} \\ &= V^{\mu'} \frac{\partial x^\mu}{\partial x^{\mu'}} \partial_\mu , \end{aligned} \tag{2.18}$$

and hence, since the matrix $\partial x^{\mu'}/\partial x^\mu$ is the inverse of the matrix $\partial x^\mu/\partial x^{\mu'}$,

$$V^{\mu'} = \frac{\partial x^{\mu'}}{\partial x^\mu} V^\mu . \tag{2.19}$$

Since the basis vectors are usually not written explicitly, the rule (2.19) for transforming components is what we call the "vector transformation law." We notice that it is compatible with the transformation of vector components in special relativity under Lorentz transformations, $V^{\mu'} = \Lambda^{\mu'}{}_\mu V^\mu$, since a Lorentz transformation is a special kind of coordinate transformation, with $x^{\mu'} = \Lambda^{\mu'}{}_\mu x^\mu$. But (2.19) is much more general, as it encompasses the behavior of vectors under arbitrary changes of coordinates (and therefore bases), not just linear transformations. As usual, we are trying to emphasize a somewhat subtle ontological distinction—in principle, tensor components need not change when we change coordinates, they change when we change the basis in the tangent space, but we have decided to use the coordinates to define our basis. Therefore a change of coordinates induces a change of basis, as indicated in Figure 2.20.

Since a vector at a point can be thought of as a directional derivative operator along a path through that point, it should be clear that a vector *field* defines a map from smooth functions to smooth functions all over the manifold, by taking a

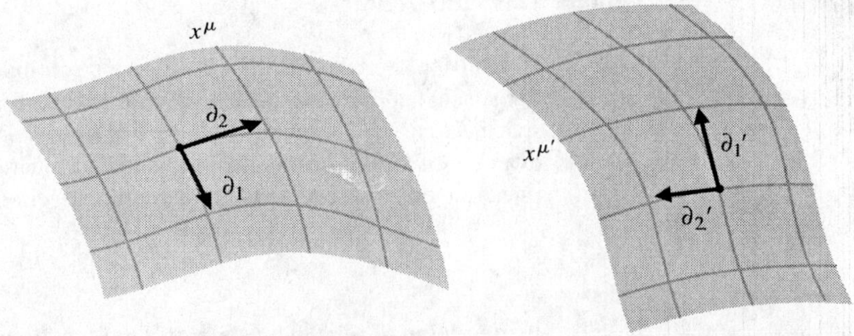

FIGURE 2.20 A change of coordinates $x^\mu \rightarrow x^{\mu'}$ induces a change of basis in the tangent space.

derivative at each point. Given two vector fields X and Y, we can therefore define their **commutator** $[X, Y]$ by its action on a function $f(x^\mu)$:

$$[X, Y](f) \equiv X(Y(f)) - Y(X(f)). \tag{2.20}$$

The virtue of the abstract point of view is that, clearly, this operator is independent of coordinates. In fact, the commutator of two vector fields is itself a vector field: if f and g are functions and a and b are real numbers, the commutator is linear,

$$[X, Y](af + bg) = a[X, Y](f) + b[X, Y](g), \tag{2.21}$$

and obeys the Leibniz rule,

$$[X, Y](fg) = f[X, Y](g) + g[X, Y](f). \tag{2.22}$$

Both properties are straightforward to check, which is a useful exercise to do. An equally interesting exercise is to derive an explicit expression for the components of the vector field $[X, Y]^\mu$, which turns out to be

$$\boxed{[X, Y]^\mu = X^\lambda \partial_\lambda Y^\mu - Y^\lambda \partial_\lambda X^\mu.} \tag{2.23}$$

By construction this is a well-defined tensor; but you should be slightly worried by the appearance of the partial derivatives, since partial derivatives of vectors are not well-defined tensors (as we discuss in the next section). Yet another fascinating exercise is to perform explicitly a coordinate transformation on the expression (2.23), to verify that all potentially nontensorial pieces cancel and the result transforms like a vector field. The commutator is a special case of the Lie derivative, discussed in Appendix B; it is sometimes referred to as the **Lie bracket**. Note that since partials commute, the commutator of the vector fields given by the partial derivatives of coordinate functions, $\{\partial_\mu\}$, always vanishes.

2.4 ■ TENSORS AGAIN

Having explored the world of vectors, we continue to retrace the steps we took in flat space, and now consider dual vectors (one-forms). Once again the cotangent space T_p^* can be thought of as the set of linear maps $\omega : T_p \to \mathbf{R}$. The canonical example of a one-form is the gradient of a function f, denoted $\mathrm{d}f$, as in (1.52). Its action on a vector $d/d\lambda$ is exactly the directional derivative of the function:

$$\mathrm{d}f\left(\frac{d}{d\lambda}\right) = \frac{df}{d\lambda}. \tag{2.24}$$

It's tempting to ask, "why shouldn't the function f itself be considered the one-form, and $df/d\lambda$ its action?" The point is that a one-form, like a vector, exists only at the point it is defined, and does not depend on information at other points on M. If you know a function in some neighborhood of a point, you can take its derivative, but not just from knowing its value at the point; the gradient, on the other hand, encodes precisely the information necessary to take the directional derivative along any curve through p, fulfilling its role as a dual vector.

You may have noticed that we defined vectors using structures intrinsic to the manifold (directional derivatives along curves), and used that definition to define one-forms in terms of the dual vector space. This might lead to the impression that vectors are somehow more fundamental; in fact, however, we could just as well have begun with an intrinsic definition of one-forms and used that to define vectors as the dual space. Roughly speaking, the space of one-forms at p is equivalent to the space of all functions that vanish at p and have the same second partial derivatives. In fact, doing it that way is more fundamental, if anything, since we can provide intrinsic definitions of all q-forms (totally antisymmetric tensors with q lower indices), which we will discuss in Section 2.9 (although we will not delve into the specifics of the intrinsic definitions).

Just as the partial derivatives along coordinate axes provide a natural basis for the tangent space, the gradients of the coordinate functions x^μ provide a natural basis for the cotangent space. Recall that in flat space we constructed a basis for T_p^* by demanding that $\hat{\theta}^{(\mu)}(\hat{e}_{(\nu)}) = \delta^\mu_\nu$. Continuing the same philosophy on an arbitrary manifold, we find that (2.24) leads to

$$\mathrm{d}x^\mu(\partial_\nu) = \frac{\partial x^\mu}{\partial x^\nu} = \delta^\mu_\nu. \tag{2.25}$$

Therefore the gradients $\{\mathrm{d}x^\mu\}$ are an appropriate set of basis one-forms; an arbitrary one-form is expanded into components as $\omega = \omega_\mu \, \mathrm{d}x^\mu$.

The transformation properties of basis dual vectors and components follow from what is by now the usual procedure. We obtain, for basis one-forms,

$$\mathrm{d}x^{\mu'} = \frac{\partial x^{\mu'}}{\partial x^\mu} \mathrm{d}x^\mu \tag{2.26}$$

and for components,

$$\omega_{\mu'} = \frac{\partial x^{\mu}}{\partial x^{\mu'}} \omega_{\mu}. \tag{2.27}$$

We will usually write the components ω_μ when we speak about a one-form ω.

Just as in flat space, a (k, l) tensor is a multilinear map from a collection of k dual vectors and l vectors to $\mathbf{R}$. Its components in a coordinate basis can be obtained by acting the tensor on basis one-forms and vectors,

$$T^{\mu_1 \cdots \mu_k}{}_{\nu_1 \cdots \nu_l} = T(\mathrm{d}x^{\mu_1}, \ldots, \mathrm{d}x^{\mu_k}, \partial_{\nu_1}, \ldots, \partial_{\nu_l}). \tag{2.28}$$

This is equivalent to the expansion

$$T = T^{\mu_1 \cdots \mu_k}{}_{\nu_1 \cdots \nu_l} \partial_{\mu_1} \otimes \cdots \otimes \partial_{\mu_k} \otimes \mathrm{d}x^{\nu_1} \otimes \cdots \otimes \mathrm{d}x^{\nu_l}. \tag{2.29}$$

The transformation law for general tensors follows the same pattern of replacing the Lorentz transformation matrix used in flat space with a matrix representing more general coordinate transformations:

$$\boxed{T^{\mu'_1 \cdots \mu'_k}{}_{\nu'_1 \cdots \nu'_l} = \frac{\partial x^{\mu'_1}}{\partial x^{\mu_1}} \cdots \frac{\partial x^{\mu'_k}}{\partial x^{\mu_k}} \frac{\partial x^{\nu_1}}{\partial x^{\nu'_1}} \cdots \frac{\partial x^{\nu_l}}{\partial x^{\nu'_l}} T^{\mu_1 \cdots \mu_k}{}_{\nu_1 \cdots \nu_l}.} \tag{2.30}$$

This tensor transformation law is straightforward to remember, since there really isn't anything else it could be, given the placement of indices.

Actually, however, it is often easier to transform a tensor by taking the identity of basis vectors and one-forms as partial derivatives and gradients at face value, and simply substituting in the coordinate transformation. As an example, consider a symmetric $(0, 2)$ tensor S on a two-dimensional manifold, whose components in a coordinate system $(x^1 = x, x^2 = y)$ are given by

$$S_{\mu\nu} = \begin{pmatrix} 1 & 0 \\ 0 & x^2 \end{pmatrix}. \tag{2.31}$$

This can be written equivalently as

$$\begin{aligned} S &= S_{\mu\nu}(\mathrm{d}x^{\mu} \otimes \mathrm{d}x^{\nu}) \\ &= (\mathrm{d}x)^2 + x^2(\mathrm{d}y)^2, \end{aligned} \tag{2.32}$$

where in the last line the tensor product symbols are suppressed for brevity (as will become our custom). Now consider new coordinates

$$\begin{aligned} x' &= \frac{2x}{y} \\ y' &= \frac{y}{2} \end{aligned} \tag{2.33}$$

(valid, for example, when $x > 0$, $y > 0$). These can be immediately inverted to obtain

$$\begin{aligned} x &= x'y' \\ y &= 2y'. \end{aligned} \tag{2.34}$$

Instead of using the tensor transformation law, we can simply use the fact that we know how to take derivatives to express $\mathrm{d}x^\mu$ in terms of $\mathrm{d}x^{\mu'}$. We have

$$\begin{aligned} \mathrm{d}x &= y'\,\mathrm{d}x' + x'\,\mathrm{d}y' \\ \mathrm{d}y &= 2\,\mathrm{d}y'. \end{aligned} \tag{2.35}$$

We need only plug these expressions directly into (2.32) to obtain (remembering that tensor products don't commute, so $\mathrm{d}x'\,\mathrm{d}y' \neq \mathrm{d}y'\,\mathrm{d}x'$):

$$S = (y')^2(\mathrm{d}x')^2 + x'y'(\mathrm{d}x'\,\mathrm{d}y' + \mathrm{d}y'\,\mathrm{d}x') + [(x')^2 + 4(x'y')^2](\mathrm{d}y')^2, \tag{2.36}$$

or

$$S_{\mu'\nu'} = \begin{pmatrix} (y')^2 & x'y' \\ x'y' & (x')^2 + 4(x'y')^2 \end{pmatrix}. \tag{2.37}$$

Notice that it is still symmetric. We did not use the transformation law (2.30) directly, but doing so would have yielded the same result, as you can check.

For the most part the various tensor operations we defined in flat space are unaltered in a more general setting: contraction, symmetrization, and so on. There are three important exceptions: partial derivatives, the metric, and the Levi–Civita tensor. Let's look at the partial derivative first.

Unfortunately, the partial derivative of a tensor is not, in general, a new tensor. The gradient, which is the partial derivative of a scalar, is an honest (0, 1) tensor, as we have seen. But the partial derivative of higher-rank tensors is not tensorial, as we can see by considering the partial derivative of a one-form, $\partial_\mu W_\nu$, and changing to a new coordinate system:

$$\begin{aligned} \frac{\partial}{\partial x^{\mu'}} W_{\nu'} &= \frac{\partial x^\mu}{\partial x^{\mu'}} \frac{\partial}{\partial x^\mu} \left(\frac{\partial x^\nu}{\partial x^{\nu'}} W_\nu \right) \\ &= \frac{\partial x^\mu}{\partial x^{\mu'}} \frac{\partial x^\nu}{\partial x^{\nu'}} \left(\frac{\partial}{\partial x^\mu} W_\nu \right) + W_\nu \frac{\partial x^\mu}{\partial x^{\mu'}} \frac{\partial}{\partial x^\mu} \frac{\partial x^\nu}{\partial x^{\nu'}}. \end{aligned} \tag{2.38}$$

The second term in the last line should not be there if $\partial_\mu W_\nu$ were to transform as a (0, 2) tensor. As you can see, it arises because the derivative of the transformation matrix does not vanish, as it did for Lorentz transformations in flat space.

Differentiation is obviously an important tool in physics, so we will have to invent new tensorial operations to take the place of the partial derivative. In fact we will invent several: the exterior derivative, the covariant derivative, and the Lie derivative.

2.5 ■ THE METRIC

The metric tensor is such an important object in curved space that it is given a new symbol, $g_{\mu\nu}$ (while $\eta_{\mu\nu}$ is reserved specifically for the Minkowski metric). There are few restrictions on the components of $g_{\mu\nu}$, other than that it be a symmetric (0, 2) tensor. It is usually, though not always, taken to be nondegenerate, meaning that the determinant $g = |g_{\mu\nu}|$ doesn't vanish. This allows us to define the inverse metric $g^{\mu\nu}$ via

$$g^{\mu\nu} g_{\nu\sigma} = g_{\lambda\sigma} g^{\lambda\mu} = \delta^{\mu}_{\sigma}. \tag{2.39}$$

The symmetry of $g_{\mu\nu}$ implies that $g^{\mu\nu}$ is also symmetric. Just as in special relativity, the metric and its inverse may be used to raise and lower indices on tensors. You may be familiar with the notion of a "metric" used in the study of topology, where we also demand that the metric be positive–definite (no negative eigenvalues). The metric we use in general relativity cannot be used to define a topology, but it will have other uses.

It will take some time to fully appreciate the role of the metric in all of its glory, but for purposes of inspiration [following Sachs and Wu (1977)] we can list the various uses to which $g_{\mu\nu}$ will be put: (1) the metric supplies a notion of "past" and "future"; (2) the metric allows the computation of path length and proper time; (3) the metric determines the "shortest distance" between two points, and therefore the motion of test particles; (4) the metric replaces the Newtonian gravitational field ϕ; (5) the metric provides a notion of locally inertial frames and therefore a sense of "no rotation"; (6) the metric determines causality, by defining the speed of light faster than which no signal can travel; (7) the metric replaces the traditional Euclidean three-dimensional dot product of Newtonian mechanics. Obviously these ideas are not all completely independent, but we get some sense of the importance of this tensor.

In our discussion of path lengths in special relativity we (somewhat handwavingly) introduced the line element as $ds^2 = \eta_{\mu\nu} dx^\mu dx^\nu$, which was used to get the length of a path. Of course now that we know that $\mathrm{d}x^\mu$ is really a basis dual vector, it becomes natural to use the terms "metric" and "line element" interchangeably, and write

$$ds^2 = g_{\mu\nu}\, \mathrm{d}x^\mu\, \mathrm{d}x^\nu. \tag{2.40}$$

To be perfectly consistent we should write this as "g," and sometimes will, but more often than not g is used for the determinant $|g_{\mu\nu}|$. For example, we know that the Euclidean line element in a three-dimensional space with Cartesian coordinates is

$$ds^2 = (\mathrm{d}x)^2 + (\mathrm{d}y)^2 + (\mathrm{d}z)^2. \tag{2.41}$$

We can now change to any coordinate system we choose. For example, in spherical coordinates we have

$$x = r \sin\theta \cos\phi$$
$$y = r \sin\theta \sin\phi$$
$$z = r \cos\theta, \tag{2.42}$$

which leads directly to

$$ds^2 = \mathrm{d}r^2 + r^2\,\mathrm{d}\theta^2 + r^2 \sin^2\theta\,\mathrm{d}\phi^2. \tag{2.43}$$

Obviously the components of the metric look different than those in Cartesian coordinates, but all of the properties of the space remain unaltered.

Most references are not sufficiently picky to distinguish between "dx," the informal notion of an infinitesimal displacement, and "$\mathrm{d}x$," the rigorous notion of a basis one-form given by the gradient of a coordinate function. (They also tend to neglect the fact that tensor products don't commute, and write expressions like $\mathrm{d}x\mathrm{d}y + \mathrm{d}y\mathrm{d}x$ as $2\mathrm{d}x\mathrm{d}y$; it should be clear what is meant from the context.) In fact our notation "ds^2" does not refer to the differential of anything, or the square of anything; it's just conventional shorthand for the metric tensor, a multilinear map from two vectors to the real numbers. Thus, we have a set of equivalent expressions for the inner product of two vectors V^μ and W^ν:

$$g_{\mu\nu}V^\mu W^\nu = g(V, W) = ds^2(V, W). \tag{2.44}$$

Meanwhile, "$(\mathrm{d}x)^2$" refers specifically to the honest (0, 2) tensor $\mathrm{d}x \otimes \mathrm{d}x$.

A good example of a non-Euclidean manifold is the two-sphere, which can be thought of as the locus of points in $\mathbf{R}^3$ at distance 1 from the origin. The metric in the (θ, ϕ) coordinate system can be derived by setting $r = 1$ and $\mathrm{d}r = 0$ in (2.43):

$$ds^2 = \mathrm{d}\theta^2 + \sin^2\theta\,\mathrm{d}\phi^2. \tag{2.45}$$

This is completely consistent with the interpretation of ds as an infinitesimal length, as illustrated in Figure 2.21. Anyone paying attention should at this point be asking, "What in the world does it mean to set $\mathrm{d}r = 0$? We know that $\mathrm{d}r$ is a well-defined nonvanishing one-form field." As occasionally happens, we are using sloppy language to motivate a step that is actually quite legitimate; see Appendix A for a discussion of how submanifolds inherit metrics from the spaces in which they are embedded.

As we shall see, the metric tensor contains all the information we need to describe the curvature of the manifold (at least in what is called Riemannian geometry; we will get into some of the subtleties in the next chapter). In Minkowski space we can choose coordinates in which the components of the metric are constant; but it should be clear that the existence of curvature is more subtle than having the metric depend on the coordinates, since in the example above we showed how the metric in flat Euclidean space in spherical coordinates is a function of r and θ. Later, we shall see that constancy of the metric components is sufficient for a space to be flat, and in fact there always exists a coordinate system on any

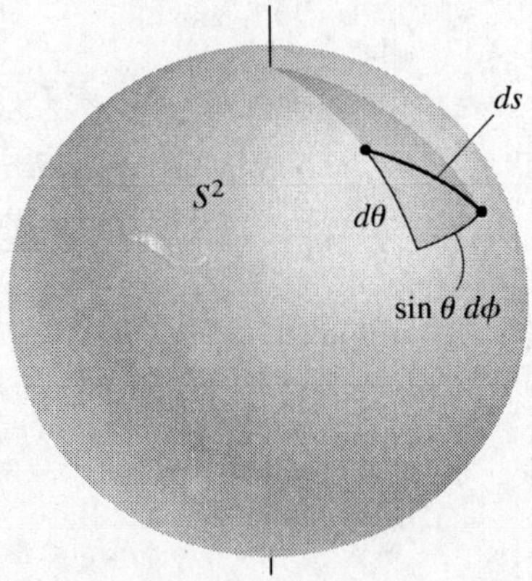

FIGURE 2.21 The line element on a two-dimensional sphere.

flat space in which the metric is constant. But we might not know how to find such a coordinate system, and there are many ways for a space to deviate from flatness; we will therefore want a more precise characterization of the curvature, which will be introduced later.

A useful characterization of the metric is obtained by putting $g_{\mu\nu}$ into its **canonical form**. In this form the metric components become

$$g_{\mu\nu} = \text{diag}\,(-1, -1, \ldots, -1, +1, +1, \ldots, +1, 0, 0, \ldots, 0), \tag{2.46}$$

where "diag" means a diagonal matrix with the given elements. The **signature** of the metric is the number of both positive and negative eigenvalues; we speak of "a metric with signature minus-plus-plus-plus" for Minkowski space, for example. If any of the eigenvalues are zero, the metric is "degenerate," and the inverse metric will not exist; if the metric is continuous and nondegenerate, its signature will be the same at every point. We will always deal with continuous, nondegenerate metrics. If all of the signs are positive, the metric is called **Euclidean** or **Riemannian** (or just positive definite), while if there is a single minus it is called **Lorentzian** or **pseudo-Riemannian**, and any metric with some $+1$'s and some -1's is called indefinite. (So the word Euclidean sometimes means that the space is flat, and sometimes doesn't, but it always means that the canonical form is strictly positive; the terminology is unfortunate but standard.) The spacetimes of interest in general relativity have Lorentzian metrics.

We haven't yet demonstrated that it is always possible to put the metric into canonical form. In fact it is always possible to do so at some point $p \in M$, but in general it will only be possible at that single point, not in any neighborhood of p. Actually we can do slightly better than this; it turns out that at any point p there exists a coordinate system $x^{\hat{\mu}}$ in which $g_{\hat{\mu}\hat{\nu}}$ takes its canonical form and the first derivatives $\partial_{\hat{\sigma}} g_{\hat{\mu}\hat{\nu}}$ all vanish (while the second derivatives $\partial_{\hat{\rho}}\partial_{\hat{\sigma}} g_{\hat{\mu}\hat{\nu}}$ cannot be made to all vanish):

$$g_{\hat{\mu}\hat{\nu}}(p) = \eta_{\hat{\mu}\hat{\nu}}, \quad \partial_{\hat{\sigma}} g_{\hat{\mu}\hat{\nu}}(p) = 0. \tag{2.47}$$

Such coordinates are known as **locally inertial coordinates**, and the associated basis vectors constitute a **local Lorentz frame**; we often put hats on the indices when we are in these special coordinates. Notice that in locally inertial coordinates the metric at p looks like that of flat space to first order. This is the rigorous notion of the idea that "small enough regions of spacetime look like flat (Minkowski) space." Also, there is no difficulty in simultaneously constructing sets of *basis vectors* at every point in M such that the metric takes its canonical form; the problem is that in general there will not be a *coordinate system* from which this basis can be derived. Bases of this sort are discussed in Appendix J.

We will delay a discussion of how to construct locally inertial coordinates until Chapter 3. It is useful, however, to sketch a proof of their existence for the specific case of a Lorentzian metric in four dimensions. The idea is to consider the transformation law for the metric

$$g_{\hat{\mu}\hat{\nu}} = \frac{\partial x^{\mu}}{\partial x^{\hat{\mu}}} \frac{\partial x^{\nu}}{\partial x^{\hat{\nu}}} g_{\mu\nu}, \tag{2.48}$$

and expand both sides in Taylor series in the sought-after coordinates $x^{\hat{\mu}}$. The expansion of the old coordinates x^{μ} looks like

$$x^{\mu} = \left(\frac{\partial x^{\mu}}{\partial x^{\hat{\mu}}}\right)_p x^{\hat{\mu}} + \frac{1}{2}\left(\frac{\partial^2 x^{\mu}}{\partial x^{\hat{\mu}_1} \partial x^{\hat{\mu}_2}}\right)_p x^{\hat{\mu}_1} x^{\hat{\mu}_2} + \frac{1}{6}\left(\frac{\partial^3 x^{\mu}}{\partial x^{\hat{\mu}_1} \partial x^{\hat{\mu}_2} \partial x^{\hat{\mu}_3}}\right)_p x^{\hat{\mu}_1} x^{\hat{\mu}_2} x^{\hat{\mu}_3} + \cdots, \tag{2.49}$$

with the other expansions proceeding along the same lines. [For simplicity we have set $x^{\mu}(p) = x^{\hat{\mu}}(p) = 0$.] Then, using some extremely schematic notation, the expansion of (2.48) to second order is

$$\begin{aligned}
&(\hat{g})_p + (\hat{\partial}\hat{g})_p\, \hat{x} + (\hat{\partial}\hat{\partial}\hat{g})_p\, \hat{x}\hat{x} \\
&= \left(\frac{\partial x}{\partial \hat{x}} \frac{\partial x}{\partial \hat{x}} g\right)_p + \left(\frac{\partial x}{\partial \hat{x}} \frac{\partial^2 x}{\partial \hat{x} \partial \hat{x}} g + \frac{\partial x}{\partial \hat{x}} \frac{\partial x}{\partial \hat{x}} \hat{\partial} g\right)_p \hat{x} \\
&+ \left(\frac{\partial x}{\partial \hat{x}} \frac{\partial^3 x}{\partial \hat{x} \partial \hat{x} \partial \hat{x}} g + \frac{\partial^2 x}{\partial \hat{x} \partial \hat{x}} \frac{\partial^2 x}{\partial \hat{x} \partial \hat{x}} g + \frac{\partial x}{\partial \hat{x}} \frac{\partial^2 x}{\partial \hat{x} \partial \hat{x}} \hat{\partial} g + \frac{\partial x}{\partial \hat{x}} \frac{\partial x}{\partial \hat{x}} \hat{\partial}\hat{\partial} g\right)_p \hat{x}\hat{x}.
\end{aligned} \tag{2.50}$$

We can set terms of equal order in $\hat{x}$ on each side equal to each other. Therefore, the components $g_{\hat{\mu}\hat{\nu}}(p)$, 10 numbers in all (to describe a symmetric two-index tensor), are determined by the matrix $(\partial x^{\mu}/\partial x^{\hat{\mu}})_p$. This is a 4×4

matrix with no constraints; thus, we are free to choose 16 numbers. Clearly this is enough freedom to put the 10 numbers of $g_{\hat{\mu}\hat{\nu}}(p)$ into canonical form, at least as far as having enough degrees of freedom is concerned. (In fact there are some limitations—if you go through the procedure carefully, you find for example that you cannot change the signature.) The six remaining degrees of freedom can be interpreted as exactly the six parameters of the Lorentz group; we know that these leave the canonical form unchanged. At first order we have the derivatives $\partial_{\hat{\sigma}} g_{\hat{\mu}\hat{\nu}}(p)$, four derivatives of ten components for a total of 40 numbers. But looking at the right-hand side of (2.50) we see that we now have the additional freedom to choose $(\partial^2 x^\mu/\partial x^{\hat{\mu}_1}\partial x^{\hat{\mu}_2})_p$. In this set of numbers there are 10 independent choices of the indices $\hat{\mu}_1$ and $\hat{\mu}_2$ (it's symmetric, since partial derivatives commute) and four choices of μ, for a total of 40 degrees of freedom. This is precisely the number of choices we need to determine all of the first derivatives of the metric, which we can therefore set to zero. At second order, however, we are concerned with $\partial_{\hat{\rho}}\partial_{\hat{\sigma}} g_{\hat{\mu}\hat{\nu}}(p)$; this is symmetric in $\hat{\rho}$ and $\hat{\sigma}$ as well as $\hat{\mu}$ and $\hat{\nu}$, for a total of $10 \times 10 = 100$ numbers. Our ability to make additional choices is contained in $(\partial^3 x^\mu/\partial x^{\hat{\mu}_1}\partial x^{\hat{\mu}_2}\partial x^{\hat{\mu}_3})_p$. This is symmetric in the three lower indices, which gives 20 possibilities, times four for the upper index gives us 80 degrees of freedom—20 fewer than we require to set the second derivatives of the metric to zero. So in fact we cannot make the second derivatives vanish; the deviation from flatness must therefore be measured by the 20 degrees of freedom representing the second derivatives of the metric tensor field. We will see later how this comes about, when we characterize curvature using the Riemann tensor, which will turn out to have 20 independent components in four dimensions.

Locally inertial coordinates are unbelievably useful. Best of all, their usefulness does not generally require that we actually do the work of constructing such coordinates (although we will give a recipe for doing so in the next chapter), but simply that we know that they do exist. The usual trick is to take a question of physical interest, answer it in the context of locally inertial coordinates, and then express that answer in a coordinate-independent form. Take a very simple example, featuring an observer with four-velocity $U^{\hat{\mu}}$ and a rocket flying past with four-velocity $V^{\hat{\mu}}$. What does the observer measure as the ordinary three-velocity of the rocket? In special relativity the answer is straightforward. Work in inertial coordinates (globally, not just locally) such that the observer is in the rest frame and the rocket is moving along the x-axis. Then the four-velocity of the observer is $U^{\hat{\mu}} = (1, 0, 0, 0)$ and the four-velocity of the rocket is $V^{\hat{\mu}} = (\gamma, v\gamma, 0, 0)$, where v is the three-velocity and $\gamma = 1/\sqrt{1-v^2}$, so that $v = \sqrt{1-\gamma^{-2}}$. Since we are in flat spacetime (for the moment), we have

$$\gamma = -\eta_{\hat{\mu}\hat{\nu}} U^{\hat{\mu}} V^{\hat{\nu}} = -U_{\hat{\mu}} V^{\hat{\mu}}, \tag{2.51}$$

since $\eta_{00} = -1$. The flat-spacetime answer would therefore be

$$v = \sqrt{1 - (U_{\hat{\mu}} V^{\hat{\mu}})^{-2}}. \tag{2.52}$$

Now we can go back to curved spacetime, where the metric is no longer flat. But at the point where the measurement is being done, we are free to use locally inertial coordinates, in which case the components of $g_{\hat{\mu}\hat{\nu}}$ are precisely those of $\eta_{\hat{\mu}\hat{\nu}}$. So (2.52) is still true in curved spacetime in this particular coordinate system. But (2.52) is a completely tensorial equation, which doesn't care what coordinate system we are in; therefore it is true in complete generality. This kind of procedure will prove its value over and over.

2.6 ■ AN EXPANDING UNIVERSE

A simple example of a nontrivial Lorentzian geometry is provided by a four-dimensional cosmological spacetime with metric

$$ds^2 = -dt^2 + a^2(t)[dx^2 + dy^2 + dz^2]. \tag{2.53}$$

This describes a universe for which "space at a fixed moment of time" is a flat three-dimensional Euclidean space, which is expanding as a function of time. Worldlines that remain at constant spatial coordinates x^i are said to be comoving; similarly, we denote a region of space that expands along with boundaries defined by fixed spatial coordinates as a "comoving volume." Since the metric describes (distance)^2, the relative distance between comoving points is growing as $a(t)$ in this spacetime; the function a is called the scale factor. This is a special case of a Robertson–Walker metric, one in which spatial slices are geometrically flat; there are other cases for which spatial slices are curved (as we will discuss in Chapter 8). But our interest right now is not in where this metric came from, but in using it as a playground to illustrate some of the ideas we have developed.

Typical solutions for the scale factor are power laws,

$$a(t) = t^q, \qquad 0 < q < 1. \tag{2.54}$$

Actually there are all sorts of solutions, but these are some particularly simple and relevant ones. A matter-dominated flat universe satisfies $q = \frac{2}{3}$, while a radiation-dominated flat universe satisfies $q = \frac{1}{2}$. An obvious feature is that the scale factor goes to zero as $t \to 0$, and along with it the spatial components of the metric. This is a coordinate-dependent statement, and in principle there might be another coordinate system in which everything looks finite; in this case, however, $t = 0$ represents a true singularity of the geometry (the "Big Bang"), and should be excluded from the manifold. The range of the t coordinate is therefore

$$0 < t < \infty. \tag{2.55}$$

Our spacetime comes to an end at $t = 0$.

Light cones in this curved geometry are defined by null paths, those for which $ds^2 = 0$. We can draw a spacetime diagram by considering null paths for which

y and z are held constant; then

$$0 = -\mathrm{d}t^2 + t^{2q}\,\mathrm{d}x^2, \tag{2.56}$$

which implies

$$\frac{dx}{dt} = \pm t^{-q}. \tag{2.57}$$

You might worry that, after all that fuss about $\mathrm{d}x^\mu$ being a basis one-form and not a differential, we have sloppily "divided by $\mathrm{d}t^2$" to go from (2.56) to (2.57). The truth is much more respectable. What we actually did was to take the (0, 2) tensor defined by (2.56), which takes two vectors and returns a real number, and act it on two copies of the vector $V = (dx^\mu/d\lambda)\partial_\mu$, the tangent vector to a curve $x^\mu(\lambda)$. Consider just the $\mathrm{d}t^2$ piece acting on V:

$$\mathrm{d}t^2(V, V) \equiv (\mathrm{d}t \otimes \mathrm{d}t)(V, V) = \mathrm{d}t(V) \cdot \mathrm{d}t(V), \tag{2.58}$$

where the notation $\mathrm{d}t(V)$ refers to a real number that we compute as

$$\begin{aligned}\mathrm{d}t(V) &= \mathrm{d}t\left(\frac{dx^\mu}{d\lambda}\partial_\mu\right)\\ &= \frac{dx^\mu}{d\lambda}\mathrm{d}t\,(\partial_\mu)\\ &= \frac{dx^\mu}{d\lambda}\frac{\partial t}{\partial x^\mu}\\ &= \frac{dt}{d\lambda},\end{aligned} \tag{2.59}$$

where in the third line we have invoked (2.25). Following the same procedure with $\mathrm{d}x^2$, we find that (2.56) implies

$$0 = -\left(\frac{dt}{d\lambda}\right)^2 + t^{2q}\left(\frac{dx}{d\lambda}\right)^2, \tag{2.60}$$

from which (2.57) follows via the one-dimensional chain rule,

$$\frac{dx}{dt} = \frac{dx}{d\lambda}\frac{d\lambda}{dt}. \tag{2.61}$$

The lesson should be clear: expressions such as (2.56) describe well-defined tensors, but manipulation of the basis one-forms as if they were simply "differentials" does get you the right answer. (At least, most of the time; it's a good idea to keep the more formal definitions in mind.)

We can solve (2.57) to obtain

$$t = (1-q)^{1/(1-q)}(\pm x - x_0)^{1/(1-q)}, \tag{2.62}$$

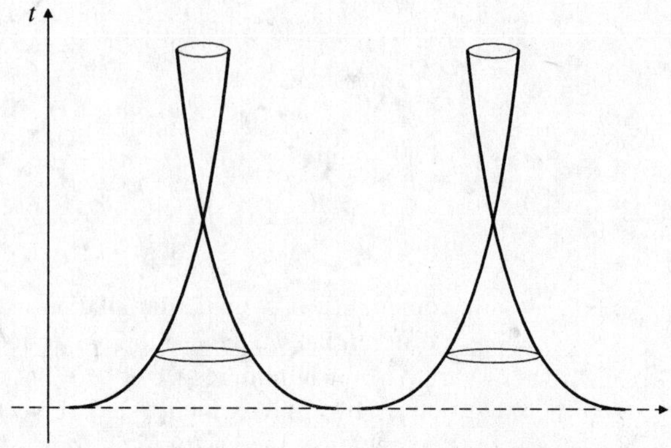

FIGURE 2.22 Spacetime diagram for a flat Robertson–Walker universe with $a(t) \propto t^q$, for $0 < q < 1$. The dashed line at the bottom of the figure represents the singularity at $t = 0$. Since light cones are tangent to the singularity, the pasts of two points may be nonoverlapping.

where x_0 is a constant of integration. These curves define the light cones of our expanding universe, as plotted in Figure 2.22. Since we have assumed $0 < q < 1$, the light cones are tangent to the singularity at $t = 0$. A crucial feature of this geometry is that the light cones of two points need not intersect in the past; this is in contrast to Minkowski space, for which the light cones of any two points always intersect in both the past and future. We say that every event defines an "horizon," outside of which there exist worldlines that can have had no influence on what happens at that event. This is because, since nothing can travel faster than light, each point can only be influenced by events that are either on, or in the interior of, its past light cone (indeed, we refer to the past light cone plus its interior as simply "the past" of an event). Two events outside each others' horizons are said to be "out of causal contact." These notions will be explored more carefully in the next section, as well as in Chapters 4 and 8.

2.7 ■ CAUSALITY

Many physical questions can be cast as an initial-value problem: given the state of a system at some moment in time, what will be the state at some later time? The fact that such questions have definite answers is due to causality, the idea that future events can be understood as consequences of initial conditions plus the laws of physics. Initial-value problems are as common in GR as in Newtonian physics or special relativity; however, the dynamical nature of the spacetime background introduces new ways in which an initial-value formulation could break down. Here we very briefly introduce some of the concepts used in understanding how causality works in GR.

We will look at the problem of evolving matter fields on a fixed background spacetime, rather than the evolution of the metric itself. Our guiding principle will be that no signals can travel faster than the speed of light; therefore information will only flow along timelike or null trajectories (not necessarily geodesics). Since it is sometimes useful to distinguish between purely timelike paths and ones that are merely non-spacelike, we define a **causal curve** to be one which is timelike or null everywhere. Then, given any subset S of a manifold M, we define the **causal future** of S, denoted $J^+(S)$, to be the set of points that can be reached from S by following a future-directed causal curve; the **chronological future** $I^+(S)$ is the set of points that can be reached by following a future-directed timelike curve. Note that a curve of zero length is causal but not chronal; therefore, a point p will always be in its own causal future $J^+(p)$, but not necessarily in its own chronological future $I^+(p)$ (although it could be, as we mention below). The causal past J^- and chronological past I^- are defined analogously.

A subset $S \subset M$ is called **achronal** if no two points in S are connected by a timelike curve; for example, any edgeless spacelike hypersurface in Minkowski spacetime is achronal. Given a closed achronal set S, we define the **future domain of dependence** of S, denoted $D^+(S)$, as the set of all points p such that *every* past-moving inextendible causal curve through p must intersect S. (Inextendible just means that the curve goes on forever, not ending at some finite point; closed means that the complement of the set is an open set.) Elements of S itself are elements of $D^+(S)$. The past domain of dependence $D^-(S)$ is defined by replacing future with past. Generally speaking, some points in M will be in one of the domains of dependence, and some will be outside; we define the boundary of $D^+(S)$ to be the **future Cauchy horizon** $H^+(S)$, and likewise the boundary of $D^-(S)$ to be the past Cauchy horizon $H^-(S)$. You can convince yourself that they are both null surfaces. The domains of dependence and Cauchy horizons are illustrated in Figure 2.23, in which S is taken to be a connected subset of an achronal surface Σ.

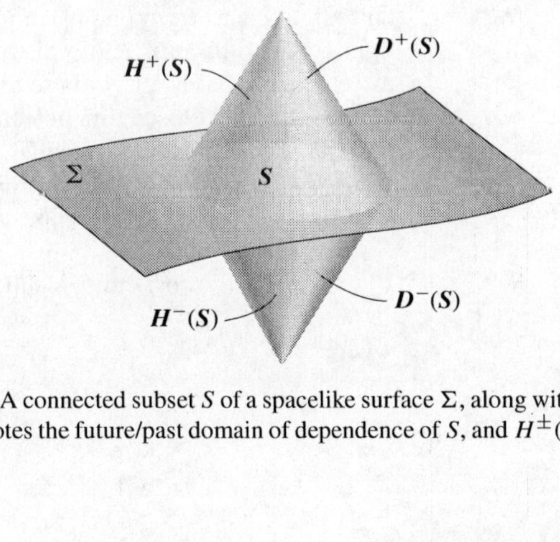

FIGURE 2.23 A connected subset S of a spacelike surface Σ, along with its causal structure. $D^\pm(S)$ denotes the future/past domain of dependence of S, and $H^\pm(S)$ the future/past Cauchy horizon.

The usefulness of these definitions should be apparent; if nothing moves faster than light, signals cannot propagate outside the light cone of any point p. Therefore, if every curve that remains inside this light cone must intersect S, then information specified on S should be sufficient to predict what the situation is at p; that is, initial data for matter fields given on S can be used to solve for the value of the fields at p. The set of all points for which we can predict what happens by knowing what happens on S is the union $D(S) = D^+(S) \cup D^-(S)$, called simply the domain of dependence. A closed achronal surface Σ is said to be a **Cauchy surface** if the domain of dependence $D(\Sigma)$ is the entire manifold; from information given on a Cauchy surface, we can predict what happens throughout all of spacetime. If a spacetime has a Cauchy surface (which it may not), it is said to be **globally hyperbolic**.

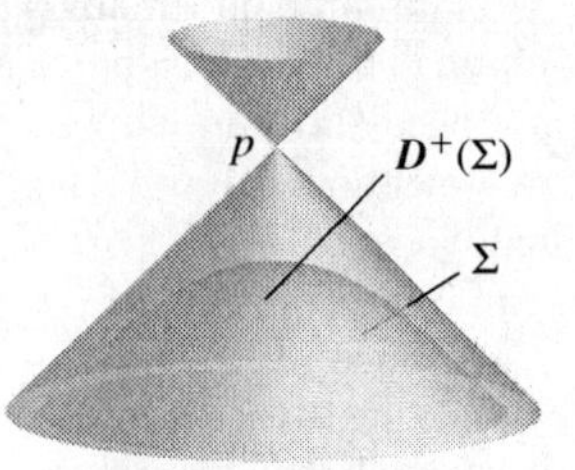

FIGURE 2.24 The surface Σ is everywhere spacelike but lies in the past of the past light cone of the point p; its domain of dependence is not all of the spacetime.

Any set Σ that is closed, achronal, and has no edge, is called a **partial Cauchy surface**. A partial Cauchy surface can fail to be an actual Cauchy surface either through its own fault, or through a fault of the spacetime. One possibility is that we have just chosen a "bad" hypersurface (although it is hard to give a general prescription for when a hypersurface is bad in this sense). Consider Minkowski space, and an edgeless spacelike hypersurface Σ, which remains to the past of the light cone of some point, as in Figure 2.24. In this case Σ is an achronal surface, but it is clear that $D^+(\Sigma)$ ends at the light cone, and we cannot use information on Σ to predict what happens throughout Minkowski space. Of course, there are other surfaces we could have picked for which the domain of dependence would have been the entire manifold, so this doesn't worry us too much.

A somewhat more nontrivial way for a Cauchy horizon to arise is through the appearance of **closed timelike curves**. In Newtonian physics, causality is enforced by the relentless forward march of an absolute notion of time. In special relativity things are even more restrictive; not only must you move forward in time, but the speed of light provides a limit on how swiftly you may move through space (you must stay within your forward light cone). In general relativity it remains true that you must stay within your forward light cone; however, this becomes strictly a local notion, as globally the curvature of spacetime might "tilt" light cones from one place to another. It becomes possible in principle for light cones to be sufficiently distorted that an observer can move on a forward-directed path that is everywhere timelike and yet intersects itself at a point in its "past"—this is a closed timelike curve.

As a simple example, consider a two-dimensional geometry with coordinates $\{t, x\}$, such that points with coordinates (t, x) and $(t, x + 1)$ are identified. The topology is thus $\mathbf{R} \times S^1$. We take the metric to be

$$ds^2 = -\cos(\lambda)dt^2 - \sin(\lambda)[dt\, dx + dx\, dt] + \cos(\lambda)dx^2, \tag{2.63}$$

where

$$\lambda = \cot^{-1} t, \tag{2.64}$$

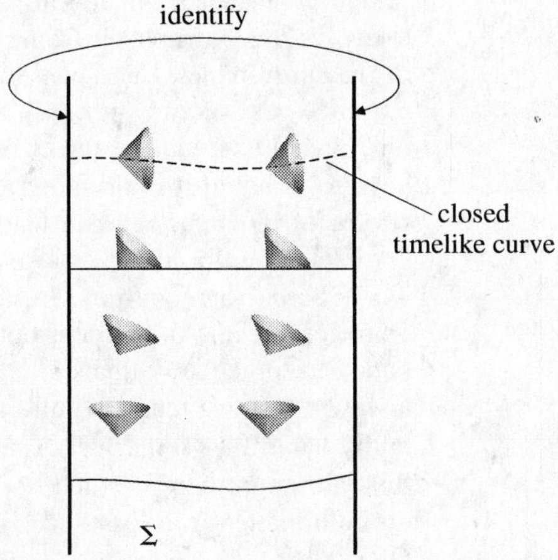

FIGURE 2.25 A cylindrical spacetime with closed timelike curves. The light cones progressively tilt, such that the domain of dependence of the surface Σ fills the lower part of the spacetime, but comes to an end when the closed timelike curves come into existence.

which goes from $\lambda(t = -\infty) = 0$ to $\lambda(t = \infty) = \pi$. This metric doesn't represent any special famous solution to general relativity, it was just cooked up to provide an interesting example of closed timelike curves; but there is a well-known example known as Misner space, with similar properties. In the spacetime defined by (2.63), the light cones progressively tilt as you go forward in time, as shown in Figure 2.25. For $t < 0$, the light cones point forward, and causality is maintained. Once $t > 0$, however, x becomes the timelike coordinate, and it is possible to travel on a timelike trajectory that wraps around the S^1 and comes back to itself; this is a closed timelike curve. If we had specified a surface Σ to this past of this point, then none of the points in the region containing closed timelike curves are in the domain of dependence of Σ, since the closed timelike curves themselves do not intersect Σ. There is thus necessarily a Cauchy horizon at the surface $t = 0$. This is obviously a worse problem than the previous one, since a well-defined initial value problem does not seem to exist in this spacetime.

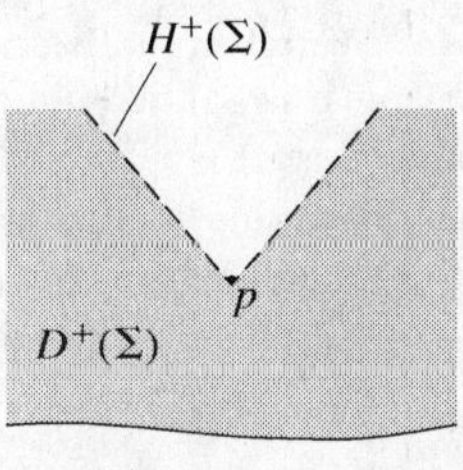

FIGURE 2.26 A singularity at p removes any points in its future from the domain of dependence of a surface Σ in its past.

A final example is provided by the existence of singularities, points that are not in the manifold even though they can be reached by traveling along a geodesic for a finite distance. Typically these occur when the curvature becomes infinite at some point; if this happens, the point can no longer be said to be part of the spacetime. Such an occurrence can lead to the emergence of a Cauchy horizon, as depicted in Figure 2.26—a point p, which is in the future of a singularity, cannot be in the domain of dependence of a hypersurface to the past of the singularity, because there will be curves from p that simply end at the singularity.

These obstacles can also arise in the initial value problem for GR, when we try to evolve the metric itself from initial data. However, they are of different degrees of troublesomeness. The possibility of picking a "bad" initial hypersurface does not arise very often, especially since most solutions are found globally (by solving Einstein's equation throughout spacetime). The one situation in which you have to be careful is in numerical solution of Einstein's equation, where a bad choice of hypersurface can lead to numerical difficulties, even if in principle a complete solution exists. Closed timelike curves seem to be something that GR works hard to avoid—there are certainly solutions that contain them, but evolution from generic initial data does not usually produce them. Singularities, on the other hand, are practically unavoidable. The simple fact that the gravitational force is always attractive tends to pull matter together, increasing the curvature, and generally leading to some sort of singularity. Apparently we must learn to live with this, although there is some hope that a well-defined theory of quantum gravity will eliminate (or at least teach us how to deal with) the singularities of classical GR.

2.8 ■ TENSOR DENSITIES

Tensors possess a compelling beauty and simplicity, but there are times when it is useful to consider nontensorial objects. Recall that in Chapter 1 we introduced the completely antisymmetric Levi–Civita symbol, defined as

$$\tilde{\epsilon}_{\mu_1\mu_2\cdots\mu_n} = \begin{cases} +1 & \text{if } \mu_1\mu_2\cdots\mu_n \text{ is an even permutation of } 01\cdots(n-1), \\ -1 & \text{if } \mu_1\mu_2\cdots\mu_n \text{ is an odd permutation of } 01\cdots(n-1), \\ 0 & \text{otherwise.} \end{cases} \tag{2.65}$$

By definition, the Levi–Civita symbol has the components specified above *in any coordinate system* (at least, in any right-handed coordinate system; switching the handedness multiplies the components of $\tilde{\epsilon}_{\mu_1\mu_2\cdots\mu_n}$ by an overall minus sign). This is called a "symbol," of course, because it is not a tensor; it is defined not to change under coordinate transformations. We were only able to treat it as a tensor in inertial coordinates in flat spacetime, since Lorentz transformations would have left the components invariant anyway. Its behavior can be related to that of an ordinary tensor by first noting that, given any $n \times n$ matrix $M^{\mu}{}_{\mu'}$, the determinant $|M|$ obeys

$$\tilde{\epsilon}_{\mu'_1\mu'_2\cdots\mu'_n}|M| = \tilde{\epsilon}_{\mu_1\mu_2\cdots\mu_n} M^{\mu_1}{}_{\mu'_1} M^{\mu_2}{}_{\mu'_2} \cdots M^{\mu_n}{}_{\mu'_n}. \tag{2.66}$$

This is just a streamlined expression for the determinant of any matrix, completely equivalent to the usual formula in terms of matrices of cofactors. (You can check it for yourself for 2×2 or 3×3 matrices.) It follows that, setting $M^{\mu}{}_{\mu'} = \partial x^{\mu}/\partial x^{\mu'}$, we have

$$\tilde{\epsilon}_{\mu_1'\mu_2'\cdots\mu_n'} = \left|\frac{\partial x^{\mu'}}{\partial x^{\mu}}\right| \tilde{\epsilon}_{\mu_1\mu_2\cdots\mu_n} \frac{\partial x^{\mu_1}}{\partial x^{\mu_1'}} \frac{\partial x^{\mu_2}}{\partial x^{\mu_2'}} \cdots \frac{\partial x^{\mu_n}}{\partial x^{\mu_n'}}, \tag{2.67}$$

where we have also used the facts that the matrix $\partial x^{\mu'}/\partial x^{\mu}$ is the inverse of $\partial x^{\mu}/\partial x^{\mu'}$, and that the determinant of an inverse matrix is the inverse of the determinant, $|M^{-1}| = |M|^{-1}$. So the Levi–Civita symbol transforms in a way close to the tensor transformation law, except for the determinant out front. Objects transforming in this way are known as **tensor densities**. Another example is given by the determinant of the metric, $g = |g_{\mu\nu}|$. It's easy to check, by taking the determinant of both sides of (2.48), that under a coordinate transformation we get

$$g(x^{\mu'}) = \left|\frac{\partial x^{\mu'}}{\partial x^{\mu}}\right|^{-2} g(x^{\mu}). \tag{2.68}$$

Therefore g is also not a tensor; it transforms in a way similar to the Levi–Civita symbol, except that the Jacobian is raised to the -2 power. The power to which the Jacobian is raised is known as the **weight** of the tensor density; the Levi–Civita symbol is a density of weight 1, while g is a (scalar) density of weight -2.

However, we don't like tensor densities as much as we like tensors. There is a simple way to convert a density into an honest tensor—multiply by $|g|^{w/2}$, where w is the weight of the density (the absolute value signs are there because $g < 0$ for Lorentzian metrics). The result will transform according to the tensor transformation law. Therefore, for example, we can define the **Levi–Civita tensor** as

$$\boxed{\epsilon_{\mu_1\mu_2\cdots\mu_n} = \sqrt{|g|}\,\tilde{\epsilon}_{\mu_1\mu_2\cdots\mu_n}.} \tag{2.69}$$

Since this is a real tensor, we can raise indices and so on. Sometimes people define a version of the Levi–Civita symbol with upper indices, $\tilde{\epsilon}^{\mu_1\mu_2\cdots\mu_n}$, whose components are numerically equal to $\mathrm{sgn}(g)\tilde{\epsilon}_{\mu_1\mu_2\cdots\mu_n}$, where $\mathrm{sgn}(g)$ is the sign of the metric determinant. This turns out to be a density of weight -1, and is related to the tensor with upper indices (obtained by using $g^{\mu\nu}$ to raise indices on $\epsilon_{\mu_1\mu_2\cdots\mu_n}$) by

$$\epsilon^{\mu_1\mu_2\cdots\mu_n} = \frac{1}{\sqrt{|g|}}\tilde{\epsilon}^{\mu_1\mu_2\cdots\mu_n}. \tag{2.70}$$

Something you often end up doing is contracting p indices on $\epsilon^{\mu_1\mu_2\cdots\mu_n}$ with $\epsilon_{\mu_1\mu_2\cdots\mu_n}$; the result can be expressed in terms of an antisymmetrized product of Kronecker deltas as

$$\epsilon^{\mu_1\mu_2\cdots\mu_p\alpha_1\cdots\alpha_{n-p}}\epsilon_{\mu_1\mu_2\cdots\mu_p\beta_1\cdots\beta_{n-p}} = (-1)^s p!(n-p)!\delta^{[\alpha_1}_{\beta_1}\cdots\delta^{\alpha_{n-p}]}_{\beta_{n-p}}, \tag{2.71}$$

where s is the number of negative eigenvalues of the metric (for Lorentzian signature with our conventions, $s = 1$). The most common example is $p = n - 1$, for which we have

$$\epsilon^{\mu_1\mu_2\cdots\mu_{n-1}\alpha}\epsilon_{\mu_1\mu_2\cdots\mu_{n-1}\beta} = (-1)^s(n-1)!\,\delta^\alpha_\beta. \qquad (2.72)$$

2.9 ■ DIFFERENTIAL FORMS

Let us now introduce a special class of tensors, known as **differential forms** (or just forms). A differential p-form is simply a $(0, p)$ tensor that is completely antisymmetric. Thus, scalars are automatically 0-forms, and dual vectors are automatically one-forms (thus explaining this terminology from before). We also have the 4-form $\epsilon_{\mu\nu\rho\sigma}$. The space of all p-forms is denoted Λ^p, and the space of all p-form fields over a manifold M is denoted $\Lambda^p(M)$. A semi-straightforward exercise in combinatorics reveals that the number of linearly independent p-forms on an n-dimensional vector space is $n!/(p!(n-p)!)$. So at a point on a four-dimensional spacetime there is one linearly independent 0-form, four 1-forms, six 2-forms, four 3-forms, and one 4-form. There are no p-forms for $p > n$, since all of the components will automatically be zero by antisymmetry.

Why should we care about differential forms? This question is hard to answer without some more work, but the basic idea is that forms can be both differentiated and integrated, without the help of any additional geometric structure. We will glance briefly at both of these operations.

Given a p-form A and a q-form B, we can form a $(p+q)$-form known as the **wedge product** $A \wedge B$ by taking the antisymmetrized tensor product:

$$(A \wedge B)_{\mu_1\cdots\mu_{p+q}} = \frac{(p+q)!}{p!\,q!} A_{[\mu_1\cdots\mu_p} B_{\mu_{p+1}\cdots\mu_{p+q}]}. \qquad (2.73)$$

Thus, for example, the wedge product of two 1-forms is

$$(A \wedge B)_{\mu\nu} = 2A_{[\mu}B_{\nu]} = A_\mu B_\nu - A_\nu B_\mu. \qquad (2.74)$$

Note that

$$A \wedge B = (-1)^{pq} B \wedge A, \qquad (2.75)$$

so you can alter the order of a wedge product if you are careful with signs. We are free to suppress indices when using forms, since we know that all of the indices are downstairs and the tensors are completely antisymmetric.

The **exterior derivative** d allows us to differentiate p-form fields to obtain $(p+1)$-form fields. It is defined as an appropriately normalized, antisymmetrized partial derivative:

$$(\mathrm{d}A)_{\mu_1\cdots\mu_{p+1}} = (p+1)\partial_{[\mu_1} A_{\mu_2\cdots\mu_{p+1}]}. \qquad (2.76)$$

The simplest example is the gradient, which is the exterior derivative of a 0-form:

$$(\mathrm{d}\phi)_\mu = \partial_\mu \phi. \qquad (2.77)$$

Exterior derivatives obey a modified version of the Leibniz rule when applied to the product of a p-form ω and a q-form η:

$$\mathrm{d}(\omega \wedge \eta) = (\mathrm{d}\omega) \wedge \eta + (-1)^p \omega \wedge (\mathrm{d}\eta). \tag{2.78}$$

You are encouraged to prove this yourself.

The reason why the exterior derivative deserves special attention is that *it is a tensor*, even in curved spacetimes, unlike its cousin the partial derivative. For $p = 1$ we can see this from the transformation law for the partial derivative of a one form, (2.38); the offending nontensorial term can be written

$$W_\nu \frac{\partial x^\mu}{\partial x^{\mu'}} \frac{\partial}{\partial x^\mu} \frac{\partial x^\nu}{\partial x^{\nu'}} = W_\nu \frac{\partial^2 x^\nu}{\partial x^{\mu'} \partial x^{\nu'}}. \tag{2.79}$$

This expression is symmetric in μ' and ν', since partial derivatives commute. But the exterior derivative is defined to be the antisymmetrized partial derivative, so this term vanishes (the antisymmetric part of a symmetric expression is zero). We are then left with the correct tensor transformation law; extension to arbitrary p is straightforward. So the exterior derivative is a legitimate tensor operator; it is not, however, an adequate substitute for the partial derivative, since it is only defined on forms. In the next chapter we will define a covariant derivative, which is closer to what we might think of as the extension of the partial derivative to arbitrary manifolds.

Another interesting fact about exterior differentiation is that, for any form A,

$$\mathrm{d}(\mathrm{d}A) = 0, \tag{2.80}$$

which is often written $\mathrm{d}^2 = 0$. This identity is a consequence of the definition of d and the fact that partial derivatives commute, $\partial_\alpha \partial_\beta = \partial_\beta \partial_\alpha$ (acting on anything). This leads us to the following mathematical aside, just for fun. We define a p-form A to be **closed** if $\mathrm{d}A = 0$, and **exact** if $A = \mathrm{d}B$ for some $(p-1)$-form B. Obviously, all exact forms are closed, but the converse is not necessarily true. On a manifold M, closed p-forms comprise a vector space $Z^p(M)$, and exact forms comprise a vector space $B^p(M)$. Define a new vector space, consisting of elements called cohomology classes, as the closed forms modulo the exact forms:

$$H^p(M) = \frac{Z^p(M)}{B^p(M)}. \tag{2.81}$$

That is, two closed forms [elements of $Z^p(M)$] define the same cohomology class [elements of $H^p(M)$] if they differ by an exact form [an element of $B^p(M)$]. Miraculously, the dimensionality of the cohomology spaces $H^p(M)$ depends only on the topology of the manifold M. Minkowski space is topologically equivalent to $\mathbf{R}^4$, which is uninteresting, so that all of the $H^p(M)$ vanish for $p > 0$; for $p = 0$ we have $H^0(M) = \mathbf{R}$. Therefore in Minkowski space all closed forms are exact except for zero-forms; zero-forms can't be exact since there are no -1-

forms for them to be the exterior derivative of. It is striking that information about the topology can be extracted in this way, which essentially involves the solutions to differential equations.

The final operation on differential forms we will introduce is **Hodge duality**. We define the *Hodge star operator* on an n-dimensional manifold as a map from p-forms to $(n-p)$-forms,

$$(*A)_{\mu_1\cdots\mu_{n-p}} = \frac{1}{p!}\epsilon^{\nu_1\cdots\nu_p}{}_{\mu_1\cdots\mu_{n-p}}A_{\nu_1\cdots\nu_p}, \tag{2.82}$$

mapping A to "A dual." Unlike our other operations on forms, the Hodge dual does depend on the metric of the manifold [which should be obvious, since we had to raise some indices on the Levi–Civita tensor in order to define (2.82)]. Applying the Hodge star twice returns either plus or minus the original form:

$$**A = (-1)^{s+p(n-p)}A, \tag{2.83}$$

where s is the number of minus signs in the eigenvalues of the metric.

Two facts on the Hodge dual: First, "duality" in the sense of Hodge is distinct from the relationship between vectors and dual vectors. The idea of "duality" is that of a transformation from one space to another with the property that doing the transformation twice gets you back to the original space. It should be clear that this holds true for both the duality between vectors and one-forms, and the Hodge duality between p-forms and $(n-p)$-forms. A requirement of dualities between vector spaces is that the original and transformed spaces have the same dimensionality; this is true of the spaces of p- and $(n-p)$-forms.

The second fact concerns differential forms in three-dimensional Euclidean space. The Hodge dual of the wedge product of two 1-forms gives another 1-form:

$$*(U\wedge V)_i = \epsilon_i{}^{jk}U_jV_k. \tag{2.84}$$

(All of the prefactors cancel.) Since 1-forms in Euclidean space are just like vectors, we have a map from two vectors to a single vector. You should convince yourself that this is just the conventional cross product, and that the appearance of the Levi–Civita tensor explains why the cross product changes sign under parity (interchange of two coordinates, or equivalently basis vectors). This is why the cross product only exists in three dimensions—because only in three dimensions do we have an interesting map from two dual vectors to a third dual vector.

Electrodynamics provides an especially compelling example of the use of differential forms. From the definition of the exterior derivative, it is clear that equation (1.97) can be concisely expressed as closure of the two-form $F_{\mu\nu}$:

$$dF = 0. \tag{2.85}$$

Does this mean that F is also exact? Yes; as we've noted, Minkowski space is topologically trivial, so all closed forms are exact. There must therefore be a one-form A_μ such that

$$F = \mathrm{d}A. \tag{2.86}$$

This one-form is the familiar **vector potential** of electromagnetism, with the 0 component given by the scalar potential, $A_0 = \Phi$, as we discussed in Chapter 1. Gauge invariance is expressed by the observation that the theory is invariant under $A \to A + \mathrm{d}\lambda$ for some scalar (zero-form) λ, and this is also immediate from the relation (2.86). The other one of Maxwell's equations, (1.96), can be expressed as an equation between three-forms:

$$\mathrm{d}(*F) = *J, \tag{2.87}$$

where the current one-form J is just the current four-vector with index lowered. Filling in the details is left for you, as good practice converting from differential-form notation to ordinary index notation.

Hodge duality is intimately related to a fascinating feature of certain field theories: duality between strong and weak coupling. It's hard not to notice that the equations (2.85) and (2.87) look very similar. Indeed, if we set $J_\mu = 0$, the equations are invariant under the "duality transformations"

$$\begin{aligned} F &\to *F, \\ *F &\to -F. \end{aligned} \tag{2.88}$$

We therefore say that the vacuum Maxwell's equations are duality invariant, while the invariance is spoiled in the presence of charges. We might imagine that magnetic as well as electric monopoles existed in nature; then we could add a magnetic current term $*J_M$ to the right-hand side of (2.85), and the equations would be invariant under duality transformations plus the additional replacement $J \leftrightarrow J_M$. (Of course a nonzero right-hand side to (2.85) is inconsistent with $F = \mathrm{d}A$, so this idea only works if A_μ is not a fundamental variable.) Dirac considered the idea of magnetic monopoles and showed that a necessary condition for their existence is that the fundamental monopole charge be inversely proportional to the fundamental electric charge. Now, the fundamental electric charge is a small number; electrodynamics is *weakly coupled*, which is why perturbation theory is so remarkably successful in quantum electrodynamics (QED). But Dirac's condition on magnetic charges implies that a duality transformation takes a theory of weakly coupled electric charges to a theory of strongly coupled magnetic monopoles (and vice-versa). Unfortunately monopoles don't fit easily into ordinary electromagnetism, so these ideas aren't directly applicable; but some sort of duality symmetry may exist in certain theories (such as supersymmetric nonabelian gauge theories). If it did, we would have the opportunity to analyze a theory that looked strongly coupled (and therefore hard to solve) by looking at the weakly coupled dual version; this is exactly what happens in certain theories. The hope is that these techniques will allow us to explore various phenomena that we know exist in strongly coupled quantum field theories, such as confinement of quarks in hadrons.

2.10 ■ INTEGRATION

An important appearance of both tensor densities and differential forms is in integration on manifolds. You have probably been exposed to the fact that in ordinary calculus on $\mathbf{R}^n$ the volume element $d^n x$ picks up a factor of the Jacobian under change of coordinates:

$$d^n x' = \left| \frac{\partial x^{\mu'}}{\partial x^{\mu}} \right| d^n x. \tag{2.89}$$

There is actually a beautiful explanation of this formula from the point of view of differential forms, which arises from the following fact: *on an n-dimensional manifold M, the integrand is properly understood as an n-form.* In other words, an integral over an n-dimensional region $\Sigma \subset M$ is a map from an n-form field ω to the real numbers:

$$\int_{\Sigma} : \omega \to \mathbf{R}. \tag{2.90}$$

Such a statement may seem strange, but it certainly looks familiar in the context of line integrals. In one dimension any one-form can be written $\omega = \omega(x)\mathrm{d}x$, where the first ω is a one-form and $\omega(x)$ denotes the (single) component function. And indeed, we write integrals in one dimension as $\int \omega(x)\mathrm{d}x$; you may be used to thinking of the symbol $\mathrm{d}x$ as an infinitesimal distance, but it is more properly a differential form.

To make this more clear, consider more than one dimension. If we are claiming that the integrand is an n-form, we need to explain in what sense it is antisymmetric, and for that matter why it is a $(0, n)$ tensor (a linear map from a set of n vectors to $\mathbf{R}$) at all. We all agree that integrals can be written as $\int f(x)\, d\mu$, where $f(x)$ is a scalar function on the manifold and $d\mu$ is the volume element, or measure. The role of the volume element is to assign to every (infinitesimal) region an (infinitesimal) real number, namely the volume of that region. A nice feature of infinitesimal regions (as opposed to ones of finite size) is that they can be taken to be rectangular parallelepipeds—in the presence of curvature we have no clear sense of what a "rectangular parallelepiped" is supposed to mean, but the effects of curvature can be neglected when we work in infinitesimal regions. Clearly we are not being rigorous here, but our present purpose is exclusively motivational.

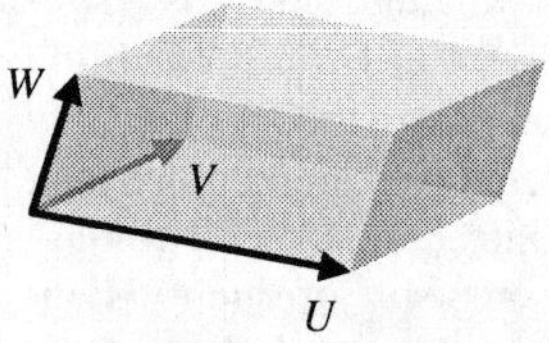

FIGURE 2.27 An infinitesimal n-dimensional region, represented as a parallelepiped, is defined by an ordered set of n vectors, shown here as U, V, and W.

As shown in Figure 2.27 (in which we take our manifold to be three-dimensional for purposes of illustration), a parallelepiped is specified by n vectors that define its edges. Our volume element, then, should be a map from n vectors to the real numbers: $d\mu(U, V, W) \in \mathbf{R}$. (Actually it should be a map from infinitesimal vectors to infinitesimal numbers, but such a map also will take finite vectors to finite numbers.) It's also clear that it should be linearly scalable by real numbers; if we change the length of any of the defining vectors, the volume changes accordingly: $d\mu(aU, bV, cW) = abc\, d\mu(U, V, W)$. Linearity with respect to adding vectors is not so obvious, but you can convince yourself by drawing pictures.

Therefore our volume element is an honest $(0, n)$ tensor. Why antisymmetric? Because we are defining an oriented element; if two of the vectors are interchanged we should get a volume of the same magnitude but opposite sign. (If this is not obvious, you should at least convince yourself that the volume should vanish when two vectors are collinear.) Thus, volume elements in n dimensions are in a very real sense n-forms.

To actually do calculations, we need to make these ideas more concrete, which turns out to be straightforward. The essential insight is to identify the naive volume element $d^n x$ as an antisymmetric tensor density constructed with wedge products:

$$d^n x = \mathrm{d}x^0 \wedge \cdots \wedge \mathrm{d}x^{n-1}. \tag{2.91}$$

The expression on the right-hand side can be misleading, because it looks like a tensor (an n-form, actually) but is really a density. Certainly if we have two functions f and g on M, then $\mathrm{d}f$ and $\mathrm{d}g$ are one-forms, and $\mathrm{d}f \wedge \mathrm{d}g$ is a two-form. But the functions appearing in (2.91) are the coordinate functions themselves, so when we change coordinates we replace the one-forms $\mathrm{d}x^\mu$ with a new set $\mathrm{d}x^{\mu'}$. You see the funny business—ordinarily a coordinate transformation changes components, but not one-forms themselves. The right-hand side of (2.91) is a coordinate-dependent object (a tensor density, to be precise) which, in the x^μ coordinate system, acts like $\mathrm{d}x^0 \wedge \cdots \wedge \mathrm{d}x^{n-1}$. Let's see this in action. First notice that the definition of the wedge product allows us to write

$$\mathrm{d}x^0 \wedge \cdots \wedge \mathrm{d}x^{n-1} = \frac{1}{n!}\tilde{\epsilon}_{\mu_1\cdots\mu_n}\, \mathrm{d}x^{\mu_1} \wedge \cdots \wedge \mathrm{d}x^{\mu_n}, \tag{2.92}$$

since both the wedge product and the Levi–Civita symbol are completely antisymmetric. (The factor of $1/n!$ takes care of the overcounting introduced by summing over permutations of the indices.) Under a coordinate transformation $\tilde{\epsilon}_{\mu_1\cdots\mu_n}$ stays the same, while the one-forms change according to (2.26), leading to

$$\begin{aligned} \tilde{\epsilon}_{\mu_1\cdots\mu_n}\, \mathrm{d}x^{\mu_1} \wedge \cdots \wedge \mathrm{d}x^{\mu_n} &= \tilde{\epsilon}_{\mu_1\cdots\mu_n} \frac{\partial x^{\mu_1}}{\partial x^{\mu'_1}} \cdots \frac{\partial x^{\mu_n}}{\partial x^{\mu'_n}}\, \mathrm{d}x^{\mu'_1} \wedge \cdots \wedge \mathrm{d}x^{\mu'_n} \\ &= \left|\frac{\partial x^\mu}{\partial x^{\mu'}}\right| \tilde{\epsilon}_{\mu'_1\cdots\mu'_n}\, \mathrm{d}x^{\mu'_1} \wedge \cdots \wedge \mathrm{d}x^{\mu'_n}. \end{aligned} \tag{2.93}$$

Multiplying by the Jacobian on both sides and using (2.91) and (2.92) recovers (2.89).

It is clear that the naive volume element $d^n x$ transforms as a density, not a tensor, but it is straightforward to construct an invariant volume element by multiplying by $\sqrt{|g|}$:

$$\sqrt{|g'|}\, \mathrm{d}x^{0'} \wedge \cdots \wedge \mathrm{d}x^{(n-1)'} = \sqrt{|g|}\, \mathrm{d}x^0 \wedge \cdots \wedge \mathrm{d}x^{n-1}, \tag{2.94}$$

which is of course just $(n!)^{-1}\epsilon_{\mu_1\cdots\mu_n}\, \mathrm{d}x^{\mu_1} \wedge \cdots \wedge \mathrm{d}x^{\mu_n}$. In the interest of simplicity we will usually write the volume element as $\sqrt{|g|}\, d^n x$, rather than as the explicit wedge product:

$$\sqrt{|g|}\, d^n x \equiv \sqrt{|g|}\, dx^0 \wedge \cdots \wedge dx^{n-1} ; \tag{2.95}$$

it will be enough to keep in mind that it's supposed to be an n-form. In fact, the volume element is no more or less than the Levi–Civita tensor $\epsilon_{\mu_1 \cdots \mu_n}$; restoring the explicit basis one-forms, we see

$$\begin{aligned} \epsilon &\equiv \epsilon_{\mu_1 \cdots \mu_n} dx^{\mu_1} \otimes \cdots \otimes dx^{\mu_n} \\ &= \frac{1}{n!} \epsilon_{\mu_1 \cdots \mu_n} dx^{\mu_1} \wedge \cdots \wedge dx^{\mu_n} \\ &= \frac{1}{n!} \sqrt{|g|}\, \tilde{\epsilon}_{\mu_1 \cdots \mu_n} dx^{\mu_1} \wedge \cdots \wedge dx^{\mu_n} \\ &= \sqrt{|g|}\, dx^0 \wedge \cdots \wedge dx^{n-1} \\ &\equiv \sqrt{|g|}\, d^n x . \end{aligned} \tag{2.96}$$

Notice that the combinatorial factors introduced by the epsilon tensor precisely cancel those from switching from tensor products to wedge products, which is only allowed because the epsilon tensor automatically antisymmetrizes.

The punch line, then, is simple: the integral I of a scalar function ϕ over an n-manifold is written as

$$\boxed{I = \int \phi(x) \sqrt{|g|}\, d^n x .} \tag{2.97}$$

Given explicit forms for $\phi(x)$ and $\sqrt{|g|}$, such an integral can be directly evaluated by the usual methods of multivariable calculus. The metric determinant serves to automatically take care of the correct transformation properties. You will sometimes see the more abstract notation

$$I = \int \phi(x) \epsilon ; \tag{2.98}$$

given (2.96), these two versions convey the same content.

2.11 ■ EXERCISES

1. Just because a manifold is topologically nontrivial doesn't necessarily mean it can't be covered with a single chart. In contrast to the circle S^1, show that the infinite cylinder $\mathbf{R} \times S^1$ can be covered with just one chart, by explicitly constructing the map.

2. By clever choice of coordinate charts, can we make $\mathbf{R}^2$ look like a one-dimensional manifold? Can we make $\mathbf{R}^1$ look like a two-dimensional manifold? If so, explicitly construct an appropriate atlas, and if not, explain why not. The point of this problem

is to provoke you to think deeply about what a manifold is; it can't be answered rigorously without going into more details about topological spaces. In particular, you might have to forget that you already know a definition of "open set" in the original $\mathbf{R}^2$ or $\mathbf{R}^1$, and define them as being appropriately inherited from the $\mathbf{R}^1$ or $\mathbf{R}^2$ to which they are being mapped.

3. Show that the two-dimensional torus T^2 is a manifold, by explicitly constructing an appropriate atlas. (Not a maximal one, obviously.)

4. Verify the claims made about the commutator of two vector fields at the end of Section 2.3 (linearity, Leibniz, component formula, transformation as a vector field).

5. Give an example of two linearly independent, nowhere-vanishing vector fields in $\mathbf{R}^2$ whose commutator does not vanish. Notice that these fields provide a basis for the tangent space at each point, but it cannot be a coordinate basis since the commutator doesn't vanish.

6. Consider $\mathbf{R}^3$ as a manifold with the flat Euclidean metric, and coordinates $\{x, y, z\}$. Introduce spherical polar coordinates $\{r, \theta, \phi\}$ related to $\{x, y, z\}$ by

$$
\begin{aligned}
x &= r \sin\theta \cos\phi \\
y &= r \sin\theta \sin\phi \\
z &= r \cos\theta,
\end{aligned}
\tag{2.99}
$$

so that the metric takes the form

$$ds^2 = \mathrm{d}r^2 + r^2 \mathrm{d}\theta^2 + r^2 \sin^2\theta \, \mathrm{d}\phi^2. \tag{2.100}$$

(a) A particle moves along a parameterized curve given by

$$x(\lambda) = \cos\lambda, \quad y(\lambda) = \sin\lambda, \quad z(\lambda) = \lambda. \tag{2.101}$$

Express the path of the curve in the $\{r, \theta, \phi\}$ system.

(b) Calculate the components of the tangent vector to the curve in both the Cartesian and spherical polar coordinate systems.

7. Prolate spheroidal coordinates can be used to simplify the Kepler problem in celestial mechanics. They are related to the usual cartesian coordinates (x, y, z) of Euclidean three-space by

$$
\begin{aligned}
x &= \sinh\chi \; \sin\theta \; \cos\phi, \\
y &= \sinh\chi \; \sin\theta \; \sin\phi, \\
z &= \cosh\chi \; \cos\theta.
\end{aligned}
$$

Restrict your attention to the plane $y = 0$ and answer the following questions.

(a) What is the coordinate transformation matrix $\partial x^\mu / \partial x^{\nu'}$ relating (x, z) to (χ, θ)?

(b) What does the line element ds^2 look like in prolate spheroidal coordinates?

8. Verify (2.78): for the exterior derivative of a product of a p-form ω and a q-form η, we have

$$\mathrm{d}(\omega \wedge \eta) = (\mathrm{d}\omega) \wedge \eta + (-1)^p \omega \wedge (\mathrm{d}\eta). \tag{2.102}$$

9. In Minkowski space, suppose $*F = q \sin\theta \, \mathrm{d}\theta \wedge \mathrm{d}\phi$.

(a) Evaluate $\mathrm{d} * F = *J$.

(b) What is the two-form F equal to?

(c) What are the electric and magnetic fields equal to for this solution?

(d) Evaluate $\int_V \mathrm{d}{*F}$, where V is a ball of radius R in Euclidean three space at a fixed moment of time.

10. Consider Maxwell's equations, $\mathrm{d}F = 0$, $\mathrm{d}{*F} = *J$, in 2-dimensional spacetime. Explain why one of the two sets of equations can be discarded. Show that the electromagnetic field can be expressed in terms of a scalar field. Write out the field equations for this scalar field in component form.

11. There are a lot of motivational words attached here to what is a very simple problem; don't get too distracted. In ordinary electromagnetism with point particles, the part of the action which represents the coupling of the gauge-potential one-form $A^{(1)}$ to a charged particle can be written $S = \int_\gamma A^{(1)}$, where γ is the particle worldline. (The superscript on $A^{(1)}$ is just to remind you that it is a one-form.) For this problem you will consider a theory related to ordinary electromagnetism, but this time in 11 spacetime dimensions, with a three-form gauge potential $A^{(3)}$ and four-form field strength $F^{(4)} = \mathrm{d}A^{(3)}$. Note that the field strength is invariant under a gauge transformation $A^{(3)} \to A^{(3)} + \mathrm{d}\lambda^{(2)}$ for any two-form $\lambda^{(2)}$.

(a) What would be the number of spatial dimensions of an object to which this gauge field would naturally couple (for example, ordinary E+M couples to zero-dimensional objects—point particles)?

(b) The electric charge of an ordinary electron is given by the integral of the dual of the two-form gauge field strength over a two-sphere surrounding the particle. How would you define the "charge" of the object to which $A^{(3)}$ couples? Argue that it is conserved if $\mathrm{d} * F^{(4)} = 0$.

(c) Imagine there is a "dual gauge potential" $\widetilde{A}$ that satisfies $\mathrm{d}(\widetilde{A}) = *F^{(4)}$. To what dimensionality object does it naturally couple?

(d) The action for the gauge field itself (as opposed to its coupling to other things) will be an integral over the entire 11-dimensional spacetime. What are the terms that would be allowed in such an action that are invariant under "local" gauge transformations, for instance, gauge transformations specified by a two-form $\lambda^{(2)}$ that vanishes at infinity? Restrict yourself to terms of first, second, or third order in $A^{(3)}$ and its first derivatives (no second derivatives, no higher-order terms). You may use the exterior derivative, wedge product, and Hodge dual, but not any explicit appearance of the metric.

More background: "Supersymmetry" is a hypothetical symmetry relating bosons (particles with integral spin) and fermions (particles with spin $\frac{1}{2}$, $\frac{3}{2}$, etc.). An interesting feature is that supersymmetric theories are only well-defined in 11 dimensions or less—in larger numbers of dimensions, supersymmetry would require the existence of particles with spins greater than 2, which cannot be consistently quantized. Eleven-dimensional supersymmetry is a unique theory, which naturally includes a three-form gauge potential (not to mention gravity). Recent work has shown that it also includes the various higher-dimensional objects alluded to in this problem (although we've cut some corners here). This theory turns out to be a well-defined limit of something called M-theory, which has as other limits various 10-dimensional superstring theories.

CHAPTER 3

Curvature

3.1 ■ OVERVIEW

We all know what curvature means, at least informally, and in the first two chapters of this book we have felt free to refer on occasion to the concept of curvature without giving it a careful definition. Clearly curvature depends somehow on the metric, which defines the geometry of our manifold; but it is not immediately clear how we should attribute curvature to any given metric (since, as we have seen, even the metric of a flat space can look arbitrarily complicated in a sufficiently extravagant coordinate system). As is often the case in mathematics, we require quite a bit of care to formalize our intuition about a concept into a usable mathematical structure; formalizing what we think of as "curvature" is the subject of this chapter.

The techniques we are about to develop are absolutely crucial to the subject; it is safe to say that there is a higher density of useful formulas per page in this chapter than in any of the others. Let's quickly summarize the most important ones, to provide a roadmap for the formalism to come.

All the ways in which curvature manifests itself rely on something called a "connection," which gives us a way of relating vectors in the tangent spaces of nearby points. There is a unique connection that we can construct from the metric, and it is encapsulated in an object called the *Christoffel symbol*, given by

$$\Gamma^{\lambda}_{\mu\nu} = \tfrac{1}{2} g^{\lambda\sigma}(\partial_\mu g_{\nu\sigma} + \partial_\nu g_{\sigma\mu} - \partial_\sigma g_{\mu\nu}). \tag{3.1}$$

The notation makes $\Gamma^{\lambda}_{\mu\nu}$ look like a tensor, but in fact it is not; this is why we call it an "object" or "symbol." The fundamental use of a connection is to take a *covariant derivative* ∇_μ (a generalization of the partial derivative); the covariant derivative of a vector field V^ν is given by

$$\nabla_\mu V^\nu = \partial_\mu V^\nu + \Gamma^{\nu}_{\mu\sigma} V^\sigma, \tag{3.2}$$

and covariant derivatives of other sorts of tensors are given by similar expressions. The connection also appears in the definition of *geodesics* (a generalization of the notion of a straight line). A parameterized curve $x^\mu(\lambda)$ is a geodesic if it obeys

$$\frac{d^2x^\mu}{d\lambda^2} + \Gamma^{\mu}_{\rho\sigma}\frac{dx^\rho}{d\lambda}\frac{dx^\sigma}{d\lambda} = 0, \tag{3.3}$$

known as the geodesic equation.

Finally, the technical expression of curvature is contained in the Riemann tensor, a (1, 3) tensor obtained from the connection by

$$R^{\rho}{}_{\sigma\mu\nu} = \partial_\mu \Gamma^{\rho}_{\nu\sigma} - \partial_\nu \Gamma^{\rho}_{\mu\sigma} + \Gamma^{\rho}_{\mu\lambda}\Gamma^{\lambda}_{\nu\sigma} - \Gamma^{\rho}_{\nu\lambda}\Gamma^{\lambda}_{\mu\sigma}. \tag{3.4}$$

Everything we want to know about the curvature of a manifold is given to us by the Riemann tensor; it will vanish if and only if the metric is perfectly flat. Einstein's equation of general relativity relates certain components of this tensor to the energy-momentum tensor.

These four equations are all of primary importance in the study of curved manifolds. We will now see how they arise from a careful consideration of how familiar notions of geometry in flat space adapt to this more general context.

3.2 ■ COVARIANT DERIVATIVES

In our discussion of manifolds, it became clear that there were various notions we could talk about as soon as the manifold was defined: we could define functions, take their derivatives, consider parameterized paths, set up tensors, and so on. Other concepts, such as the volume of a region or the length of a path, required some additional piece of structure, namely the introduction of a metric. It would be natural to think of the notion of curvature as something that depends exclusively on the metric. In a more careful treatment, however, we find that curvature depends on a connection, and connections may or may not depend on the metric. Nevertheless, we will also show how the existence of a metric implies a certain unique connection, whose curvature may be thought of as that of the metric. This is the connection used in general relativity, so in this particular context it is legitimate to think of curvature as characterizing the metric, without introducing any additional structures.

The connection becomes necessary when we attempt to address the problem of the partial derivative not being a good tensor operator. What we would like is a covariant derivative, that is, an operator that reduces to the partial derivative in flat space with inertial coordinates, but transforms as a tensor on an arbitrary manifold. It is conventional to spend a certain amount of time motivating the introduction of a covariant derivative, but in fact the need is obvious; equations such as $\partial_\mu T^{\mu\nu} = 0$ must be generalized to curved space somehow. So let's agree that a covariant derivative would be a good thing to have, and go about setting it up.

In flat space in inertial coordinates, the partial derivative operator ∂_μ is a map from (k, l) tensor fields to $(k, l+1)$ tensor fields, which acts linearly on its arguments and obeys the Leibniz rule on tensor products. All of this continues to be true in the more general situation we would now like to consider, but the map provided by the partial derivative depends on the coordinate system used. We would therefore like to define a **covariant derivative** operator ∇ to perform the functions of the partial derivative, but in a way independent of coordinates. Rather than simply postulating the answer (which would be perfectly acceptable), let's motivate it by thinking carefully about what properties a covariant generalization of the partial derivative *should* have—mathematical structures are, after all, invented by human beings, not found lying on sidewalks. We begin by requiring that ∇ be a map from (k, l) tensor fields, to $(k, l+1)$ tensor fields which has these two properties:

1. linearity: $\nabla(T + S) = \nabla T + \nabla S$;
2. Leibniz (product) rule: $\nabla(T \otimes S) = (\nabla T) \otimes S + T \otimes (\nabla S)$.

If ∇ is going to obey the Leibniz rule, it can always be written as the partial derivative plus some linear transformation. That is, to take the covariant derivative we first take the partial derivative, and then apply a correction to make the result covariant. [We aren't going to prove this reasonable-sounding statement; see Wald (1984) if you are interested.] Let's consider what this means for the covariant derivative of a vector V^ν. It means that, for each direction μ, the covariant derivative ∇_μ will be given by the partial derivative ∂_μ plus a correction specified by a set of n matrices $(\Gamma_\mu)^\rho{}_\sigma$ (one $n \times n$ matrix, where n is the dimensionality of the manifold, for each μ). In fact the parentheses are usually dropped and we write these matrices, known as the **connection coefficients**, with haphazard index placement as $\Gamma^\rho_{\mu\sigma}$. We therefore have

$$\boxed{\nabla_\mu V^\nu = \partial_\mu V^\nu + \Gamma^\nu_{\mu\lambda} V^\lambda.} \tag{3.5}$$

Notice that in the second term the index originally on V has moved to the Γ, and a new index is summed over. If this is the expression for the covariant derivative of a vector in terms of the partial derivative, we should be able to determine the transformation properties of $\Gamma^\nu_{\mu\lambda}$ by demanding that the left-hand side be a (1, 1) tensor. That is, we want the transformation law to be

$$\nabla_{\mu'} V^{\nu'} = \frac{\partial x^\mu}{\partial x^{\mu'}} \frac{\partial x^{\nu'}}{\partial x^\nu} \nabla_\mu V^\nu. \tag{3.6}$$

Let's look at the left side first; we can expand it using (3.5) and then transform the parts that we understand (which is everything except $\Gamma^{\nu'}_{\mu'\lambda'}$):

$$\begin{aligned}\nabla_{\mu'} V^{\nu'} &= \partial_{\mu'} V^{\nu'} + \Gamma^{\nu'}_{\mu'\lambda'} V^{\lambda'} \\ &= \frac{\partial x^\mu}{\partial x^{\mu'}} \frac{\partial x^{\nu'}}{\partial x^\nu} \partial_\mu V^\nu + \frac{\partial x^\mu}{\partial x^{\mu'}} V^\nu \frac{\partial}{\partial x^\mu} \frac{\partial x^{\nu'}}{\partial x^\nu} + \Gamma^{\nu'}_{\mu'\lambda'} \frac{\partial x^{\lambda'}}{\partial x^\lambda} V^\lambda.\end{aligned} \tag{3.7}$$

On the right-hand side we can also expand $\nabla_\mu V^\nu$:

$$\frac{\partial x^\mu}{\partial x^{\mu'}} \frac{\partial x^{\nu'}}{\partial x^\nu} \nabla_\mu V^\nu = \frac{\partial x^\mu}{\partial x^{\mu'}} \frac{\partial x^{\nu'}}{\partial x^\nu} \partial_\mu V^\nu + \frac{\partial x^\mu}{\partial x^{\mu'}} \frac{\partial x^{\nu'}}{\partial x^\nu} \Gamma^\nu_{\mu\lambda} V^\lambda. \tag{3.8}$$

These last two expressions are to be equated; the first terms in each are identical and therefore cancel, so we have

$$\Gamma^{\nu'}_{\mu'\lambda'}\frac{\partial x^{\lambda'}}{\partial x^{\lambda}}V^{\lambda}+\frac{\partial x^{\mu}}{\partial x^{\mu'}}V^{\lambda}\frac{\partial}{\partial x^{\mu}}\frac{\partial x^{\nu'}}{\partial x^{\lambda}}=\frac{\partial x^{\mu}}{\partial x^{\mu'}}\frac{\partial x^{\nu'}}{\partial x^{\nu}}\Gamma^{\nu}_{\mu\lambda}V^{\lambda}, \tag{3.9}$$

where we have changed a dummy index from ν to λ. This equation must be true for any vector V^{λ}, so we can eliminate that on both sides. Then the connection coefficients in the primed coordinates may be isolated by multiplying by $\partial x^{\lambda}/\partial x^{\sigma'}$ and relabeling $\sigma' \to \lambda'$. The result is

$$\Gamma^{\nu'}_{\mu'\lambda'}=\frac{\partial x^{\mu}}{\partial x^{\mu'}}\frac{\partial x^{\lambda}}{\partial x^{\lambda'}}\frac{\partial x^{\nu'}}{\partial x^{\nu}}\Gamma^{\nu}_{\mu\lambda}+\frac{\partial x^{\mu}}{\partial x^{\mu'}}\frac{\partial x^{\lambda}}{\partial x^{\lambda'}}\frac{\partial^2 x^{\nu'}}{\partial x^{\mu}\partial x^{\lambda}}. \tag{3.10}$$

This is not, of course, the tensor transformation law; the second term on the right spoils it. That's okay, because *the connection coefficients are not the components of a tensor.* They are purposefully constructed to be nontensorial, but in such a way that the combination (3.5) transforms as a tensor—the extra terms in the transformation of the partials and the Γ's exactly cancel. This is why we are not so careful about index placement on the connection coefficients; they are not a tensor, and therefore you should try not to raise and lower their indices.

What about the covariant derivatives of other sorts of tensors? By similar reasoning to that used for vectors, the covariant derivative of a one-form can also be expressed as a partial derivative plus some linear transformation. But there is no reason as yet that the matrices representing this transformation should be related to the coefficients $\Gamma^{\nu}_{\mu\lambda}$. In general we could write something like

$$\nabla_{\mu}\omega_{\nu}=\partial_{\mu}\omega_{\nu}+\widetilde{\Gamma}^{\lambda}_{\mu\nu}\omega_{\lambda}, \tag{3.11}$$

where $\widetilde{\Gamma}^{\lambda}_{\mu\nu}$ is a new set of matrices for each μ. Pay attention to where all of the various indices go. It is straightforward to derive that the transformation properties of $\widetilde{\Gamma}$ must be similar to those of Γ, since otherwise $\nabla_{\mu}\omega_{\nu}$ wouldn't transform as a tensor, but otherwise no relationship has been established. To do so, we need to introduce two new properties that we would like our covariant derivative to have, in addition to the two above:

3. commutes with contractions: $\nabla_{\mu}(T^{\lambda}{}_{\lambda\rho})=(\nabla T)_{\mu}{}^{\lambda}{}_{\lambda\rho}$,
4. reduces to the partial derivative on scalars: $\nabla_{\mu}\phi=\partial_{\mu}\phi$.

There is no way to "derive" these properties; we are simply demanding that they be true as part of the definition of a covariant derivative. Note that property 3 is equivalent to saying that the Kronecker delta (the identity map) is covariantly constant, $\nabla_{\mu}\delta^{\lambda}_{\sigma}=0$; this is certainly a reasonable thing to ask.

Let's see what these new properties imply. Given some one-form field ω_{μ} and vector field V^{μ}, we can take the covariant derivative of the scalar defined by $\omega_{\lambda}V^{\lambda}$ to get

$$\nabla_\mu(\omega_\lambda V^\lambda) = (\nabla_\mu \omega_\lambda)V^\lambda + \omega_\lambda(\nabla_\mu V^\lambda)$$
$$= (\partial_\mu \omega_\lambda)V^\lambda + \widetilde{\Gamma}^\sigma_{\mu\lambda}\omega_\sigma V^\lambda + \omega_\lambda(\partial_\mu V^\lambda) + \omega_\lambda \Gamma^\lambda_{\mu\rho} V^\rho. \tag{3.12}$$

But since $\omega_\lambda V^\lambda$ is a scalar, this must also be given by the partial derivative:

$$\nabla_\mu(\omega_\lambda V^\lambda) = \partial_\mu(\omega_\lambda V^\lambda)$$
$$= (\partial_\mu \omega_\lambda)V^\lambda + \omega_\lambda(\partial_\mu V^\lambda). \tag{3.13}$$

This can only be true if the terms in (3.12) with connection coefficients cancel each other; that is, rearranging dummy indices, we must have

$$0 = \widetilde{\Gamma}^\sigma_{\mu\lambda}\omega_\sigma V^\lambda + \Gamma^\sigma_{\mu\lambda}\omega_\sigma V^\lambda. \tag{3.14}$$

But both ω_σ and V^λ are completely arbitrary, so

$$\widetilde{\Gamma}^\sigma_{\mu\lambda} = -\Gamma^\sigma_{\mu\lambda}. \tag{3.15}$$

The two extra conditions we have imposed therefore allow us to express the covariant derivative of a one-form using the same connection coefficients as were used for the vector, but now with a minus sign (and indices matched up somewhat differently):

$$\boxed{\nabla_\mu \omega_\nu = \partial_\mu \omega_\nu - \Gamma^\lambda_{\mu\nu}\omega_\lambda.} \tag{3.16}$$

It should come as no surprise that the connection coefficients encode all the information necessary to take the covariant derivative of a tensor of arbitrary rank. The formula is quite straightforward; for each upper index you introduce a term with a single $+\Gamma$, and for each lower index a term with a single $-\Gamma$:

$$\begin{aligned}\nabla_\sigma T^{\mu_1\mu_2\cdots\mu_k}{}_{\nu_1\nu_2\cdots\nu_l} = {}& \partial_\sigma T^{\mu_1\mu_2\cdots\mu_k}{}_{\nu_1\nu_2\cdots\nu_l} \\ & + \Gamma^{\mu_1}_{\sigma\lambda} T^{\lambda\mu_2\cdots\mu_k}{}_{\nu_1\nu_2\cdots\nu_l} + \Gamma^{\mu_2}_{\sigma\lambda} T^{\mu_1\lambda\cdots\mu_k}{}_{\nu_1\nu_2\cdots\nu_l} + \cdots \\ & - \Gamma^{\lambda}_{\sigma\nu_1} T^{\mu_1\mu_2\cdots\mu_k}{}_{\lambda\nu_2\cdots\nu_l} - \Gamma^{\lambda}_{\sigma\nu_2} T^{\mu_1\mu_2\cdots\mu_k}{}_{\nu_1\lambda\cdots\nu_l} - \cdots .\end{aligned} \tag{3.17}$$

This is the general expression for the covariant derivative. You can check it yourself; it comes from the set of axioms we have established, and the usual requirements that tensors of various sorts be coordinate-independent entities. Sometimes an alternative notation is used; just as commas are used for partial derivatives, semicolons are used for covariant ones:

$$\nabla_\sigma T^{\mu_1\mu_2\cdots\mu_k}{}_{\nu_1\nu_2\cdots\nu_l} \equiv T^{\mu_1\mu_2\cdots\mu_k}{}_{\nu_1\nu_2\cdots\nu_l;\sigma}. \tag{3.18}$$

Once again, in this book we will stick to "∇_σ."

To define a covariant derivative, then, we need to put a connection on our manifold, which is specified in some coordinate system by a set of coefficients $\Gamma^{\lambda}_{\mu\nu}$ ($n^3 = 64$ independent components in $n = 4$ dimensions) that transform according to (3.10). (The name *connection* comes from the fact that it is used to transport vectors from one tangent space to another, as we will soon see; it is sometimes used to refer to the operator ∇, and sometimes to the coefficients $\Gamma^{\lambda}_{\mu\nu}$.) Evidently, we could define a large number of connections on any manifold, and each of them implies a distinct notion of covariant differentiation. In general relativity this freedom is not a big concern, because it turns out that every metric defines a unique connection, which is the one used in GR. Let's see how that works.

The first thing to notice is that the difference of two connections is a tensor. Imagine we have defined two different kinds of covariant derivative, ∇_μ and $\widehat{\nabla}_\mu$, with associated connection coefficients $\Gamma^{\lambda}_{\mu\nu}$ and $\widehat{\Gamma}^{\lambda}_{\mu\nu}$. Then the difference

$$S^{\lambda}{}_{\mu\nu} = \Gamma^{\lambda}_{\mu\nu} - \widehat{\Gamma}^{\lambda}_{\mu\nu} \tag{3.19}$$

is a $(1, 2)$ tensor. (Notice that we had to choose a convention for index placement.) We could show this by brute force, plugging in the transformation laws for the connection coefficients, but let's be a little more slick. Given an arbitrary vector field V^λ, we know that both $\nabla_\mu V^\lambda$ and $\widehat{\nabla}_\mu V^\lambda$ are tensors, so their difference must also be. This difference is simply

$$\begin{aligned}\nabla_\mu V^\lambda - \widehat{\nabla}_\mu V^\lambda &= \partial_\mu V^\lambda + \Gamma^{\lambda}_{\mu\nu} V^\nu - \partial_\mu V^\lambda - \widehat{\Gamma}^{\lambda}_{\mu\nu} V^\nu \\ &= S^{\lambda}{}_{\mu\nu} V^\nu .\end{aligned} \tag{3.20}$$

Since V^λ was arbitrary, and the left-hand side is a tensor, $S^{\lambda}{}_{\mu\nu}$ must be a tensor. As a trivial consequence, we learn that any set of connection coefficients can be expressed as some fiducial connection plus a tensorial correction,

$$\Gamma^{\lambda}_{\mu\nu} = \widehat{\Gamma}^{\lambda}_{\mu\nu} + S^{\lambda}{}_{\mu\nu} . \tag{3.21}$$

Next notice that, given a connection specified by $\Gamma^{\lambda}_{\mu\nu}$, we can immediately form another connection simply by permuting the lower indices. That is, the set of coefficients $\Gamma^{\lambda}_{\nu\mu}$ will also transform according to (3.10) (since the partial derivatives appearing in the last term can be commuted), so they determine a distinct connection. There is thus a tensor we can associate with any given connection, known as the **torsion tensor**, defined by

$$T^{\lambda}{}_{\mu\nu} = \Gamma^{\lambda}_{\mu\nu} - \Gamma^{\lambda}_{\nu\mu} = 2\Gamma^{\lambda}_{[\mu\nu]} . \tag{3.22}$$

It is clear that the torsion is antisymmetric in its lower indices, and a connection that is symmetric in its lower indices is known as "torsion-free."

We can now define a unique connection on a manifold with a metric $g_{\mu\nu}$ by introducing two additional properties:

- torsion-free: $\Gamma^{\lambda}_{\mu\nu} = \Gamma^{\lambda}_{(\mu\nu)}$.
- metric compatibility: $\nabla_\rho g_{\mu\nu} = 0$.

A connection is **metric compatible** if the covariant derivative of the metric with respect to that connection is everywhere zero. This implies a couple of nice properties. First, it's easy to show that both the Levi–Civita tensor and the inverse metric also have zero covariant derivative,

$$\begin{aligned} \nabla_\lambda \epsilon_{\mu\nu\rho\sigma} &= 0 \\ \nabla_\rho g^{\mu\nu} &= 0. \end{aligned} \tag{3.23}$$

Second, a metric-compatible covariant derivative commutes with raising and lowering of indices. Thus, for some vector field V^λ,

$$g_{\mu\lambda}\nabla_\rho V^\lambda = \nabla_\rho(g_{\mu\lambda}V^\lambda) = \nabla_\rho V_\mu. \tag{3.24}$$

With non-metric-compatible connections we would have to be very careful about index placement when taking a covariant derivative.

Our claim is therefore that there is exactly one torsion-free connection on a given manifold that is compatible with some given metric on that manifold. We do not want to make these two requirements part of the definition of a covariant derivative; they simply single out one of the many possible ones.

We can demonstrate both existence and uniqueness by deriving a manifestly unique expression for the connection coefficients in terms of the metric. To accomplish this, we expand out the equation of metric compatibility for three different permutations of the indices:

$$\begin{aligned} \nabla_\rho g_{\mu\nu} &= \partial_\rho g_{\mu\nu} - \Gamma^{\lambda}_{\rho\mu} g_{\lambda\nu} - \Gamma^{\lambda}_{\rho\nu} g_{\mu\lambda} = 0 \\ \nabla_\mu g_{\nu\rho} &= \partial_\mu g_{\nu\rho} - \Gamma^{\lambda}_{\mu\nu} g_{\lambda\rho} - \Gamma^{\lambda}_{\mu\rho} g_{\nu\lambda} = 0 \\ \nabla_\nu g_{\rho\mu} &= \partial_\nu g_{\rho\mu} - \Gamma^{\lambda}_{\nu\rho} g_{\lambda\mu} - \Gamma^{\lambda}_{\nu\mu} g_{\rho\lambda} = 0. \end{aligned} \tag{3.25}$$

We subtract the second and third of these from the first, and use the symmetry of the connection to obtain

$$\partial_\rho g_{\mu\nu} - \partial_\mu g_{\nu\rho} - \partial_\nu g_{\rho\mu} + 2\Gamma^{\lambda}_{\mu\nu} g_{\lambda\rho} = 0. \tag{3.26}$$

It is straightforward to solve this for the connection by multiplying by $g^{\sigma\rho}$. The result is

$$\boxed{\Gamma^{\sigma}_{\mu\nu} = \frac{1}{2} g^{\sigma\rho}(\partial_\mu g_{\nu\rho} + \partial_\nu g_{\rho\mu} - \partial_\rho g_{\mu\nu}).} \tag{3.27}$$

This formula is one of the most important in this subject; commit it to memory. Of course, we have only proved that if a metric-compatible and torsion-free connection exists, it must be of the form (3.27); you can check for yourself that the right-hand side of (3.27) transforms like a connection.

This connection we have derived from the metric is the one on which conventional general relativity is based. It is known by different names: sometimes the **Christoffel** connection, sometimes the **Levi–Civita** connection, sometimes the **Riemannian** connection. The associated connection coefficients are sometimes called **Christoffel symbols** and written as $\left\{{}^{\sigma}_{\mu\nu}\right\}$; we will sometimes call them Christoffel symbols, but we won't use the funny notation. The study of manifolds with metrics and their associated connections is called Riemannian geometry, or sometimes pseudo-Riemannian when the metric has a Lorentzian signature.

Before putting our covariant derivatives to work, we should mention some miscellaneous properties. First, note that in ordinary flat space there is an implicit connection we use all the time—the Christoffel connection constructed from the flat metric. The coefficients of the Christoffel connection in flat space vanish in Cartesian coordinates, but not in curvilinear coordinate systems. Consider for example the plane in polar coordinates, with metric

$$ds^2 = dr^2 + r^2 d\theta^2. \tag{3.28}$$

The nonzero components of the inverse metric are readily found to be $g^{rr} = 1$ and $g^{\theta\theta} = r^{-2}$. Notice that we use r and θ as indices in an obvious notation. We can compute a typical connection coefficient:

$$\begin{aligned}
\Gamma^r_{rr} &= \tfrac{1}{2} g^{r\rho}(\partial_r g_{r\rho} + \partial_r g_{\rho r} - \partial_\rho g_{rr}) \\
&= \tfrac{1}{2} g^{rr}(\partial_r g_{rr} + \partial_r g_{rr} - \partial_r g_{rr}) \\
&\quad + \tfrac{1}{2} g^{r\theta}(\partial_r g_{r\theta} + \partial_r g_{\theta r} - \partial_\theta g_{rr}) \\
&= \tfrac{1}{2}(1)(0+0-0) + \tfrac{1}{2}(0)(0+0-0) \\
&= 0.
\end{aligned} \tag{3.29}$$

Sadly, it vanishes. But not all of them do:

$$\begin{aligned}
\Gamma^r_{\theta\theta} &= \tfrac{1}{2} g^{r\rho}(\partial_\theta g_{\theta\rho} + \partial_\theta g_{\rho\theta} - \partial_\rho g_{\theta\theta}) \\
&= \tfrac{1}{2} g^{rr}(\partial_\theta g_{\theta r} + \partial_\theta g_{r\theta} - \partial_r g_{\theta\theta}) \\
&= \tfrac{1}{2}(1)(0+0-2r) \\
&= -r.
\end{aligned} \tag{3.30}$$

Continuing to turn the crank, we eventually find

$$\begin{aligned}
\Gamma^r_{\theta r} &= \Gamma^r_{r\theta} = 0 \\
\Gamma^\theta_{rr} &= 0 \\
\Gamma^\theta_{r\theta} &= \Gamma^\theta_{\theta r} = \frac{1}{r} \\
\Gamma^\theta_{\theta\theta} &= 0.
\end{aligned} \tag{3.31}$$

From these and similar expressions, we can derive formulas for the divergence, gradient, and curl in curvilinear coordinate systems.

Contrariwise, even in a curved space it is still possible to make the Christoffel symbols vanish at any one point. This is because, as we argued in the last chapter, we can always make the first derivative of the metric vanish at a point; so by (3.27) the connection coefficients derived from this metric will also vanish. Of course this can only be established at a point, not in some neighborhood of the point. We will discuss this more fully in Section 3.4.

Another useful property is that the formula for the divergence of a vector (with respect to the Christoffel connection) has a simplified form. The covariant divergence of V^μ is given by

$$\nabla_\mu V^\mu = \partial_\mu V^\mu + \Gamma^\mu_{\mu\lambda} V^\lambda. \tag{3.32}$$

It is straightforward to show that the Christoffel connection satisfies

$$\Gamma^\mu_{\mu\lambda} = \frac{1}{\sqrt{|g|}} \partial_\lambda \sqrt{|g|}, \tag{3.33}$$

and we therefore obtain

$$\nabla_\mu V^\mu = \frac{1}{\sqrt{|g|}} \partial_\mu (\sqrt{|g|} V^\mu). \tag{3.34}$$

There are also formulas for the divergences of higher-rank tensors, but they are generally not such a great simplification.

We use the Christoffel covariant derivative in the curved-space version of Stokes's theorem (see Appendix E). If V^μ is a vector field over a region Σ with boundary $\partial\Sigma$, Stokes's theorem is

$$\boxed{\int_\Sigma \nabla_\mu V^\mu \sqrt{|g|}\, d^n x = \int_{\partial\Sigma} n_\mu V^\mu \sqrt{|\gamma|}\, d^{n-1}x,} \tag{3.35}$$

where n_μ is normal to $\partial\Sigma$, and γ_{ij} is the induced metric on $\partial\Sigma$. If the connection weren't metric-compatible or torsion-free, there would be additional terms in this equation.

The last thing we need to mention is that converting partial derivatives into covariant derivatives is not always necessary in order to construct well-defined tensors; in particular, the exterior derivative and the vector-field commutator are both well-defined in terms of partials, essentially because both involve an antisymmetrization that cancels the nontensorial piece of the partial derivative transformation law. The same feature implies that they could, on the other hand, be equally well-defined in terms of (torsion-free) covariant derivatives; the antisymmetrization causes the connection coefficient terms to vanish. Thus, if ∇ is the Christoffel connection, ω_μ is a one-form, and X^μ and Y^μ are vector fields, we can write

$$(d\omega)_{\mu\nu} = 2\partial_{[\mu}\omega_{\nu]} = 2\nabla_{[\mu}\omega_{\nu]} \tag{3.36}$$

and

$$[X, Y]^\mu = X^\lambda \partial_\lambda Y^\mu - Y^\lambda \partial_\lambda X^\mu = X^\lambda \nabla_\lambda Y^\mu - Y^\lambda \nabla_\lambda X^\mu. \tag{3.37}$$

If the connection is not torsion-free, the last equalities in these expressions are no longer true; the more fundamental definitions of the exterior derivative and the commutator are those in terms of the partial derivative.

Before moving on, let's review the process by which we have been adding structures to our mathematical constructs. We started with the basic notion of a set, which you were presumed to be familiar with (informally, if not rigorously). We introduced the concept of open subsets of our set; this is equivalent to introducing a topology, and promoted the set to a topological space. Then by demanding that each open set look like a region of $\mathbf{R}^n$ (with n the same for each set) and that the coordinate charts be smoothly sewn together, the topological space became a manifold. A manifold is simultaneously a very flexible and powerful structure, and comes equipped naturally with a tangent bundle, tensor bundles of various ranks, the ability to take exterior derivatives, and so forth. We then proceeded to put a metric on the manifold, resulting in a manifold with metric (sometimes Riemannian manifold). Independently of the metric we found we could introduce a connection, allowing us to take covariant derivatives. Once we have a metric, however, there is automatically a unique torsion-free metric-compatible connection. Likewise we could introduce an independent volume form, although one is automatically determined by the metric. In principle there is nothing to stop us from introducing more than one connection, or volume form, or metric, on any given manifold. In general relativity we do have a physical metric, which determines volumes and the covariant derivative, and the independence of these notions is not a crucial feature.

3.3 ■ PARALLEL TRANSPORT AND GEODESICS

Now that we know how to take covariant derivatives, let's step back and put this in the context of differentiation more generally. We think of a derivative as a way of quantifying how fast something is changing. In the case of tensors, the crucial issue is "changing with respect to what?" An ordinary function defines a number at each point in spacetime, and it is straightforward to compare two different numbers, so we shouldn't be surprised that the partial derivative of functions remained valid on arbitrary manifolds. But a tensor is a map from vectors and dual vectors to the real numbers, and it's not clear how to compare such maps at different points in spacetime. Since we have successfully constructed a covariant derivative, can we think of it as somehow measuring the rate of change of tensors? The answer turns out to be yes: the covariant derivative quantifies the instantaneous rate of

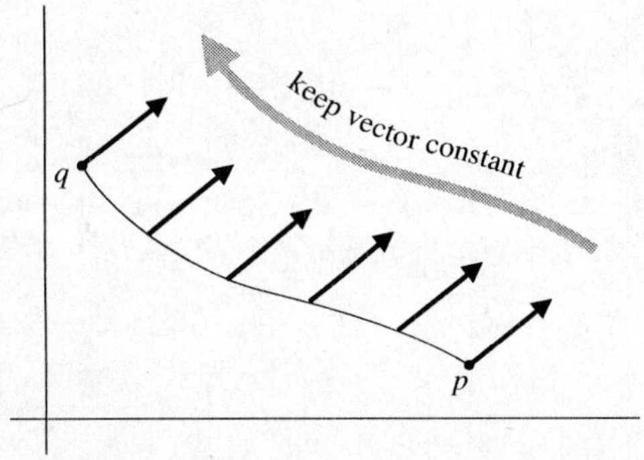

FIGURE 3.1 In flat space, we can parallel transport a vector by simply keeping its Cartesian components constant.

change of a tensor field in comparison to what the tensor would be if it were "parallel transported." In other words, a connection defines a specific way of keeping a tensor constant (along some path), on the basis of which we can compare nearby tensors.

It turns out that the concept of parallel transport is interesting in its own right, and worth spending some time thinking about. Recall that in flat space it is unnecessary to be very careful about the fact that vectors are elements of tangent spaces defined at individual points; it is actually very natural to compare vectors at different points (where by "compare" we mean add, subtract, take the dot product, and so on). The reason why it is natural is because it makes sense, in flat space, to move a vector from one point to another while keeping it constant, as illustrated in Figure 3.1. Then, once we get the vector from one point to another, we can do the usual operations allowed in a vector space.

This concept of moving a vector along a path, keeping constant all the while, is known as *parallel transport*. Parallel transport requires a connection to be well-defined; the intuitive manipulation of vectors in flat space makes implicit use of the Christoffel connection on this space. The crucial difference between flat and curved spaces is that, in a curved space, *the result of parallel transporting a vector from one point to another will depend on the path taken between the points.* Without yet assembling the complete mechanism of parallel transport, we can use our intuition about the two-sphere to see that this is the case. Start with a vector on the equator, pointing along a line of constant longitude. Parallel transport it up to the north pole along a line of longitude in the obvious way. Then take the original vector, parallel transport it along the equator by an angle θ, and then move it up to the north pole as before. As shown in Figure 3.2, it should be clear that the vector, parallel transported along two paths, arrived at the same destination with two different values (rotated by θ).

It therefore appears as if there is no natural way to uniquely move a vector from one tangent space to another; we can always parallel-transport it, but the result depends on the path, and there is no natural choice of which path to take.

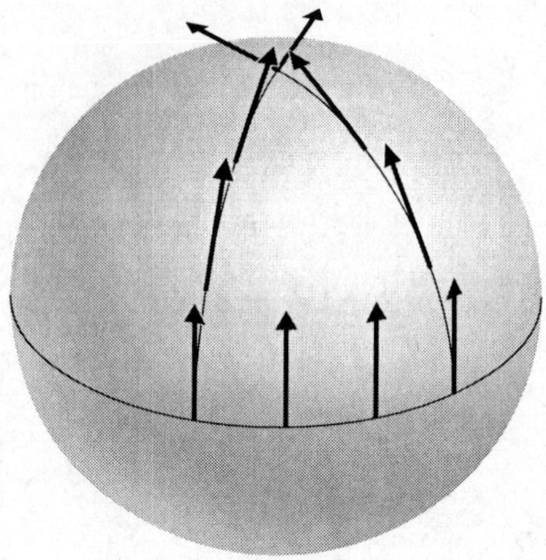

FIGURE 3.2 Parallel transport on a two-sphere. On a curved manifold, the result of parallel transport can depend on the path taken.

Unlike some of the problems we have encountered, *there is no solution to this one*—we simply must learn to live with the fact that two vectors can only be compared in a natural way if they are elements of the same tangent space. For example, two particles passing by each other have a well-defined relative velocity, which cannot be greater than the speed of light. But two particles at different points on a curved manifold do not have any well-defined notion of relative velocity—the concept simply makes no sense. Of course, in certain special situations it is still useful to talk as if it did make sense, but occasional usefulness is not a substitute for rigorous definition. In cosmology, for example, the light from distant galaxies is redshifted with respect to the frequencies we would observe from a nearby stationary source. Since this phenomenon bears such a close resemblance to the conventional Doppler effect due to relative motion, we are very tempted to say that the galaxies are "receding away from us" at a speed defined by their redshift. At a rigorous level this is nonsense, what Wittgenstein would call a "grammatical mistake"—the galaxies are not receding, since the notion of their velocity with respect to us is not well-defined. What is actually happening is that the metric of spacetime between us and the galaxies has changed (the universe has expanded) along the path of the photon from here to there, leading to an increase in the wavelength of the light. As an example of how you can go wrong, naive application of the Doppler formula to the redshift of galaxies implies that some of them are receding faster than light, in apparent contradiction with relativity. The resolution of this apparent paradox is simply that the very notion of their recession should not be taken literally.

Enough about what we cannot do; let's see what we can. Parallel transport is supposed to be the curved-space generalization of the concept of "keeping the vector constant" as we move it along a path; similarly for a tensor of arbitrary rank.

Given a curve $x^\mu(\lambda)$, the requirement of constancy of a tensor $T^{\mu_1\mu_2\cdots\mu_k}{}_{\nu_1\nu_2\cdots\nu_l}$ along this curve in flat space is simply that the components be constant:

$$\frac{d}{d\lambda}T^{\mu_1\mu_2\cdots\mu_k}{}_{\nu_1\nu_2\cdots\nu_l} = \frac{dx^\mu}{d\lambda}\frac{\partial}{\partial x^\mu}T^{\mu_1\mu_2\cdots\mu_k}{}_{\nu_1\nu_2\cdots\nu_l} = 0.$$

To make this properly tensorial we simply replace this partial derivative by a covariant one, and define the **directional covariant derivative** to be

$$\frac{D}{d\lambda} = \frac{dx^\mu}{d\lambda}\nabla_\mu. \tag{3.38}$$

This is a map, defined only along the path, from (k, l) tensors to (k, l) tensors. We then define **parallel transport** of the tensor T along the path $x^\mu(\lambda)$ to be the requirement that the covariant derivative of T along the path vanishes:

$$\left(\frac{D}{d\lambda}T\right)^{\mu_1\mu_2\cdots\mu_k}{}_{\nu_1\nu_2\cdots\nu_l} \equiv \frac{dx^\sigma}{d\lambda}\nabla_\sigma T^{\mu_1\mu_2\cdots\mu_k}{}_{\nu_1\nu_2\cdots\nu_l} = 0. \tag{3.39}$$

This is a well-defined tensor equation (since both the tangent vector $dx^\mu/d\lambda$ and the covariant derivative ∇T are tensors), known as the **equation of parallel transport**. For a vector it takes the form

$$\frac{d}{d\lambda}V^\mu + \Gamma^\mu_{\sigma\rho}\frac{dx^\sigma}{d\lambda}V^\rho = 0. \tag{3.40}$$

We can look at the parallel transport equation as a first-order differential equation defining an initial-value problem: given a tensor at some point along the path, there will be a unique continuation of the tensor to other points along the path such that the continuation solves (3.39). We say that such a tensor is parallel-transported.

The notion of parallel transport is obviously dependent on the connection, and different connections lead to different answers. If the connection is metric-compatible, the metric is always parallel transported with respect to it:

$$\frac{D}{d\lambda}g_{\mu\nu} = \frac{dx^\sigma}{d\lambda}\nabla_\sigma g_{\mu\nu} = 0. \tag{3.41}$$

It follows that the inner product of two parallel-transported vectors is preserved. That is, if V^μ and W^ν are parallel-transported along a curve $x^\sigma(\lambda)$, we have

$$\begin{aligned}\frac{D}{d\lambda}(g_{\mu\nu}V^\mu W^\nu) &= \left(\frac{D}{d\lambda}g_{\mu\nu}\right)V^\mu W^\nu + g_{\mu\nu}\left(\frac{D}{d\lambda}V^\mu\right)W^\nu + g_{\mu\nu}V^\mu\left(\frac{D}{d\lambda}W^\nu\right)\\ &= 0.\end{aligned} \tag{3.42}$$

This means that parallel transport with respect to a metric-compatible connection preserves the norm of vectors, the sense of orthogonality, and so on.

With parallel transport defined, the next logical step is to discuss geodesics. A geodesic is the curved-space generalization of the notion of a straight line in Euclidean space. We all know what a straight line is: it's the path of shortest distance between two points. But there is an equally good definition—a straight line is a path that parallel-transports its own tangent vector. It will turn out that these two concepts coincide if and only if the connection is the Christoffel connection.

We'll start with the second definition (a geodesic is a curve along which the tangent vector is parallel-transported), since it is computationally much more straightforward. The tangent vector to a path $x^\mu(\lambda)$ is $dx^\mu/d\lambda$. The condition that it be parallel transported is thus

$$\frac{D}{d\lambda}\frac{dx^\mu}{d\lambda} = 0, \tag{3.43}$$

or alternatively

$$\boxed{\frac{d^2x^\mu}{d\lambda^2} + \Gamma^\mu_{\rho\sigma}\frac{dx^\rho}{d\lambda}\frac{dx^\sigma}{d\lambda} = 0.} \tag{3.44}$$

This is the **geodesic equation**, another one you should memorize. We can easily see that it reproduces the usual notion of straight lines if the connection coefficients are the Christoffel symbols in Euclidean space; in that case we can choose Cartesian coordinates in which $\Gamma^\mu_{\rho\sigma} = 0$, and the geodesic equation is just $d^2x^\mu/d\lambda^2 = 0$, which is the equation for a straight line.

That was embarrassingly simple; let's turn to the more nontrivial case of the shortest-distance definition. As we know, various subtleties are involved in the definition of distance in a Lorentzian spacetime; for null paths the distance is zero, for timelike paths it's more convenient to use the proper time. So in the name of simplicity let's do the calculation just for a timelike path—the resulting equation will turn out to be good for any path, so we are not losing any generality. We therefore consider the proper time functional,

$$\tau = \int \left(-g_{\mu\nu}\frac{dx^\mu}{d\lambda}\frac{dx^\nu}{d\lambda}\right)^{1/2} d\lambda, \tag{3.45}$$

where the integral is over the path. To search for shortest-distance paths, we could do the usual calculus-of-variations treatment to seek critical points of this functional. They will turn out to be curves of *maximum* proper time, consistent with our discussion of the twin paradox in Chapter 1. However, we can simplify the algebra by means of a trick. The integral (3.45) is of the form $\tau = \int \sqrt{-f}\, d\lambda$, where $f = g_{\mu\nu}(dx^\mu/d\lambda)(dx^\nu/d\lambda)$. The variation looks like

$$\begin{aligned}\delta\tau &= \int \delta\sqrt{-f}\, d\lambda \\ &= -\int \frac{1}{2}(-f)^{-1/2}\,\delta f\, d\lambda.\end{aligned} \tag{3.46}$$

It makes things easier if we now specify that our parameter is the proper time τ itself, rather than the arbitrary parameter λ, so that the tangent vector is the four-velocity U^μ. This fixes the value of f,

$$f = g_{\mu\nu}\frac{dx^\mu}{d\tau}\frac{dx^\nu}{d\tau} = g_{\mu\nu}U^\mu U^\nu = -1. \tag{3.47}$$

From (3.46) we then have

$$\delta\tau = -\frac{1}{2}\int \delta f\, d\tau. \tag{3.48}$$

Stationary points of (3.45)—paths for which $\delta\tau = 0$—are therefore equivalent to stationary points (with fixed parameterization) of the simpler integral

$$I = \frac{1}{2}\int f\, d\tau = \frac{1}{2}\int g_{\mu\nu}\frac{dx^\mu}{d\tau}\frac{dx^\nu}{d\tau}\, d\tau. \tag{3.49}$$

(The $\frac{1}{2}$ is by no means necessary, but will make things nicer later on.) Taking variations of this expression is thus a shortcut to finding shortest-distance paths, one that we will wisely follow.

Stationary points of I will of course obey the Euler–Lagrange equations (1.128), but evaluating them involves repeated application of the chain rule, and it is just as simple to directly consider the change in the integral under infinitesimal variations of the path,

$$\begin{aligned} x^\mu &\rightarrow x^\mu + \delta x^\mu \\ g_{\mu\nu} &\rightarrow g_{\mu\nu} + (\partial_\sigma g_{\mu\nu})\delta x^\sigma. \end{aligned} \tag{3.50}$$

The second line comes from Taylor expansion in curved spacetime, which as you can see, uses the partial derivative, not the covariant derivative. This is because we are simply thinking of the components $g_{\mu\nu}$ as functions on spacetime in some specific coordinate system. Plugging this into (3.49) and keeping only terms first-order in δx^μ, we get

$$\delta I = \frac{1}{2}\int \left[\partial_\sigma g_{\mu\nu}\frac{dx^\mu}{d\tau}\frac{dx^\nu}{d\tau}\delta x^\sigma + g_{\mu\nu}\frac{d(\delta x^\mu)}{d\tau}\frac{dx^\nu}{d\tau} + g_{\mu\nu}\frac{dx^\mu}{d\tau}\frac{d(\delta x^\nu)}{d\tau}\right] d\tau. \tag{3.51}$$

The last two terms can be integrated by parts; for example,

$$\begin{aligned} \frac{1}{2}\int \left[g_{\mu\nu}\frac{dx^\mu}{d\tau}\frac{d(\delta x^\nu)}{d\tau}\right] d\tau &= -\frac{1}{2}\int \left[g_{\mu\nu}\frac{d^2x^\mu}{d\tau^2} + \frac{dg_{\mu\nu}}{d\tau}\frac{dx^\mu}{d\tau}\right]\delta x^\nu\, d\tau \\ &= -\frac{1}{2}\int \left[g_{\mu\nu}\frac{d^2x^\mu}{d\tau^2} + \partial_\sigma g_{\mu\nu}\frac{dx^\sigma}{d\tau}\frac{dx^\mu}{d\tau}\right]\delta x^\nu\, d\tau, \end{aligned} \tag{3.52}$$

where we have neglected boundary terms, which vanish because we take our variation δx^μ to vanish at the endpoints of the path. In the second line we have used the chain rule on the derivative of $g_{\mu\nu}$. The variation (3.51) then becomes, after rearranging some dummy indices,

$$\delta I = -\int \left[g_{\mu\sigma} \frac{d^2 x^\mu}{d\tau^2} + \frac{1}{2}\left(\partial_\mu g_{\nu\sigma} + \partial_\nu g_{\sigma\mu} - \partial_\sigma g_{\mu\nu}\right) \frac{dx^\mu}{d\tau}\frac{dx^\nu}{d\tau} \right] \delta x^\sigma \, d\tau. \tag{3.53}$$

Since we are searching for stationary points, we want δI to vanish for any variation δx^σ; this implies

$$g_{\mu\sigma} \frac{d^2 x^\mu}{d\tau^2} + \frac{1}{2}\left(\partial_\mu g_{\nu\sigma} + \partial_\nu g_{\sigma\mu} - \partial_\sigma g_{\mu\nu}\right) \frac{dx^\mu}{d\tau}\frac{dx^\nu}{d\tau} = 0, \tag{3.54}$$

and multiplying by the inverse metric $g^{\rho\sigma}$ finally leads to

$$\frac{d^2 x^\rho}{d\tau^2} + \frac{1}{2} g^{\rho\sigma}\left(\partial_\mu g_{\nu\sigma} + \partial_\nu g_{\sigma\mu} - \partial_\sigma g_{\mu\nu}\right) \frac{dx^\mu}{d\tau}\frac{dx^\nu}{d\tau} = 0. \tag{3.55}$$

We see that this is precisely the geodesic equation (3.40), but with the specific choice of Christoffel connection (3.27). Thus, on a manifold with metric, extremals of the length functional are curves that parallel transport their tangent vector with respect to the Christoffel connection associated with that metric. It doesn't matter if any other connection is defined on the same manifold. Of course, in GR the Christoffel connection is the only one used, so the two notions are the same.

The variational principle provides a convenient way to actually calculate the Christoffel symbols for a given metric. Rather than simply plugging into (3.27), it is often less work to explicitly vary the integral (3.49), with the metric of interest substituted in for $g_{\mu\nu}$. An example of this procedure is shown in Section 3.5.

3.4 ■ PROPERTIES OF GEODESICS

The primary usefulness of geodesics in general relativity is that they are the paths followed by unaccelerated test particles. A **test particle** is a body that does not itself influence the geometry through which it moves—never perfectly true, but often an excellent approximation. This concept allows us to explore, for example, the properties of the gravitational field around the Sun, without worrying about the field of the planet whose motion we are considering. The geodesic equation can be thought of as the generalization of Newton's law $\mathbf{f} = m\mathbf{a}$, for the case $\mathbf{f} = 0$, to curved spacetime. It is also possible to introduce forces by adding terms to the right-hand side; in fact, looking back to the expression (1.106) for the Lorentz force in special relativity, it is natural to guess that

$$\frac{d^2 x^\mu}{d\tau^2} + \Gamma^\mu_{\rho\sigma} \frac{dx^\rho}{d\tau}\frac{dx^\sigma}{d\tau} = \frac{q}{m} F^\mu{}_\nu \frac{dx^\nu}{d\tau}. \tag{3.56}$$

We will talk about this more later, but in fact your guess would be correct.

We should say some more careful words about the parameterization of a geodesic path. When we presented the geodesic equation as the requirement that the tangent vector be parallel-transported, (3.44), we parameterized our path with some parameter λ, whereas when we found the formula (3.55) for the extremal of the spacetime interval, we wound up with a very specific parameterization, the proper time. Of course from the form of (3.55) it is clear that a transformation,

$$\tau \rightarrow \lambda = a\tau + b, \tag{3.57}$$

for some constants a and b, leaves the equation invariant. Any parameter related to the proper time in this way is called an **affine parameter**, and is just as good as the proper time for parameterizing a geodesic. What was hidden in our derivation of (3.44) was that *the demand that the tangent vector be parallel-transported actually constrains the parameterization of the curve*, specifically to one related to the proper time by (3.57). In other words, if you start at some point and with some initial direction, and then construct a curve by beginning to walk in that direction and keeping your tangent vector parallel transported, you will not only define a path in the manifold but also (up to linear transformations) define the parameter along the path.

Of course, there is nothing to stop you from using any other parameterization you like, but then (3.44) will not be satisfied. More generally you will satisfy an equation of the form

$$\frac{d^2x^\mu}{d\alpha^2} + \Gamma^\mu_{\rho\sigma}\frac{dx^\rho}{d\alpha}\frac{dx^\sigma}{d\alpha} = f(\alpha)\frac{dx^\mu}{d\alpha}, \tag{3.58}$$

for some parameter $\alpha(\lambda)$, where $f(\alpha)$ is related to the affine parameter by

$$f(\alpha) = -\left(\frac{d^2\alpha}{d\lambda^2}\right)\left(\frac{d\alpha}{d\lambda}\right)^{-2}. \tag{3.59}$$

Conversely, if (3.58) is satisfied along a curve you can always find an affine parameter $\lambda(\alpha)$ for which the geodesic equation (3.44) will be satisfied.

For timelike paths, we can write the geodesic equation in terms of the four-velocity $U^\mu = dx^\mu/d\tau$ as

$$U^\lambda \nabla_\lambda U^\mu = 0. \tag{3.60}$$

Similarly, in terms of the four-momentum $p^\mu = mU^\mu$, the geodesic equation is simply

$$p^\lambda \nabla_\lambda p^\mu = 0. \tag{3.61}$$

This relation expresses the idea that freely-falling particles keep moving in the direction in which their momenta are pointing.

For null paths, the proper time vanishes and τ is not an appropriate affine parameter. Nevertheless, it is still perfectly well-defined to ask whether a parameterized path $x^\mu(\lambda)$ satisfies the geodesic equation (3.44). If a null path is a geodesic for some parameter λ, it will also be a geodesic for any other affine parameter of the form $a\lambda + b$. However, there is no preferred choice among these parameters like the proper time is for timelike paths. Once we choose a parameter at some point along the path, of course, there is a unique continuation to the rest of the path if we want to solve the geodesic equation. It is often convenient to choose the normalization of the affine parameter λ along a null geodesic such that $dx^\mu/d\lambda$ is equal to the momentum four-vector:

$$p^\mu = \frac{dx^\mu}{d\lambda}. \tag{3.62}$$

This is in contrast to timelike paths, where $dx^\mu/d\tau$ is the momentum per unit mass. Then an observer with four-velocity U^μ measures the energy of the particle (or equivalently the frequency, since we are setting $\hbar = 1$) to be

$$E = -p_\mu U^\mu. \tag{3.63}$$

This expression always tells us the energy of a particle with momentum p^μ as measured by an observer with four-velocity U^μ, whether p^μ is null or timelike; you can check it by going to locally inertial coordinates. (A caveat: this expression for E does not include potential energy, only the intrinsic energy from motion and inertia. In a general spacetime there will not be a well-defined notion of gravitational potential energy, although in special cases it does exist.)

An important property of geodesics in a spacetime with Lorentzian metric is that the character (timelike/null/spacelike) of the geodesic, relative to a metric-compatible connection, never changes. This is simply because parallel transport preserves inner products, and the character is determined by the inner product of the tangent vector with itself. This is why we were consistent to consider purely timelike paths when we derived (3.55); for spacelike paths we would have derived the same equation, since the only difference is an overall minus sign in the final answer.

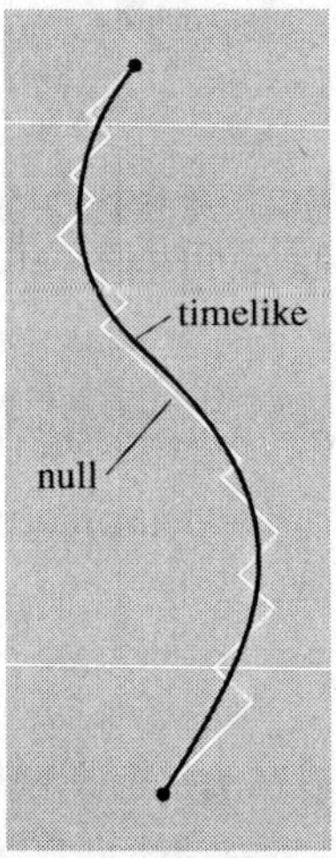

FIGURE 3.3 We can always approximate a timelike path by a sequence of null paths with a total path length of zero. Hence, timelike geodesics must be maxima of the proper time rather than minima.

Let's now explain the earlier remark that timelike geodesics are maxima of the proper time. The reason we know this is true is that, given any timelike curve (geodesic or not), we can approximate it to arbitrary accuracy by a null curve. To do this all we have to do is to consider "jagged" null curves that follow the timelike one, as portrayed in Figure 3.3. As we increase the number of sharp corners, the null curve comes closer and closer to the timelike curve while still having zero path length. Timelike geodesics cannot therefore be curves of minimum proper time, since they are always infinitesimally close to curves of less proper time (zero, in fact); actually they maximize the proper time. This is how you can remember which twin in the twin paradox ages more—the one who stays home is basically on a geodesic, and therefore experiences more proper time. Of course even this is being a little cavalier; actually every time we say "maximize" or "minimize" we should add the modifier "locally." Often the case is that between two points on a manifold there is more than one geodesic. For instance, on S^2 we can draw a great circle through any two points, and imagine traveling between them either the short way or the long way around. One of these is obviously longer than the other, although both are stationary points of the length functional.

Geodesics provide a convenient way of mapping the tangent space T_p of a point p to a region of the manifold that contains p, called the **exponential map**. This map in turn defines a set of coordinates for this region that are automatically the locally inertial coordinates discussed in Section 2.5 [coordinates $x^{\hat{\mu}}$ around a point p such that $g_{\hat{\mu}\hat{\nu}}(p) = \eta_{\hat{\mu}\hat{\nu}}$ and $\partial_{\hat{\sigma}} g_{\hat{\mu}\hat{\nu}}(p) = 0$]. We begin by noticing that any vector $k \in T_p$ defines a unique geodesic passing through it, for which k is the tangent vector at p, and $\lambda(p) = 0$:

$$\frac{dx^\mu}{d\lambda}(\lambda = 0) = k^\mu. \tag{3.64}$$

Uniqueness follows from the fact that the geodesic equation is a second-order differential equation, and specifying initial data in the form $x^\mu(p)$ and $k^\mu = (dx^\mu/d\lambda)(p)$ completely determines a solution. On this geodesic there will be a unique point in M for which $\lambda = 1$. The exponential map at p, $\exp_p : T_p \to M$, is then defined as

$$\exp_p(k) = x^\nu(\lambda = 1), \tag{3.65}$$

where $x^\nu(\lambda)$ solves the geodesic equation subject to (3.64), as shown in Figure 3.4.

For some set of tangent vectors k^μ near the zero vector, this map will be well-defined, and in fact invertible. Depending on the geometry, however, different geodesics emanating from a single point may eventually cross, at which point $\exp_p : T_p \to M$ is no longer one-to-one. Furthermore, the range of the exponential map is not necessarily the whole manifold, and the domain is not necessarily the whole tangent space. The range can fail to be all of M simply because there can be two points that are not connected by any geodesic. An example is given by anti-de Sitter space, discussed in Chapter 8. The domain can fail to be all of T_p because a geodesic may run into a singularity, which we think of as "the edge of the manifold." Manifolds that have such singularities are known as **geodesically incomplete**. In a more careful discussion it would actually be the

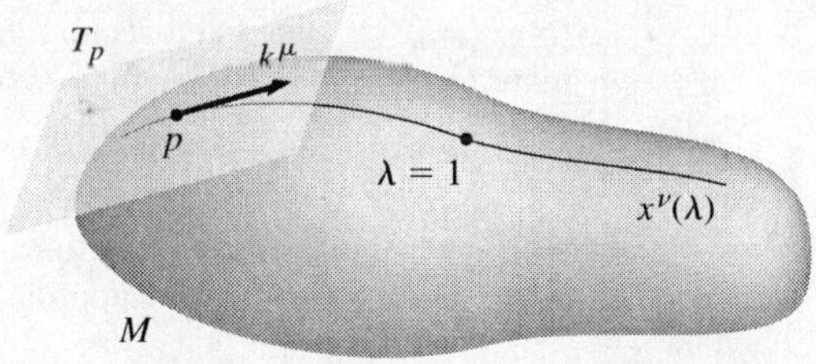

FIGURE 3.4 The exponential map takes a vector in T_p to a point in M that lies at unit affine parameter along the geodesic to which the vector is tangent.

other way around: the best way we have of defining a singularity is as a place where geodesics appear to "end," after we remove trivial cases in which a part of the manifold is artificially excluded by hand. See Wald (1984) or Hawking and Ellis (1973). This problem is not merely technical; the singularity theorems of Hawking and Penrose state that, for certain matter content, spacetimes in general relativity are almost guaranteed to be geodesically incomplete. As examples, two of the most useful spacetimes in GR—the Schwarzschild solution describing black holes and the Friedmann–Robertson–Walker solutions describing homogeneous, isotropic cosmologies—both feature important singularities; these will be discussed in later chapters.

We now use the exponential map to construct locally inertial coordinates. The easy part is to find basis vectors $\{\hat{e}_{(\hat{\mu})}\}$ for T_p such that the components of the metric are those of the canonical form:

$$g_{\hat{\mu}\hat{\nu}} = g(\hat{e}_{(\hat{\mu})}, \hat{e}_{(\hat{\nu})}) = \eta_{\hat{\mu}\hat{\nu}}. \tag{3.66}$$

Here $g(\,,\,)$ denotes the metric, thought of as a multilinear map from $T_p \times T_p$ to $\mathbf{R}$. And the hats have different meanings: over e they remind us that it's a basis vector, and over the indices they remind us that we are in locally inertial coordinates (as we shall see). This is easy because it's just linear algebra, not yet referring to coordinates; starting with any set of components for $g_{\mu\nu}$, we can always diagonalize this matrix and then rescale our basis vectors to satisfy (3.66). The hard part, we would expect, is finding a coordinate system $x^{\hat{\mu}}$ for which the basis vectors $\{\hat{e}_{(\hat{\mu})}\}$ comprise the coordinate basis, $\hat{e}_{(\hat{\mu})} = \partial_{\hat{\mu}}$, and such that the first partial derivatives of $g_{\hat{\mu}\hat{\nu}}$ vanish. In fact, however, the exponential map achieves this automatically. For any point q sufficiently close to p, there is a unique geodesic path connecting p to q, and a unique parameterization λ such that $\lambda(p) = 0$ and $\lambda(q) = 1$. At p the tangent vector k to this geodesic can be written as a linear combination of our basis vectors, $k = k^{\hat{\mu}}\hat{e}_{(\hat{\mu})}$. We define the sought-after coordinates $x^{\hat{\mu}}$ simply to be these components: $x^{\hat{\mu}}(q) = k^{\hat{\mu}}$. In other words, we have defined the coordinates $x^{\hat{\mu}}(q)$ to be the components (with respect to our normalized basis $\{\hat{e}_{(\hat{\mu})}\}$) of the tangent vector k that gets mapped to q by $\exp_p$. Coordinates constructed in this way are known as **Riemann normal coordinates** at p.

We still need to verify that these Riemann normal coordinates satisfy $\partial_{\hat{\sigma}} g_{\hat{\mu}\hat{\nu}}(p) = 0$. Note that a ray in the tangent space (a parameterized set of vectors of the form $\lambda k^{\hat{\mu}}$, for some fixed vector $k^{\hat{\mu}}$) gets mapped to a geodesic by the exponential map. Therefore, in Riemann normal coordinates, a curve $x^{\hat{\mu}}(\lambda)$ of the form

$$x^{\hat{\mu}}(\lambda) = \lambda k^{\hat{\mu}} \tag{3.67}$$

will solve the geodesic equation. Indeed, *any* geodesic through p may be expressed this way, for some appropriate vector $k^{\hat{\mu}}$. We therefore have

$$\frac{d^2 x^{\hat{\mu}}}{d\lambda^2} = 0 \tag{3.68}$$

along any geodesic through p in this coordinate system. But, by the geodesic equation, we also have

$$\frac{d^2x^{\hat{\mu}}}{d\lambda^2}(p) = -\Gamma^{\hat{\mu}}_{\hat{\rho}\hat{\sigma}}(p)k^{\hat{\rho}}k^{\hat{\sigma}}, \tag{3.69}$$

where $k^{\hat{\rho}} = (dx^{\hat{\rho}}/d\lambda)(p)$. Since this holds for arbitrary $k^{\hat{\rho}}$, we conclude that

$$\Gamma^{\hat{\mu}}_{\hat{\rho}\hat{\sigma}}(p) = 0. \tag{3.70}$$

Now apply metric compatibility:

$$\begin{aligned} 0 &= \nabla_{\hat{\sigma}} g_{\hat{\mu}\hat{\nu}} \\ &= \partial_{\hat{\sigma}} g_{\hat{\mu}\hat{\nu}} - \Gamma^{\hat{\lambda}}_{\hat{\sigma}\hat{\mu}} g_{\hat{\lambda}\hat{\nu}} - \Gamma^{\hat{\lambda}}_{\hat{\sigma}\hat{\nu}} g_{\hat{\mu}\hat{\lambda}} \\ &= \partial_{\hat{\sigma}} g_{\hat{\mu}\hat{\nu}}, \end{aligned} \tag{3.71}$$

where all quantities are evaluated at p. We see that Riemann normal coordinates provide a realization of the locally inertial coordinates discussed in Section 2.5. They are not unique; there are an infinite number of non-Riemann-normal coordinate systems in which $g_{\hat{\mu}\hat{\nu}}(p) = \eta_{\hat{\mu}\hat{\nu}}$ and $\partial_{\hat{\sigma}} g_{\hat{\mu}\hat{\nu}}(p) = 0$, but in an expansion around p they will differ from the Riemann normal coordinates only at third order in $x^{\hat{\mu}}$.

3.5 ■ THE EXPANDING UNIVERSE REVISITED

Let's put some of the technology we have developed to work in understanding a simple metric. Recall the expanding-universe metric we studied in Chapter 2,

$$\begin{aligned} ds^2 &= -dt^2 + a^2(t)[dx^2 + dy^2 + dz^2] \\ &= -dt^2 + a^2(t)\delta_{ij}dx^i dx^j. \end{aligned} \tag{3.72}$$

This metric describes a universe consisting of flat spatial sections expanding as a function of time, with the relative distance between particles at fixed spatial coordinates growing proportionally to the scale factor $a(t)$.

Faced with a metric, the first thing we do is to calculate the Christoffel symbols. As mentioned at the end of Section 3.3, the easiest technique for doing so is actually to vary explicitly an integral of the form (3.49). Plugging in the metric under consideration, we have

$$I = \frac{1}{2}\int \left[-\left(\frac{dt}{d\tau}\right)^2 + a^2(t)\delta_{ij}\frac{dx^i}{d\tau}\frac{dx^j}{d\tau} \right] d\tau. \tag{3.73}$$

The technique is to consider variations $x^\mu \rightarrow x^\mu + \delta x^\mu$ and demand that δI vanish. We get n equations on an n-dimensional manifold (in this case $n = 4$), one for each μ; each equation corresponds to a component of the geodesic equation

(3.44). In the equation derived by varying with respect to x^μ, the coefficient of $(dx^\rho/d\tau)(dx^\sigma/d\tau)$ will be $\Gamma^\mu_{\rho\sigma}$.

For the metric (3.72), we need to consider separately variations with respect to $x^0 = t$ and one of the x^i's (it doesn't matter which one, since the results for each spacelike direction will be equivalent). Let's start with $t \to t + \delta t$. The nontrivial time dependence comes from the scale factor, for which, to first order,

$$a(t + \delta t) = a(t) + \dot{a}\delta t, \tag{3.74}$$

where $\dot{a} = da/dt$. We therefore have

$$\begin{aligned} \delta I &= \frac{1}{2}\int \left[-2\frac{dt}{d\tau}\frac{d(\delta t)}{d\tau} + 2a\dot{a}\delta_{ij}\frac{dx^i}{d\tau}\frac{dx^j}{d\tau}\delta t \right] d\tau \\ &= \int \left[\frac{d^2 t}{d\tau^2} + a\dot{a}\delta_{ij}\frac{dx^i}{d\tau}\frac{dx^j}{d\tau} \right] \delta t \, d\tau, \end{aligned} \tag{3.75}$$

where we have dropped a boundary term after integrating by parts (as always). Setting the coefficient of δt equal to zero implies

$$\frac{d^2 t}{d\tau^2} + a\dot{a}\delta_{ij}\frac{dx^i}{d\tau}\frac{dx^j}{d\tau} = 0, \tag{3.76}$$

which is supposed to be equivalent to the geodesic equation (with $\mu = 0$)

$$\frac{d^2 x^0}{d\tau^2} + \Gamma^0_{\rho\sigma}\frac{dx^\rho}{d\tau}\frac{dx^\sigma}{d\tau} = 0. \tag{3.77}$$

Comparison of these two equations implies

$$\begin{aligned} \Gamma^0_{00} &= 0, \\ \Gamma^0_{i0} = \Gamma^0_{0i} &= 0, \\ \Gamma^0_{ij} &= a\dot{a}\,\delta_{ij}. \end{aligned} \tag{3.78}$$

We can repeat this procedure for a spatial coordinate, $x^i \to x^i + \delta x^i$. The variation is then

$$\begin{aligned} \delta I &= \frac{1}{2}\int a^2 \left(2\delta_{ij}\frac{dx^i}{d\tau}\frac{d(\delta x^j)}{d\tau} \right) d\tau \\ &= -\int \left(a^2 \frac{d^2 x^i}{d\tau^2} + 2a\frac{da}{d\tau}\frac{dx^i}{d\tau} \right) \delta_{ij}\delta x^j \, d\tau. \end{aligned} \tag{3.79}$$

We can express $da/d\tau$ in terms of $\dot{a}$ by using the chain rule,

$$\frac{da}{d\tau} = \dot{a}\frac{dt}{d\tau}. \tag{3.80}$$

Then setting the coefficient of δx^j equal to zero in (3.79) implies

$$\frac{d^2x^i}{d\tau^2} + 2\frac{\dot{a}}{a}\frac{dt}{d\tau}\frac{dx^i}{d\tau} = 0. \tag{3.81}$$

Comparing to the geodesic equation, we find that the Christoffel symbols must satisfy

$$\Gamma^i_{\rho\sigma}\frac{dx^\rho}{d\tau}\frac{dx^\sigma}{d\tau} = 2\frac{\dot{a}}{a}\frac{dt}{d\tau}\frac{dx^i}{d\tau}. \tag{3.82}$$

The Christoffel symbols are therefore given by

$$\begin{aligned}\Gamma^i_{00} &= 0\\ \Gamma^i_{j0} = \Gamma^i_{0j} &= \frac{\dot{a}}{a}\delta^i_j\\ \Gamma^i_{jk} &= 0.\end{aligned} \tag{3.83}$$

Together, (3.78) and (3.83) are all of the connection coefficients for the metric (3.72). These are, of course, necessary both for studying geodesics of the spacetime and for taking covariant derivatives; in fact, (3.76) and (3.81) together *are* the geodesic equation. Let's put this to work by solving for null geodesics, those followed by massless particles such as photons, for which we have to use λ rather than τ as a parameter. Without loss of generality we can consider paths along the x-direction, for which $x^\mu(\lambda) = \{t(\lambda), x(\lambda), 0, 0\}$. It is trivial to solve for null paths of this sort, using $ds^2 = 0$. We have

$$0 = -dt^2 + a^2(t)dx^2, \tag{3.84}$$

which implies

$$\frac{dx}{d\lambda} = \frac{1}{a}\frac{dt}{d\lambda}. \tag{3.85}$$

In Section 2.6 we solved this for $a = t^q$, but here we will remain more general. Also, we have chosen to consider paths moving in the positive x-direction, which determines the sign of $dx/d\lambda$. We must distinguish, however, between "null paths" and "null geodesics": the latter are a much more restrictive class, and to show that these paths are geodesics, we need to solve for the coordinates t and x in terms of the parameter λ.

Let's solve for $dt/d\lambda$, which will turn out to be the quantity in which we are most interested. Plugging the null condition (3.85) into the $\mu = 0$ component of the geodesic equation (3.76), and remembering to replace $\tau \to \lambda$, we get

$$\frac{d^2t}{d\lambda^2} + \frac{\dot{a}}{a}\left(\frac{dt}{d\lambda}\right)^2 = 0. \tag{3.86}$$

It is straightforward to verify that this is solved by

$$\frac{dt}{d\lambda} = \frac{\omega_0}{a}, \tag{3.87}$$

where ω_0 is a constant. For a given $a(t)$, this could be instantly integrated to yield $t(\lambda)$. But more interesting is to consider the energy E of the photon as it would be measured by a comoving observer (one at fixed spatial coordinates), who would have four-velocity

$$U^\mu = (1, 0, 0, 0). \tag{3.88}$$

Don't get tricked into thinking that the timelike component of the four-velocity of a particle at rest will always equal unity; we need to satisfy the normalization condition $g_{\mu\nu}U^\mu U^\nu = -1$, which in the rest frame ($U^i = 0$) implies $U^0 = \sqrt{-g_{00}}$. According to (3.63), and using $p^\mu = dx^\mu/d\lambda$, we have

$$\begin{aligned} E &= -p_\mu U^\mu \\ &= -g_{00}\frac{dx^0}{d\lambda}U^0 \\ &= \frac{\omega_0}{a}. \end{aligned} \tag{3.89}$$

We see why the notation ω_0 was chosen for the constant of proportionality in (3.87): ω_0 is simply the frequency of the photon when $a = 1$. Recall that $E = \hbar\omega$, and we are using units in which $\hbar = 1$.

We have uncovered a profound phenomenon: the **cosmological redshift**. A photon emitted with energy E_1 at scale factor a_1 and observed with energy E_2 at scale factor a_2 will have

$$\frac{E_2}{E_1} = \frac{a_1}{a_2}. \tag{3.90}$$

This is called a "redshift" because the wavelength of the photon is inversely proportional to the frequency, and in an expanding universe the wavelength therefore grows with time. As a practical matter this provides an easy way to measure the change in the scale factor between us and distant galaxies, and also serves as a proxy for the distance: since the universe has been monotonically expanding, a greater redshift implies a greater distance. In conventional notation, the amount of redshift is denoted by

$$z = \frac{\omega_1 - \omega_2}{\omega_2} = \frac{a_2}{a_1} - 1, \tag{3.91}$$

so that z vanishes if there has been no expansion, for instance, if the emitter and observer are so close that there hasn't been enough time for the universe to appreciably expand.

As mentioned in Section 3.3, the cosmological redshift is *not* a Doppler shift (despite the understandable temptation to refer to the "velocity" of receding galaxies). Now we can understand this statement quantitatively. You might imagine that, as far as the behavior of emitted photons is concerned, there is little difference between two galaxies physically moving apart in a flat spacetime and two galaxies at fixed comoving coordinates in an expanding spacetime. But let's consider a specific (unrealistic, but educational) example. Start with flat spacetime, and imagine that our two galaxies are initially not moving apart, but are at rest in some globally inertial coordinate system. One emits a photon toward the other; while the photon is traveling, we quickly move the two galaxies apart until they are twice their original separation, then leave them stationary at that distance; and then the photon is absorbed by the second galaxy. Clearly there will be no Doppler shift, since the galaxies were at rest both at emission and absorption. Now consider the analogous phenomenon in an expanding spacetime, with the galaxies stuck at fixed comoving coordinates. We begin with the scale factor constant (the universe is not expanding). One galaxy emits a photon, and we imagine that during the photon's journey the universe starts expanding until the scale factor is twice its original size, and then stops expanding before the photon is absorbed. In this case there certainly will be a redshift, despite the fact that there was no "relative motion" (an ill-defined concept in any case) at either absorption or emission; the photon's wavelength will have doubled as the scale factor doubled, so we observe a redshift $z = 1$. This demonstrates the conceptual distinction between the cosmological redshift and the conventional Doppler effect.

Beyond the geodesic equation, covariant derivatives will play a role in generalizing laws of physics from the flat spacetime of special relativity to the curved geometry of general relativity. As we will discuss in more detail in the next chapter, a simple rule of thumb is simply to replace all partial derivatives by covariant derivatives, and all appearances of the flat spacetime metric $\eta_{\mu\nu}$ by the curved metric $g_{\mu\nu}$. For example, the energy-momentum conservation equation of special relativity, $\partial_\mu T^{\mu\nu} = 0$, where $T^{\mu\nu}$ is the energy-momentum tensor, becomes

$$\nabla_\mu T^{\mu\nu} = 0. \tag{3.92}$$

In cosmology, we typically model the matter filling the universe as a perfect fluid; the corresponding energy-momentum tensor comes from generalizing (1.114) to curved spacetime,

$$T^{\mu\nu} = (\rho + p)U^\mu U^\nu + p g^{\mu\nu}. \tag{3.93}$$

Recall that ρ is the energy density, p is the pressure, and U^μ is the four-velocity of the fluid. For the metric (3.72) the components of the inverse metric are

$$g^{\mu\nu} = \begin{pmatrix} -1 & & & \\ & a^{-2} & & \\ & & a^{-2} & \\ & & & a^{-2} \end{pmatrix}. \tag{3.94}$$

We can take the fluid to be in its rest frame in these coordinates, so that the components of the four-velocity are $U^\mu = (1, 0, 0, 0)$. In fact the fluid would have to be in its rest frame for this particular metric to solve Einstein's equation, as we will later see. The energy-momentum tensor therefore takes the form

$$T^{\mu\nu} = \begin{pmatrix} \rho & & & \\ & a^{-2}p & & \\ & & a^{-2}p & \\ & & & a^{-2}p \end{pmatrix}. \tag{3.95}$$

Note that these components are specific to the metric (3.72), and will generally look different for other metrics.

Let's see what the energy-momentum conservation equation $\nabla_\mu T^{\mu\nu} = 0$ implies for a perfect fluid in an expanding universe. The rule for covariant derivatives (3.17) implies

$$\nabla_\mu T^{\mu\nu} = \partial_\mu T^{\mu\nu} + \Gamma^\mu_{\mu\lambda} T^{\lambda\nu} + \Gamma^\nu_{\mu\lambda} T^{\mu\lambda} = 0. \tag{3.96}$$

This equation has four components, one for each μ, although the three $\mu = i \in \{1, 2, 3\}$ are equivalent. Let's first look at the $\nu = 0$ component, piece by piece. The first term is straightforward,

$$\partial_\mu T^{\mu 0} = \partial_0 T^{00} = \dot{\rho}. \tag{3.97}$$

The second term is

$$\Gamma^\mu_{\mu\lambda} T^{\lambda 0} = \Gamma^\mu_{\mu 0} T^{00} = 3\frac{\dot{a}}{a}\rho, \tag{3.98}$$

and the third term is

$$\Gamma^0_{\mu\lambda} T^{\mu\lambda} = \Gamma^0_{00} T^{00} + \Gamma^0_{11} T^{11} + \Gamma^0_{22} T^{22} + \Gamma^0_{33} T^{33} = 3\frac{\dot{a}}{a}p. \tag{3.99}$$

In each of these sets of equations, we have first invoked the fact that $T^{\mu\nu}$ is diagonal, and then used the explicit formulae for the energy-momentum tensor and the connection coefficients in this metric. All together, then, we find

$$\dot{\rho} = -3\frac{\dot{a}}{a}(\rho + p). \tag{3.100}$$

Now let's look at one of the spatial components, choosing $\nu = 1$ for definiteness. Once again working piece by piece, we have for the first term in (3.96),

$$\partial_\mu T^{\mu 1} = \partial_1 T^{11} = a^{-2}\partial_x p. \tag{3.101}$$

The second and third terms are

$$\Gamma^\mu_{\mu\lambda} T^{\lambda 1} = \Gamma^\mu_{\mu 1} T^{11} = 0, \tag{3.102}$$

and

$$\Gamma^1_{\mu\lambda}T^{\mu\lambda} = \Gamma^1_{00}T^{00} + \Gamma^1_{11}T^{11} + \Gamma^1_{22}T^{22} + \Gamma^1_{33}T^{33} = 0. \tag{3.103}$$

Equivalent results will hold for $\nu = 2$ and $\nu = 3$. So the spatial components of the energy-momentum conservation equation simply amount to

$$\partial_i p = 0. \tag{3.104}$$

It is illuminating to compare these results to those we would obtain in Minkowski spacetime, which can be found by simply setting $a = 1$, $\dot{a} = 0$. The pressure-gradient equation (3.104) is unaffected, so there is no effect of curvature on the spatial components: for a fluid that is motionless as measured by a comoving observer, the pressure must be constant throughout space. For the timelike component, on the other hand, the expansion of the universe introduces a nonzero right-hand side to (3.100). To understand the consequences of this new feature, let us consider equations of state of the form

$$p = w\rho, \tag{3.105}$$

where w is some constant. Then (3.100) becomes

$$\frac{\dot{\rho}}{\rho} = -3(1+w)\frac{\dot{a}}{a}, \tag{3.106}$$

which can be solved to yield

$$\rho \propto a^{-3(1+w)}. \tag{3.107}$$

In Chapter 1 we mentioned three kinds of perfect fluid with equations of state of the form (3.105): dust, with $w = 0$; radiation, with $w = \frac{1}{3}$; and vacuum, with $w = -1$. A set of nonrelativistic, noninteracting particles behaves like dust; a set of photons or other massless particles behaves like radiation; and a nonzero constant energy density throughout spacetime acts like vacuum. From (3.107) we see that the equation of state determines how the energy density evolves as the universe expands:

$$\begin{array}{lll} \text{matter} & p = 0 & \rho \propto a^{-3} \\ \text{radiation} & p = \frac{1}{3}\rho & \rho \propto a^{-4} \\ \text{vacuum} & p = -\rho & \rho = \text{constant}. \end{array} \tag{3.108}$$

We will explore these behaviors more thoroughly in Chapter 8; for right now let's simply note that they make sense. For dust, the energy density comes from the rest mass of each particle; if all the particles have mass m, the energy density is simply $\rho = nm$, where n is the number density. Since the number density goes down as a^{-3} (the physical volume of a comoving region goes up, while the total number of particles stays constant), while the masses remain unchanged, we expect that the energy density obeys $\rho \propto a^{-3}$. For radiation, meanwhile, the energy of each particle (such as a photon) redshifts away as a^{-1} as the universe expands; since the number density still obeys $n \propto a^{-3}$, we expect that $\rho \propto a^{-4}$. Finally, the vacuum energy density is an intrinsic and unchanging amount of energy in any physical volume; it doesn't redshift at all as the universe expands, so we get $\rho =$ constant.

This example brings to life the differences between flat and curved spacetimes. For example, consider what we might be tempted to call the "energy," the integral over space of the energy density: $E = \int \rho a^3 \, d^3x$, where the boundaries are at fixed comoving coordinates, so the region expands along with the universe, and the factor of a^3 comes from the square root of the determinant of the spatial metric $a^2\delta_{ij}$. This number is clearly not conserved in general. For dust, since $\rho \propto a^{-3}$, E remains constant as the universe expands; but for radiation it decreases, and for vacuum energy it increases. This is upsetting, since conservation of energy is one of the more cherished principles of physics. What has happened? One way of thinking about this is from the viewpoint of Noether's theorem, which states that every symmetry implies a conserved quantity. Energy is the conserved quantity that derives from invariance under time translations. Clearly, in an expanding universe, the energy-momentum tensor is defined on a background that is changing with time; therefore there is no reason to believe that the energy should be conserved. ("There is no timelike Killing vector," in the language to be introduced in Section 3.8.) Nevertheless, we continue to refer to $\nabla_\mu T^{\mu\nu} = 0$ as the energy momentum conservation equation. It conveys the idea that there is a definite law obeyed by the energy-momentum tensor, even if there is no integral corresponding to a conserved energy. The transition from flat to curved spacetime induces the additional Christoffel-symbol terms in (3.96); these terms serve, roughly speaking, to allow transfer of energy between the matter fields (comprising $T^{\mu\nu}$) and the gravitational field. This notion is not very formal, however, and you shouldn't push it too far—it turns out to be difficult to associate a local energy density to the gravitational field, although it is possible in certain circumstances.

Of course there is also a notion of time-translation invariance that refers not to the background spacetime, but to the theory itself (that is, to the equations that define the theory rather than a specific solution to them). We haven't yet developed the dynamical equations of general relativity, but they will turn out to be invariant under time translations, as well as under any other sort of coordinate transformations, as indeed they must be. This general coordinate invariance leads to a set of constraints on allowed configurations of the theory, and generally requires a more subtle analysis.

In the end, you should come to accept that there is a profound difference between flat and curved spacetimes, and some of our favorite notions from flat-spacetime physics will be seriously modified in this more general context. This is not a sign of any flaw in general relativity, but a natural consequence of discarding the rigid spacetime geometry we learn to take for granted.

3.6 ■ THE RIEMANN CURVATURE TENSOR

Having set up the machinery of covariant derivatives and parallel transport, we are at last prepared to discuss curvature proper. The curvature is quantified by the Riemann tensor, which is derived from the connection. The idea behind this measure of curvature is that we know what we mean by "flatness" of a connection—the conventional (and usually implicit) Christoffel connection associated with a Euclidean or Minkowskian metric has a number of properties that can be thought of as different manifestations of flatness. These include the fact that parallel transport around a closed loop leaves a vector unchanged, that covariant derivatives of tensors commute, and that initially parallel geodesics remain parallel. As we shall see, the Riemann tensor arises when we study how any of these properties are altered in more general contexts.

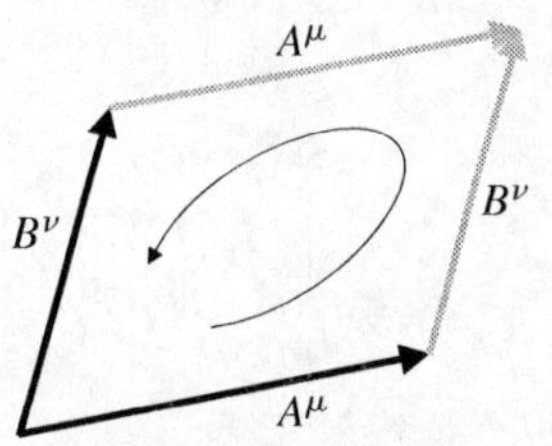

FIGURE 3.5 An infinitesimal loop defined by two vectors A^μ and B^ν.

We have already argued, using the two-sphere as an example, that parallel transport of a vector around a closed loop in a curved space will lead to a transformation of the vector. The resulting transformation depends on the total curvature enclosed by the loop; it would be more useful to have a local description of the curvature at each point, which is what the Riemann tensor is supposed to provide. One conventional way to introduce the Riemann tensor, therefore, is to consider parallel transport around an infinitesimal loop. We are not going to do that here, but take a more direct route. Nevertheless, even without working through the details, it is possible to see what form the answer should take. Since spacetime looks flat in sufficiently small regions, our loop will be specified by two (infinitesimal) vectors A^μ and B^ν. We imagine parallel transporting a vector V^μ by first moving it in the direction of A^μ, then along B^ν, then backward along A^μ and B^ν to return to the starting point, as shown in Figure 3.5. We know the action of parallel transport is independent of coordinates, so there should be some tensor that tells us how the vector changes when it comes back to its starting point; it will be a linear transformation on a vector, and therefore involve one upper and one lower index. But it will also depend on the two vectors A and B that define the loop; therefore there should be two additional lower indices to contract with A^μ and B^ν. Furthermore, the tensor should be antisymmetric in these two indices, since interchanging the vectors corresponds to traversing the loop in the opposite direction, and should give the inverse of the original answer. This is consistent with the fact that the transformation should vanish if A and B are the same vector. We therefore expect that the expression for the change δV^ρ experienced by this vector when parallel transported around the loop should be of the form

$$\delta V^\rho = R^\rho{}_{\sigma\mu\nu} V^\sigma A^\mu B^\nu, \tag{3.109}$$

where $R^\rho{}_{\sigma\mu\nu}$ is a $(1, 3)$ tensor known as the **Riemann tensor** (or simply curvature tensor). It is antisymmetric in the last two indices:

$$R^\rho{}_{\sigma\mu\nu} = -R^\rho{}_{\sigma\nu\mu}. \tag{3.110}$$

Of course, if (3.109) is taken as a definition of the Riemann tensor, a convention needs to be chosen for the ordering of the indices. There is no agreement at all on what this convention should be, so be careful.

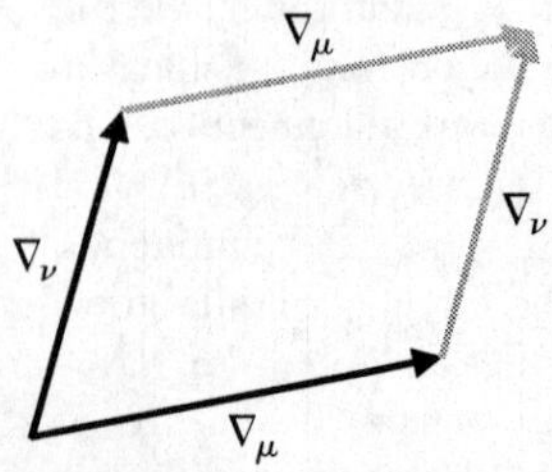

FIGURE 3.6 The commutator of two covariant derivatives.

Knowing what we do about parallel transport, we could very carefully perform the necessary manipulations to see what happens to the vector under this operation, and the result would be a formula for the curvature tensor in terms of the connection coefficients. It is much quicker, however, to consider a related operation, the commutator of two covariant derivatives. The relationship between this and parallel transport around a loop should be evident; the covariant derivative of a tensor in a certain direction measures how much the tensor changes relative to what it would have been if it had been parallel transported, since the covariant derivative of a tensor in a direction along which it is parallel transported is zero. The commutator of two covariant derivatives, then, measures the difference between parallel transporting the tensor first one way and then the other, versus the opposite ordering, as shown in Figure 3.6.

The actual computation is very straightforward. Considering a vector field V^ρ, we take

$$\begin{aligned}
[\nabla_\mu, \nabla_\nu]V^\rho &= \nabla_\mu \nabla_\nu V^\rho - \nabla_\nu \nabla_\mu V^\rho \\
&= \partial_\mu(\nabla_\nu V^\rho) - \Gamma^\lambda_{\mu\nu}\nabla_\lambda V^\rho + \Gamma^\rho_{\mu\sigma}\nabla_\nu V^\sigma - (\mu \leftrightarrow \nu) \\
&= \partial_\mu \partial_\nu V^\rho + (\partial_\mu \Gamma^\rho_{\nu\sigma})V^\sigma + \Gamma^\rho_{\nu\sigma}\partial_\mu V^\sigma - \Gamma^\lambda_{\mu\nu}\partial_\lambda V^\rho - \Gamma^\lambda_{\mu\nu}\Gamma^\rho_{\lambda\sigma}V^\sigma \\
&\quad + \Gamma^\rho_{\mu\sigma}\partial_\nu V^\sigma + \Gamma^\rho_{\mu\sigma}\Gamma^\sigma_{\nu\lambda}V^\lambda - (\mu \leftrightarrow \nu) \\
&= (\partial_\mu \Gamma^\rho_{\nu\sigma} - \partial_\nu \Gamma^\rho_{\mu\sigma} + \Gamma^\rho_{\mu\lambda}\Gamma^\lambda_{\nu\sigma} - \Gamma^\rho_{\nu\lambda}\Gamma^\lambda_{\mu\sigma})V^\sigma - 2\Gamma^\lambda_{[\mu\nu]}\nabla_\lambda V^\rho .
\end{aligned} \tag{3.111}$$

In the last step we have relabeled some dummy indices and eliminated some terms that cancel when antisymmetrized. We recognize that the antisymmetrized connection coefficients in the last term are simply one-half times the torsion tensor, and that the left hand side is manifestly a tensor; therefore the expression in parentheses must be a tensor itself. We write

$$[\nabla_\mu, \nabla_\nu]V^\rho = R^\rho{}_{\sigma\mu\nu}V^\sigma - T^\lambda{}_{\mu\nu}\nabla_\lambda V^\rho , \tag{3.112}$$

where the Riemann tensor is identified as

$$\boxed{R^\rho{}_{\sigma\mu\nu} = \partial_\mu \Gamma^\rho_{\nu\sigma} - \partial_\nu \Gamma^\rho_{\mu\sigma} + \Gamma^\rho_{\mu\lambda}\Gamma^\lambda_{\nu\sigma} - \Gamma^\rho_{\nu\lambda}\Gamma^\lambda_{\mu\sigma} .} \tag{3.113}$$

Notice a number of things about the derivation of this expression:

- Of course we have not demonstrated that (3.113) is actually the same tensor that appeared in (3.109), but in fact it's true. You are asked to show this in the Exercises.
- It is perhaps surprising that the commutator $[\nabla_\mu, \nabla_\nu]$, which appears to be a differential operator, has an action on vector fields that (in the absence of

torsion, at any rate) is a simple multiplicative transformation. The Riemann tensor measures that part of the commutator of covariant derivatives that is proportional to the vector field, while the torsion tensor measures the part that is proportional to the covariant derivative of the vector field; the second derivative doesn't enter at all.

- Notice that the expression (3.113) is constructed from nontensorial elements; you can check that the transformation laws all work out to make this particular combination a legitimate tensor.
- The antisymmetry of $R^\rho{}_{\sigma\mu\nu}$ in its last two indices is immediate from this formula and its derivation.
- We constructed the curvature tensor completely from the connection (no mention of the metric was made). We were sufficiently careful that the above expression is true for any connection, whether or not it is metric compatible or torsion free.
- Using what are by now our usual methods, the action of $[\nabla_\rho, \nabla_\sigma]$ can be computed on a tensor of arbitrary rank. The answer is

$$\begin{aligned}
[\nabla_\rho, \nabla_\sigma] X^{\mu_1\cdots\mu_k}{}_{\nu_1\cdots\nu_l} \\
&= -T^\lambda{}_{\rho\sigma} \nabla_\lambda X^{\mu_1\cdots\mu_k}{}_{\nu_1\cdots\nu_l} \\
&\quad + R^{\mu_1}{}_{\lambda\rho\sigma} X^{\lambda\mu_2\cdots\mu_k}{}_{\nu_1\cdots\nu_l} + R^{\mu_2}{}_{\lambda\rho\sigma} X^{\mu_1\lambda\cdots\mu_k}{}_{\nu_1\cdots\nu_l} + \cdots \\
&\quad - R^\lambda{}_{\nu_1\rho\sigma} X^{\mu_1\cdots\mu_k}{}_{\lambda\nu_2\cdots\nu_l} - R^\lambda{}_{\nu_2\rho\sigma} X^{\mu_1\cdots\mu_k}{}_{\nu_1\lambda\cdots\nu_l} - \cdots.
\end{aligned} \tag{3.114}$$

Both the torsion tensor and the Riemann tensor, thought of as multilinear maps, have elegant expressions in terms of the vector-field commutator. Thinking of the torsion as a map from two vector fields to a third vector field, we have

$$T(X, Y) = \nabla_X Y - \nabla_Y X - [X, Y], \tag{3.115}$$

and thinking of the Riemann tensor as a map from three vector fields to a fourth one, we have (in funny-looking but standard notation)

$$R(X, Y)Z = \nabla_X \nabla_Y Z - \nabla_Y \nabla_X Z - \nabla_{[X,Y]} Z. \tag{3.116}$$

In these expressions, the notation ∇_X refers to the covariant derivative along the vector field X; in components, $\nabla_X = X^\mu \nabla_\mu$. So, for example, (3.116) is equivalent to

$$\begin{aligned}
R^\rho{}_{\sigma\mu\nu} X^\mu Y^\nu Z^\sigma &= X^\lambda \nabla_\lambda (Y^\eta \nabla_\eta Z^\rho) - Y^\lambda \nabla_\lambda (X^\eta \nabla_\eta Z^\rho) \\
&\quad - (X^\lambda \partial_\lambda Y^\eta - Y^\lambda \partial_\lambda X^\eta) \nabla_\eta Z^\rho,
\end{aligned} \tag{3.117}$$

which you can check is equivalent to (3.113). Note that the two vectors X and Y in (3.116) correspond to the last two indices in the component form of the Riemann tensor. The last term in (3.116), involving the commutator $[X, Y]$, vanishes when X and Y are taken to be the coordinate basis vector fields (since $[\partial_\mu, \partial_\nu] = 0$), which is why this term did not arise when we originally took the commutator of two covariant derivatives. We will not use this notation extensively, but you might see it in the literature, so you should be able to decode it.

Having defined the curvature tensor as something that characterizes the connection, let us now admit that in GR we are most concerned with the Christoffel connection. In this case the connection is derived from the metric, and the associated curvature may be thought of as that of the metric itself. This identification allows us to finally make sense of our informal notion that spaces for which the metric looks Euclidean or Minkowskian are flat. In fact it works both ways:

- If a coordinate system exists in which the components of the metric are constant, the Riemann tensor will vanish.
- If the Riemann tensor vanishes, we can always construct a coordinate system in which the metric components are constant.

Technically, these statements should be restricted to a region of the manifold that is simply-connected (all loops in the region can be smoothly deformed to a point without leaving the region); we will implicitly assume this condition below.

The first of these is easy to show. If we are in some coordinate system such that $\partial_\sigma g_{\mu\nu} = 0$ everywhere, not just at a point, then $\Gamma^\rho_{\mu\nu} = 0$ and $\partial_\sigma \Gamma^\rho_{\mu\nu} = 0$; thus $R^\rho{}_{\sigma\mu\nu} = 0$ by (3.113). But this is a tensor equation, and if it is true in one coordinate system it must be true in any coordinate system. Therefore, the statement that the Riemann tensor vanishes is a necessary condition for it to be possible to find coordinates in which the components of $g_{\mu\nu}$ are constant everywhere.

The second claim, that $R^\rho{}_{\sigma\mu\nu} = 0$ everywhere implies we can find a coordinate system in which the metric components are constant everywhere, is harder to prove (but not very hard). Consider as a warm-up a one-form $\omega = \omega_\mu \mathrm{d}x^\mu$, defined at some point p. For any path $x^\mu(\lambda)$ that includes p, we can construct a unique one-form field along the path by demanding that ω_μ be parallel-transported:

$$\frac{dx^\mu}{d\lambda}\nabla_\mu \omega_\nu = 0. \tag{3.118}$$

In general, if we performed this procedure for distinct paths that started at p and passed through some other point q, the value of ω_μ at q would depend on the path. However, if the Riemann tensor vanishes everywhere, the parallel-transport will be path-independent, and we can define a unique one-form field throughout the manifold. Therefore (3.118) must be true for arbitrary $dx^\mu/d\lambda$; this can only be true if ω_μ is covariantly constant:

$$\nabla_\mu \omega_\nu = 0. \tag{3.119}$$

On an arbitrary manifold there will be no solutions to this equation; it is only possible here because we are assuming that the curvature vanishes. We can take the antisymmetric part of (3.119), and from (3.36) we know that this is just the exterior derivative:

$$\nabla_{[\mu}\omega_{\nu]} = \partial_{[\mu}\omega_{\nu]} = 0, \tag{3.120}$$

or, in index-free notation,

$$\mathrm{d}\omega = 0. \tag{3.121}$$

In other words, ω is closed. It is also exact (there exists a scalar function α such that $\omega = \mathrm{d}\alpha$), since we have restricted the topology of the region in which we are working. In components we have

$$\omega_\mu = \partial_\mu \alpha. \tag{3.122}$$

There is nothing special about the one-form ω, so we can repeat this procedure with a set of one-forms $\hat{\theta}^{(a)}$, where $a \in \{1 \dots n\}$ on an n-dimensional manifold. We may choose our one-forms to comprise a normalized basis for the dual space T_p^*, such that the components of the metric with respect to this basis are those of the canonical form; in other words,

$$ds^2(p) = \eta_{ab}\, \hat{\theta}^{(a)} \otimes \hat{\theta}^{(b)}. \tag{3.123}$$

Here we use η_{ab} in a generalized sense, as a matrix with either $+1$ or -1 for each diagonal element and zeroes elsewhere. The actual arrangement of the $+1$'s and -1's depends on the canonical form of the metric, but is irrelevant for the present argument. Now let us parallel transport the entire set of basis forms all over the manifold; the vanishing of the Riemann tensor ensures that the result will be independent of the path taken. Since the metric is always automatically parallel-transported with respect to a metric-compatible connection, the metric components will remain unchanged,

$$ds^2(\text{anywhere}) = \eta_{ab}\, \hat{\theta}^{(a)} \otimes \hat{\theta}^{(b)}. \tag{3.124}$$

We therefore have specified a set of one-form fields, which everywhere define a basis in which the metric components are constant. This is completely unimpressive; it can be done on any manifold, regardless of what the curvature is. What we would like to show is that this is a *coordinate* basis (which will only be possible if the curvature vanishes). However, by the same arguments that led to (3.122), we know that all of the $\hat{\theta}^{(a)}$'s are exact forms, so that there exists a set of functions y^a such that the one-form fields are their gradients,

$$\hat{\theta}^{(a)} = \mathrm{d}y^a. \tag{3.125}$$

These n functions are precisely the sought-after coordinates; all over the manifold the metric takes the form

$$ds^2 = \eta_{ab}\, \mathrm{d}y^a \mathrm{d}y^b. \tag{3.126}$$

At this point you are welcome to switch from using a, b as indices to μ, ν if you prefer.

We have thus verified that the Riemann tensor provides us with an answer to the question of whether some horrible-looking metric is secretly that of flat space in a perverse coordinate system. If we calculate the Riemann tensor of such a metric and find that it vanishes, we know that the metric is flat; if it doesn't vanish, there is curvature.

3.7 ■ PROPERTIES OF THE RIEMANN TENSOR

The Riemann tensor, with four indices, naively has n^4 independent components in an n-dimensional space. In fact the antisymmetry property (3.110) means that there are only $n(n-1)/2$ independent values these last two indices can take on, leaving us with $n^3(n-1)/2$ independent components. When we consider the Christoffel connection, however, a number of other symmetries reduce the number of independent components further. Let's consider these now.

The simplest way to derive these additional symmetries is to examine the Riemann tensor with all lower indices,

$$R_{\rho\sigma\mu\nu} = g_{\rho\lambda} R^{\lambda}{}_{\sigma\mu\nu}. \tag{3.127}$$

Let us further consider the components of this tensor in locally inertial coordinates $x^{\hat{\mu}}$ established at a point p. Then the Christoffel symbols themselves will vanish, although their derivatives will not. We therefore have

$$\begin{aligned} R_{\hat{\rho}\hat{\sigma}\hat{\mu}\hat{\nu}}(p) &= g_{\hat{\rho}\hat{\lambda}}(\partial_{\hat{\mu}}\Gamma^{\hat{\lambda}}_{\hat{\nu}\hat{\sigma}} - \partial_{\hat{\nu}}\Gamma^{\hat{\lambda}}_{\hat{\mu}\hat{\sigma}}) \\ &= \tfrac{1}{2} g_{\hat{\rho}\hat{\lambda}} g^{\hat{\lambda}\hat{\tau}}(\partial_{\hat{\mu}}\partial_{\hat{\nu}} g_{\hat{\sigma}\hat{\tau}} + \partial_{\hat{\mu}}\partial_{\hat{\sigma}} g_{\hat{\tau}\hat{\nu}} - \partial_{\hat{\mu}}\partial_{\hat{\tau}} g_{\hat{\nu}\hat{\sigma}} - \partial_{\hat{\nu}}\partial_{\hat{\mu}} g_{\hat{\sigma}\hat{\tau}} \\ &\qquad - \partial_{\hat{\nu}}\partial_{\hat{\sigma}} g_{\hat{\tau}\hat{\mu}} + \partial_{\hat{\nu}}\partial_{\hat{\tau}} g_{\hat{\mu}\hat{\sigma}}) \\ &= \tfrac{1}{2}(\partial_{\hat{\mu}}\partial_{\hat{\sigma}} g_{\hat{\rho}\hat{\nu}} - \partial_{\hat{\mu}}\partial_{\hat{\rho}} g_{\hat{\nu}\hat{\sigma}} - \partial_{\hat{\nu}}\partial_{\hat{\sigma}} g_{\hat{\rho}\hat{\mu}} + \partial_{\hat{\nu}}\partial_{\hat{\rho}} g_{\hat{\mu}\hat{\sigma}}). \end{aligned} \tag{3.128}$$

In the first line we have used $\Gamma^{\hat{\tau}}_{\hat{\mu}\hat{\nu}}(p) = 0$, in the second line we have used $\partial_{\hat{\mu}} g^{\hat{\lambda}\hat{\tau}} = 0$ in Riemann normal coordinates, and the fact that partials commute in the third line. From this expression we can notice immediately three properties of $R_{\rho\sigma\mu\nu}$: it is antisymmetric in its first two indices,

$$\boxed{R_{\rho\sigma\mu\nu} = -R_{\sigma\rho\mu\nu},} \tag{3.129}$$

it is antisymmetric in its last two indices [which we already knew from (3.110)],

$$\boxed{R_{\rho\sigma\mu\nu} = -R_{\rho\sigma\nu\mu},} \tag{3.130}$$

and it is invariant under interchange of the first pair of indices with the second:

$$\boxed{R_{\rho\sigma\mu\nu} = R_{\mu\nu\rho\sigma}.} \tag{3.131}$$

With a little more work, which is left to your imagination, we can see that the sum of cyclic permutations of the last three indices vanishes:

$$\boxed{R_{\rho\sigma\mu\nu} + R_{\rho\mu\nu\sigma} + R_{\rho\nu\sigma\mu} = 0.} \tag{3.132}$$

Given (3.130), it's easy to see that this last property is equivalent to the vanishing of the antisymmetric part of the last three indices:

$$\boxed{R_{\rho[\sigma\mu\nu]} = 0.} \tag{3.133}$$

All of these properties have been derived in a special coordinate system, but they are all tensor equations; therefore they will be true in any coordinates (so we haven't bothered with hats on the indices). Not all of them are independent; with some effort, you can show that (3.129), (3.130), and (3.133) together imply (3.131). The logical interdependence of the equations is usually less important than the fact that they are true.

Given these relationships between the different components of the Riemann tensor, how many independent quantities remain? Let's begin with the facts that $R_{\rho\sigma\mu\nu}$ is antisymmetric in the first two indices, antisymmetric in the last two indices, and symmetric under interchange of these two pairs. This means that we can think of it as a symmetric matrix $R_{[\rho\sigma][\mu\nu]}$, where the pairs $\rho\sigma$ and $\mu\nu$ are thought of as individual indices. An $m \times m$ symmetric matrix has $m(m+1)/2$ independent components, while an $n \times n$ antisymmetric matrix has $n(n-1)/2$ independent components. We therefore have

$$\tfrac{1}{2}\left[\tfrac{1}{2}n(n-1)\right]\left[\tfrac{1}{2}n(n-1)+1\right] = \tfrac{1}{8}(n^4 - 2n^3 + 3n^2 - 2n) \tag{3.134}$$

independent components. We still have to deal with the additional symmetry (3.133). An immediate consequence of (3.133) is that the totally antisymmetric part of the Riemann tensor vanishes,

$$R_{[\rho\sigma\mu\nu]} = 0. \tag{3.135}$$

In fact, this equation plus the other symmetries (3.129), (3.130), and (3.131), are enough to imply (3.133), as can be easily shown by expanding (3.135) and manipulating the resulting terms. Therefore imposing the additional constraint of (3.135) is equivalent to imposing (3.133), once the other symmetries have been accounted for. How many independent restrictions does this represent? Let us imagine decomposing

$$R_{\rho\sigma\mu\nu} = X_{\rho\sigma\mu\nu} + R_{[\rho\sigma\mu\nu]}. \tag{3.136}$$

It is easy to see that any totally antisymmetric 4-index tensor is automatically antisymmetric in its first and last indices, and symmetric under interchange of the two pairs. Therefore these properties are independent restrictions on $X_{\rho\sigma\mu\nu}$, unrelated to the requirement (3.135). Now a totally antisymmetric 4-index tensor has $n(n-1)(n-2)(n-3)/4!$ terms, and therefore (3.135) reduces the number of independent components by this amount. We are left with

$$\tfrac{1}{8}(n^4 - 2n^3 + 3n^2 - 2n) - \tfrac{1}{24}n(n-1)(n-2)(n-3) = \tfrac{1}{12}n^2(n^2-1) \tag{3.137}$$

independent components of the Riemann tensor.

In four dimensions, therefore, the Riemann tensor has 20 independent components. (In one dimension it has none.) These twenty functions are precisely the 20 degrees of freedom in the second derivatives of the metric that we could not set to zero by a clever choice of coordinates when we first discussed locally inertial coordinates in Chapter 2. This should reinforce your confidence that the Riemann tensor is an appropriate measure of curvature.

In addition to the algebraic symmetries of the Riemann tensor (which constrain the number of independent components at any point), it obeys a differential identity, which constrains its relative values at different points. Consider the covariant derivative of the Riemann tensor, evaluated in locally inertial coordinates:

$$\begin{aligned}\nabla_{\hat\lambda} R_{\hat\rho\hat\sigma\hat\mu\hat\nu} &= \partial_{\hat\lambda} R_{\hat\rho\hat\sigma\hat\mu\hat\nu} \\ &= \tfrac{1}{2}\partial_{\hat\lambda}(\partial_{\hat\mu}\partial_{\hat\sigma} g_{\hat\rho\hat\nu} - \partial_{\hat\mu}\partial_{\hat\rho} g_{\hat\nu\hat\sigma} - \partial_{\hat\nu}\partial_{\hat\sigma} g_{\hat\rho\hat\mu} + \partial_{\hat\nu}\partial_{\hat\rho} g_{\hat\mu\hat\sigma}).\end{aligned} \tag{3.138}$$

It may seem illegitimate to take the derivative of an expression that is only true at a point, but the terms we are neglecting are all proportional to $\partial_{\hat\sigma} g_{\hat\mu\hat\nu}$, and therefore vanish. We would like to consider the sum of cyclic permutations of the first three indices:

$$\begin{aligned}&\nabla_{\hat\lambda} R_{\hat\rho\hat\sigma\hat\mu\hat\nu} + \nabla_{\hat\rho} R_{\hat\sigma\hat\lambda\hat\mu\hat\nu} + \nabla_{\hat\sigma} R_{\hat\lambda\hat\rho\hat\mu\hat\nu} \\ &\quad = \tfrac{1}{2}(\partial_{\hat\lambda}\partial_{\hat\mu}\partial_{\hat\sigma} g_{\hat\rho\hat\nu} - \partial_{\hat\lambda}\partial_{\hat\mu}\partial_{\hat\rho} g_{\hat\nu\hat\sigma} - \partial_{\hat\lambda}\partial_{\hat\nu}\partial_{\sigma} g_{\hat\rho\hat\mu} + \partial_{\hat\lambda}\partial_{\hat\nu}\partial_{\hat\rho} g_{\hat\mu\hat\sigma} \\ &\qquad + \partial_{\hat\rho}\partial_{\hat\mu}\partial_{\hat\lambda} g_{\hat\sigma\hat\nu} - \partial_{\hat\rho}\partial_{\hat\mu}\partial_{\hat\sigma} g_{\hat\nu\hat\lambda} - \partial_{\hat\rho}\partial_{\hat\nu}\partial_{\hat\lambda} g_{\hat\sigma\hat\mu} + \partial_{\hat\rho}\partial_{\hat\nu}\partial_{\hat\sigma} g_{\hat\mu\hat\lambda} \\ &\qquad + \partial_{\hat\sigma}\partial_{\hat\mu}\partial_{\hat\rho} g_{\hat\lambda\hat\nu} - \partial_{\hat\sigma}\partial_{\hat\mu}\partial_{\hat\lambda} g_{\hat\nu\hat\rho} - \partial_{\hat\sigma}\partial_{\hat\nu}\partial_{\hat\rho} g_{\hat\lambda\hat\mu} + \partial_{\hat\sigma}\partial_{\hat\nu}\partial_{\hat\lambda} g_{\hat\mu\hat\rho}) \\ &\quad = 0.\end{aligned} \tag{3.139}$$

Once again, since this is an equation between tensors it is true in any coordinate system, even though we derived it in a particular one. We recognize by now that the antisymmetry $R_{\rho\sigma\mu\nu} = -R_{\sigma\rho\mu\nu}$ allows us to write this result as

$$\nabla_{[\lambda} R_{\rho\sigma]\mu\nu} = 0. \tag{3.140}$$

This is known as the **Bianchi identity**. For a general connection there would be additional terms involving the torsion tensor. It is closely related to the Jacobi identity, since (recalling the definition of the Riemann tensor in terms of the commutator of covariant derivatives) it expresses

$$[[\nabla_\lambda, \nabla_\rho], \nabla_\sigma] + [[\nabla_\rho, \nabla_\sigma], \nabla_\lambda] + [[\nabla_\sigma, \nabla_\lambda], \nabla_\rho] = 0. \tag{3.141}$$

The Riemann tensor has four indices. At times it is useful to express a tensor as a sum of various pieces that are individually easier to handle and may have direct physical interpretations. The trick is to do this in a coordinate-invariant way. For example, we could decompose the Riemann tensor into $R^\rho{}_{\sigma ij}$ and $R^\rho{}_{\sigma i0}$, from which we could reconstruct the entire tensor (since $R^\rho{}_{\sigma 00}$ vanishes). But clearly this decomposition is not invariant under change of basis; we want to find a decomposition that is preserved when we change coordinates. What we are really doing is considering representations of the Lorentz group. We have two fundamental tricks at our disposal: taking contractions, and taking symmetric/antisymmetric parts. For example, given an arbitrary (0, 2) tensor $X_{\mu\nu}$, we can decompose it into its symmetric and antisymmetric pieces,

$$X_{\mu\nu} = X_{(\mu\nu)} + X_{[\mu\nu]}, \tag{3.142}$$

and the symmetric part can be further decomposed into its trace $X = g^{\mu\nu}X_{(\mu\nu)}$ and a trace-free part $\widehat{X}_{\mu\nu} = X_{(\mu\nu)} - \frac{1}{n}Xg_{\mu\nu}$, so that

$$X_{\mu\nu} = \frac{1}{n}Xg_{\mu\nu} + \widehat{X}_{\mu\nu} + X_{[\mu\nu]}. \tag{3.143}$$

(Note that $X_{[\mu\nu]}$ is automatically traceless.) When we change coordinates, the different pieces $Xg_{\mu\nu}$, $\widehat{X}_{\mu\nu}$, and $X_{[\mu\nu]}$ are rotated into themselves, not into each other; we say that they define "invariant subspaces" of the space of (0, 2) tensors. For more complicated tensors the equivalent decomposition might not be so simple.

For the Riemann tensor, our first step is to take a contraction to form the **Ricci tensor**:

$$\boxed{R_{\mu\nu} = R^\lambda{}_{\mu\lambda\nu}.} \tag{3.144}$$

For the curvature tensor formed from an arbitrary (not necessarily Christoffel) connection, there are a number of independent contractions to take. Our primary concern is with the Christoffel connection, for which (3.144) is the only independent contraction; all others either vanish, or are related to this one. The Ricci tensor associated with the Christoffel connection is automatically symmetric,

$$\boxed{R_{\mu\nu} = R_{\nu\mu},} \tag{3.145}$$

as a consequence of the symmetries of the Riemann tensor. The trace of the Ricci tensor is the **Ricci scalar** (or **curvature scalar**):

$$\boxed{R = R^{\mu}{}_{\mu} = g^{\mu\nu} R_{\mu\nu}.} \tag{3.146}$$

We could also form the trace-free part $\widehat{R}_{\mu\nu} = R_{\mu\nu} - \frac{1}{n} R g_{\mu\nu}$, but this turns out not to be especially useful; it is more common to express things in terms of $R_{\mu\nu}$ and R.

The Ricci tensor and scalar contain all of the information about traces of the Riemann tensor, leaving us the trace-free parts. These are captured by the **Weyl tensor**, which is basically the Riemann tensor with all of its contractions removed. It is given in n dimensions by

$$\begin{aligned} C_{\rho\sigma\mu\nu} = R_{\rho\sigma\mu\nu} &- \frac{2}{(n-2)} \left(g_{\rho[\mu} R_{\nu]\sigma} - g_{\sigma[\mu} R_{\nu]\rho} \right) \\ &+ \frac{2}{(n-1)(n-2)} g_{\rho[\mu} g_{\nu]\sigma} R \end{aligned} \tag{3.147}$$

This messy formula is designed so that all possible contractions of $C_{\rho\sigma\mu\nu}$ vanish, while it retains the symmetries of the Riemann tensor:

$$\begin{aligned} C_{\rho\sigma\mu\nu} &= C_{[\rho\sigma][\mu\nu]}, \\ C_{\rho\sigma\mu\nu} &= C_{\mu\nu\rho\sigma}, \\ C_{\rho[\sigma\mu\nu]} &= 0. \end{aligned} \tag{3.148}$$

The Weyl tensor is only defined in three or more dimensions, and in three dimensions it vanishes identically. One of the most important properties of the Weyl tensor is that it is invariant under conformal transformations (discussed in Appendix G). This means that if you compute $C^{\rho}{}_{\sigma\mu\nu}$ (note that the first index is upstairs) for some metric $g_{\mu\nu}$, and then compute it again for a metric given by $\omega^2(x) g_{\mu\nu}$, where $\omega(x)$ is an arbitrary nonvanishing function of spacetime, you get the same answer. For this reason it is often known as the *conformal tensor*.

An especially useful form of the Bianchi identity comes from contracting twice on (3.139):

$$\begin{aligned} 0 &= g^{\nu\sigma} g^{\mu\lambda} (\nabla_\lambda R_{\rho\sigma\mu\nu} + \nabla_\rho R_{\sigma\lambda\mu\nu} + \nabla_\sigma R_{\lambda\rho\mu\nu}) \\ &= \nabla^\mu R_{\rho\mu} - \nabla_\rho R + \nabla^\nu R_{\rho\nu}, \end{aligned} \tag{3.149}$$

or

$$\nabla^\mu R_{\rho\mu} = \tfrac{1}{2} \nabla_\rho R. \tag{3.150}$$

Notice that, unlike the partial derivative, it makes sense to raise an index on the covariant derivative, due to metric compatibility. We define the **Einstein tensor** as

$$G_{\mu\nu} = R_{\mu\nu} - \tfrac{1}{2}Rg_{\mu\nu}. \tag{3.151}$$

In four dimensions the Einstein tensor can be thought of as a trace-reversed version of the Ricci tensor. We then see that the twice-contracted Bianchi identity (3.150) is equivalent to

$$\nabla^{\mu}G_{\mu\nu} = 0. \tag{3.152}$$

The Einstein tensor, which is symmetric due to the symmetry of the Ricci tensor and the metric, will be of great importance in general relativity.

We should pause at this point to contrast the formalism we have developed with our intuitive notion of curvature. Our intuition is unfortunately contaminated by the fact that we are used to thinking about one- and two-dimensional spaces embedded in the (almost) Euclidean space in which we live. We think, for example, of a straight line as having no curvature, while a circle (S^1) is curved. However, according to (3.137), in one, two, three, and four dimensions there are 0, 1, 6 and 20 independent components of the Riemann tensor, respectively. (Everything we say about the curvature in these examples refers to the curvature associated with the Christoffel connection, and therefore the metric.) Therefore it is impossible for a one-dimensional space such as S^1 to have any curvature as we have defined it. The apparent contradiction stems from the fact that our intuitive notion of curvature depends on the extrinsic geometry of the manifold, which characterizes how a space is embedded in some larger space, while the Riemann curvature is a property of the intrinsic geometry of a space, which could be measured by observers confined to the manifold. Beings that lived on a circle and had no access to the larger world would necessarily think that they lived in a flat geometry—for example, there is no possibility of a nondegenerate infinitesimal loop around which we could parallel-transport a vector and have it come back rotated from its original position. Extrinsic curvature, discussed in Appendix D, is occasionally useful in GR when we wish to describe submanifolds of spacetime; but most often we are interested in the intrinsic geometry of spacetime itself, which does not rely on any embeddings.

We can illustrate the intrinsic/extrinsic difference further with an example from two dimensions, where the curvature has one independent component. In fact, all of the information about the curvature is contained in the single component of the Ricci scalar. Consider a torus, portrayed in Figure 3.7, which can be thought of as a square region of the plane with opposite sides identified (topologically, $S^1 \times S^1$). Although a torus embedded in three dimensions looks curved from our point of view, it should be clear that we can put a metric on the torus whose components are constant in an appropriate coordinate system—simply unroll it and use the Euclidean metric of the plane, $ds^2 = dx^2 + dy^2$. In this metric, the torus is flat.

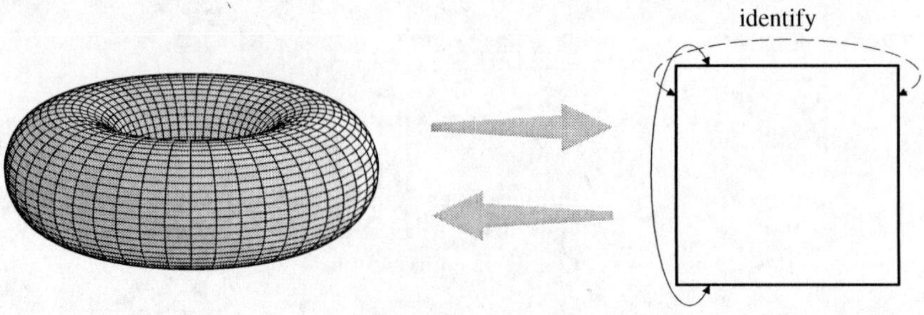

FIGURE 3.7 A torus thought of as a square in flat space with opposite sides identified.

There is also nothing to stop us from introducing a different metric in which the torus is not flat, but the point we are trying to emphasize is that it can be made flat in some metric. Every time we embed a manifold in a larger space, the manifold inherits an "induced metric" from the background in which it is embedded, as discussed in the Appendix A. Our point here is that a torus embedded in a flat three-dimensional Euclidean space will have an induced metric that is curved, but we can nevertheless choose to put a different metric on it so that the intrinsic geometry is flat.

Let's turn to a simple example where the curvature does not vanish. We have already talked about the two-sphere S^2, with metric

$$ds^2 = a^2(d\theta^2 + \sin^2\theta\, d\phi^2), \tag{3.153}$$

where a is the radius of the sphere. It will actually be the radius if our sphere is embedded in $\mathbf{R}^3$, but we can call it the radius even in the absence of any embedding. Two-dimensional people living on the sphere could calculate a by measuring the area of the sphere, dividing by 4π, and taking the square root; using the word "radius" to refer to this quantity is merely a convenience. We should also point out that the notion of a sphere is sometimes used in the weaker topological sense, without any particular metric being assumed; the metric we are using is called the *round metric*. Without going through the details, the nonzero connection coefficients for (3.153) are

$$\begin{aligned} \Gamma^\theta_{\phi\phi} &= -\sin\theta\cos\theta \\ \Gamma^\phi_{\theta\phi} = \Gamma^\phi_{\phi\theta} &= \cot\theta. \end{aligned} \tag{3.154}$$

Let's compute a promising component of the Riemann tensor:

$$\begin{aligned} R^\theta{}_{\phi\theta\phi} &= \partial_\theta\Gamma^\theta_{\phi\phi} - \partial_\phi\Gamma^\theta_{\theta\phi} + \Gamma^\theta_{\theta\lambda}\Gamma^\lambda_{\phi\phi} - \Gamma^\theta_{\phi\lambda}\Gamma^\lambda_{\theta\phi} \\ &= (\sin^2\theta - \cos^2\theta) - (0) + (0) - (-\sin\theta\cos\theta)(\cot\theta) \\ &= \sin^2\theta. \end{aligned} \tag{3.155}$$

The notation is obviously imperfect, since the Greek letter λ is a dummy index that is summed over, while the Greek letters θ and ϕ represent specific coordinates. Lowering an index, we have

$$\begin{aligned} R_{\theta\phi\theta\phi} &= g_{\theta\lambda} R^{\lambda}{}_{\phi\theta\phi} \\ &= g_{\theta\theta} R^{\theta}{}_{\phi\theta\phi} \\ &= a^2 \sin^2\theta. \end{aligned} \tag{3.156}$$

It is easy to check that all of the components of the Riemann tensor either vanish or are related to this one by symmetry. We can go on to compute the Ricci tensor via $R_{\mu\nu} = g^{\alpha\beta} R_{\alpha\mu\beta\nu}$. We obtain

$$\begin{aligned} R_{\theta\theta} &= g^{\phi\phi} R_{\phi\theta\phi\theta} = 1 \\ R_{\theta\phi} &= R_{\phi\theta} = 0 \\ R_{\phi\phi} &= g^{\theta\theta} R_{\theta\phi\theta\phi} = \sin^2\theta. \end{aligned} \tag{3.157}$$

The Ricci scalar is similarly straightforward:

$$R = g^{\theta\theta} R_{\theta\theta} + g^{\phi\phi} R_{\phi\phi} = \frac{2}{a^2}. \tag{3.158}$$

Therefore the Ricci scalar, which for a two-dimensional manifold completely characterizes the curvature, is a constant over the two-sphere. If we had perturbed the metric (corresponding physically to bumps on the sphere), this would no longer have been the case. Note that the scalar curvature decreases as the radius of the sphere increases. Even in more general contexts, we will sometimes refer to the "radius of curvature" of a manifold as providing a length scale over which the curvature varies; the larger the radius of curvature, the smaller the curvature itself.

3.8 ■ SYMMETRIES AND KILLING VECTORS

The real world is a messy place, and we have no hope of finding a metric that describes our actual universe, or even any small part thereof, with perfect precision. Instead, we model spacetime via various approximations appropriate to the physical situation being studied. For example, the geometry outside a star or planet may be approximated, to some order of precision, as being spherically symmetric, even if the real situation includes small deviations from this symmetry—these may be added in later as perturbations.

General relativity is no different from other fields of physics, then, in being especially interested in solutions with symmetry. In fact, such properties may be even more crucial in GR than in, say, electromagnetism, since the nonlinear nature of Einstein's equation (discussed in the next chapter) makes it hard to find any exact solutions at all. In the context of curved spacetime, however, we need to be more careful than usual about what exactly is meant by "symmetry." In this section we develop some useful tools for studying symmetry; a deeper investigation can be found in Appendix B.

We think of a manifold M as possessing a symmetry if the geometry is invariant under a certain transformation that maps M to itself; that is, if the metric is the same, in some sense, from one point to another. In fact different tensor fields may possess different symmetries; symmetries of the metric are called **isometries**. Sometimes the existence of isometries is obvious; consider, for example, four-dimensional Minkowski space,

$$ds^2 = \eta_{\mu\nu}dx^\mu dx^\nu = -dt^2 + dx^2 + dy^2 + dz^2. \tag{3.159}$$

We know of several isometries of this space; these include translations ($x^\mu \rightarrow x^\mu + a^\mu$, with a^μ fixed) and Lorentz transformations ($x^\mu \rightarrow \Lambda^\mu{}_\nu x^\nu$, with $\Lambda^\mu{}_\nu$ a Lorentz-transformation matrix). The fact that the metric is invariant under translations is made immediately apparent by the simple fact that the metric coefficients $\eta_{\mu\nu}$ are independent of the individual coordinate functions x^μ. Indeed, whenever $\partial_{\sigma_*} g_{\mu\nu} = 0$ for some fixed σ_* (but for all μ and ν), there will be a symmetry under translations along x^{σ_*}:

$$\partial_{\sigma_*} g_{\mu\nu} = 0 \quad \Rightarrow \quad x^{\sigma_*} \rightarrow x^{\sigma_*} + a^{\sigma_*} \text{ is a symmetry.} \tag{3.160}$$

The careful reader will have noticed that we still haven't precisely defined what we mean by symmetry; roughly we imagine that the metric is invariant under some transformation, but the precise meaning is only developed in Appendix B. Also, the implication arrow in (3.160) only goes one way, and it would be nice to have a clean criterion for deciding when a given transformation counts as a symmetry; this will come soon.

Isometries of the form (3.160) have immediate consequences for the motion of test particles as described by the geodesic equation. Recall from (3.61) that the geodesic equation can be written in terms of the four-momentum $p^\mu = mU^\mu$ (valid for timelike paths, at least) as

$$p^\lambda \nabla_\lambda p^\mu = 0. \tag{3.161}$$

By metric compatibility we are free to lower the index μ, and then we may expand the covariant derivative to obtain

$$p^\lambda \partial_\lambda p_\mu - \Gamma^\sigma_{\lambda\mu} p^\lambda p_\sigma = 0. \tag{3.162}$$

The first term tells us how the momentum components change along the path,

$$p^\lambda \partial_\lambda p_\mu = m\frac{dx^\lambda}{d\tau}\partial_\lambda p_\mu = m\frac{dp_\mu}{d\tau}, \tag{3.163}$$

while the second term is

$$\Gamma^{\sigma}_{\lambda\mu} p^{\lambda} p_{\sigma} = \tfrac{1}{2} g^{\sigma\nu} (\partial_{\lambda} g_{\mu\nu} + \partial_{\mu} g_{\nu\lambda} - \partial_{\nu} g_{\lambda\mu}) p^{\lambda} p_{\sigma} \tag{3.164}$$

$$= \tfrac{1}{2} (\partial_{\lambda} g_{\mu\nu} + \partial_{\mu} g_{\nu\lambda} - \partial_{\nu} g_{\lambda\mu}) p^{\lambda} p^{\nu} \tag{3.165}$$

$$= \tfrac{1}{2} (\partial_{\mu} g_{\nu\lambda}) p^{\lambda} p^{\nu}, \tag{3.166}$$

where we have used the symmetry of $p^{\lambda} p^{\nu}$ to go from the second line to the third. So, without yet making any assumptions about symmetry, we see that the geodesic equation can be written as

$$m \frac{dp_{\mu}}{d\tau} = \frac{1}{2} (\partial_{\mu} g_{\nu\lambda}) p^{\lambda} p^{\nu}. \tag{3.167}$$

Therefore, if all of the metric coefficients are independent of the coordinate x^{σ_*}, we find that this isometry implies that the momentum component p_{σ_*} is a conserved quantity of the motion:

$$\partial_{\sigma_*} g_{\mu\nu} = 0 \quad \Rightarrow \quad \frac{dp_{\sigma_*}}{d\tau} = 0. \tag{3.168}$$

This will hold along any geodesic, even though we only derived it for timelike ones. The conserved quantities implied by isometries are extremely useful in studying the motion of test particles in curved backgrounds.

Of course, even though independence of the metric components on one or more coordinates implies the existence of isometries, the converse does not necessarily hold. Symmetry under Lorentz transformations, for example, is not manifest as independence of $\eta_{\mu\nu}$ on any coordinates; indeed, in four dimensions, there are four types of translations and six types of Lorentz transformations, for a total of ten, which is obviously larger than the number of dimensions the metric could possibly be independent of. What is more, it would be simple enough to transform to a complicated coordinate system where not even the translational symmetries were obvious. Such a coordinate transformation would change the metric components, but not the underlying geometry, which is what the symmetry is really characterizing. Clearly a more systematic procedure is called for.

We can develop such a procedure by casting the right-hand equation of (3.168), expressing constancy of one of the components of the momentum, in a more manifestly covariant form. If x^{σ_*} is the coordinate which $g_{\mu\nu}$ is independent of, let us consider the vector ∂_{σ_*}, which we label as K:

$$K = \partial_{\sigma_*}, \tag{3.169}$$

which is equivalent in component notation to

$$K^{\mu} = (\partial_{\sigma_*})^{\mu} = \delta^{\mu}_{\sigma_*}. \tag{3.170}$$

We say that the vector K^{μ} generates the isometry; this means that the transformation under which the geometry is invariant is expressed infinitesimally as a motion in the direction of K^μ. Again, the notion is developed more fully in Appendix B. In terms of this vector, the noncovariant-looking quantity p_{σ_*} is simply

$$p_{\sigma_*} = K^\nu p_\nu = K_\nu p^\nu. \tag{3.171}$$

Meanwhile, the constancy of this (scalar) quantity along the path is equivalent to the statement that its directional derivative along the geodesic vanishes:

$$\frac{dp_{\sigma_*}}{d\tau} = 0 \quad \leftrightarrow \quad p^\mu \nabla_\mu (K_\nu p^\nu) = 0. \tag{3.172}$$

Expanding the expression on the right, we obtain

$$\begin{aligned} p^\mu \nabla_\mu (K_\nu p^\nu) &= p^\mu K_\nu \nabla_\mu p^\nu + p^\mu p^\nu \nabla_\mu K_\nu \\ &= p^\mu p^\nu \nabla_\mu K_\nu \\ &= p^\mu p^\nu \nabla_{(\mu} K_{\nu)}, \end{aligned} \tag{3.173}$$

where in the second line we have invoked the geodesic equation ($p^\mu \nabla_\mu p^\nu = 0$). In the third line we have used the fact that $p^\mu p^\nu$ is automatically symmetric in μ and ν, so only the symmetric part of $\nabla_\mu K_\nu$ could possibly contribute. We therefore conclude that any vector K_μ that satisfies $\nabla_{(\mu} K_{\nu)} = 0$ implies that $K_\nu p^\nu$ is conserved along a geodesic trajectory:

$$\boxed{\nabla_{(\mu} K_{\nu)} = 0 \quad \Rightarrow \quad p^\mu \nabla_\mu (K_\nu p^\nu) = 0.} \tag{3.174}$$

The equation on the left is known as **Killing's equation**, and vector fields that satisfy it are known as **Killing vector fields** (or simply Killing vectors). You can verify for yourself that, if the metric is independent of some coordinate x^{σ_*}, the vector ∂_{σ_*} will satisfy Killing's equation. In fact, if a vector K^μ satisfies Killing's equation, it will always be possible to find a coordinate system in which $K = \partial_{\sigma_*}$; but in general we cannot find coordinates in which all the Killing vectors are simultaneously of this form, nor is this form necessary for the vector to satisfy Killing's equation.

As we investigate in Appendix B, Killing vector fields on a manifold are in one-to-one correspondence with continuous symmetries of the metric on that manifold. Every Killing vector implies the existence of conserved quantities associated with geodesic motion. This can be understood physically: by definition the metric is unchanging along the direction of the Killing vector. Loosely speaking, therefore, a free particle will not feel any forces in this direction, and the component of its momentum in that direction will consequently be conserved. In fact, the same kind of logic by which we showed that $K_\nu p^\nu$ is conserved along a geodesic if $\nabla_{(\mu} K_{\nu)} = 0$ generalizes to additional indices: a **Killing tensor** is a symmetric l-index tensor $K_{\nu_1 \cdots \nu_l}$ that satisfies the obvious generalization of Killing's equation, and correspondingly leads to conserved quantities by contracting with l copies of the momentum:

$$\nabla_{(\mu} K_{\nu_1 \cdots \nu_l)} = 0 \quad \Rightarrow \quad p^\mu \nabla_\mu (K_{\nu_1 \cdots \nu_l} p^{\nu_1} \cdots p^{\nu_l}) = 0. \tag{3.175}$$

Simple examples of Killing tensors are the metric itself, and symmetrized tensor products of Killing vectors. Killing tensors are not related in a simple way to symmetries of the spacetime, but they will simplify our analysis of rotating black holes and expanding universes.

Derivatives of Killing vectors can be related to the Riemann tensor by

$$\nabla_\mu \nabla_\sigma K^\rho = R^\rho{}_{\sigma\mu\nu} K^\nu, \tag{3.176}$$

as you are asked to prove in the exercises. Contracting this expression yields

$$\nabla_\mu \nabla_\sigma K^\mu = R_{\sigma\nu} K^\nu. \tag{3.177}$$

These relations, along with the Bianchi identity and Killing's equation, suffice to show that the directional derivative of the Ricci scalar along a Killing vector field will vanish,

$$K^\lambda \nabla_\lambda R = 0. \tag{3.178}$$

This last fact is another reflection of the idea that the geometry is not changing along a Killing vector field.

Besides leading to conserved quantities for the motion of individual particles, the existence of a timelike Killing vector allows us to define a conserved energy for the entire spacetime. Given a Killing vector K_ν and a conserved energy-momentum tensor $T_{\mu\nu}$, we can construct a current

$$J_T^\mu = K_\nu T^{\mu\nu} \tag{3.179}$$

that is automatically conserved,

$$\begin{aligned} \nabla_\mu J_T^\mu &= (\nabla_\mu K_\nu) T^{\mu\nu} + K_\nu (\nabla_\mu T^{\mu\nu}) \\ &= 0. \end{aligned} \tag{3.180}$$

The first term vanishes by virtue of Killing's equation (since the symmetry of the upper indices serves to automatically symmetrize the lower indices), and the second term vanishes by conservation of $T_{\mu\nu}$. If K_ν is timelike, we can integrate over a spacelike hypersurface Σ to define the total energy,

$$E_T = \int_\Sigma J_T^\mu n_\mu \sqrt{\gamma}\, d^3x, \tag{3.181}$$

where γ_{ij} is the induced metric on Σ and n_μ is the normal vector to Σ. In Appendix E we discuss integration over hypersurfaces, and in particular Stokes's theorem; as explained there, E_T will be the same when integrated over any spacelike hypersurface, and is therefore conserved. This result fits nicely with our discussion in Section 3.5, where we found that the total energy is not typically conserved in an expanding universe; expansion means that the metric is changing with time, so there is no isometry in this direction. When there is a timelike Killing vector, we can write the metric in a form where it is independent of the timelike coordinate, and Noether's theorem implies a conserved energy. Similarly, spacelike Killing vectors may be used to construct conserved momenta (or angular momenta).

Although it may or may not be simple to actually solve Killing's equation in any given spacetime, it is frequently possible to write down some Killing vectors by inspection. (Of course a generic metric has no Killing vectors at all, but to keep things simple we often deal with metrics with high degrees of symmetry.) For example, in $\mathbf{R}^3$ with metric $ds^2 = dx^2 + dy^2 + dz^2$, independence of the metric components with respect to x, y, and z immediately yields three Killing vectors:

$$\begin{aligned} X^\mu &= (1, 0, 0) \\ Y^\mu &= (0, 1, 0) \\ Z^\mu &= (0, 0, 1). \end{aligned} \tag{3.182}$$

These clearly represent the three translations. There are also three rotational symmetries in $\mathbf{R}^3$, which are not quite as simple. To find them, imagine first going to polar coordinates,

$$\begin{aligned} x &= r \sin\theta \cos\phi \\ y &= r \sin\theta \sin\phi \\ z &= r \cos\theta, \end{aligned} \tag{3.183}$$

where the metric takes the form

$$ds^2 = dr^2 + r^2 d\theta^2 + r^2 \sin^2\theta \, d\phi^2. \tag{3.184}$$

Now the metric (the *same* metric, just in a different coordinate system) is manifestly independent of ϕ. We therefore know that $R = \partial_\phi$ is a Killing vector. Transforming back to Cartesian coordinates, this becomes

$$R = -y\partial_x + x\partial_y. \tag{3.185}$$

The Cartesian components R^μ are therefore $(-y, x, 0)$. Since this represents a rotation about the z-axis, it is straightforward to guess the components of all three rotational Killing vectors:

$$\begin{aligned} R^\mu &= (-y, \ x, \ 0) \\ S^\mu &= (\ z, \ 0, -x) \\ T^\mu &= (\ 0, -z, \ y), \end{aligned} \tag{3.186}$$

representing rotations about the z, y, and x-axes, respectively. You can check for yourself that these actually do solve Killing's equation. The overall signs don't matter, since minus a Killing vector is still a Killing vector.

This exercise leads directly to the Killing vectors for the two-sphere S^2 with metric

$$ds^2 = d\theta^2 + \sin^2\theta\, d\phi^2. \tag{3.187}$$

Since this sphere can be thought of as the locus of points at unit distance from the origin in $\mathbf{R}^3$, and the rotational Killing vectors all rotate such a sphere into itself, they also represent symmetries of S^2. To get explicit coordinate-basis representations for these vectors, we first transform the three-dimensional vectors (3.186) to polar coordinates $x^{\mu'} = (r, \theta, \phi)$. A straightforward calculation reveals

$$\begin{aligned} R &= \partial_\phi \\ S &= \cos\phi\, \partial_\theta - \cot\theta \sin\phi\, \partial_\phi \\ T &= -\sin\phi\, \partial_\theta - \cot\theta \cos\phi\, \partial_\phi . \end{aligned} \tag{3.188}$$

Notice that there are no components along ∂_r, which makes sense for a rotational isometry. Therefore the expressions (3.188) for the three rotational Killing vectors in $\mathbf{R}^3$ are exactly the same as those of S^2 in spherical polar coordinates.

In $n \geq 2$ dimensions, there can be more Killing vectors than dimensions. This is because a set of Killing vector fields can be linearly independent, even though at any one point on the manifold the vectors at that point are linearly dependent. It is trivial to show (so you should do it yourself) that a linear combination of Killing vectors with *constant* coefficients is still a Killing vector (in which case the linear combination does not count as an independent Killing vector), but this is not generally true with coefficients that vary over the manifold. You can also show that the commutator of two Killing vector fields is a Killing vector field; this is very useful to know, but it may be the case that the commutator gives you a vector field that is not linearly independent (or it may simply vanish). The problem of finding all of the Killing vectors of a metric is therefore somewhat tricky, as it is not always clear when to stop looking.

3.9 ■ MAXIMALLY SYMMETRIC SPACES

How symmetric can a space possibly be? An example of a space with the highest possible degree of symmetry is $\mathbf{R}^n$ with the flat Euclidean metric. Consider the isometries of this space, which we know to be translations and rotations in n dimensions, from the perspective of what they do in the neighborhood of some fixed point p. The translations are those transformations that move the point; there are n independent axes along which it can be moved, and hence n total translations. The rotations, centered at p, are those transformations that leave p invariant; they can be thought of as moving one of the axes through p into one of the others. There are n axes, and for each axis there are $n - 1$ other axes into which it can be rotated, but we shouldn't count a rotation of y into x as separate from a rotation of x into y, so the total number of independent rotations is $\frac{1}{2}n(n-1)$. We therefore have

$$n + \tfrac{1}{2}n(n-1) = \tfrac{1}{2}n(n+1) \tag{3.189}$$

independent symmetries of $\mathbf{R}^n$. But our counting argument only referred to the behavior of the symmetry in a neighborhood of p, not globally all over the manifold; so even in the presence of curvature the counting should be the same. If the metric signature is not Euclidean, some of the rotations will actually be boosts, but again the counting will be the same. The number of isometries is, of course, the number of linearly independent Killing vector fields. We therefore refer to an n-dimensional manifold with $\frac{1}{2}n(n+1)$ Killing vectors as a **maximally symmetric space**. The most familiar examples of maximally symmetric spaces are n-dimensional Euclidean spaces $\mathbf{R}^n$ and the n-dimensional spheres S^n. For an n-dimensional sphere we usually think of the isometries as consisting of $\frac{1}{2}n(n+1)$ independent rotations, rather than as some collection of both rotations and translations. However, if we consider the action of these rotations on some fixed point p, a moment's thought convinces us that the entire set can be decomposed into $\frac{1}{2}n(n-1)$ rotations around the point (keeping p fixed), and another n that move p along each direction, just as in $\mathbf{R}^n$.

If a manifold is maximally symmetric, the curvature is the same everywhere (as expressed by translation-like isometries) and the same in every direction (as expressed by rotation-like isometries). Hence, if we know the curvature of a maximally symmetric space at one point, we know it everywhere. Indeed, there are only a small number of possible maximally symmetric spaces; they are classified by the curvature scalar R (which will be constant everywhere), the dimensionality n, the metric signature, and perhaps some discrete pieces of information relating to the global topology (distinguishing, for example, an n-torus from $\mathbf{R}^n$, and tori of different sizes from each other). It follows that we should be able to reconstruct the entire Riemann tensor of such a space from the Ricci scalar R; let's see how this works.

The basic idea is simply that, since the geometry looks the same in all directions, the curvature tensor should look the same in all directions. What might this mean? First choose locally inertial coordinates at some point p, so that $g_{\hat{\mu}\hat{\nu}} = \eta_{\hat{\mu}\hat{\nu}}$. Of course, locally inertial coordinates are not unique; for example, we can perform a Lorentz transformation at p and the metric components will remain those of $\eta_{\hat{\mu}\hat{\nu}}$. (By "doing a Lorentz transformation" we really are referring to a change of basis vectors in T_p; in a curved spacetime, this only makes sense at a single point, not over a region.) Since the geometry is maximally symmetric, we want the same to be true of the Riemann tensor; that is, the components of $R_{\hat{\rho}\hat{\sigma}\hat{\mu}\hat{\nu}}$ should not change under a Lorentz transformation either, since there is no preferred direction in spacetime. But there are unique tensors that do not change their components under Lorentz transformations—the metric, the Kronecker delta, and the Levi–Civita tensor. This means that, in these coordinates and at this point, the components of $R_{\hat{\rho}\hat{\sigma}\hat{\mu}\hat{\nu}}$ will be proportional to a tensor constructed from these invariant tensors. Attempting to match the symmetries of the Riemann tensor reveals that there is a unique possibility:

$$R_{\hat{\rho}\hat{\sigma}\hat{\mu}\hat{\nu}} \propto g_{\hat{\rho}\hat{\mu}}g_{\hat{\sigma}\hat{\nu}} - g_{\hat{\rho}\hat{\nu}}g_{\hat{\sigma}\hat{\mu}}. \tag{3.190}$$

But this is a completely tensorial relation, so it must be true in any coordinate system. We have argued in favor of this relation at a single point p, but in a maximally symmetric space all points are created equal, so it must also be true at any other point as well. The proportionality constant is easily fixed by contracting both sides twice [the left-hand side becomes R, and the right-hand side is $n(n-1)$]. We end up with an equation true in any maximally symmetric space, at any point, in any coordinate system:

$$R_{\rho\sigma\mu\nu} = \frac{R}{n(n-1)}(g_{\rho\mu}g_{\sigma\nu} - g_{\rho\nu}g_{\sigma\mu}). \tag{3.191}$$

Likewise, if the Riemann tensor satisfies this condition (with R a constant over the manifold), the metric will be maximally symmetric. In two dimensions, finding that R is a constant is sufficient to prove that a space is maximally symmetric, since there is only one independent component of the curvature. In higher dimensions you have to work harder.

Locally, then (ignoring questions of global topology), a maximally symmetric space of given dimension and signature is fully specified by R. The basic classification of such spaces is simply whether R is positive, zero, or negative, since the magnitude of R represents an overall scaling of the size of the space. For Euclidean signatures, the flat maximally symmetric spaces are planes or appropriate higher-dimensional generalizations, while the positively curved ones are spheres. Maximally symmetric Euclidean spaces of negative curvature are hyperboloids, denoted H^n. These are less familiar because even a two-dimensional hyperboloid cannot be isometrically embedded in $\mathbf{R}^3$. Let's examine this two-dimensional hyperboloid briefly.

There are a number of ways of representing H^2, which has the same topology as $\mathbf{R}^2$. One simple way is as the **Poincaré half-plane**, which is the region $y > 0$ of a two-dimensional region with coordinates $\{x, y\}$ and metric

$$ds^2 = \frac{a^2}{y^2}(dx^2 + dy^2). \tag{3.192}$$

The geometry of the Poincaré half-plane is of course different from that of the upper half of $\mathbf{R}^2$, despite the use of similar coordinates. For example, we can compute the length of a line segment stretching vertically ($x =$ constant) from y_1 to y_2:

$$\begin{aligned}\Delta s &= \int_{y_1}^{y_2} \sqrt{g_{\mu\nu}\frac{dx^\mu}{dy}\frac{dx^\nu}{dy}}\,dy \\ &= a\int_{y_1}^{y_2}\frac{dy}{y} \\ &= a\ln\left(\frac{y_2}{y_1}\right). \end{aligned} \tag{3.193}$$

This is not at all the result $\Delta s = y_2 - y_1$ we would expect in Euclidean space. In particular, notice that the path length becomes infinite for paths that approach the boundary $y = 0$. In other words, it's not really a boundary at all; it's infinitely far away, as far as anyone living on the hyperboloid is concerned.

The nonvanishing Christoffel symbols for (3.192) are

$$\begin{aligned}\Gamma^x_{xy} = \Gamma^x_{yx} &= -y^{-1} \\ \Gamma^y_{xx} &= y^{-1} \\ \Gamma^y_{yy} &= -y^{-1}. \end{aligned} \tag{3.194}$$

From these it is straightforward to show that geodesics satisfy

$$(x - x_0)^2 + y^2 = l^2, \tag{3.195}$$

for some constants x_0 and l. Curves of this form are semicircles with centers located on the x-axis, as shown in Figure 3.8. In the limit as $x_0 \to \infty$ and $l \to \infty$ with $l - x_0$ fixed, we get a straight vertical line. Following our discussion of S^2 at the end of Section 3.7, we calculate a representative component of the Riemann tensor to be

$$R^x{}_{yxy} = -y^{-2}. \tag{3.196}$$

As with the two-sphere, all other components are either vanishing or related to this by symmetries. This is simply a reflection of the fact that we are in two di-

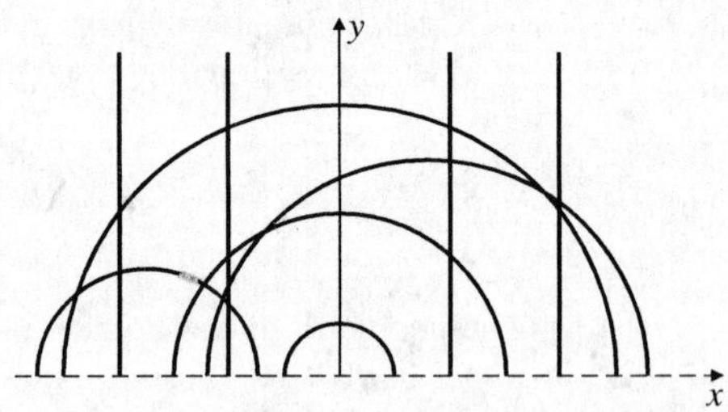

FIGURE 3.8 The upper half plane with a negatively curved metric. Geodesics are semicircles and straight lines that intersect the x-axis vertically.

mensions, with only one independent component of curvature. Turning the crank yields the Ricci tensor,

$$\begin{aligned} R_{xx} &= -y^{-2} \\ R_{xy} &= 0 \\ R_{yy} &= -y^{-2}, \end{aligned} \tag{3.197}$$

and the curvature scalar,

$$R = -\frac{2}{a^2}. \tag{3.198}$$

We see that it matches that of S^2 with the opposite sign, and in particular that it is a constant. Since we are in two dimensions, this is enough to ensure that our metric really is maximally symmetric. Of course there are coordinates in which H^2 looks very different; one is introduced in the Exercises.

Locally, then, a maximally symmetric space of Euclidean signature is either a plane, a sphere, or a hyperboloid, depending on the sign of R. Globally, any maximally symmetric space (of Euclidean signature) can be constructed by taking a carefully chosen region of one of these three spaces and identifying different sides, as the flat torus can be constructed from $\mathbf{R}^2$. As an aside, let's briefly mention a connection between local geometry and global topology, encompassed by the Gauss–Bonnet theorem. For a two-dimensional compact boundaryless orientable manifold, this reads

$$\chi(M) = \frac{1}{4\pi}\int_M R\sqrt{|g|}\, d^n x, \tag{3.199}$$

where $\chi(M)$ is a topological invariant of the space, known as the Euler characteristic. In general it can be calculated from the cohomology spaces mentioned in Chapter 2; in two dimensions, however, it is simply given by

$$\chi(M) = 2(1-g), \tag{3.200}$$

where g is the genus of the surface (zero for a sphere, and equal to the number of handles of a torus or Riemann surface). The Gauss–Bonnet theorem holds whether or not the curvature R is a constant; when it is, however, we see that all Riemann surfaces of genus $g \geq 2$ must have negative curvature, just as a sphere must be positively curved and a torus must be flat.

Continuing our aside, think for the moment about string theory, which claims that the fundamental objects comprising the universe are small one-dimensional loops of string. Such strings have two-dimensional "world-sheets" rather than one-dimensional worldlines. Doing perturbation theory in string theory (the equivalent of calculating Feynman diagrams in quantum field theory) involves summing over all world-sheet geometries (generally, for technical reasons, Euclidean geometries). This sounds like a lot of geometries, but in two dimensions any metric can be written as some fiducial metric times a conformal factor. This should be plausible, since there is only one curvature component; you are asked to prove it in the Exercises. The fiducial metric can be chosen differently for each world-sheet topology, and we can make our lives easier by choosing it to be (locally) a metric of maximal symmetry—the round sphere for genus zero, the plane for genus one, and the hyperboloid for higher genera. Even more fortunately, the string theories of greatest physical interest are the so-called critical string theories, for which the conformal factor itself doesn't matter. This is one of the things that makes doing calculations in perturbative string theory possible; we only have to sum over a discrete set of topologies, with a finite number of modular parameters for each topology (such as the parameters telling us the sizes of the different directions in a torus).

We close this section with one last point. We have explored the maximally symmetric spaces of Euclidean signatures; there are, of course, corresponding spacetimes with Lorentzian signatures. We know that the maximally symmetric spacetime with $R = 0$ is simply Minkowski space. The positively curved maximally symmetric spacetime is called de Sitter space, while that with negative curvature is imaginatively labeled anti-de Sitter space. These spacetimes will be more thoroughly discussed in Chapter 8.

It should be clear by now that the Appendices flesh out these ideas in important ways. Impatient readers may skip over them, but it would be a shame to do so.

3.10 ■ GEODESIC DEVIATION

The Riemann tensor shows up as a consequence of curvature in one more way: geodesic deviation. You have undoubtedly heard that the defining property of Euclidean (flat) geometry is the parallel postulate: initially parallel lines remain parallel forever. Of course in a curved space this is not true; on a sphere, certainly, initially parallel geodesics will eventually cross. We would like to quantify this behavior for an arbitrary curved space.

The problem is that the notion of "parallel" does not extend naturally from flat to curved spaces. The best we can do is to consider geodesic curves that might be initially parallel, and see how they behave as we travel down the geodesics. To this end we consider a one-parameter family of geodesics, $\gamma_s(t)$. That is, for each $s \in \mathbf{R}$, γ_s is a geodesic parameterized by the affine parameter t. The collection of these curves defines a smooth two-dimensional surface (embedded in a manifold M of arbitrary dimensionality). The coordinates on this surface may be chosen to be s and t, provided we have chosen a family of geodesics that do not cross. The entire surface is the set of points $x^\mu(s,t) \in M$. We have two natural vector fields: the tangent vectors to the geodesics,

$$T^\mu = \frac{\partial x^\mu}{\partial t}, \tag{3.201}$$

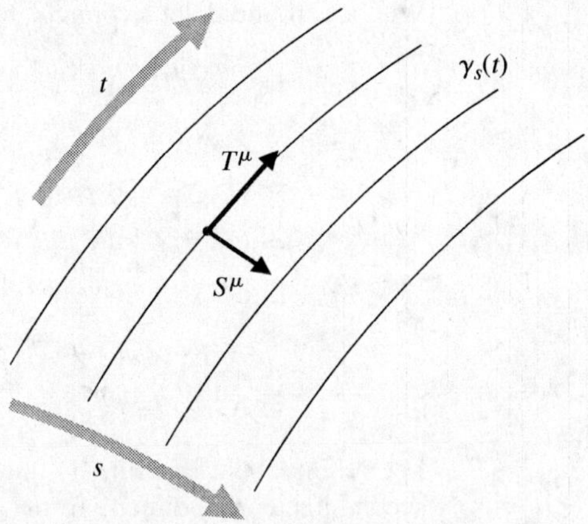

FIGURE 3.9 A set of geodesics $\gamma_s(t)$, with tangent vectors T^μ. The vector field S^μ measures the deviation between nearby geodesics.

and the deviation vectors

$$S^\mu = \frac{\partial x^\mu}{\partial s}. \tag{3.202}$$

This name derives from the informal notion that S^μ points from one geodesic toward the neighboring ones.

The idea that S^μ points from one geodesic to the next inspires us to define the "relative velocity of geodesics,"

$$V^\mu = (\nabla_T S)^\mu = T^\rho \nabla_\rho S^\mu, \tag{3.203}$$

and the "relative acceleration of geodesics,"

$$A^\mu = (\nabla_T V)^\mu = T^\rho \nabla_\rho V^\mu. \tag{3.204}$$

You should take the names with a grain of salt, but these vectors are certainly well-defined. This notion of *relative* acceleration between geodesics should be distinguished from the acceleration of a path away from being a geodesic, which would be given (when t is the proper time) by $a^\mu = T^\sigma \nabla_\sigma T^\mu$.

Since S and T are basis vectors adapted to a coordinate system, their commutator vanishes:

$$[S, T] = 0. \tag{3.205}$$

From (3.37) we then have

$$S^\rho \nabla_\rho T^\mu = T^\rho \nabla_\rho S^\mu. \tag{3.206}$$

With this in mind, let's compute the acceleration:

$$
\begin{aligned}
A^\mu &= T^\rho\nabla_\rho(T^\sigma\nabla_\sigma S^\mu) \\
&= T^\rho\nabla_\rho(S^\sigma\nabla_\sigma T^\mu) \\
&= (T^\rho\nabla_\rho S^\sigma)(\nabla_\sigma T^\mu) + T^\rho S^\sigma\nabla_\rho\nabla_\sigma T^\mu \\
&= (S^\rho\nabla_\rho T^\sigma)(\nabla_\sigma T^\mu) + T^\rho S^\sigma(\nabla_\sigma\nabla_\rho T^\mu + R^\mu{}_{\nu\rho\sigma}T^\nu) \\
&= (S^\rho\nabla_\rho T^\sigma)(\nabla_\sigma T^\mu) + S^\sigma\nabla_\sigma(T^\rho\nabla_\rho T^\mu) - (S^\sigma\nabla_\sigma T^\rho)\nabla_\rho T^\mu \\
&\quad + R^\mu{}_{\nu\rho\sigma}T^\nu T^\rho S^\sigma \\
&= R^\mu{}_{\nu\rho\sigma}T^\nu T^\rho S^\sigma .
\end{aligned}
\tag{3.207}
$$

Let's think about this line by line. The first line is the definition of A^μ, and the second line comes directly from (3.206). The third line is simply the Leibniz rule. The fourth line replaces a double covariant derivative by the derivatives in the opposite order plus the Riemann tensor. In the fifth line we use Leibniz again (in the opposite order from usual), and then we cancel two identical terms and notice that the term involving $T^\rho\nabla_\rho T^\mu$ vanishes because T^μ is the tangent vector to a geodesic. The result,

$$
\boxed{A^\mu = \frac{D^2}{dt^2}S^\mu = R^\mu{}_{\nu\rho\sigma}T^\nu T^\rho S^\sigma ,}
\tag{3.208}
$$

is the **geodesic deviation equation**. It expresses something that we might have expected: the relative acceleration between two neighboring geodesics is proportional to the curvature.

The geodesic deviation equation characterizes the behavior of a one-parameter family of neighboring geodesics. We will sometimes be interested in keeping track of the behavior of a multi-dimensional set of neighboring geodesics, perhaps representing a bundle of photons or a distribution of massive test particles. Such a set of geodesics forms a congruence; in Appendix F we derive equations that describe the evolution of such congruences.

Physically, of course, the acceleration of neighboring geodesics is interpreted as a manifestation of gravitational tidal forces. In the next chapter we explore in more detail how properties of curved spacetime are reflected by physics in a gravitational field.

3.11 ■ EXERCISES

1. Verify these consequences of metric compatibility ($\nabla_\sigma g_{\mu\nu} = 0$):

$$
\begin{aligned}
\nabla_\sigma g^{\mu\nu} &= 0 \\
\nabla_\lambda \epsilon_{\mu\nu\rho\sigma} &= 0.
\end{aligned}
\tag{3.209}
$$

2. You are familiar with the operations of gradient ($\nabla\phi$), divergence ($\nabla \cdot \mathbf{V}$) and curl ($\nabla \times \mathbf{V}$) in ordinary vector analysis in three-dimensional Euclidean space. Using covariant derivatives, derive formulae for these operations in spherical polar coordinates $\{r, \theta, \phi\}$ defined by

$$x = r \sin\theta \cos\phi \tag{3.210}$$

$$y = r \sin\theta \sin\phi \tag{3.211}$$

$$z = r \cos\theta. \tag{3.212}$$

Compare your results to those in Jackson (1999) or an equivalent text. Are they identical? Should they be?

3. Imagine we have a *diagonal* metric $g_{\mu\nu}$. Show that the Christoffel symbols are given by

$$\Gamma^{\lambda}_{\mu\nu} = 0 \tag{3.213}$$

$$\Gamma^{\lambda}_{\mu\mu} = -\tfrac{1}{2}(g_{\lambda\lambda})^{-1}\partial_\lambda g_{\mu\mu} \tag{3.214}$$

$$\Gamma^{\lambda}_{\mu\lambda} = \partial_\mu \left(\ln\sqrt{|g_{\lambda\lambda}|}\right) \tag{3.215}$$

$$\Gamma^{\lambda}_{\lambda\lambda} = \partial_\lambda \left(\ln\sqrt{|g_{\lambda\lambda}|}\right) \tag{3.216}$$

In these expressions, $\mu \neq \nu \neq \lambda$, and repeated indices are *not* summed over.

4. In Euclidean three-space, we can define paraboloidal coordinates (u, v, ϕ) via

$$x = uv\cos\phi \quad y = uv\sin\phi \quad z = \tfrac{1}{2}(u^2 - v^2).$$

(a) Find the coordinate transformation matrix between paraboloidal and Cartesian coordinates $\partial x^\alpha/\partial x^{\beta'}$ and the inverse transformation. Are there any singular points in the map?

(b) Find the basis vectors and basis one-forms in terms of Cartesian basis vectors and forms.

(c) Find the metric and inverse metric in paraboloidal coordinates.

(d) Calculate the Christoffel symbols.

(e) Calculate the divergence $\nabla_\mu V^\mu$ and Laplacian $\nabla_\mu \nabla^\mu f$.

5. Consider a 2-sphere with coordinates (θ, ϕ) and metric

$$ds^2 = d\theta^2 + \sin^2\theta\, d\phi^2. \tag{3.217}$$

(a) Show that lines of constant longitude (ϕ = constant) are geodesics, and that the only line of constant latitude (θ = constant) that is a geodesic is the equator ($\theta = \pi/2$).

(b) Take a vector with components $V^\mu = (1, 0)$ and parallel-transport it once around a circle of constant latitude. What are the components of the resulting vector, as a function of θ?

6. A good approximation to the metric outside the surface of the Earth is provided by

$$ds^2 = -(1 + 2\Phi)dt^2 + (1 - 2\Phi)dr^2 + r^2(d\theta^2 + \sin^2\theta\, d\phi^2), \tag{3.218}$$

where

$$\Phi = -\frac{GM}{r} \tag{3.219}$$

may be thought of as the familiar Newtonian gravitational potential. Here G is Newton's constant and M is the mass of the earth. For this problem Φ may be assumed to be small.

(a) Imagine a clock on the surface of the Earth at distance R_1 from the Earth's center, and another clock on a tall building at distance R_2 from the Earth's center. Calculate the time elapsed on each clock as a function of the coordinate time t. Which clock moves faster?

(b) Solve for a geodesic corresponding to a circular orbit around the equator of the Earth ($\theta = \pi/2$). What is $d\phi/dt$?

(c) How much proper time elapses while a satellite at radius R_1 (skimming along the surface of the earth, neglecting air resistance) completes one orbit? You can work to first order in Φ if you like. Plug in the actual numbers for the radius of the Earth and so on (don't forget to restore the speed of light) to get an answer in seconds. How does this number compare to the proper time elapsed on the clock stationary on the surface?

7. For this problem you will use the parallel propagator introduced in Appendix I to see how the Riemann tensor arises from parallel transport around an infinitesimal loop. Consider the following loop:

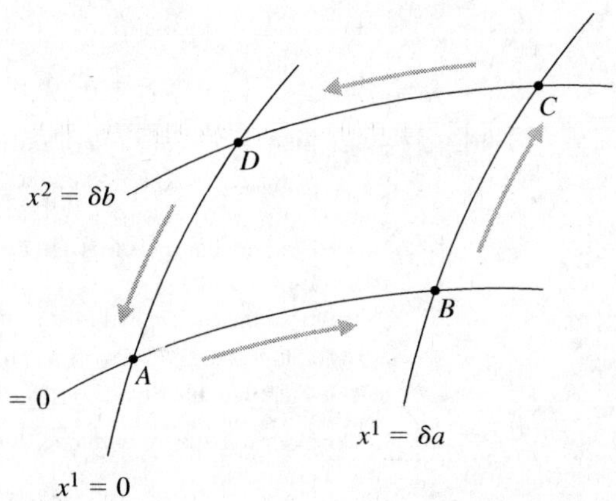

Using the infinite series expression for the parallel propagator, compute to lowest nontrivial order in δa and δb the transformation induced on a vector that is parallel transported around this loop from A to B to C to D and back to A, and show it is proportional to the appropriate components of the Riemann tensor. To make things easy, you can use x^1 and x^2 as parameters on the appropriate legs of the journey.

8. The metric for the three-sphere in coordinates $x^\mu = (\psi, \theta, \phi)$ can be written

$$ds^2 = d\psi^2 + \sin^2\psi\,(d\theta^2 + \sin^2\theta\, d\phi^2). \tag{3.220}$$

(a) Calculate the Christoffel connection coefficients. Use whatever method you like, but it is good practice to get the connection coefficients by varying the integral (3.49).

(b) Calculate the Riemann tensor, Ricci tensor, and Ricci scalar.

(c) Show that (3.191) is obeyed by this metric, confirming that the three-sphere is a maximally symmetric space (as you would expect).

9. Show that the Weyl tensor $C^{\mu}{}_{\nu\rho\sigma}$ is left invariant by a conformal transformation.

10. Show that, for $n \geq 4$, the Weyl tensor satisfies a version of the Bianchi identity,

$$\nabla_\rho C^\rho{}_{\sigma\mu\nu} = 2\frac{(n-3)}{(n-2)}\left(\nabla_{[\mu} R_{\nu]\sigma} + \frac{1}{2(n-1)} g_{\sigma[\mu}\nabla_{\nu]} R\right). \tag{3.221}$$

11. Since the Poincare half-plane with metric (3.192) is maximally symmetric, we might expect that it is rotationally symmetric around any point, although this certainly isn't evident in the $\{x, y\}$ coordinates. If that is so, it should be possible to put the metric in a form where the rotational symmetry is manifest, such as

$$ds^2 = f^2(r)[dr^2 + r^2 d\theta^2]. \tag{3.222}$$

To show that this works, calculate the curvature scalar for this metric and solve for the function $f(r)$ subject to the condition $R = -2/a^2$ everywhere. What is the range of the coordinate r?

12. Show that any Killing vector K^μ satisfies the relations mentioned in the text:

$$\nabla_\mu \nabla_\sigma K^\rho = R^\rho{}_{\sigma\mu\nu} K^\nu$$
$$K^\lambda \nabla_\lambda R = 0. \tag{3.223}$$

13. Find explicit expressions for a complete set of Killing vector fields for the following spaces:

(a) Minkowski space, with metric $ds^2 = -dt^2 + dx^2 + dy^2 + dz^2$.

(b) A spacetime with coordinates $\{u, v, x, y\}$ and metric

$$ds^2 = -(du dv + dv du) + a^2(u)dx^2 + b^2(u)dy^2, \tag{3.224}$$

where a and b are unspecified functions of u. This represents a gravitational wave spacetime. (*Hints*, which you need not show: there are five Killing vectors in all, and all of them have a vanishing u component K^u.)

Be careful, in all of these cases, about the distinction between upper and lower indices.

14. Consider the three Killing vectors of the the two-sphere, (3.188). Show that their commutators satisfy the following algebra:

$$[R, S] = T$$
$$[S, T] = R$$
$$[T, R] = S. \tag{3.225}$$

15. Use Raychaudhuri's equation, discussed in Appendix F, to show that, if a fluid is flowing on geodesics through spacetime with zero shear and expansion, then spacetime must have a timelike Killing vector.

16. Consider again the metric on a three-sphere,

$$ds^2 = d\psi^2 + \sin^2\psi\,(d\theta^2 + \sin^2\theta\, d\phi^2). \tag{3.226}$$

In this problem we make use of noncoordinate bases, discussed in Appendix J. In an orthonormal frame of one-forms $\hat{\theta}^{(a)}$ the metric would become

$$ds^2 = \hat{\theta}^{(1)} \otimes \hat{\theta}^{(1)} + \hat{\theta}^{(2)} \otimes \hat{\theta}^{(2)} + \hat{\theta}^{(3)} \otimes \hat{\theta}^{(3)}. \tag{3.227}$$

(a) Find such an orthonormal frame of one-forms, such that the matrix e^μ_a is diagonal. Don't worry about covering the entire manifold.

(b) Compute the components of the spin connection by solving $de^a + \omega^a{}_b \wedge e^b = 0$.

(c) Compute the components of the Riemann tensor $R^\rho{}_{\sigma\mu\nu}$ in the coordinate basis adapted to x^μ by computing the components of the curvature two-form $R^a{}_{b\mu\nu}$ and then converting.

CHAPTER

4 Gravitation

4.1 ■ PHYSICS IN CURVED SPACETIME

Having paid our mathematical dues, we are now prepared to examine the physics of gravitation as described by general relativity. This subject falls naturally into two pieces: how the gravitational field influences the behavior of matter, and how matter determines the gravitational field. In Newtonian gravity, these two elements consist of the expression for the acceleration of a body in a gravitational potential Φ,

$$\mathbf{a} = -\nabla\Phi, \tag{4.1}$$

and Poisson's differential equation for the potential in terms of the matter density ρ and Newton's gravitational constant G:

$$\nabla^2\Phi = 4\pi G\rho. \tag{4.2}$$

In general relativity, the analogous statements will describe how the curvature of spacetime acts on matter to manifest itself as gravity, and how energy and momentum influence spacetime to create curvature. In either case it would be legitimate to start at the top, by stating outright the laws governing physics in curved spacetime and working out their consequences. Instead, we will try to be a little more motivational, starting with basic physical principles and attempting to argue that these lead naturally to an almost unique physical theory.

In Chapter 2 we motivated our discussion of manifolds by introducing the Einstein Equivalence Principle, or EEP: "In small enough regions of spacetime, the laws of physics reduce to those of special relativity; it is impossible to detect the existence of a gravitational field by means of local experiments." The EEP arises from the idea that gravity is *universal*; it affects all particles (and indeed all forms of energy-momentum) in the same way. This feature of universality led Einstein to propose that what we experience as gravity is a manifestation of the curvature of spacetime. The idea is simply that something so universal as gravitation could be most easily described as a fundamental feature of the background on which matter fields propagate, as opposed to as a conventional force. At the same time, the identification of spacetime as a curved manifold is supported by the similarity between the undetectability of gravity in local regions and our ability to find locally inertial coordinates ($g_{\hat{\mu}\hat{\nu}} = \eta_{\hat{\mu}\hat{\nu}}$, $\partial_{\hat{\rho}} g_{\hat{\mu}\hat{\nu}} = 0$ at a point p) on a manifold.

Best of all, this abstract philosophizing translates directly into a simple recipe for generalizing laws of physics to the curved-spacetime context, known as the **minimal-coupling principle**. In its baldest form, this recipe may be stated as follows:

1. Take a law of physics, valid in inertial coordinates in flat spacetime.
2. Write it in a coordinate-invariant (tensorial) form.
3. Assert that the resulting law remains true in curved spacetime.

It may seem somewhat melodramatic to take such a simple idea and spread it out into a three-part procedure. We hope only to make clear that there is nothing very complicated going on. Operationally, this recipe usually amounts to taking an agreed-upon law in flat space and replacing the Minkowski metric $\eta_{\mu\nu}$ by the more general metric $g_{\mu\nu}$, and replacing partial derivatives ∂_μ by covariant derivatives ∇_μ. For this reason, this recipe is sometimes known as the "Comma-Goes-to-Semicolon Rule," by those who use commas and semicolons to denote partial and covariant derivatives.

As a straightforward example, we can consider the motion of freely-falling (unaccelerated) particles. In flat space such particles move in straight lines; in equations, this is expressed as the vanishing of the second derivative of the parameterized path $x^\mu(\lambda)$:

$$\frac{d^2x^\mu}{d\lambda^2} = 0. \tag{4.3}$$

This is not, in general coordinates, a tensorial equation; although $dx^\mu/d\lambda$ are the components of a well-defined vector, the second derivative components $d^2x^\mu/d\lambda^2$ are not. You might really think that this is a tensorial-looking equation; however, you can readily check that it's not even true in polar coordinates, unless you expect free particles to move in circles. We can use the chain rule to write

$$\frac{d^2x^\mu}{d\lambda^2} = \frac{dx^\nu}{d\lambda}\partial_\nu\frac{dx^\mu}{d\lambda}. \tag{4.4}$$

Now it is clear how to generalize this to curved space—simply replace the partial derivative by a covariant one,

$$\frac{dx^\nu}{d\lambda}\partial_\nu\frac{dx^\mu}{d\lambda} \quad\rightarrow\quad \frac{dx^\nu}{d\lambda}\nabla_\nu\frac{dx^\mu}{d\lambda} = \frac{d^2x^\mu}{d\lambda^2} + \Gamma^\mu_{\rho\sigma}\frac{dx^\rho}{d\lambda}\frac{dx^\sigma}{d\lambda}. \tag{4.5}$$

We recognize, then, that the appropriate general-relativistic version of the Newtonian relation (4.3) is simply the geodesic equation,

$$\frac{d^2x^\mu}{d\lambda^2} + \Gamma^\mu_{\rho\sigma}\frac{dx^\rho}{d\lambda}\frac{dx^\sigma}{d\lambda} = 0. \tag{4.6}$$

In general relativity, therefore, free particles move along geodesics; we have mentioned this before, but now you have a slightly better idea why it is true.

As an even more straightforward example, and one that we have referred to already, we have the law of energy-momentum conservation in flat spacetime:

$$\partial_\mu T^{\mu\nu} = 0. \tag{4.7}$$

Plugging into our recipe reveals the appropriate generalization to curved spacetime:

$$\nabla_\mu T^{\mu\nu} = 0. \tag{4.8}$$

It really is just that simple—sufficiently so that we felt quite comfortable using this equation in Chapter 3, without any detailed justification. Of course, this simplicity should not detract from the profound consequences of the generalization to curved spacetime, as illustrated in the example of the expanding universe.

It is one thing to generalize an equation from flat to curved spacetime; it is something altogether different to argue that the result describes gravity. To do so, we can show how the usual results of Newtonian gravity fit into the picture. We define the Newtonian limit by three requirements: the particles are moving slowly (with respect to the speed of light), the gravitational field is weak (so that it can be considered as a perturbation of flat space), and the field is also static (unchanging with time). Let us see what these assumptions do to the geodesic equation, taking the proper time τ as an affine parameter. "Moving slowly" means that

$$\frac{dx^i}{d\tau} \ll \frac{dt}{d\tau}, \tag{4.9}$$

so the geodesic equation becomes

$$\frac{d^2 x^\mu}{d\tau^2} + \Gamma^\mu_{00}\left(\frac{dt}{d\tau}\right)^2 = 0. \tag{4.10}$$

Since the field is static ($\partial_0 g_{\mu\nu} = 0$), the relevant Christoffel symbols Γ^μ_{00} simplify:

$$\begin{aligned} \Gamma^\mu_{00} &= \tfrac{1}{2} g^{\mu\lambda}(\partial_0 g_{\lambda 0} + \partial_0 g_{0\lambda} - \partial_\lambda g_{00}) \\ &= -\tfrac{1}{2} g^{\mu\lambda}\partial_\lambda g_{00}. \end{aligned} \tag{4.11}$$

Finally, the weakness of the gravitational field allows us to decompose the metric into the Minkowski form plus a small perturbation:

$$g_{\mu\nu} = \eta_{\mu\nu} + h_{\mu\nu}, \quad |h_{\mu\nu}| \ll 1. \tag{4.12}$$

We are working in inertial coordinates, so $\eta_{\mu\nu}$ is the canonical form of the metric. The "smallness condition" on the metric perturbation $h_{\mu\nu}$ doesn't really make sense in arbitrary coordinates. From the definition of the inverse metric, $g^{\mu\nu} g_{\nu\sigma} = \delta^\mu_\sigma$, we find that to first order in h,

$$g^{\mu\nu} = \eta^{\mu\nu} - h^{\mu\nu}, \tag{4.13}$$

where $h^{\mu\nu} = \eta^{\mu\rho}\eta^{\nu\sigma}h_{\rho\sigma}$. In fact, we can use the Minkowski metric to raise and lower indices on an object of any definite order in h, since the corrections would only contribute at higher orders. If you like, think of $h_{\mu\nu}$ as a symmetric (0, 2) tensor field propagating in Minkowski space and interacting with other fields.

Putting it all together, to first order in $h_{\mu\nu}$ we find

$$\Gamma^{\mu}_{00} = -\tfrac{1}{2}\eta^{\mu\lambda}\partial_\lambda h_{00}. \tag{4.14}$$

The geodesic equation (4.10) is therefore

$$\frac{d^2x^\mu}{d\tau^2} = \frac{1}{2}\eta^{\mu\lambda}\partial_\lambda h_{00}\left(\frac{dt}{d\tau}\right)^2. \tag{4.15}$$

Using $\partial_0 h_{00} = 0$, the $\mu = 0$ component of this is just

$$\frac{d^2t}{d\tau^2} = 0. \tag{4.16}$$

That is, $dt/d\tau$ is constant. To examine the spacelike components of (4.15), recall that the spacelike components of $\eta^{\mu\nu}$ are just those of a 3×3 identity matrix. We therefore have

$$\frac{d^2x^i}{d\tau^2} = \frac{1}{2}\left(\frac{dt}{d\tau}\right)^2 \partial_i h_{00}. \tag{4.17}$$

Dividing both sides by $(dt/d\tau)^2$ has the effect of converting the derivative on the left-hand side from τ to t, leaving us with

$$\frac{d^2x^i}{dt^2} = \frac{1}{2}\partial_i h_{00}. \tag{4.18}$$

This begins to look a great deal like Newton's theory of gravitation. In fact, if we compare this equation to (4.1), we find that they are the same once we identify

$$h_{00} = -2\Phi, \tag{4.19}$$

or in other words

$$g_{00} = -(1 + 2\Phi). \tag{4.20}$$

Therefore, we have shown that the curvature of spacetime is indeed sufficient to describe gravity in the Newtonian limit, as long as the metric takes the form (4.20). It remains, of course, to find field equations for the metric that imply this is the form taken, and that for a single gravitating body we recover the Newtonian formula

$$\Phi = -\frac{GM}{r}, \tag{4.21}$$

but that will come soon enough.

The straightforward procedure we have outlined for generalizing laws of physics to curved spacetime does have some subtleties, which we address in Section 4.7. But it's more than good enough for our present purposes, so let's not delay our pursuit of the second half of our task, obtaining the field equation for the metric in general relativity.

4.2 ■ EINSTEIN'S EQUATION

Just as Maxwell's equations govern how the electric and magnetic fields respond to charges and currents, Einstein's field equation governs how the metric responds to energy and momentum. Ultimately the field equation must be postulated and tested against experiment, not derived from any bedrock principles; however, we can motivate it on the basis of plausibility arguments. We will actually do this in two ways: first by some informal reasoning by analogy, close to what Einstein himself was thinking, and then by starting with an action and deriving the corresponding equations of motion.

The informal argument begins with the realization that we would like to find an equation that supersedes the Poisson equation for the Newtonian potential:

$$\nabla^2 \Phi = 4\pi G\rho, \tag{4.22}$$

where $\nabla^2 = \delta^{ij}\partial_i\partial_j$ is the Laplacian in space and ρ is the mass density. [The explicit form of Φ given in (4.21) is one solution of (4.22), for the case of a pointlike mass distribution.] What characteristics should our sought-after equation possess? On the left-hand side of (4.22) we have a second-order differential operator acting on the gravitational potential, and on the right-hand side a measure of the mass distribution. A relativistic generalization should take the form of an equation between tensors. We know what the tensor generalization of the mass density is; it's the energy-momentum tensor $T_{\mu\nu}$. The gravitational potential, meanwhile, should get replaced by the metric tensor, because in (4.20) we had to relate a perturbation of the metric to the Newtonian potential to successfully reproduce gravity. We might therefore guess that our new equation will have $T_{\mu\nu}$ set proportional to some tensor, which is second-order in derivatives of the metric; something along the lines of

$$[\nabla^2 g]_{\mu\nu} \propto T_{\mu\nu}, \tag{4.23}$$

but of course we want it to be completely tensorial.

The left-hand side of (4.23) is not a sensible tensor; it's just a suggestive notation to indicate that we would like a symmetric (0, 2) tensor that is second-order in derivatives of the metric. The first choice might be to act the d'Alembertian $\Box = \nabla^\mu\nabla_\mu$ on the metric $g_{\mu\nu}$, but this is automatically zero by metric compatibility. Fortunately, there is an obvious quantity which is not zero and is constructed from second derivatives (and first derivatives) of the metric: the Riemann tensor $R^\rho{}_{\sigma\mu\nu}$. Recall that the Riemann tensor is constructed from the Christoffel symbols and their first derivatives, and the Christoffel symbols are constructed from the metric and its first derivatives, so $R^\rho{}_{\sigma\mu\nu}$ contains second derivatives of $g_{\mu\nu}$. It doesn't have the right number of indices, but we can contract it to form the Ricci tensor $R_{\mu\nu}$, which does (and is symmetric to boot). It is therefore tempting to guess that the gravitational field equations are

$$R_{\mu\nu} = \kappa T_{\mu\nu}, \tag{4.24}$$

for some constant κ. In fact, Einstein did suggest this equation at one point. There is a problem, unfortunately, with conservation of energy. If we want to preserve

$$\nabla^\mu T_{\mu\nu} = 0, \tag{4.25}$$

by (4.24) we would have

$$\nabla^\mu R_{\mu\nu} = 0. \tag{4.26}$$

This is certainly not true in an arbitrary geometry; we have seen from the Bianchi identity (3.150) that

$$\nabla^\mu R_{\mu\nu} = \tfrac{1}{2}\nabla_\nu R. \tag{4.27}$$

But our proposed field equation implies that $R = \kappa g^{\mu\nu} T_{\mu\nu} = \kappa T$, so taking these together we have

$$\nabla_\mu T = 0. \tag{4.28}$$

The covariant derivative of a scalar is just the partial derivative, so (4.28) is telling us that T is constant throughout spacetime. This is highly implausible, since $T = 0$ in vacuum while $T \neq 0$ in matter. We have to try harder.

Of course we don't have to try much harder, since we already know of a symmetric (0, 2) tensor, constructed from the Ricci tensor, which is automatically conserved: the Einstein tensor

$$G_{\mu\nu} = R_{\mu\nu} - \tfrac{1}{2} R g_{\mu\nu}, \tag{4.29}$$

which always obeys $\nabla^\mu G_{\mu\nu} = 0$. We are therefore led to propose

$$G_{\mu\nu} = \kappa T_{\mu\nu} \tag{4.30}$$

as a field equation for the metric. (Actually it is probably more common to write out $R_{\mu\nu} - \frac{1}{2} R g_{\mu\nu}$, rather than use the abbreviation $G_{\mu\nu}$.) This equation satisfies all of the obvious requirements: the right-hand side is a covariant expression of the energy and momentum density in the form of a symmetric and conserved (0, 2) tensor, while the left-hand side is a symmetric and conserved (0, 2) tensor constructed from the metric and its first and second derivatives. It only remains to fix the proportionality constant κ, and to see whether the result actually reproduces gravity as we know it. In other words, does this equation predict the Poisson equation for the gravitational potential in the Newtonian limit?

To answer this, note that contracting both sides of (4.30) yields (in four dimensions)

$$R = -\kappa T, \tag{4.31}$$

and using this we can rewrite (4.30) as

$$R_{\mu\nu} = \kappa(T_{\mu\nu} - \tfrac{1}{2}Tg_{\mu\nu}). \tag{4.32}$$

This is the same equation, just written slightly differently. We would like to see if it predicts Newtonian gravity in the weak-field, time-independent, slowly-moving-particles limit. We consider a perfect-fluid source of energy-momentum, for which

$$T_{\mu\nu} = (\rho + p)U_\mu U_\nu + pg_{\mu\nu}, \tag{4.33}$$

where U^μ is the fluid four-velocity and ρ and p are the rest-frame energy and momentum densities. In fact for the Newtonian limit we may neglect the pressure; roughly speaking, the pressure of a body becomes important when its constituent particles are traveling at speeds close to that of light, which we exclude from the Newtonian limit by hypothesis. So we are actually considering the energy-momentum tensor of dust:

$$T_{\mu\nu} = \rho U_\mu U_\nu. \tag{4.34}$$

The "fluid" we are considering is some massive body, such as the Earth or the Sun. We will work in the fluid rest frame, in which

$$U^\mu = (U^0, 0, 0, 0). \tag{4.35}$$

The timelike component can be fixed by appealing to the normalization condition $g_{\mu\nu}U^\mu U^\nu = -1$. In the weak-field limit we write, in accordance with (4.12) and (4.13),

$$\begin{aligned} g_{00} &= -1 + h_{00}, \\ g^{00} &= -1 - h_{00}. \end{aligned} \tag{4.36}$$

Then to first order in $h_{\mu\nu}$ we get

$$U^0 = 1 + \tfrac{1}{2}h_{00}. \tag{4.37}$$

In fact, however, this is needlessly careful, as we are going to plug the four-velocity into (4.34), and the energy density ρ is already considered small (spacetime will be flat as ρ is taken to zero). So to our level of approximation, we can simply take $U^0 = 1$, and correspondingly $U_0 = -1$. Then

$$T_{00} = \rho, \tag{4.38}$$

and all other components vanish. In this limit the rest energy $\rho = T_{00}$ will be much larger than the other terms in $T_{\mu\nu}$, so we want to focus on the $\mu = 0$, $\nu = 0$ component of (4.32). The trace, to lowest nontrivial order, is

$$T = g^{00}T_{00} = -T_{00} = -\rho. \tag{4.39}$$

We plug this into the 00 component of our proposed gravitational field equation (4.32), to get

$$R_{00} = \tfrac{1}{2}\kappa\rho. \tag{4.40}$$

This is an equation relating derivatives of the metric to the energy density. To find the explicit expression in terms of the metric, we need to evaluate $R_{00} = R^{\lambda}{}_{0\lambda 0}$. In fact we only need $R^{i}{}_{0i0}$, since $R^{0}{}_{000} = 0$. We have

$$R^{i}{}_{0j0} = \partial_j\Gamma^{i}_{00} - \partial_0\Gamma^{i}_{j0} + \Gamma^{i}_{j\lambda}\Gamma^{\lambda}_{00} - \Gamma^{i}_{0\lambda}\Gamma^{\lambda}_{j0}. \tag{4.41}$$

The second term here is a time derivative, which vanishes for static fields. The third and fourth terms are of the form $(\Gamma)^2$, and since Γ is first-order in the metric perturbation these contribute only at second order, and can be neglected. We are left with $R^{i}{}_{0j0} = \partial_j\Gamma^{i}_{00}$. From this we get

$$\begin{aligned} R_{00} &= R^{i}{}_{0i0} \\ &= \partial_i\left[\tfrac{1}{2}g^{i\lambda}(\partial_0 g_{\lambda 0} + \partial_0 g_{0\lambda} - \partial_\lambda g_{00})\right] \\ &= -\tfrac{1}{2}\delta^{ij}\partial_i\partial_j h_{00} \\ &= -\tfrac{1}{2}\nabla^2 h_{00}. \end{aligned} \tag{4.42}$$

Comparing to (4.40), we see that the 00 component of (4.30) in the Newtonian limit predicts

$$\nabla^2 h_{00} = -\kappa\rho. \tag{4.43}$$

Since (4.19) sets $h_{00} = -2\Phi$, this is precisely the Poisson equation (4.22), if we set $\kappa = 8\pi G$.

So our guess, (4.30), seems to have worked out. With the normalization chosen so as to correctly recover the Newtonian limit, we can present **Einstein's equation** for general relativity:

$$\boxed{R_{\mu\nu} - \tfrac{1}{2}Rg_{\mu\nu} = 8\pi G T_{\mu\nu}.} \tag{4.44}$$

This tells us how the curvature of spacetime reacts to the presence of energy-momentum. G is of course Newton's constant of gravitation; it has nothing to do with the trace of $G_{\mu\nu}$. Einstein, you may have heard, thought that the left-hand side was nice and geometrical, while the right-hand side was somewhat less compelling.

It is sometimes useful to rewrite Einstein's equation in a slightly different form. Following (4.31) and (4.32), we can take the trace of (4.44) to find that $R = -8\pi GT$. Plugging this in and moving the trace term to the right-hand side, we obtain

$$\boxed{R_{\mu\nu} = 8\pi G\left(T_{\mu\nu} - \tfrac{1}{2}Tg_{\mu\nu}\right).} \tag{4.45}$$

The difference between this and (4.44) is purely cosmetic; in substance they are precisely the same. We will often be interested in the Einstein's equation in vacuum, where $T_{\mu\nu} = 0$ (for example, outside a star or planet). Then of course the right-hand side of (4.45) vanishes. Therefore the vacuum Einstein equation is simply

$$\boxed{R_{\mu\nu} = 0.} \tag{4.46}$$

This is both slightly less formidable, and of considerable physical usefulness.

4.3 ■ LAGRANGIAN FORMULATION

An alternative route to Einstein's equation is through the principle of least action, as we discussed for classical field theories in flat spacetime at the end of Chapter 1. Let's spend a moment to generalize those results to curved spacetime, and then see what kind of Lagrangian is appropriate for general relativity. We'll work in n dimensions, since our results will not depend on the dimensionality; we will, however, assume that our metric has Lorentzian signature.

Consider a field theory in which the dynamical variables are a set of fields $\Phi^i(x)$. The classical solutions to such a theory will be those that are critical points of an action S, generally expressed as an integral over space of a Lagrange density $\mathcal{L}$,

$$S = \int \mathcal{L}(\Phi^i, \nabla_\mu \Phi^i)\, d^n x. \tag{4.47}$$

Note that we are now imagining that the Lagrangian is a function of the fields and their covariant (rather than partial) derivatives, as is appropriate in curved space. Note also that, since $d^n x$ is a density rather than a tensor, $\mathcal{L}$ is also a density (since their product must be a well-defined tensor); we typically write

$$\mathcal{L} = \sqrt{-g}\,\widehat{\mathcal{L}}, \tag{4.48}$$

where $\widehat{\mathcal{L}}$ is indeed a scalar. You might think it would be sensible to forget about what we are calling $\mathcal{L}$ and just focus on $\widehat{\mathcal{L}}$, but in fact both quantities are useful in different circumstances; it is $\mathcal{L}$ that will matter whenever we are varying with respect to the metric itself. The associated Euler–Lagrange equations make use of the scalar $\widehat{\mathcal{L}}$, and are otherwise like those in flat space, but with covariant instead of partial derivatives:

$$\frac{\partial\widehat{\mathcal{L}}}{\partial\Phi} - \nabla_\mu\left(\frac{\partial\widehat{\mathcal{L}}}{\partial(\nabla_\mu\Phi)}\right) = 0. \tag{4.49}$$

In deriving these equations, we make use of Stokes's theorem (3.35),

$$\int_\Sigma \nabla_\mu V^\mu \sqrt{|g|}\, d^n x = \int_{\partial\Sigma} n_\mu V^\mu \sqrt{|\gamma|}\, d^{n-1}x, \tag{4.50}$$

and set the variation equal to zero at infinity (the boundary). Integration by parts therefore takes the form

$$\int A^\mu(\nabla_\mu B)\sqrt{-g}\, d^n x = -\int (\nabla_\mu A^\mu) B\sqrt{-g}\, d^n x + \text{boundary terms}. \tag{4.51}$$

For example, the curved-spacetime generalization of the action for a single scalar field ϕ considered in Chapter 1 would be

$$S_\phi = \int \left[-\frac{1}{2}g^{\mu\nu}(\nabla_\mu\phi)(\nabla_\nu\phi) - V(\phi)\right]\sqrt{-g}\, d^n x, \tag{4.52}$$

which would lead to an equation of motion

$$\Box\phi - \frac{dV}{d\phi} = 0, \tag{4.53}$$

where the covariant d'Alembertian is

$$\Box = \nabla^\mu\nabla_\mu = g^{\mu\nu}\nabla_\mu\nabla_\nu. \tag{4.54}$$

Just as in flat spacetime, the combination $g^{\mu\nu}(\nabla_\mu\phi)(\nabla_\nu\phi)$ is often abbreviated as $(\nabla\phi)^2$. Of course, the covariant derivatives are equivalent to partial derivatives when acting on scalars, but it is wise to use the ∇_μ notation still; you never know when you might integrate by parts and suddenly be acting on a vector.

With that as a warm-up, we turn to the construction of an action for general relativity. Our dynamical variable is now the metric $g_{\mu\nu}$; what scalars can we make out of the metric to serve as a Lagrangian? Since we know that the metric can be set equal to its canonical form and its first derivatives set to zero at any one point, any nontrivial scalar must involve at least second derivatives of the metric. The Riemann tensor is of course made from second derivatives of the metric, and we argued earlier that the only independent scalar we could construct from the Riemann tensor was the Ricci scalar R. What we did not show, but is nevertheless true, is that any nontrivial tensor made from products of the metric and its first and second derivatives can be expressed in terms of the metric and the Riemann tensor. Therefore, the *only* independent scalar constructed from the metric, which is no higher than second order in its derivatives, is the Ricci scalar. Hilbert figured that this was therefore the simplest possible choice for a Lagrangian, and proposed

$$S_H = \int \sqrt{-g}R\, d^n x, \tag{4.55}$$

known as the **Hilbert action** (or sometimes the Einstein–Hilbert action). As we shall see, he was right.

The equation of motion should come from varying the action with respect to the metric. Unfortunately the action isn't quite in the form (4.47), since it can't be written in terms of covariant derivatives of $g_{\mu\nu}$ (which would simply vanish). Therefore, instead of simply plugging into the Euler–Lagrange equations, we will consider directly the behavior of S_H under small variations of the metric. In fact it is more convenient to vary with respect to the inverse metric $g^{\mu\nu}$. Since $g^{\mu\lambda}g_{\lambda\nu} = \delta^\mu_\nu$, and the Kronecker delta is unchanged under any variation, it is straightforward to express variations of the metric and inverse metric in terms of each other:

$$\delta g_{\mu\nu} = -g_{\mu\rho}g_{\nu\sigma}\delta g^{\rho\sigma}, \tag{4.56}$$

so stationary points with respect to variations in $g^{\mu\nu}$ are equivalent to those with respect to variations in $g_{\mu\nu}$. Using $R = g^{\mu\nu}R_{\mu\nu}$, we have

$$\delta S_H = (\delta S)_1 + (\delta S)_2 + (\delta S)_3, \tag{4.57}$$

where

$$\begin{aligned}
(\delta S)_1 &= \int d^n x\, \sqrt{-g}g^{\mu\nu}\delta R_{\mu\nu} \\
(\delta S)_2 &= \int d^n x\, \sqrt{-g}R_{\mu\nu}\delta g^{\mu\nu} \\
(\delta S)_3 &= \int d^n x\, R\delta\sqrt{-g}.
\end{aligned} \tag{4.58}$$

The second term $(\delta S)_2$ is already in the form of some expression multiplied by $\delta g^{\mu\nu}$; let's examine the others more closely.

Recall that the Ricci tensor is the contraction of the Riemann tensor, which is given by

$$R^\rho{}_{\mu\lambda\nu} = \partial_\lambda \Gamma^\rho_{\nu\mu} + \Gamma^\rho_{\lambda\sigma}\Gamma^\sigma_{\nu\mu} - (\lambda \leftrightarrow \nu). \tag{4.59}$$

The variation of the Riemann tensor with respect to the metric can be found by first varying the connection with respect to the metric, and then substituting into this expression. However, let us consider arbitrary variations of the connection by replacing

$$\Gamma^\rho_{\nu\mu} \to \Gamma^\rho_{\nu\mu} + \delta\Gamma^\rho_{\nu\mu}. \tag{4.60}$$

The variation $\delta\Gamma^{\rho}_{\nu\mu}$ is the difference of two connections, and therefore is itself a tensor. We can thus take its covariant derivative,

$$\nabla_\lambda(\delta\Gamma^{\rho}_{\nu\mu}) = \partial_\lambda(\delta\Gamma^{\rho}_{\nu\mu}) + \Gamma^{\rho}_{\lambda\sigma}\delta\Gamma^{\sigma}_{\nu\mu} - \Gamma^{\sigma}_{\lambda\nu}\delta\Gamma^{\rho}_{\sigma\mu} - \Gamma^{\sigma}_{\lambda\mu}\delta\Gamma^{\rho}_{\nu\sigma}. \tag{4.61}$$

Here and elsewhere, the covariant derivatives are taken with respect to $g_{\mu\nu}$, not $g_{\mu\nu} + \delta g_{\mu\nu}$. Given this expression and a small amount of labor, it is easy to show that, to first order in the variation,

$$\delta R^{\rho}{}_{\mu\lambda\nu} = \nabla_\lambda(\delta\Gamma^{\rho}_{\nu\mu}) - \nabla_\nu(\delta\Gamma^{\rho}_{\lambda\mu}). \tag{4.62}$$

You are encouraged check this yourself. Therefore, the contribution of the first term in (4.58) to δS can be written

$$\begin{aligned}(\delta S)_1 &= \int d^n x\sqrt{-g}\, g^{\mu\nu}\left[\nabla_\lambda(\delta\Gamma^{\lambda}_{\nu\mu}) - \nabla_\nu(\delta\Gamma^{\lambda}_{\lambda\mu})\right]\\ &= \int d^n x\sqrt{-g}\,\nabla_\sigma\left[g^{\mu\nu}(\delta\Gamma^{\sigma}_{\mu\nu}) - g^{\mu\sigma}(\delta\Gamma^{\lambda}_{\lambda\mu})\right],\end{aligned} \tag{4.63}$$

where we have used metric compatibility and relabeled some dummy indices. We can now plug in the expression for $\delta\Gamma^{\sigma}_{\mu\nu}$ in terms of $\delta g^{\mu\nu}$, which works out to be

$$\delta\Gamma^{\sigma}_{\mu\nu} = -\tfrac{1}{2}\left[g_{\lambda\mu}\nabla_\nu(\delta g^{\lambda\sigma}) + g_{\lambda\nu}\nabla_\mu(\delta g^{\lambda\sigma}) - g_{\mu\alpha}g_{\nu\beta}\nabla^\sigma(\delta g^{\alpha\beta})\right], \tag{4.64}$$

leading to

$$(\delta S)_1 = \int d^n x\sqrt{-g}\,\nabla_\sigma[g_{\mu\nu}\nabla^\sigma(\delta g^{\mu\nu}) - \nabla_\lambda(\delta g^{\sigma\lambda})], \tag{4.65}$$

as you are also welcome to check. But (4.63) [or (4.65)] is an integral with respect to the natural volume element of the covariant divergence of a vector; by Stokes's theorem, this is equal to a boundary contribution at infinity, which we can set to zero by making the variation vanish at infinity. Therefore this term contributes nothing to the total variation. Although to be honest, we have cheated. The boundary term will include not only the metric variation, but also its first derivative, which is not traditionally set to zero. For our present purposes it doesn't matter, but in principle we might care about what happens at the boundary, and would have to include an additional term in the action to take care of this subtlety.

To make sense of the $(\delta S)_3$ term we need to use the following fact, true for any square matrix M with nonvanishing determinant:

$$\ln(\det M) = \mathrm{Tr}(\ln M). \tag{4.66}$$

Here, $\ln M$ is defined by $\exp(\ln M) = M$. For numbers this is obvious, for matrices it's a little less straightforward. The variation of this identity yields

$$\frac{1}{\det M}\delta(\det M) = \mathrm{Tr}(M^{-1}\delta M). \tag{4.67}$$

We have used the cyclic property of the trace to allow us to ignore the fact that M^{-1} and δM may not commute. Taking the matrix M to be the metric $g_{\mu\nu}$, so that $\det M = \det g_{\mu\nu} = g$, we get

$$\begin{aligned} \delta g &= g(g^{\mu\nu}\delta g_{\mu\nu}) \\ &= -g(g_{\mu\nu}\delta g^{\mu\nu}). \end{aligned} \tag{4.68}$$

In the last step we converted from $\delta g_{\mu\nu}$ to $\delta g^{\mu\nu}$ using (4.56). Now we can just plug in to get

$$\begin{aligned} \delta\sqrt{-g} &= -\frac{1}{2\sqrt{-g}}\delta g \\ &= \frac{1}{2}\frac{g}{\sqrt{-g}}g_{\mu\nu}\delta g^{\mu\nu} \\ &= -\frac{1}{2}\sqrt{-g}g_{\mu\nu}\delta g^{\mu\nu}. \end{aligned} \tag{4.69}$$

Hearkening back to (4.58), and remembering that $(\delta S)_1$ does not contribute, we find

$$\delta S_H = \int d^n x\sqrt{-g}\left[R_{\mu\nu} - \frac{1}{2}Rg_{\mu\nu}\right]\delta g^{\mu\nu}. \tag{4.70}$$

Recall that the functional derivative of the action satisfies

$$\delta S = \int \sum_i \left(\frac{\delta S}{\delta\Phi^i}\delta\Phi^i\right) d^n x, \tag{4.71}$$

where $\{\Phi^i\}$ is a complete set of fields being varied (in our case, it's just $g^{\mu\nu}$). Stationary points are those for which each $\delta S/\delta\Phi^i = 0$, so we recover Einstein's equation in vacuum:

$$\frac{1}{\sqrt{-g}}\frac{\delta S_H}{\delta g^{\mu\nu}} = R_{\mu\nu} - \frac{1}{2}Rg_{\mu\nu} = 0. \tag{4.72}$$

The advantage of the Lagrangian approach is manifested by the fact that our very first guess (which was practically unique) gave the right answer, in contrast with our previous trial-and-error method. This is a reflection of two elegant features of this technique: First, the Lagrangian is a scalar, rather than a tensor, and therefore more restricted; second, the symmetries of the theory are straightforwardly imposed (in this case, we automatically derived a tensor with vanishing divergence, which is related to diffeomorphism invariance, as discussed in Appendix B).

We derived Einstein's equation "in vacuum" because we only included the gravitational part of the action, not additional terms for matter fields. What we would really like, however, is to get the nonvacuum field equation as well. That means we consider an action of the form

$$S = \frac{1}{16\pi G} S_H + S_M, \tag{4.73}$$

where S_M is the action for matter, and we have presciently normalized the gravitational action so that we get the right answer. Following through the same procedure as above leads to

$$\frac{1}{\sqrt{-g}} \frac{\delta S}{\delta g^{\mu\nu}} = \frac{1}{16\pi G} \left(R_{\mu\nu} - \frac{1}{2} R g_{\mu\nu} \right) + \frac{1}{\sqrt{-g}} \frac{\delta S_M}{\delta g^{\mu\nu}} = 0. \tag{4.74}$$

We now boldly define the energy-momentum tensor to be

$$T_{\mu\nu} = -2 \frac{1}{\sqrt{-g}} \frac{\delta S_M}{\delta g^{\mu\nu}}. \tag{4.75}$$

This allows us to recover the complete Einstein's equation,

$$R_{\mu\nu} - \tfrac{1}{2} R g_{\mu\nu} = 8\pi G T_{\mu\nu}, \tag{4.76}$$

or equivalently, $G_{\mu\nu} = 8\pi G T_{\mu\nu}$.

Why should we think that (4.75) is really the energy-momentum tensor? In some sense it is only because it is a symmetric, conserved, (0, 2) tensor with dimensions of energy density; if you prefer to call it by some other name, go ahead. But it also accords with our preconceived expectations. Consider again the action for a scalar field, (4.52). Now vary this action with respect, not to ϕ, but to the inverse metric:

$$\delta S_\phi = \int d^n x \left[\sqrt{-g} \left(-\frac{1}{2} \delta g^{\mu\nu} \nabla_\mu \phi \nabla_\nu \phi \right) + \delta\sqrt{-g} \left(-\frac{1}{2} g^{\mu\nu} \nabla_\mu \phi \nabla_\nu \phi - V(\phi) \right) \right] \tag{4.77}$$

$$= \int d^n x \sqrt{-g}\, \delta g^{\mu\nu} \left[-\frac{1}{2} \nabla_\mu \phi \nabla_\nu \phi + \left(-\frac{1}{2} g_{\mu\nu} \right) \left(-\frac{1}{2} g^{\rho\sigma} \nabla_\rho \phi \nabla_\sigma \phi - V(\phi) \right) \right]. \tag{4.78}$$

We therefore have

$$\begin{aligned} T^{(\phi)}_{\mu\nu} &= -2 \frac{1}{\sqrt{-g}} \frac{\delta S_\phi}{\delta g^{\mu\nu}} \\ &= \nabla_\mu \phi \nabla_\nu \phi - \frac{1}{2} g_{\mu\nu} g^{\rho\sigma} \nabla_\rho \phi \nabla_\sigma \phi - g_{\mu\nu} V(\phi). \end{aligned} \tag{4.79}$$

In flat spacetime this reduces to what we had asserted, in Chapter 1, was the correct energy-momentum tensor for a scalar field.

On the other hand, in Minkowski space there is an alternative definition for the energy-momentum tensor, which is sometimes given in books on electromagnetism or field theory. In this context energy-momentum conservation arises as a consequence of symmetry of the Lagrangian under spacetime translations. **Noether's theorem** states that every symmetry of a Lagrangian implies the existence of a conservation law; invariance under the four spacetime translations leads to a tensor $S^{\mu\nu}$, which obeys $\partial_\mu S^{\mu\nu} = 0$ (four relations, one for each value of ν). The details can be found in Wald (1984) or Peskin and Schroeder (1995). Applying Noether's procedure to a Lagrangian that depends on some fields Φ^i and their first derivatives $\partial_\mu \Phi^i$ (in flat spacetime), we obtain

$$S^{\mu\nu} = \frac{\delta \mathcal{L}}{\delta(\partial_\mu \Phi^i)} \partial^\nu \Phi^i - \eta^{\mu\nu} \mathcal{L}, \tag{4.80}$$

where a sum over i is implied. You can check that this tensor is conserved by virtue of the equations of motion of the matter fields. $S^{\mu\nu}$ often goes by the name "canonical energy-momentum tensor"; however, there are a number of reasons why it is more convenient for us to use (4.75). First, (4.75) is in fact what appears on the right hand side of Einstein's equation when it is derived from an action, and it is not always possible to generalize (4.80) to curved spacetime. But even in flat space (4.75) has its advantages; it is manifestly symmetric, and also guaranteed to be gauge invariant, neither of which is true for (4.80). We will therefore stick with (4.75) as the definition of the energy-momentum tensor.

Now that Einstein's equation has been derived, the rest of this chapter is devoted to exploring some of its properties. These discussions are fascinating but not strictly necessary; if you like, you can jump right to the applications discussed in subsequent chapters.

4.4 ■ PROPERTIES OF EINSTEIN'S EQUATION

Einstein's equation may be thought of as a set of second-order differential equations for the metric tensor field $g_{\mu\nu}$. There are really ten independent equations (since both sides are symmetric two-index tensors), which seems to be exactly right for the ten unknown functions of the metric components. However, the Bianchi identity $\nabla^\mu G_{\mu\nu} = 0$ represents four constraints on the functions $R_{\mu\nu}(x)$, so there are only six truly independent equations in (4.44). In fact this is appropriate, since if a metric is a solution to Einstein's equation in one coordinate system x^μ it should also be a solution in any other coordinate system $x^{\mu'}$. This means that there are four unphysical degrees of freedom in $g_{\mu\nu}$, represented by the four functions $x^{\mu'}(x^\mu)$, and we should expect that Einstein's equation only constrains the six coordinate-independent degrees of freedom.

As differential equations, these are extremely complicated; the Ricci scalar and tensor are contractions of the Riemann tensor, which involves derivatives and products of the Christoffel symbols, which in turn involve the inverse metric and derivatives of the metric. Furthermore, the energy-momentum tensor $T_{\mu\nu}$ will generally involve the metric as well. The equations are also nonlinear, so that two known solutions cannot be superposed to find a third. It is therefore very difficult to solve Einstein's equation in any sort of generality, and it is usually necessary to make some simplifying assumptions. Even in vacuum, where we set the energy-momentum tensor to zero, the resulting equation (4.46) can be very difficult to solve. The most popular sort of simplifying assumption is that the metric has a significant degree of symmetry, and we will see later how isometries make life easier.

The nonlinearity of general relativity is worth a remark. In Newtonian gravity the potential due to two point masses is simply the sum of the potentials for each mass, but clearly this does not carry over to general relativity outside the weak-field limit. There is a physical reason for this, namely that in GR the gravitational field couples to itself. This can be thought of as a consequence of the equivalence principle—if gravitation did not couple to itself, a gravitational atom (two particles bound by their mutual gravitational attraction) would have a different inertial mass than gravitational mass (due to the negative binding energy). The nonlinearity of Einstein's equation is a reflection of the back-reaction of gravity on itself.

A nice way to think about this is provided by Feynman diagrams. These are used in quantum field theory to calculate the amplitudes for scattering processes, which can be obtained by summing the various contributions from different interactions, each represented by its own diagram. Even if we don't go so far as to quantize gravity and calculate scattering cross-sections (see the end of this section), we can still draw Feynman diagrams as a simple way of keeping track of which interactions exist and which do not. A simple example is provided by the electromagnetic interaction between two electrons; this can be thought of as due to exchange of a virtual photon, as shown in Figure 4.1.

In contrast, there is no diagram in which two photons exchange another photon between themselves, because electromagnetism is linear (there is no back-reaction). The gravitational interaction, meanwhile, can be thought of as deriving from the exchange of a virtual graviton (a quantized perturbation of the metric). The nonlinearity manifests itself as the fact that both electrons and gravi-

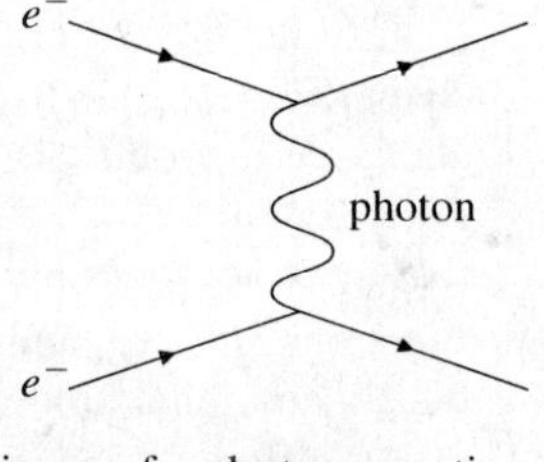

FIGURE 4.1 A Feynman diagram for electromagnetism. In quantum field theory, such diagrams are used to calculate amplitudes for scattering processes; here, just think of it as a cartoon representing a certain interaction. The point of this particular diagram is that the coupling of photons to electrons is what causes the electromagnetic interaction between them. In contrast, there is no coupling of photons to other photons, and no analogous diagram in which photons interact.

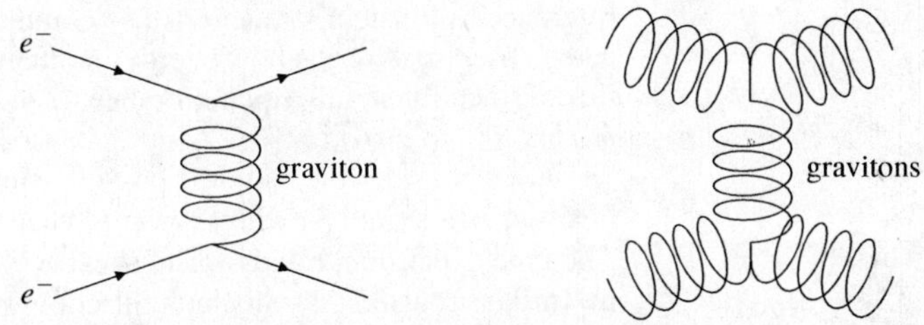

FIGURE 4.2 Feynman diagrams for gravity. Upon quantization, Einstein's equation predicts spin-two particles called gravitons. We don't know how to carry out such a quantization consistently, but the existence of gravitons is sufficiently robust that it is expected to be a feature of any well-defined scheme. Since gravity couples to energy-momentum, gravitons interact with every kind of particle, including other gravitons. This provides a way of thinking about the nonlinearity of Einstein's theory.

tons can exchange virtual gravitons, and therefore exert a gravitational force, as shown in Figure 4.2. There is nothing unique about this feature of gravity; it is shared by most gauge theories, such as quantum chromodynamics, the theory of the strong interactions. Electromagnetism is actually the exception; the linearity can be traced to the fact that the relevant gauge group, U(1), is abelian. But nonlinearity does represent a departure from the Newtonian theory. This difference is experimentally detectable; the reason why (as we shall see) the orbit of Mercury is different in GR versus Newtonian gravity is that the gravitational field influences itself, and the closer we get to the Sun, the more noticeable that influence is.

Beyond the fact that it is complicated and nonlinear, it is worth thinking a bit about what Einstein's equation is actually telling us. Clearly it relates the energy-momentum distribution to components of the curvature tensor; but from a physical point of view, precisely what kind of gravitational field is generated by a given kind of source? One way to answer this question is to consider the evolution of the *expansion* θ of a family of neighboring timelike geodesics. We imagine a small ball of free test particles moving along geodesics with four-velocities U^μ, and follow their evolution; the expansion $\theta = \nabla_\mu U^\mu$ tells us how the volume of the ball is growing (or shrinking, if $\theta < 0$) at any one moment of time. Clearly the value of the expansion will depend on the initial conditions for our test particles. The effects of gravity, on the other hand, are encoded in the *evolution* of the expansion, which is governed by Raychaudhuri's equation. This equation, discussed in Appendix F, tells us that the derivative of the expansion with respect to the proper time τ along the geodesics is given by the following expression:

$$\frac{d\theta}{d\tau} = 2\omega^2 - 2\sigma^2 - \frac{1}{3}\theta^2 - R_{\mu\nu}U^\mu U^\nu. \tag{4.81}$$

The terms on the right-hand side are explained carefully in Appendix F; ω encodes the rotation of the geodesics, σ encodes the shear, and $R_{\mu\nu}$ is of course the Ricci tensor. Raychaudhuri's equation is a purely geometric relation, making no reference to Einstein's equation. The combination of the two equations, however, can be used to describe how energy-momentum influences the motion of test particles, since Einstein's equation relates $T_{\mu\nu}$ to $R_{\mu\nu}$ and Raychaudhuri's equation relates $R_{\mu\nu}$ to $d\theta/d\tau$.

Let us consider the simplest possible situation, where we start with all of the nearby particles at rest with respect to each other in a small region of spacetime. Then the expansion, rotation, and shear will all vanish at this initial moment. Let us further construct locally inertial coordinates $x^{\hat{\mu}}$, in which $U^{\hat{\mu}}$ is in its rest frame, so that $U^{\hat{\mu}} = (1, 0, 0, 0)$ and $R_{\hat{\mu}\hat{\nu}}U^{\hat{\mu}}U^{\hat{\nu}} = R_{\hat{0}\hat{0}}$. We therefore have (in these coordinates, at this point)

$$\frac{d\theta}{d\tau} = -R_{\hat{0}\hat{0}}. \tag{4.82}$$

Now we can turn to Einstein's equation, in the form

$$R_{\mu\nu} = 8\pi G\left(T_{\mu\nu} - \tfrac{1}{2}Tg_{\mu\nu}\right). \tag{4.83}$$

Since we are in locally inertial coordinates, we have

$$g_{\hat{\mu}\hat{\nu}} = \eta_{\hat{\mu}\hat{\nu}} \tag{4.84}$$

$$T = g^{\hat{\mu}\hat{\nu}}T_{\hat{\mu}\hat{\nu}} = -\rho + p_x + p_y + p_z, \tag{4.85}$$

where $\rho = T_{\hat{0}\hat{0}}$ is the rest-frame energy density and $p_k = T_{\hat{k}\hat{k}}$ is the pressure in the $x^{\hat{k}}$ direction. Thus, (4.82) becomes

$$\frac{d\theta}{d\tau} = -4\pi G(\rho + p_x + p_y + p_z). \tag{4.86}$$

This equation is telling us that energy and pressure create a gravitational field that works to decrease the volume of our initially stationary ball of test particles (if ρ and the p_i's are all positive). In other words, gravity is attractive.

Of course, from (4.86) we see that gravity is not *necessarily* attractive; we could imagine sources for which $\rho+p_x+p_y+p_z$ were a negative number. Clearly, the role of pressure bears noting. For one thing, it represents an unambiguous departure from Newtonian theory, in which the pressure does not influence gravity (it doesn't appear in Poisson's equation, $\nabla^2\Phi = 4\pi G\rho$). The difference is hard to notice in our Solar System, since the pressure in the Sun and planets is much less than the energy density, which is dominated by the rest masses of the constituent particles. For another thing, notice that the *gravitational* effect of the pressure is opposite to that of the *direct* effect with which we are more familiar, namely that positive pressure works to push things apart. In most circumstances the direct effect of pressure is much more noticeable. However, the pressure can only act directly when there is a pressure gradient (for example, a change in pressure between the interior and exterior of a piston), whereas the gravitational effect depends only on the value of the pressure locally. If there were a perfectly smooth pressure, it would only be detectable through its gravitational effect; an example is provided by vacuum energy, discussed in Section 4.5.

As a final comment on (4.86), let's point out that it is completely equivalent to Einstein's equation—they convey identical information. This very specific relation will hold for any set of initially motionless test particles; the only way this can happen is if all of the components of Einstein's equation are true. If we like, then, we can state Einstein's equation in words[1] as follows: "The expansion of the volume of any set of particles initially at rest is proportional to (minus) the sum of the energy density and the three components of pressure."

So Einstein's equation tells us that energy density and pressure affect the Ricci tensor in such a way as to attract particles together when ρ and p are positive. What about the components of the Riemann tensor that are not included in the Ricci tensor? In Chapter 3 we found that these components were described by the Weyl tensor (expressed here in four dimensions),

$$C_{\rho\sigma\mu\nu} = R_{\rho\sigma\mu\nu} + \tfrac{1}{3}g_{\rho[\mu}g_{\nu]\sigma}R - g_{\rho[\mu}R_{\nu]\sigma} + g_{\sigma[\mu}R_{\nu]\rho}. \tag{4.87}$$

The Ricci tensor is the trace of the Riemann tensor, while the Weyl tensor describes the trace-free part; together they provide a complete characterization of the curvature. Clearly, given some specified energy-momentum distribution, there is still some freedom in the choice of Weyl curvature, since there is no analogue of Einstein's equation to relate $C^{\rho}{}_{\sigma\mu\nu}$ algebraically to $T_{\mu\nu}$. This is exactly as it should be. Imagine for example a spacetime that is vacuum everywhere, $R_{\mu\nu} = 0$. Flat Minkowski space is a possible solution in such a case, but so is a gravitational wave propagating through empty spacetime (as we will discuss in Chapter 7).

Since only $R_{\mu\nu}$ enters Einstein's equation, it might appear that the components of $C_{\rho\sigma\mu\nu}$ are completely unconstrained. But recall that we are not permitted to arbitrarily specify the components of the curvature tensor throughout a manifold; they are related by the Bianchi identity,

$$\nabla_{[\lambda}R_{\rho\sigma]\mu\nu} = 0. \tag{4.88}$$

As you showed in Exercise 10 of Chapter 3, this identity implies a differential relation for the Weyl tensor of the form

$$\nabla^{\rho}C_{\rho\sigma\mu\nu} = \nabla_{[\mu}R_{\nu]\sigma} + \tfrac{1}{6}g_{\sigma[\mu}\nabla_{\nu]}R. \tag{4.89}$$

On the right-hand side, the Riemann tensor only appears via its contractions the Ricci scalar and tensor, which can be related to $T_{\mu\nu}$ by Einstein's equation; we therefore have

$$\nabla^{\rho}C_{\rho\sigma\mu\nu} = 8\pi G\left(\nabla_{[\mu}T_{\nu]\sigma} + \tfrac{1}{3}g_{\sigma[\mu}\nabla_{\nu]}T\right). \tag{4.90}$$

So, while $R_{\mu\nu}$ and $T_{\mu\nu}$ are related algebraically through Einstein's equation, $C_{\rho\sigma\mu\nu}$ and $T_{\mu\nu}$ are related by this first-order differential equation. There will be

[1] J.C Baez, "The Meaning of Einstein's Equation," http://arXiv.org/abs/gr-qc/0103044.

a number of possible solutions for a given energy-momentum distribution, each specified by certain boundary conditions. This equation can be thought of as a propagation equation for gravitational waves, in close analogy with Maxwell's equations $\nabla_\mu F^{\nu\mu} = J^\nu$.

Having listed all of these lovely properties of Einstein's equation, it seems only fair that we should mention one distressing feature: the well-known difficulty of reconciling general relativity with quantum mechanics. GR is a classical field theory: the dynamical variable is a field (the metric) defined on spacetime, and coordinate-invariant quantities constructed from this field (such as the curvature scalar) can in principle be specified and measured to arbitrary accuracy. In the case of other field theories, such as electromagnetism, there are well-understood procedures for beginning with the classical theory and quantizing it, to obtain the dynamics of operators acting on wave functions living in a Hilbert space. For GR, the usual procedures run into both technical and conceptual difficulties, a description of which is beyond the scope of this book. One aspect of the technical difficulties is that GR is not "renormalizable" in the way that the Standard Model of particle physics is; when considering higher-order quantum effects, infinities appear that cannot be absorbed in any finite number of parameters. Nonrenormalizability does not mean that theory is fundamentally incorrect, but is a strong suggestion that it should only be taken seriously up to a certain energy scale.

Fortunately, the regime in which observable effects of quantum gravity are expected to become important is far from our everyday experience (or, for that matter, any conditions we can produce in the lab). Way back in 1899 Planck noticed that his constant h, for which nowadays we more often substitute $\hbar = h/2\pi = 1.05 \times 10^{-27}$ cm^2 g/sec, could be combined with Newton's constant $G = 6.67 \times 10^{-8}\text{cm}^3\,\text{g}^{-1}\,\text{sec}^{-2}$ and the speed of light $c = 3.00 \times 10^{10}\text{cm sec}^{-1}$ to form a basic set of dimensionful quantities: the Planck mass,

$$m_P = \left(\frac{\hbar c}{G}\right)^{1/2} = 2.18 \times 10^{-5}\,\text{g}, \tag{4.91}$$

the Planck length,

$$l_P = \left(\frac{\hbar G}{c^3}\right)^{1/2} = 1.62 \times 10^{-33}\,\text{cm}, \tag{4.92}$$

the Planck time,

$$t_P = \left(\frac{\hbar G}{c^5}\right)^{1/2} = 5.39 \times 10^{-44}\,\text{sec}, \tag{4.93}$$

and the Planck energy,

$$E_P = \left(\frac{\hbar c^5}{G}\right)^{1/2} = 1.95 \times 10^{16}\,\text{erg} \tag{4.94}$$

$$= 1.22 \times 10^{19}\,\text{GeV}. \tag{4.95}$$

A GeV is 10^9 electron volts, a common unit in particle physics, as it is approximately the mass of a proton. We usually set $\hbar = c = 1$, so that these quantities are all indistinguishable in the sense that $m_P = l_P^{-1} = t_P^{-1} = E_P$. You will hear people say things like "the Planck mass is 10^{19} GeV"; or simply refer to "the Planck scale." Another commonly used quantity is the reduced Planck scale, $\bar{m}_P = m_P/\sqrt{8\pi} = 2.43 \times 10^{18}$ GeV, which is often more convenient in equations—note that the coefficient of the curvature scalar in (4.73) is $\bar{m}_P^2/2$. Most likely, quantum gravity does not become important until we consider particle masses greater than m_P, or times shorter than t_P, or lengths smaller than l_P, or energies higher than E_P; at lower scales, classical GR should suffice. Since these are all far removed from observable phenomena, constructing a consistent theory of quantum gravity is more an issue of principle than of practice. On the other hand, quantum effects in curved spacetime might be important in the real world; as we will discuss in Chapter 8, they might lead to density fluctuations in the early universe, which grow into the galaxies and large-scale structure we observe today.

There is a leading contender for a fully quantum theory that would encompass GR in the appropriate limit: string theory. In string theory we imagine that the fundamental objects are not point particles like electrons or photons, but rather small one-dimensional objects called strings, which can be either closed loops or open segments. String theory was originally proposed as a model of the strong nuclear force, but it was soon realized that the theory inevitably predicted a massless spin-two particle: exactly what a quantum theory of gravity would require. String theory seems to be a consistent quantum theory, and it predicts gravity, but there is still a great deal about it that we don't understand. In particular, the way in which a classical spacetime arises out of fundamental strings is somewhat mysterious, and the connection to direct experiments is tenuous at best. Nevertheless, string theory is remarkably rich and robust, and promises to be an important part of theoretical physics for the foreseeable future.

4.5 ■ THE COSMOLOGICAL CONSTANT

A characteristic feature of general relativity is that the source for the gravitational field is the entire energy-momentum tensor. In nongravitational physics, only *changes* in energy from one state to another are measurable; the normalization of the energy is arbitrary. For example, the motion of a particle with potential energy $V(x)$ is precisely the same as that with a potential energy $V(x) + V_0$, for any constant V_0. In gravitation, however, the actual value of the energy matters, not just the differences between states.

This behavior opens up the possibility of **vacuum energy**: an energy density characteristic of empty space. One feature that we might want the vacuum to exhibit is that it not pick out a preferred direction; it will still be possible to have a nonzero energy density if the associated energy-momentum tensor is Lorentz invariant in locally inertial coordinates. Lorentz invariance implies that the corresponding energy-momentum tensor should be proportional to the metric,

$$T^{(\text{vac})}_{\hat{\mu}\hat{\nu}} = -\rho_{\text{vac}}\eta_{\hat{\mu}\hat{\nu}}, \tag{4.96}$$

since $\eta_{\hat{\mu}\hat{\nu}}$ is the only Lorentz invariant (0, 2) tensor. This generalizes straightforwardly from inertial coordinates to arbitrary coordinates as

$$T^{(\text{vac})}_{\mu\nu} = -\rho_{\text{vac}}g_{\mu\nu}. \tag{4.97}$$

Comparing to the perfect-fluid energy-momentum tensor $T_{\mu\nu} = (\rho + p)U_\mu U_\nu + pg_{\mu\nu}$, we find that the vacuum looks like a perfect fluid with an isotropic pressure opposite in sign to the energy density,

$$p_{\text{vac}} = -\rho_{\text{vac}}. \tag{4.98}$$

The energy density should be constant throughout spacetime, since a gradient would not be Lorentz invariant.

If we decompose the energy-momentum tensor into a matter piece $T^{(\text{M})}_{\mu\nu}$ and a vacuum piece $T^{(\text{vac})}_{\mu\nu} = -\rho_{\text{vac}}g_{\mu\nu}$, Einstein's equation is

$$R_{\mu\nu} - \frac{1}{2}Rg_{\mu\nu} = 8\pi G\left(T^{(\text{M})}_{\mu\nu} - \rho_{\text{vac}}g_{\mu\nu}\right). \tag{4.99}$$

Soon after inventing GR, Einstein tried to find a static cosmological model, since that was what astronomical observations of the time seemed to imply. The result was the Einstein static universe, which will be discussed in Chapter 8. In order for this static cosmology to solve the field equation with an ordinary matter source, it was necessary to add a new term called the **cosmological constant**, Λ, which enters as

$$R_{\mu\nu} - \frac{1}{2}Rg_{\mu\nu} + \Lambda g_{\mu\nu} = 8\pi GT_{\mu\nu}. \tag{4.100}$$

From comparison with (4.99), we see that the cosmological constant is precisely equivalent to introducing a vacuum energy density

$$\rho_{\text{vac}} = \frac{\Lambda}{8\pi G}. \tag{4.101}$$

The terms "cosmological constant" and "vacuum energy" are essentially interchangeable.

Is a nonzero vacuum energy something we should expect? We arrived at the Hilbert Lagrangian $\widehat{\mathcal{L}}_H = R$ by looking for the simplest possible scalar we could construct from the metric. Of course there is an even simpler one, namely a constant. Using (4.69), it is straightforward to check that

$$S = \int d^4x\,\sqrt{-g}\left[\frac{1}{16\pi G}(R - 2\Lambda) + \widehat{\mathcal{L}}_{\text{M}}\right] \tag{4.102}$$

leads to the modified equation (4.100); alternatively, the vacuum Lagrangian is simply

$$\widehat{\mathcal{L}}_{\rm vac} = -\rho_{\rm vac}. \tag{4.103}$$

So it is certainly easy to introduce vacuum energy; however, we have no insight into its expected value, since it enters as an arbitrary constant.

The vacuum energy ultimately is a constant of nature in its own right. (An exception occurs in certain theories where a spacetime symmetry such as supersymmetry or conformal invariance governs the value of the vacuum energy; here we are considering a more generic field theory.) Nevertheless, there are various distinct contributions to the vacuum energy, and it would be strange if the total value were much smaller than the individual contributions. One such contribution comes from zero-point fluctuations—the energies of quantum fields in their vacuum state.

Consider a simple harmonic oscillator, a particle moving in a one-dimensional potential $V(x) = \frac{1}{2}\omega^2 x^2$. Classically, the vacuum for this system is the state in which the particle is motionless and at the minimum of the potential ($x = 0$), for which the energy in this case vanishes. Quantum-mechanically, however, the uncertainty principle forbids us from isolating the particle both in position and momentum, and we find that the lowest energy state has an energy $E_0 = \frac{1}{2}\hbar\omega$ (where we have temporarily reintroduced explicit factors of $\hbar$ for clarity). Of course, in the absence of gravity, either system actually has a vacuum energy that is completely arbitrary; we could add any constant to the potential without changing the theory. But quantum fluctuations have changed the zero-point energy from our classical expectation.

A precisely analogous situation holds in field theory. If we take the Fourier transform of a free quantum field (one where we ignore interactions for simplicity), we find that it becomes an infinite number of harmonic oscillators in momentum space, as we discuss in Chapter 9. The frequency ω of each oscillator is $\omega = \sqrt{m^2 + k^2}$, where m is the mass of the field and k is the magnitude of the wave vector of the mode. If we set the classical vacuum energy to zero, each of these modes contributes a quantum zero-point energy of $\hbar\omega/2$. Formally, adding all of these contributions together yields an infinite result. If, however, we discard the very high-momentum modes on the grounds that we trust our theory only up to a certain ultraviolet momentum cutoff $k_{\rm max}$, we find that the resulting energy density is of the form

$$\rho_{\rm vac} \sim \hbar k_{\rm max}^4. \tag{4.104}$$

This answer could have been guessed by dimensional analysis; the numerical constants that have been neglected will depend on the precise theory under consideration. If we are confident that we can use ordinary quantum field theory all the way up to the reduced Planck scale $\bar{m}_{\rm P} = (8\pi G)^{-1/2} \sim 10^{18}$ GeV, we expect a contribution of order

$$\rho_{\rm vac} \sim (10^{18}\ {\rm GeV})^4 \sim 10^{112}\ {\rm erg/cm^3}. \tag{4.105}$$

Field theory may fail earlier, although quantum gravity is the best reason we have to believe it will fail at any specific scale.

As we will discuss in Chapter 8, cosmological observations imply

$$|\rho_\Lambda^{(\text{obs})}| \leq (10^{-12}\ \text{GeV})^4 \sim 10^{-8}\ \text{erg/cm}^3, \tag{4.106}$$

much smaller than the naive expectation just derived. The ratio of (4.105) to (4.106) is the origin of the famous discrepancy of 120 orders of magnitude between the theoretical and observational values of the cosmological constant. We are free to imagine that the bare vacuum energy is adjusted so that the net cosmological constant is consistent with the limit (4.106), except for one problem: we know of no special symmetry that could enforce a vanishing vacuum energy while remaining consistent with the known laws of physics; this conundrum is the "cosmological constant problem." We will discuss the cosmological effects of vacuum energy more in Chapter 8.[2]

4.6 ■ ENERGY CONDITIONS

Sometimes it is useful to think about Einstein's equation without specifying the theory of matter from which $T_{\mu\nu}$ is derived. This leaves us with a great deal of arbitrariness; consider for example the question, What metrics obey Einstein's equation? In the absence of some constraints on $T_{\mu\nu}$, the answer is any metric at all; simply take the metric of your choice, compute the Einstein tensor $G_{\mu\nu}$ for this metric, and then demand that $T_{\mu\nu}$ be equal to $G_{\mu\nu}$. It will automatically be conserved, by the Bianchi identity. Our real concern is with the existence of solutions to Einstein's equation in the presence of "realistic" sources of energy and momentum, whatever that means. One strategy is to consider specific kinds of sources, such as scalar fields, dust, or electromagnetic fields. However, we occasionally wish to understand properties of Einstein's equations that hold for a variety of different sources. In this circumstance it is convenient to impose *energy conditions* that limit the arbitrariness of $T_{\mu\nu}$.

Energy conditions are coordinate-invariant restrictions on the energy-momentum tensor. We must therefore construct scalars from $T_{\mu\nu}$, which is typically accomplished by contracting with arbitrary timelike vectors t^μ or null vectors ℓ^μ. For example, the weak energy condition (WEC) states that $T_{\mu\nu}t^\mu t^\nu \geq 0$ for all timelike vectors t^μ. For purposes of physical intuition, it is useful to consider the special case where the source is a perfect fluid, so that the energy-momentum tensor takes the form

$$T_{\mu\nu} = (\rho + p)U_\mu U_\nu + p g_{\mu\nu}, \tag{4.107}$$

where U^μ is the fluid four-velocity. Let's use this form to translate the WEC into physical terms. Because the pressure is isotropic, $T_{\mu\nu}t^\mu t^\nu$ will be nonnegative

[2]For more on the physics and cosmology of vacuum energy, see S.M. Carroll, *Liv. Rev. Rel.* **4**, 1 (2001), `http://arxiv.org/astro-ph/0004075`.

for all timelike vectors t^μ if both $T_{\mu\nu}U^\mu U^\nu \geq 0$ and $T_{\mu\nu}\ell^\mu\ell^\nu \geq 0$ for some null vector ℓ^μ (convince yourself of this; it's just adding vectors). We therefore evaluate

$$T_{\mu\nu}U^\mu U^\nu = \rho, \quad T_{\mu\nu}\ell^\mu\ell^\nu = (\rho + p)(U_\mu\ell^\mu)^2. \tag{4.108}$$

The WEC therefore implies $\rho \geq 0$ and $\rho + p \geq 0$. These are simply the reasonable-sounding requirements that the energy density be nonnegative and the pressure not be too large compared to the energy density. Of course we need not restrict ourselves to perfect fluids, we merely use them to gain insight into the requirements the energy conditions impose.

There are a number of different energy conditions, appropriate to different circumstances. Some of the most popular are the following:

- The **Weak Energy Condition** or WEC, as just discussed, states that $T_{\mu\nu}t^\mu t^\nu \geq 0$ for all timelike vectors t^μ, or equivalently that $\rho \geq 0$ and $\rho + p \geq 0$.

- The **Null Energy Condition** or NEC states that $T_{\mu\nu}\ell^\mu\ell^\nu \geq 0$ for all null vectors ℓ^μ, or equivalently that $\rho + p \geq 0$. It is a special case of the WEC, with the timelike vector replaced by a null vector. The energy density may now be negative, so long as there is a compensating positive pressure.

- The **Dominant Energy Condition** or DEC includes the WEC ($T_{\mu\nu}t^\mu t^\nu \geq 0$ for all timelike vectors t^μ), as well as the additional requirement that $T^{\mu\nu}t_\mu$ is a nonspacelike vector (namely, that $T_{\mu\nu}T^\nu{}_\lambda t^\mu t^\lambda \leq 0$). For a perfect fluid, these conditions together are equivalent to the simple requirement that $\rho \geq |p|$; the energy density must be nonnegative, and greater than or equal the magnitude of the pressure.

- The **Null Dominant Energy Condition** or NDEC is the DEC condition for null vectors only: for any null vector ℓ^μ, $T_{\mu\nu}\ell^\mu\ell^\nu \geq 0$ and $T^{\mu\nu}\ell_\mu$ is a nonspacelike vector. The allowed density and pressure are the same as for the DEC, except that negative densities are allowed so long as $p = -\rho$. In other words, the NDEC excludes all sources excluded by the DEC, except for a negative vacuum energy.

- The **Strong Energy Condition** or SEC states that $T_{\mu\nu}t^\mu t^\nu \geq \frac{1}{2}T^\lambda{}_\lambda t^\sigma t_\sigma$ for all timelike vectors t^μ, or equivalently that $\rho + p \geq 0$ and $\rho + 3p \geq 0$. Note that the SEC does *not* imply the WEC. It implies the NEC, along with excluding excessively large negative pressures. From (4.86) we see that it is the SEC that implies gravitation is attractive.

These conditions are illustrated in Figure 4.3. In addition we have plotted the constraint $w \geq -1$, where $w = p/\rho$ is called the **equation-of-state parameter**. This is a useful concept in cosmology, where sources often have equations of state $p = w\rho$ with w being a constant (of course, w is defined whether it is constant

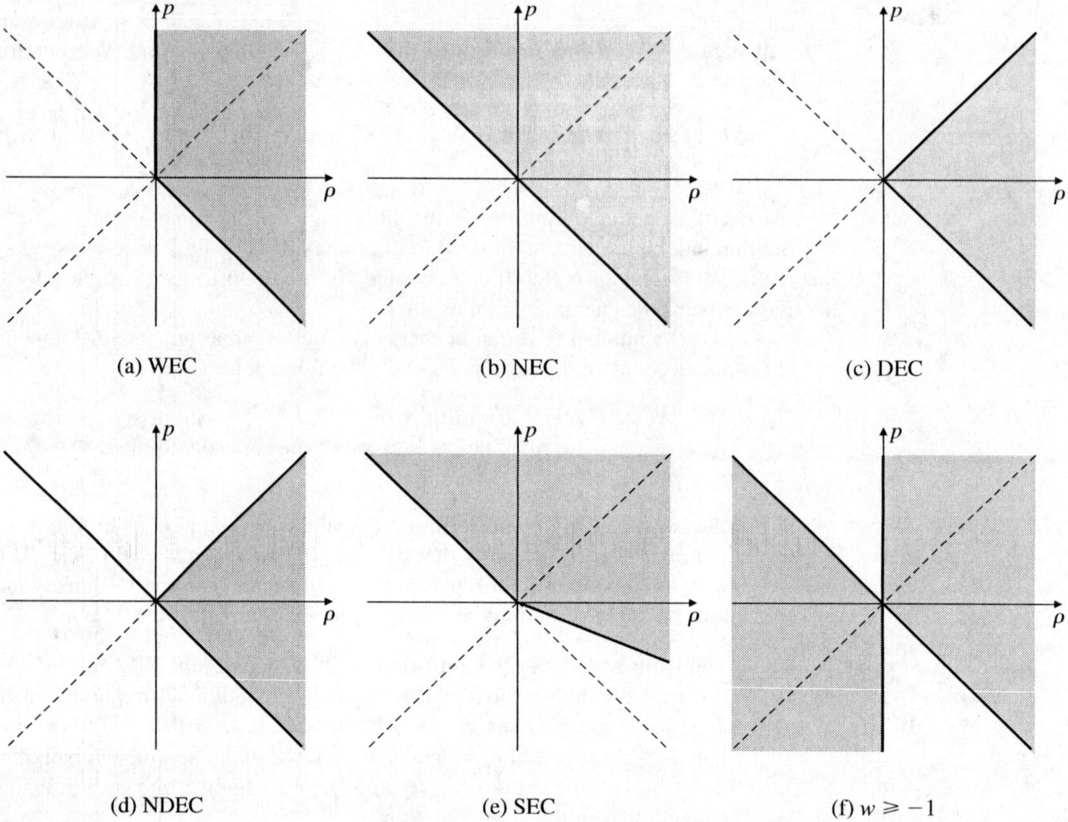

FIGURE 4.3 Energy conditions as applied to perfect fluids, expressed as allowed regions of energy density ρ and pressure p. Illustrated are the Weak Energy Condition (WEC), Null Energy Condition (NEC), Dominant Energy Condition (DEC), Null Dominant Energy Condition (NDEC), and the Strong Energy Condition (SEC). For comparison, we also have illustrated the condition $w \geq -1$, where $w = p/\rho$ is the equation-of-state parameter.

or not). If we restrict ourselves to sources with $\rho \geq 0$, then any of the energy conditions mentioned above will imply $w \geq -1$.

Most ordinary classical forms of matter, including scalar fields and electromagnetic fields, obey the DEC (see Exercises), and hence the less restrictive conditions (WEC, NEC, NDEC). The SEC is useful in the proof of some singularity theorems, but can be violated by certain forms of matter, such as a massive scalar field. It turns out that quantum fields can generically violate any of the energy conditions we have listed; there may, however, be inequalities involving integrals over regions of spacetime that are satisfied even by quantum fields. This is an area of current investigation.

The energy conditions are not, strictly speaking, related to energy conservation; the Bianchi identity guarantees that $\nabla_\mu T^{\mu\nu} = 0$ regardless of whether we impose any additional constraints on $T^{\mu\nu}$. Rather, they serve to prevent other properties that we think of as "unphysical," such as energy propagating faster than the speed of light, or empty space spontaneously decaying into compensating regions of positive and negative energy. In particular, Hawking and Ellis (1973) prove a *conservation theorem*: Essentially, if the energy-momentum tensor obeys the DEC and vanishes in some spacelike region, then it will necessarily vanish everywhere in the future domain of dependence of that region (see Section 2.7 for the definition of the future domain of dependence). Thus, energy cannot spontaneously appear from nothing, nor can it sneak outside the light cone. The theorem does not include the converse statement (that sources violating the DEC are necessarily acausal), so it pays to be careful.

4.7 ■ THE EQUIVALENCE PRINCIPLE REVISITED

In this section we will examine more carefully the underpinnings and consequences of the Principle of Equivalence, which we used in Section 4.1 to motivate the minimal-coupling procedure for generalizing physics to curved spacetime. We will see that the Principle of Equivalence is not a sacred physical law, nor is it even a mathematically rigorous statement; at a more fundamental level, it arises as a consequence of the nature of general relativity as an effective field theory valid at macroscopic distances, and our job is to determine which kinds of couplings between matter and the metric we would expect in such a theory.

In practice, it is common to invoke the Equivalence Principle to justify any of the following four ideas:

1. Laws of physics should be expressed (or at least be expressible) in generally covariant form.
2. There exists a metric on spacetime, the curvature of which we interpret as gravity.
3. There do not exist any other fields that resemble gravity.
4. The interactions of matter fields to curvature are minimal: they do not involve direct couplings to the Riemann tensor or its contractions.

These very different statements each have a very different status: the first is vacuous, the second is both profound and almost certainly true, the third is interesting and testable, and the fourth is just a useful approximation. Let's examine each of them in turn.

The first statement is sometimes called the Principle of Covariance. It is more or less content-free. "Generally covariant" simply means that all of the terms in an equation transform in the same way under a change of coordinates, so that the form of the equation is coordinate-invariant. Due to the universal nature of the tensor transformation law, the most straightforward way of achieving this aim is to make the equation manifestly tensorial. Certainly there is nothing wrong if a law is expressed in a form that is not generally covariant, as long as we know that it is possible to rewrite it in a coordinate-independent way. On the other hand, it is *always* possible to write laws in a coordinate-independent way, if the laws are well-defined to begin with. A physical system acting in a certain way doesn't know which coordinate system you are using to describe it; consequently, anything deserving of the name "law of physics" (as opposed to some particular statement of that law) must be independent of coordinates. An insistence on explicit coordinate-independence says nothing about the adaptation of laws to curved spacetime; as we have seen, manifestly tensorial equations take on the same form regardless of the geometry.

Consider Maxwell's equations in flat spacetime, as we wrote them in Chapter 1:

$$\partial_\mu F^{\nu\mu} = J^\nu. \tag{4.109}$$

The right-hand side is a well-defined tensor, while the left-hand side is not, due to the appearance of the partial derivative. That's okay, since we know that this equation is valid only in inertial coordinates in Minkowski space. A coordinate-invariant way of expressing the same law is

$$\nabla_\mu F^{\nu\mu} = J^\nu. \tag{4.110}$$

No physical principle needs to be invoked to conclude that this is the correct formulation in Minkowski space; it is the *unique* tensorial equation, which is equivalent to (4.109) in inertial coordinates. It is not the unique generalization to curved spacetime, since we could imagine new terms involving products of $F_{\mu\nu}$ and $R^\rho{}_{\sigma\mu\nu}$; the status of such additional terms is directly addressed by the minimal-coupling assumption, point four in the above list. By itself, however, making things "tensorial" or "generally covariant" is a simple matter of logical necessity, not a physical principle that one could imagine disproving by experiment. (Another spin on the same idea is "diffeomorphism invariance," discussed in the Appendix B.)

The second purported consequence of the Equivalence Principle from our list above is much deeper, and by no means obvious. Although he was inspired by the EP, this geometric insight was Einstein's great breakthrough. At the beginning of Chapter 2 we discussed why such an insight was warranted: the EP implies that gravity is universal, which implies in turn that gravitational fields become impossible to measure in small regions of spacetime, a feature which in turn is most directly implemented by identifying gravitation with the effects of spacetime geometry. These steps are well-motivated suggestions, not rigorously derived consequences; once we have the idea that there is a metric whose curvature gives rise to gravity, we can check its usefulness by comparing with experiment. As we've mentioned, it passes with flying colors. An accumulation of evidence (such as the gravitational redshift discussed in Chapter 2) is consistent with the idea that idealized rods and clocks behave as they should if the geometry of spacetime were curved. Still, one should not imagine proving that there really is a metric with the desired properties; we make the hypothesis, test it against ever-more precise experiments, and deduce its range of usefulness. Indeed, the demands of eventually reconciling general relativity with quantum mechanics suggest to many that the metric will ultimately be revealed as a concept derived from a more fundamental collection of degrees of freedom. For our present purposes this ultimate resolution doesn't matter; the idea of a curved metric has proven its usefulness beyond a reasonable doubt, and we work to extend our understanding of its properties until they run up against insurmountable obstacles (either theoretical or empirical).

Given our conviction that the effects of gravitation are best ascribed to the curvature of a metric on spacetime, what would we conclude if experiments were to detect an apparent violation of the Equivalence Principle? For example, we might imagine an experiment that revealed that the acceleration of test bodies in the direction of the Earth or Sun actually did depend, ever so slightly, on the composition of the test body. (The best current limits on such anomalous accelerations constrain them to be less than 10^{-12} times that due to gravity.)[3] In such a circumstance, nobody would really be tempted to declare that general relativity had been completely undermined and it was necessary to start over. Rather, we would return to the definition of "test body," which includes the proviso that the body be uncharged. An electron, for example, would not make a good test body, as it would be buffeted about by ambient electromagnetic fields as well as by gravity. Similarly, by far the most straightforward explanation of any hypothetical anomalous acceleration on purportedly neutral test bodies would be to imagine that we had discovered the existence of a new long-range field, under which our test bodies were actually charged. To have remain undetected thus far, such a field must be either very weakly coupled, or must couple almost universally, so as to mimic the effects of gravity. We could imagine, for example, scalar fields that couple to the trace of the energy-momentum tensor, or vector fields that couple to baryon number. The mass of ordinary test bodies is almost proportional to their baryon number, which counts the number of protons and neutrons in the body. It is therefore sometimes convenient to think of "tests of the Equivalence Principle" as tests of the third of our statements above—that there do not exist any other fields that resemble gravity (where a field resembles gravity if it is long-range and couples almost universally to mass). Again, detecting a violation of this hypothesis would be most directly interpreted as discovery of a new "fifth force" rather than as a repudiation of Einstein's ideas. As to whether we should expect to discover such a new field if we improve upon current experiments, it is hard to say; on the one hand, it is easy to concoct models with new long-range forces, but on the other hand, they would typically be strong enough to already have been detected. At this stage it is still worthwhile to keep an open mind.

Beyond the very existence of the metric, the heart of the Equivalence Principle lies in the fourth of our formulations, that the interactions of matter fields to curvature are minimal: they do not involve direct couplings to the Riemann tensor or its contractions. For example, we could consider the following possible alternative

[3] Y. Su et al., *Phys. Rev.* **D 50**, 3614 (1994).

to the conventional geodesic equation:

$$\frac{d^2x^\mu}{d\lambda^2} + \Gamma^\mu_{\rho\sigma}\frac{dx^\rho}{d\lambda}\frac{dx^\sigma}{d\lambda} = \alpha\,(\nabla_\sigma R)\frac{dx^\mu}{d\lambda}\frac{dx^\sigma}{d\lambda}, \tag{4.111}$$

where R is the Ricci scalar and α is a coupling constant. This equation also reduces to straight-line motion in flat spacetime, but would allow for direct detection of spacetime curvature in small regions by measurement of the coupling to $\nabla_\sigma R$. Why, then, does nature choose the simple geodesic equation? As a first step toward an answer, consider the dimensions of the coupling α. Since $c = 1$ and space and time have the same units, we can use length as our basic dimension. The metric, the inverse metric, and $dx^\mu/d\lambda$ are then dimensionless. The partial derivative operator has units of inverse length, as does the covariant derivative. The Christoffel symbols involve first derivatives of the metric, and thus have dimensions of inverse length; similarly, the Riemann tensor, Ricci tensor, and Ricci scalar have dimensions of inverse length squared:

$$\left[\frac{dx^\mu}{d\lambda}\right] = [g_{\mu\nu}] = [g^{\mu\nu}] = L^0, \quad [\nabla_\mu] = [\Gamma^\mu_{\rho\sigma}] = L^{-1}, \quad [R] = L^{-2}. \tag{4.112}$$

To be consistent, the coupling α must have dimensions of length squared:

$$[\alpha] = L^2. \tag{4.113}$$

The square root of α therefore defines a length scale; what should the length scale be? We don't know for sure, but there is every reason to believe it should be extremely small. There are two arguments for this. One is that, since the coupling represented by α is of gravitational origin, the only reasonable expectation for the relevant length scale is

$$\alpha \sim l_{\mathrm{P}}^2, \tag{4.114}$$

where l_{P} is the Planck length. Another reason is simply a more sophisticated version of this "what else could it be?" rationale. Although general relativity is a classical theory, at a deeper level we expect that it is merely an effective field theory describing an underlying quantum-mechanical structure. Even without knowing what this structure may be, a generic expectation (derived from our experience with quantum field theories we do understand) is that the effective classical limit should contain all possible interactions, but with dimensionful length parameters representing scales at which new degrees of freedom become important (recall our discussion of effective field theory at the end of Chapter 1). Thus, the Fermi theory of the weak interactions contains a length scale, which we now know to correspond to the scale of electroweak symmetry breaking where W and Z bosons become relevant. Since we do not expect new gravitational physics to arise before the Planck scale, the higher-order interactions associated with gravity should be suppressed by appropriate powers of the Planck length.

How much suppression does this represent? One measure would be to compare l_P (and thus the likely value of the parameter α) to a typical gravitational length scale near the vicinity of the Earth. The strength of gravity on Earth is characterized by the acceleration due to gravity, $a_g = 980$ cm/sec^2. To construct a quantity with dimensions of length, we define

$$l_\oplus = c^2/a_g \sim 10^{18} \text{ cm}, \tag{4.115}$$

where the symbol $\oplus$ in this context stands for the Earth (not a direct sum). So the relative strength of higher-order gravitational effects is measured by

$$\frac{l_P}{l_\oplus} \sim 10^{-51}. \tag{4.116}$$

In fact, since we expect $\alpha \sim l_P^2$, the suppression will be of order 10^{-102}. Consequently, there seems to be little need to worry about the possible role of such couplings. But dramatic departures should be kept in mind; recent ideas about large extra dimensions have opened up the possibility of observing direct gravitational interactions at particle accelerators. Ultimately, there is no way to resolve these problems by pure thought alone; only experiment can decide among the alternatives.

4.8 ■ ALTERNATIVE THEORIES

General relativity has passed a wide variety of experimental tests. Nevertheless, it is always possible that the next experiment we do will reveal a deviation from Einstein's original formulation. Let us therefore briefly consider ways in which general relativity could be modified. There are an uncountable number of such ways, but we will consider four different possibilities:

- gravitational scalar fields
- extra spatial dimensions
- higher-order terms in the action
- nonChristoffel connections

A popular set of alternative models are known as **scalar-tensor theories** of gravity, since they involve both the metric tensor, $g_{\mu\nu}$ and a scalar field, λ. In particular, the scalar field couples directly to the curvature scalar, not simply to the metric (as the Equivalence Principle would seem to imply). The action can be written as a sum of a gravitational piece, a pure-scalar piece, and a matter piece:

$$S = S_{fR} + S_\lambda + S_M, \tag{4.117}$$

where

$$S_{fR} = \int d^4x\sqrt{-g}\, f(\lambda) R, \tag{4.118}$$

$$S_\lambda = \int d^4x\sqrt{-g}\left[-\frac{1}{2}h(\lambda)g^{\mu\nu}(\partial_\mu\lambda)(\partial_\nu\lambda) - U(\lambda)\right], \tag{4.119}$$

and

$$S_{\rm M} = \int d^4x\sqrt{-g}\,\widehat{\mathcal{L}}_{\rm M}(g_{\mu\nu}, \psi_i). \tag{4.120}$$

Here, $f(\lambda)$, $h(\lambda)$ and $U(\lambda)$ are functions that define the theory, and the matter Lagrangian $\widehat{\mathcal{L}}_{\rm M}$ depends on the metric and a set of matter fields ψ_i, but not on λ. By change of variables we can always set $h(\lambda) = 1$, but we leave it here to facilitate comparison with models found in the literature.

The equations of motion for this theory include the gravitational equation (from varying with respect to the metric), and the scalar equation (from varying with respect to λ), as well as the appropriate matter equations. Let's start with the gravitational equation, which we can derive by following the same steps as for the ordinary Hilbert action (4.55). We consider perturbations of the metric,

$$g^{\mu\nu} \rightarrow g^{\mu\nu} + \delta g^{\mu\nu}. \tag{4.121}$$

Following the procedure from Section 4.3, the variation of the gravitational part of the action is

$$\delta S_{fR} = \int d^4x\sqrt{-g}\, f(\lambda)\Big[\Big(R_{\mu\nu} - \frac{1}{2}Rg_{\mu\nu}\Big)\delta g^{\mu\nu} + \nabla_\sigma\nabla^\sigma(g_{\mu\nu}\delta g^{\mu\nu}) - \nabla_\mu\nabla_\nu(\delta g^{\mu\nu})\Big]. \tag{4.122}$$

For the Hilbert action, f is a constant, so the last two terms are total derivatives, which can be converted to surface terms through integration by parts and therefore ignored. Now integration by parts (twice) picks up derivatives of f, and we obtain

$$\delta S_{fR} = \int d^4x\sqrt{-g}\,\left[f(\lambda)G_{\mu\nu} + g_{\mu\nu}\Box f - \nabla_\mu\nabla_\nu f\right]\delta g^{\mu\nu}, \tag{4.123}$$

where $G_{\mu\nu}$ is the Einstein tensor. We have discarded surface terms as usual, although there are subtleties concerning boundary contributions in this case; see Wald (1984) for a discussion. The gravitational equation of motion, including contributions from S_λ and $S_{\rm M}$, is thus

$$G_{\mu\nu} = f^{-1}(\lambda)\left(\tfrac{1}{2}T^{(\rm M)}_{\mu\nu} + \tfrac{1}{2}T^{(\lambda)}_{\mu\nu} + \nabla_\mu\nabla_\nu f - g_{\mu\nu}\Box f\right), \tag{4.124}$$

where the energy-momentum tensors are $T^{(i)}_{\mu\nu} = -2(-g)^{-1/2}\delta S_i/\delta g^{\mu\nu}$; in particular,

$$T^{(\lambda)}_{\mu\nu} = h(\lambda)\nabla_\mu\lambda\nabla_\nu\lambda - g_{\mu\nu}\left[\tfrac{1}{2}h(\lambda)g^{\rho\sigma}\nabla_\rho\lambda\nabla_\sigma\lambda + U(\lambda)\right]. \tag{4.125}$$

From looking at the coefficient of $T^{(\mathrm{M})}_{\mu\nu}$ in (4.124), we see that when the scalar field is *constant* (or practically so), we may identify $f(\lambda) = 1/(16\pi G)$, as makes sense from the original action (4.118). Meanwhile, if λ varies slightly from point to point in spacetime, it would be interpreted as a spacetime-dependent Newton's constant. The dynamics that control this variation are determined by the equation of motion for λ, which is straightforward to derive as

$$h\,\Box\lambda + \tfrac{1}{2}h' g^{\mu\nu}\nabla_\mu\lambda\nabla_\nu\lambda - U' + f'R = 0, \tag{4.126}$$

where primes denote differentiation with respect to λ. Notice that if we set $h(\lambda) = 1$ to get a conventional kinetic term for the scalar, λ obeys a conventional scalar-field equation of motion, with an additional coupling to the curvature scalar. In the real world, we don't want $f(\lambda)$ to vary too much, as it would have observable consequences in the classic experimental tests of GR in the solar system, and also in cosmological tests such as primordial nucleosynthesis. This can be ensured either by choosing $U(\lambda)$ so that there is a minimum to the potential and λ cannot deviate too far without a large input of energy—in other words, λ has a large mass—or by choosing $f(\lambda)$ and $h(\lambda)$ so that large changes in λ give rise to relatively small changes in the effective value of Newton's constant.

One of the earliest scalar-tensor models is known as Brans–Dicke theory, and corresponds in our notation to the choices

$$f(\lambda) = \frac{\lambda}{16\pi}, \quad h(\lambda) = \frac{\omega}{8\pi\lambda}, \quad U(\lambda) = 0. \tag{4.127}$$

where ω is a coupling constant. The scalar-tensor action takes the form

$$S_{\mathrm{BD}} = \int d^4x\sqrt{-g}\left[\frac{\lambda}{16\pi}R - \frac{\omega}{16\pi}g^{\mu\nu}\frac{(\partial_\mu\lambda)(\partial_\nu\lambda)}{\lambda}\right]. \tag{4.128}$$

In the Brans–Dicke theory, the scalar field is massless, but in the $\omega \to \infty$ limit the field becomes nondynamical and ordinary GR is recovered. Current bounds from Solar System tests imply $\omega > 500$, so if there is such a scalar field it must couple only weakly to the Ricci scalar.

A popular approach to dealing with scalar-tensor theories is to perform a conformal transformation to bring the theory in to a form that looks like conventional GR. We define a conformal metric

$$\tilde{g}_{\mu\nu} = 16\pi\widetilde{G}f(\lambda)g_{\mu\nu}, \tag{4.129}$$

where $\widetilde{G}$ will become Newton's constant in the conformal frame. Using formulae for conformal transformations from the Appendix G, the action S_{fR} from (4.118) becomes

$$\begin{aligned} S_{fR} &= \int d^4x\sqrt{-g}\, f(\lambda)R \\ &= \int d^4x\sqrt{-\tilde{g}}\,(16\pi\tilde{G})^{-1}\left[\tilde{R} - \frac{3}{2}\tilde{g}^{\rho\sigma} f^{-2}\left(\frac{df}{d\lambda}\right)^2 (\tilde{\nabla}_\rho\lambda)(\tilde{\nabla}_\sigma\lambda)\right], \end{aligned} \tag{4.130}$$

where as usual we have integrated by parts and discarded surface terms. In the conformal frame, therefore, the curvature scalar appears by itself, not multiplied by any function of λ. This frame is sometimes called the **Einstein frame**, since Einstein's equations for the conformal metric $\tilde{g}_{\mu\nu}$ take on their conventional form. The original frame with metric $g_{\mu\nu}$ is called the **Jordan frame**, or sometimes the **string frame**. (String theory typically predicts a scalar-tensor theory rather than ordinary GR, and the string worldsheet responds to the metric $g_{\mu\nu}$.)

Before going on with our analysis of the conformally-transformed theory, consider what happens if we choose

$$f(\lambda) = e^{\lambda/\sqrt{3}}, \quad h(\lambda) = 0, \quad U(\lambda) = 0, \tag{4.131}$$

a specific choice for $f(\lambda)$, but turning off the pure scalar terms in S_λ. Then we notice that the Einstein frame action (4.130) actually includes a conventional kinetic term for the scalar, even though it wasn't present in the Jordan frame action (4.118). Even without an explicit kinetic term for λ, the degrees of freedom of this theory include a propagating scalar as well as the metric. This should hopefully become more clear after we examine the degrees of freedom of the gravitational field in Chapter 7. There we will find that the metric $g_{\mu\nu}$ actually includes scalar (spin-0) and vector (spin-1) degrees of freedom as well as the expected tensor (spin-2) degrees of freedom; however, with the standard Hilbert action, these degrees of freedom are constrained rather than freely propagating. What we have just found is that multiplying R by a scalar in the action serves to bring the scalar degree of freedom to life, which is revealed explicitly in the Einstein frame.

If we do choose to include the pure-scalar action S_λ, we obtain

$$S_{fR} + S_\lambda = \int d^4x\sqrt{-\tilde{g}}\left[\frac{\tilde{R}}{16\pi\tilde{G}} - \frac{1}{2}K(\lambda)\tilde{g}^{\rho\sigma}(\tilde{\nabla}_\rho\lambda)(\tilde{\nabla}_\sigma\lambda) - \frac{U(\lambda)}{(16\pi\tilde{G})^2 f^2(\lambda)}\right], \tag{4.132}$$

where

$$K(\lambda) = \frac{1}{16\pi\tilde{G}f^2}\left[fh + 3(f')^2\right]. \tag{4.133}$$

We can make our action look utterly conventional by defining a new scalar field ϕ via

$$\phi = \int K^{1/2}\, d\lambda, \tag{4.134}$$

in terms of which the action becomes

$$S_{fR} + S_\lambda = \int d^4x \sqrt{-\tilde{g}} \left[\frac{\tilde{R}}{16\pi \tilde{G}} - \frac{1}{2}\tilde{g}^{\rho\sigma}(\tilde{\nabla}_\rho \phi)(\tilde{\nabla}_\sigma \phi) - V(\phi) \right], \tag{4.135}$$

where

$$V(\phi) = \frac{U(\lambda(\phi))}{(16\pi \tilde{G})^2 f^2(\lambda(\phi))}. \tag{4.136}$$

Amazingly, in the Einstein frame we have a completely ordinary theory of a scalar field in curved spacetime. So long as $f(\lambda)$ is well-behaved, the variables $(\tilde{g}_{\mu\nu}, \phi)$ can be used instead of $(g_{\mu\nu}, \lambda)$, in the sense that varying with respect to the new variables is equivalent to starting with the original equations of motion (4.124) and (4.126) and then doing the transformations (4.129) and (4.134).

Finally, we add in the matter action (4.120). Varying with respect to $\tilde{g}_{\mu\nu}$ will yield an energy-momentum tensor in the Einstein frame. In the original variables $(g_{\mu\nu}, \lambda)$, we knew that $S_{\rm M}$ was independent of λ, but now it will depend on both of the new variables $(\tilde{g}_{\mu\nu}, \phi)$; we can use the chain rule to characterize this dependence. Let us also assume that $S_{\rm M}$ depends on $g_{\mu\nu}$ only algebraically, not through derivatives. This will hold for ordinary scalar-field or gauge-field matter; things become more complicated for fermions, which we won't discuss here. We obtain

$$\begin{aligned} \tilde{T}_{\mu\nu} &\equiv -2\frac{1}{\sqrt{-\tilde{g}}}\frac{\delta S_{\rm M}}{\delta \tilde{g}^{\mu\nu}} \\ &= -2\frac{1}{\sqrt{-\tilde{g}}}\frac{\partial g^{\alpha\beta}}{\partial \tilde{g}^{\mu\nu}}\frac{\delta S_{\rm M}}{\delta g^{\alpha\beta}} \\ &= -2(16\pi \tilde{G} f)^{-1}\frac{1}{\sqrt{-g}}\delta^\alpha_\mu \delta^\beta_\nu \frac{\delta S_{\rm M}}{\delta g^{\alpha\beta}} \\ &= (16\pi \tilde{G} f)^{-1} T_{\mu\nu}. \end{aligned} \tag{4.137}$$

A similar trick works for the coupling of matter to ϕ, which comes from varying $S_{\rm M}$ with respect to ϕ, using $g^{\alpha\beta} = 16\pi \tilde{G} f \tilde{g}^{\alpha\beta}$:

$$\begin{aligned} \frac{\delta S_{\rm M}}{\delta \phi} &= \frac{\partial g^{\alpha\beta}}{\partial \phi}\frac{\delta S_{\rm M}}{\delta g^{\alpha\beta}} \\ &= \left(16\pi \tilde{G}\frac{df}{d\phi}\tilde{g}^{\alpha\beta}\right)\left(-\frac{1}{2}\sqrt{-g}T^{\rm M}_{\alpha\beta}\right) \\ &= -\frac{1}{2f}\frac{df}{d\phi}\sqrt{-\tilde{g}}\tilde{T}^{\rm M}, \end{aligned} \tag{4.138}$$

where

$$\widetilde{T}^{(\mathrm{M})} = \tilde{g}^{\alpha\beta}\widetilde{T}^{(\mathrm{M})}_{\alpha\beta} = \frac{1}{(16\pi\widetilde{G}f)^2} g^{\alpha\beta} T^{(\mathrm{M})}_{\alpha\beta} \tag{4.139}$$

is the trace of the energy-momentum tensor in the conformal frame.

Varying (4.135) with respect to $\tilde{g}_{\mu\nu}$ and ϕ returns equations of motion equivalent to Einstein's equations and an equation for ϕ. The gravitational equation is

$$\widetilde{G}_{\mu\nu} = 8\pi\widetilde{G}\left(\widetilde{T}^{(\mathrm{M})}_{\mu\nu} + \widetilde{T}^{(\phi)}_{\mu\nu}\right), \tag{4.140}$$

where

$$\widetilde{T}^{(\phi)}_{\mu\nu} = \widetilde{\nabla}_\mu\phi\widetilde{\nabla}_\nu\phi - \tilde{g}_{\mu\nu}\left[\tfrac{1}{2}\tilde{g}^{\rho\sigma}\widetilde{\nabla}_\rho\phi\widetilde{\nabla}_\sigma\phi + V(\phi)\right], \tag{4.141}$$

and the scalar field equation is

$$\widetilde{\Box}\phi - \frac{dV}{d\phi} = \frac{1}{2f}\frac{df}{d\phi}\widetilde{T}^{(\mathrm{M})}. \tag{4.142}$$

Given that (4.140) looks just like Einstein's equation with both matter and scalar-field sources, why should we even bother to call this scalar-tensor theory an alternative to GR? Isn't it the same theory, just in different variables? In fact it is not the same, because of the dependence of S_{M} on ϕ in the Einstein frame. In particular, physical test particles will move along geodesics of $g_{\mu\nu}$, which will not generally coincide with those of $\tilde{g}_{\mu\nu}$. The original metric is the one that test particles "see." So either we work in the original variables $(g_{\mu\nu}, \lambda)$, where the gravitational field equation is altered, or we use the new variables $(\tilde{g}_{\mu\nu}, \phi)$, in which the equations of motion for matter are altered; either way, there will be unambiguously measurable departures (in principle) from ordinary GR.

Another way to modify general relativity is to allow for the existence of extra spatial dimensions; in fact the physical consequences of extra dimensions turn out to be closely related to those of scalar-tensor theories. By extra dimensions we don't simply mean considering GR in higher-dimensional spaces, but rather considering models in which the spacetime appears four-dimensional on large scales even though there are really $4 + d$ total dimensions. The simplest way for this to happen is if the extra d dimensions are "compactified" on some manifold; it is this possibility we consider here.[4] Models of this kind are known as Kaluza-Klein theories.

Let G_{ab} be the metric for a $(4 + d)$-dimensional spacetime with coordinates X^a, where indices a, b run from 0 to $d + 3$.

[4]We follow the analysis of S.M. Carroll, J. Geddes, M. Hoffman, and R.M. Wald, *Phys. Rev. D* **66**, 024036 (2002); `http://arxiv.org/hep-th/0110149`. The original papers on extra dimensions are those by Kaluza and Klein: T. Kaluza, *Sitzungsber. Preuss. Akad. Wiss. Berlin (Math. Phys.)* **K1**, 966 (1921); O. Klein, *Z. Phys.* **37**, 895 (1926) [*Surveys High Energ. Phys.* **5**, 241 (1926)]; O. Klein, *Nature* **118**, 516 (1926).

$$ds^2 = G_{ab}\mathrm{d}X^a\mathrm{d}X^b = g_{\mu\nu}(x)\mathrm{d}x^\mu\mathrm{d}x^\nu + b^2(x)\gamma_{ij}(y)\mathrm{d}y^i\mathrm{d}y^j, \tag{4.143}$$

where the x^μ are coordinates in the four-dimensional spacetime and the y^i are coordinates on the extra-dimensional manifold, taken to be a maximally symmetric space with metric γ_{ij}. Of course the geometry of the extra dimensions is actually something dynamical that should be determined by solving the full equations of motion, but we are going to take (4.143) as a simplifying ansatz. (In a more complete treatment, we would expand the dynamical modes of the compactified geometry as a Fourier series, and show that the modes we are presently neglecting have larger masses than the overall-size mode we are choosing to examine.) The action is the $(4+d)$-dimensional Hilbert action plus a matter term:

$$S = \int d^{4+d}X\sqrt{-G}\left(\frac{1}{16\pi G_{4+d}}R[G_{ab}] + \widehat{\mathcal{L}}_{\mathrm{M}}\right), \tag{4.144}$$

where $\sqrt{-G}$ is the square root of minus the determinant of G_{ab}, $R[G_{ab}]$ is the Ricci scalar of G_{ab}, and $\widehat{\mathcal{L}}_{\mathrm{M}}$ is the matter Lagrange density with the metric determinant factored out.

The first step is to dimensionally reduce the action (4.144). By this we mean to actually perform the integral over the extra dimensions, which is possible because we have assumed that the extra-dimensional scale factor b is independent of y^i. Therefore we can express everything in terms of $g_{\mu\nu}$, γ_{IJ}, and $b(x)$, integrate over the extra dimensions, and arrive at an effective four-dimensional theory. From the metric (4.143) we have

$$\sqrt{-G} = b^d\sqrt{-g}\sqrt{\gamma}, \tag{4.145}$$

and we can evaluate the curvature scalar for this metric to obtain

$$\begin{aligned} R[G_{ab}] = {} & R[g_{\mu\nu}] + b^{-2}R[\gamma_{ij}] - 2db^{-1}g^{\mu\sigma}\nabla_\mu\nabla_\sigma b \\ & - d(d-1)b^{-2}g^{\mu\sigma}(\nabla_\mu b)(\nabla_\sigma b), \end{aligned} \tag{4.146}$$

where ∇_μ is the covariant derivative associated with the four-dimensional metric $g_{\mu\nu}$. We denote by $\mathcal{V}$ the volume of the extra dimensions when $b=1$; it is given by

$$\mathcal{V} = \int d^d y\sqrt{\gamma}. \tag{4.147}$$

The four-dimensional Newton's constant G_4 is determined by evaluating the coefficient of the curvature scalar in the action; we find that G_4 is related to its higher-dimensional analogue by

$$\frac{1}{16\pi G_4} = \frac{\mathcal{V}}{16\pi G_{4+d}}. \tag{4.148}$$

We are thus left with

$$S = \int d^4x\sqrt{-g}\left\{\frac{1}{16\pi G_4}\left[b^d R[g_{\mu\nu}] + d(d-1)b^{d-2}g^{\mu\nu}(\nabla_\mu b)(\nabla_\nu b)\right.\right.$$
$$\left.\left. + d(d-1)\kappa b^{d-2}\right] + \mathcal{V}b^d\widehat{\mathcal{L}}_{\rm M}\right\}, \qquad (4.149)$$

where we have integrated by parts for convenience, and introduced the curvature parameter κ of γ_{ij}, given by

$$\kappa = \frac{R[\gamma_{ij}]}{d(d-1)}. \qquad (4.150)$$

Comparing to (4.117)–(4.120), we see that the dimensionally-reduced action is precisely that of a scalar-tensor theory; the size of the extra dimensions plays the role of the scalar field. We can therefore make it look more conventional by performing a change of variables and a conformal transformation,

$$\beta(x) = \ln b,$$
$$\tilde{g}_{\mu\nu} = e^{d\beta} g_{\mu\nu}, \qquad (4.151)$$

which turns the reduced action into that of a scalar field coupled to gravity in the Einstein frame. Following the same procedure as outlined in our discussion of scalar-tensor theories yields

$$S = \int d^4x\sqrt{-\tilde{g}}\left\{\frac{1}{16\pi G_4}\left[R[\tilde{g}_{\mu\nu}] - \frac{1}{2}d(d+2)\tilde{g}^{\mu\nu}(\tilde{\nabla}_\mu\beta)(\tilde{\nabla}_\nu\beta)\right.\right.$$
$$\left.\left. + d(d-1)\kappa e^{(d+2)\beta}\right] + \mathcal{V}e^{-d\beta}\widehat{\mathcal{L}}_{\rm M}\right\}, \qquad (4.152)$$

where we have dropped terms that are total derivatives.

To turn β into a canonically normalized scalar field, we make one final change of variables, to

$$\phi = \sqrt{\frac{d(d+2)}{2}}\bar{m}_{\rm P}\beta, \qquad (4.153)$$

where the reduced Planck mass is $\bar{m}_{\rm P} = (8\pi G_4)^{-1/2}$. We are then left with

$$S = \int d^4x\sqrt{-\tilde{g}}\left\{\frac{1}{16\pi G_4}R[\tilde{g}_{\mu\nu}] - \frac{1}{2}\tilde{g}^{\mu\nu}(\tilde{\nabla}_\mu\phi)(\tilde{\nabla}_\nu\phi)\right.$$
$$\left. + \frac{1}{2}\kappa d(d-1)\bar{m}_{\rm P}^2 e^{-\sqrt{2(d+2)/d}\,\phi/\bar{m}_{\rm P}} + \mathcal{V}e^{-\sqrt{2d/(d+2)}\,\phi/\bar{m}_{\rm P}}\widehat{\mathcal{L}}_{\rm M}\right\}.$$
$$(4.154)$$

The scalar ϕ is known as the **dilaton** or **radion**, and characterizes the size of the extra-dimensional manifold.

The last two terms in (4.154) represent (minus) the potential $V(\phi)$. If we ignore the matter term $\widehat{\mathcal{L}}_{\rm M}$, the behavior of the dilaton will depend only on the sign of κ. If the extra-dimensional manifold is flat ($\kappa = 0$), the potential vanishes and we simply have a massless scalar field; this possibility runs afoul of the experimental constraints on scalar-tensor theories mentioned above. If there is curvature ($\kappa \neq 0$), the potential has no minimum; for $\kappa > 0$ the field will roll to $-\infty$, while for $\kappa < 0$ the field will roll to $+\infty$. But $\phi \propto \ln b$, so this means the scale factor $b(x)$ of the extra dimensions either shrinks to zero or becomes arbitrarily large, in either case ruining the hope for stable extra dimensions. Stability can be achieved, however, by choosing an appropriate matter Lagrangian, and an appropriate field configuration in the extra dimensions.

Let us now move on to a different kind of alternative theory, those that feature Lagrangians of more than second order in derivatives of the metric. We could imagine an action of the form

$$S = \int d^n x \sqrt{-g}(R + \alpha_1 R^2 + \alpha_2 R_{\mu\nu}R^{\mu\nu} + \alpha_3 g^{\mu\nu}\nabla_\mu R \nabla_\nu R + \cdots), \quad (4.155)$$

where the α's are coupling constants and the dots represent every other scalar we can make from the curvature tensor, its contractions, and its derivatives. Traditionally, such terms have been neglected on the reasonable grounds that they merely complicate a theory that is already both aesthetically pleasing and empirically successful. There is also, classically speaking, a more substantive objection. In conventional form, Einstein's equation leads to a well-posed initial value problem for the metric, in which coordinates and momenta specified at an initial time can be used to predict future evolution. With higher-derivative terms, we would require not only those data, but also some number of derivatives of the momenta; the character of the theory is dramatically altered.

However, there are also good reasons to consider such additional terms. As mentioned in our brief discussion of quantum gravity, one of the technical obstacles to consistent quantization of general relativity is that the theory is nonrenormalizable: Inclusion of higher-order quantum effects leads to infinite answers. With the appropriate combination of higher-order Lagrangian terms, it turns out that you can actually render the theory renormalizable, which gives some hope of constructing a consistent quantum theory.[5] Unfortunately, it turns out that renormalizability comes at too high a price; these models generally feature negative-energy field excitations (ghosts). Consequently, the purported vacuum state (empty space) would be unstable to decay into positive- and negative-energy modes, which is inconsistent with both empirical experience and theoretical prejudice.

Nevertheless, the prevailing current view is that GR is an effective theory valid at energies below the Planck scale, and we should actually include all of the pos-

[5] See, for example, K.S. Stelle, *Phys. Rev.* **D16**, 953 (1977).

sible higher-order terms; but they will be suppressed by appropriate powers of the Planck scale, just as we argued in our discussion of the Equivalence Principle in Section 4.7. They will therefore only become important when the length scale characteristic of the curvature approaches the Planck scale, which is far from any plausible experiment. Higher-order terms are therefore interesting in principle, but not in practice. On the other hand, similar reasoning would lead us to expect a huge vacuum energy term, since it is lower-order than the Hilbert action, which we know not to be true; so we should keep an open mind.

As a final alternative to general relativity, we should mention the possibility that the connection really is not derived from the metric, but in fact has an independent existence as a fundamental field. As one of the exercises you are asked to show that it is possible to consider the conventional action for general relativity but treat it as a function of both the metric $g_{\mu\nu}$ and a torsion-free connection $\Gamma^{\lambda}_{\rho\sigma}$, and the equations of motion derived from varying such an action with respect to the connection imply that $\Gamma^{\lambda}_{\rho\sigma}$ is actually the Christoffel connection associated with $g_{\mu\nu}$. We could drop the demand that the connection be torsion-free, in which case the torsion tensor could lead to additional propagating degrees of freedom. The basic reason why such theories do not receive much attention is simply because the torsion is itself a tensor; there is nothing to distinguish it from other, nongravitational tensor fields. Thus, we do not really lose any generality by considering theories of torsion-free connections (which lead to GR) plus any number of tensor fields, which we can name what we like. Similar considerations apply when we consider dropping the requirement of metric compatibility—any connection can be written as a metric-compatible connection plus a tensorial correction, so any such theory is equivalent to GR plus extra tensor fields, which wouldn't really deserve to be called an "alternative to general relativity".

4.9 ■ EXERCISES

1. The Lagrange density for electromagnetism in curved space is

$$\mathcal{L} = \sqrt{-g}\left(-\tfrac{1}{4}F^{\mu\nu}F_{\mu\nu} + A_{\mu}J^{\mu}\right), \tag{4.156}$$

where J^{μ} is the conserved current.

(a) Derive the energy-momentum tensor by functional differentiation with respect to the metric. You can assume that the $A_{\mu}J^{\mu}$ term does not contribute to the energy-momentum tensor.

(b) Consider adding a new term to the Lagrangian,

$$\mathcal{L}' = \beta R^{\mu\nu} g^{\rho\sigma} F_{\mu\rho} F_{\nu\sigma}.$$

How are Maxwell's equations altered in the presence of this term? Einstein's equation? Is the current still conserved?

2. We showed how to derive Einstein's equation by varying the Hilbert action with respect to the metric. They can also be derived by treating the metric and connection as independent degrees of freedom and varying separately with respect to them; this is known as the **Palatini formalism**. That is, we consider the action

$$S = \int d^4x \sqrt{-g}\, g^{\mu\nu} R_{\mu\nu}(\Gamma),$$

where the Ricci tensor is thought of as constructed purely from the connection, not using the metric. Variation with respect to the metric gives the usual Einstein's equations, but for a Ricci tensor constructed from a connection that has no a priori relationship to the metric. Imagining from the start that the connection is symmetric (torsion free), show that variation of this action with respect to the connection coefficients leads to the requirement that the connection be metric compatible, that is, the Christoffel connection. Remember that Stokes's theorem, relating the integral of the covariant divergence of a vector to an integral of the vector over the boundary, does not work for a general covariant derivative. The best strategy is to write the connection coefficients as a sum of the Christoffel symbols $\widetilde{\Gamma}^{\lambda}_{\mu\nu}$ and a tensor $C^{\lambda}{}_{\mu\nu}$,

$$\Gamma^{\lambda}_{\mu\nu} = \widetilde{\Gamma}^{\lambda}_{\mu\nu} + C^{\lambda}{}_{\mu\nu},$$

and then show that $C^{\lambda}{}_{\mu\nu}$ must vanish.

3. The four-dimensional δ-function on a manifold M is defined by

$$\int_M F(x^\mu) \left[\frac{\delta^{(4)}(x^\sigma - y^\sigma)}{\sqrt{-g}}\right] \sqrt{-g}\, d^4x = F(y^\sigma), \tag{4.157}$$

for an arbitrary function $F(x^\mu)$. Meanwhile, the energy-momentum tensor for a pressureless perfect fluid (dust) is

$$T^{\mu\nu} = \rho U^\mu U^\nu, \tag{4.158}$$

where ρ is the energy density and U^μ is the four-velocity. Consider such a fluid that consists of a single particle traveling on a world line $x^\mu(\tau)$, with τ the proper time. The energy-momentum tensor for this fluid is then given by

$$T^{\mu\nu}(y^\sigma) = m \int_M \left[\frac{\delta^{(4)}(y^\sigma - x^\sigma(\tau))}{\sqrt{-g}}\right] \frac{dx^\mu}{d\tau}\frac{dx^\nu}{d\tau} d\tau, \tag{4.159}$$

where m is the rest mass of the particle. Show that covariant conservation of the energy-momentum tensor, $\nabla_\mu T^{\mu\nu} = 0$, implies that $x^\mu(\tau)$ satisfies the geodesic equation.

4. Show that the energy-momentum tensors for electromagnetism and for scalar field theory satisfy the dominant energy condition, and thus also the weak, null, and null dominant conditions. Show that they also satisfy $w \geq -1$.

5. A spacetime is static if there is a timelike Killing vector that is orthogonal to spacelike hypersurfaces. (See Appendices D and F for more discussion, including a definition of Raychaudhuri's equation.)

(a) Generally speaking, if a vector field v^μ is orthogonal to a set of hypersurfaces defined by $f =$ constant, then we can write the vector as $v_\mu = h\nabla_\mu f$ (here both f and h are functions). Show that this implies

$$v_{[\sigma}\nabla_\mu v_{\nu]} = 0.$$

(b) Imagine we have a perfect fluid with zero pressure (dust), which generates a solution to Einstein's equations. Show that the metric can be static only if the fluid four-velocity is parallel to the timelike (and hypersurface-orthogonal) Killing vector.

(c) Use Raychaudhuri's equation to prove that there is no static solution to Einstein's equations if the pressure is zero and the energy density is greater than zero.

6. Let K be a Killing vector field. Show that an electromagnetic field with potential $A_\mu = K_\mu$ solves Maxwell's equations if the metric is a vacuum solution to Einstein's equations. This is a slight cheat, since you won't be in vacuum if there is a nonzero electromagnetic field strength, but we assume the field strength is small enough not to dramatically affect the geometry.

CHAPTER

5 The Schwarzschild Solution

5.1 ■ THE SCHWARZSCHILD METRIC

The most obvious application of a theory of gravity is to a spherically symmetric gravitational field. This would be the relevant situation to describe, for example, the field created by the Earth or the Sun (to a good approximation), in which apples fall or planets move. Furthermore, our first concern is with exterior solutions (empty space surrounding a gravitating body), since understanding the motion of test particles outside an object is both easier and more immediately useful than considering the relatively inaccessible interior. In addition to its practical usefulness, the answer to this problem in general relativity will lead us to remarkable solutions describing new phenomena of great interest to physicists and astronomers: black holes. In this chapter we examine the simple case of vacuum solutions with perfect spherical symmetry; in the next chapter we consider features of black holes in more general contexts.

In GR, the unique spherically symmetric vacuum solution is the **Schwarzschild metric**; it is second only to Minkowski space in the list of important spacetimes. In spherical coordinates $\{t, r, \theta, \phi\}$, the metric is given by

$$\boxed{ds^2 = -\left(1 - \frac{2GM}{r}\right) dt^2 + \left(1 - \frac{2GM}{r}\right)^{-1} dr^2 + r^2\, d\Omega^2,} \tag{5.1}$$

where $d\Omega^2$ is the metric on a unit two-sphere,

$$d\Omega^2 = d\theta^2 + \sin^2\theta\, d\phi^2. \tag{5.2}$$

The constant M is interpreted as the mass of the gravitating object (although some care is required in making this identification). In this section we will derive the Schwarzschild metric by trial and error; in the next section we will be more systematic in both the derivation of the solution and its consequences.

Since we are interested in the solution *outside* a spherical body, we care about Einstein's equation in vacuum,

$$R_{\mu\nu} = 0. \tag{5.3}$$

Our hypothesized source is static (unevolving) and spherically symmetric, so we will look for solutions that also have these properties. Rigorous definitions of both "static" and "spherically symmetric" require some care, due to subtleties of coordinate independence. For now we will interpret static to imply two conditions: that all metric components are independent of the time coordinate, and that there are no time-space cross terms ($dt dx^i + dx^i dt$) in the metric. The latter condition makes sense if we imagine performing a time inversion $t \to -t$; the dt^2 term remains invariant, as do any $dx^i dx^j$ terms, while cross terms would not. Since we hope to find a solution that is independent of time, it should be invariant under time reversal, and we therefore leave cross terms out. To impose spherical symmetry, we begin by writing the metric of Minkowski space (a spherically symmetric spacetime we know something about) in polar coordinates $x^\mu = (t, r, \theta, \phi)$:

$$ds^2_{\text{Minkowski}} = -dt^2 + dr^2 + r^2\, d\Omega^2. \tag{5.4}$$

One requirement to preserve spherical symmetry is that we maintain the form of $d\Omega^2$; that is, if we want our spheres to be perfectly round, the coefficient of the $d\phi^2$ term should be $\sin^2\theta$ times that of the $d\theta^2$ term. But we are otherwise free to multiply all of the terms by separate coefficients, so long as they are only functions of the radial coordinate r:

$$ds^2 = -e^{2\alpha(r)}\, dt^2 + e^{2\beta(r)}\, dr^2 + e^{2\gamma(r)} r^2\, d\Omega^2. \tag{5.5}$$

We've expressed our functions as exponentials so that the signature of the metric doesn't change. In a full treatment, we would allow for complete freedom and see what happens.

We can use our ability to change coordinates to make a slight simplification to the static, spherically-symmetric metric (5.5), even before imposing Einstein's equation. Unlike other theories of physics, in general relativity we simultaneously define coordinates and the metric as a function of those coordinates. In other words, we don't know ahead of time what, for example, the radial coordinate r really is; we can only interpret it once the solution is in our hands. Let us therefore imagine defining a new coordinate $\bar{r}$ via

$$\bar{r} = e^{\gamma(r)} r, \tag{5.6}$$

with an associated basis one-form

$$d\bar{r} = e^{\gamma}\, dr + e^{\gamma} r\, d\gamma = \left(1 + r\frac{d\gamma}{dr}\right) e^{\gamma}\, dr. \tag{5.7}$$

In terms of this new variable, the metric (5.5) becomes

$$ds^2 = -e^{2\alpha(r)}\, dt^2 + \left(1 + r\frac{d\gamma}{dr}\right)^{-2} e^{2\beta(r) - 2\gamma(r)} d\bar{r}^2 + \bar{r}^2\, d\Omega^2, \tag{5.8}$$

where each function of r is a function of $\bar{r}$ in the obvious way. But now let us make the following relabelings:

$$\bar{r} \to r \tag{5.9}$$

$$\left(1 + r\frac{d\gamma}{dr}\right)^{-2} e^{2\beta(r)-2\gamma(r)} \to e^{2\beta}. \tag{5.10}$$

There is nothing to stop us from doing this, as they are simply labels, with no independent external definition. If you wish you can continue to use $\bar{r}$, and set (5.10) equal to $e^{2\bar{\beta}}$, but we won't bother. Our metric (5.8) becomes

$$ds^2 = -e^{2\alpha(r)}\,dt^2 + e^{2\beta(r)}\,dr^2 + r^2\,d\Omega^2. \tag{5.11}$$

This looks exactly like (5.5), except that the $e^{2\gamma}$ factor has disappeared. We have not set $e^{2\gamma}$ equal to one, which would be a statement about the geometry; we have simply chosen our radial coordinate such that this factor doesn't exist. Thus, (5.11) is precisely as general as (5.5).

Let's now take this metric and use Einstein's equation to solve for the functions $\alpha(r)$ and $\beta(r)$. We begin by evaluating the Christoffel symbols. If we use labels (t, r, θ, ϕ) for $(0, 1, 2, 3)$ in the usual way, the Christoffel symbols are given by

$$\begin{aligned}
\Gamma^t_{tr} &= \partial_r\alpha & \Gamma^r_{tt} &= e^{2(\alpha-\beta)}\partial_r\alpha & \Gamma^r_{rr} &= \partial_r\beta \\
\Gamma^\theta_{r\theta} &= \frac{1}{r} & \Gamma^r_{\theta\theta} &= -re^{-2\beta} & \Gamma^\phi_{r\phi} &= \frac{1}{r} \\
\Gamma^r_{\phi\phi} &= -re^{-2\beta}\sin^2\theta & \Gamma^\theta_{\phi\phi} &= -\sin\theta\cos\theta & \Gamma^\phi_{\theta\phi} &= \frac{\cos\theta}{\sin\theta}.
\end{aligned} \tag{5.12}$$

Anything not written down explicitly is meant to be zero, or related to what is written by symmetries. From these we get the following nonvanishing components of the Riemann tensor:

$$\begin{aligned}
R^t{}_{rtr} &= \partial_r\alpha\,\partial_r\beta - \partial_r^2\alpha - (\partial_r\alpha)^2 \\
R^t{}_{\theta t\theta} &= -re^{-2\beta}\,\partial_r\alpha \\
R^t{}_{\phi t\phi} &= -re^{-2\beta}\sin^2\theta\;\partial_r\alpha \\
R^r{}_{\theta r\theta} &= re^{-2\beta}\,\partial_r\beta \\
R^r{}_{\phi r\phi} &= re^{-2\beta}\sin^2\theta\;\partial_r\beta \\
R^\theta{}_{\phi\theta\phi} &= (1 - e^{-2\beta})\sin^2\theta.
\end{aligned} \tag{5.13}$$

Taking the contraction as usual yields the Ricci tensor:

$$\begin{aligned}
R_{tt} &= e^{2(\alpha-\beta)}\left[\partial_r^2\alpha + (\partial_r\alpha)^2 - \partial_r\alpha\,\partial_r\beta + \tfrac{2}{r}\partial_r\alpha\right] \\
R_{rr} &= -\partial_r^2\alpha - (\partial_r\alpha)^2 + \partial_r\alpha\,\partial_r\beta + \tfrac{2}{r}\partial_r\beta \\
R_{\theta\theta} &= e^{-2\beta}[r(\partial_r\beta - \partial_r\alpha) - 1] + 1 \\
R_{\phi\phi} &= \sin^2\theta\,R_{\theta\theta},
\end{aligned} \tag{5.14}$$

and for future reference we calculate the curvature scalar,

$$R = -2e^{-2\beta}\left[\partial_r^2\alpha + (\partial_r\alpha)^2 - \partial_r\alpha\partial_r\beta + \frac{2}{r}(\partial_r\alpha - \partial_r\beta) + \frac{1}{r^2}(1 - e^{2\beta})\right]. \tag{5.15}$$

With the Ricci tensor calculated, we would like to set it equal to zero. Since R_{tt} and R_{rr} vanish independently, we can write

$$0 = e^{2(\beta-\alpha)}R_{tt} + R_{rr} = \frac{2}{r}(\partial_r\alpha + \partial_r\beta), \tag{5.16}$$

which implies $\alpha = -\beta + c$, where c is some constant. We can set this constant equal to zero by rescaling our time coordinate by $t \to e^{-c}t$, after which we have

$$\alpha = -\beta. \tag{5.17}$$

Next let us turn to $R_{\theta\theta} = 0$, which now reads

$$e^{2\alpha}(2r\partial_r\alpha + 1) = 1. \tag{5.18}$$

This is equivalent to

$$\partial_r(re^{2\alpha}) = 1. \tag{5.19}$$

We can solve this to obtain

$$e^{2\alpha} = 1 - \frac{R_S}{r}, \tag{5.20}$$

where R_S is some undetermined constant. With (5.17) and (5.20), our metric becomes

$$ds^2 = -\left(1 - \frac{R_S}{r}\right)dt^2 + \left(1 - \frac{R_S}{r}\right)^{-1}dr^2 + r^2\,d\Omega^2. \tag{5.21}$$

We now have no freedom left except for the single constant R_S, so this form had better solve the remaining equations $R_{tt} = 0$ and $R_{rr} = 0$; it is straightforward to check that it does, for any value of R_S.

The only thing left to do is to interpret the constant R_S, called the **Schwarzschild radius**, in terms of some physical parameter. Nothing could be simpler. In Chapter 4 we found that, in the weak-field limit, the tt component of the metric around a point mass satisfies

$$g_{tt} = -\left(1 - \frac{2GM}{r}\right). \tag{5.22}$$

The Schwarzschild metric should reduce to the weak-field case when $r \gg 2GM$, but for the tt component the forms are already exactly the same; we need only identify

$$R_S = 2GM. \tag{5.23}$$

This can be thought of as the definition of the parameter M.

Our final result is the Schwarzschild metric, (5.1). We have shown that it is a static, spherically symmetric vacuum solution to Einstein's equation; M functions as a parameter, which we happen to know can be interpreted as the conventional Newtonian mass that we would measure by studying orbits at large distances from the gravitating source. It won't simply be the sum of the masses of the constituents of whatever body is curving spacetime, since there will be a contribution from what we might think of as the gravitational binding energy; however, in the weak field limit, the quantities will agree. Note that as $M \to 0$ we recover Minkowski space, which is to be expected. Note also that the metric becomes progressively Minkowskian as $r \to \infty$; this property is known as **asymptotic flatness.** A more technical definition involves matching regions at infinity in a conformal diagram, as discussed in the next chapter.

5.2 ■ BIRKHOFF'S THEOREM

Birkhoff's theorem is the statement that the Schwarzschild metric is the *unique* vacuum solution with spherical symmetry (and in particular, that there are no time-dependent solutions of this form); proving it is an instructive exercise, which consists of three major steps. First, we argue that a spherically symmetric spacetime can be foliated by two-spheres—in other words, that (almost) every point lies on a unique sphere that is left invariant by the generators of spherical symmetry. Second, we show on purely geometric grounds that the metric on such a space can always (at least in a local region) be put in the form

$$ds^2 = d\tau^2(a, b) + r^2(a, b)\, d\Omega^2(\theta, \phi), \tag{5.24}$$

where (a, b) are coordinates transverse to the spheres, and r is a function of these coordinates. Third, we plug this metric into Einstein's equation in vacuum to show that Schwarzschild is the unique solution. We will argue in favor of the first two points at a level of rigor that is likely to be convincing to most physicists, although mathematicians will be uneasy; the third point is straightforward calculation. For a more careful treatment see Hawking and Ellis (1973). We will use a few concepts from Appendix C, which may be useful to read at this point. Of course, if you are more interested in exploring properties of the Schwarzschild solution than in proving its uniqueness, you are welcome to skip right to the next section.

We begin with the concept of a four-dimensional spherically symmetric spacetime M. Spherically symmetric means having the same symmetries as a sphere. (In this chapter the word sphere refers specifically to S^2, not spheres of other dimension.) The symmetries of a sphere are precisely those of ordinary rotations in three-dimensional Euclidean space; in the language of group theory, they comprise the special orthogonal group SO(3). (Recall the discussion of the Lorentz and rotation groups in Chapter 1.) In the case of a metric on a manifold, symmetries are characterized by the existence of Killing vectors. In Section 3.8 we found the three Killing vectors of S^2, labeled (R, S, T); in (θ, ϕ) coordinates they take the form

$$
\begin{aligned}
R &= \partial_\phi \\
S &= \cos\phi\,\partial_\theta - \cot\theta\sin\phi\,\partial_\phi \\
T &= -\sin\phi\,\partial_\theta - \cot\theta\cos\phi\,\partial_\phi.
\end{aligned}
\tag{5.25}
$$

A spherically symmetric manifold is one that has three Killing vector fields that are the same as those on S^2. But how do we know, in a coordinate-independent way, that a set of Killing vectors on one manifold is the same as that on some other manifold? The structure of a set of symmetry transformations is given by the commutation relations of the transformations, which express the difference between performing two infinitesimal transformations in one order versus the reversed order. In group theory these are expressed by the Lie algebra of the symmetry generators, while in differential geometry they are expressed by the commutators of the Killing vector fields. There is a deep connection here, which we don't have time to pursue; see Schutz (1980). In the Exercises for Chapter 3 you verified that the commutators of the rotational Killing vectors (R, S, T) satisfied

$$
\begin{aligned}
[R, S] &= T \\
[S, T] &= R \\
[T, R] &= S.
\end{aligned}
\tag{5.26}
$$

This algebra of Killing vectors fully characterizes the kind of symmetry we have. A manifold will be said to possess **spherical symmetry** if and only if there are three Killing fields satisfying (5.26).

In Appendix C we discuss Frobenius's theorem, which states that if you have a set of vector fields whose commutator closes—the commutator of any two fields in the set is a linear combination of other fields in the set—then the integral curves of these vector fields fit together to describe submanifolds of the manifold on which they are all defined. The dimensionality of the submanifold may be smaller than the number of vectors, or it could be equal, but obviously not larger. Vector fields that obey (5.26) will of course form 2-spheres. Since the vector fields stretch throughout the space, every point will be on exactly one of these spheres. (Actually, it's almost every point—we will show below how it can fail to be absolutely every point.) Thus, we say that a spherically symmetric manifold can be foliated into spheres.

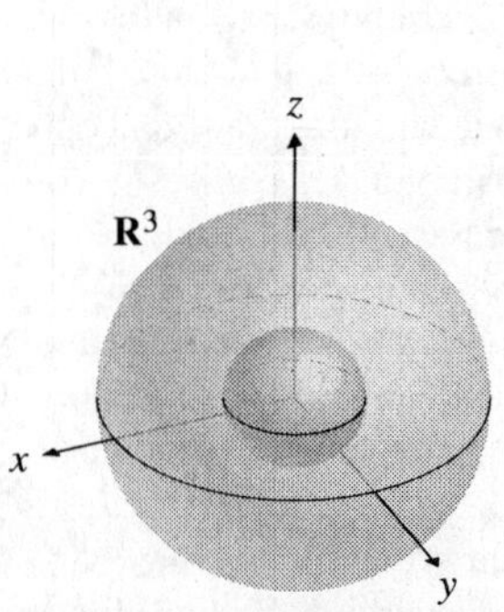

FIGURE 5.1 Foliating $\mathbf{R}^3$ (minus the origin) by two-spheres.

Let's consider some examples to bring this down to earth. The simplest example is flat three-dimensional Euclidean space. If we pick an origin, then $\mathbf{R}^3$ is clearly spherically symmetric with respect to rotations around this origin. Under such rotations (that is, under the flow of the Killing vector fields), points move into each other, but each point stays on an S^2 at a fixed distance from the origin.

These spheres foliate $\mathbf{R}^3$, as depicted in Figure 5.1. Of course, they don't really foliate all of the space, since the origin itself just stays put under rotations—it

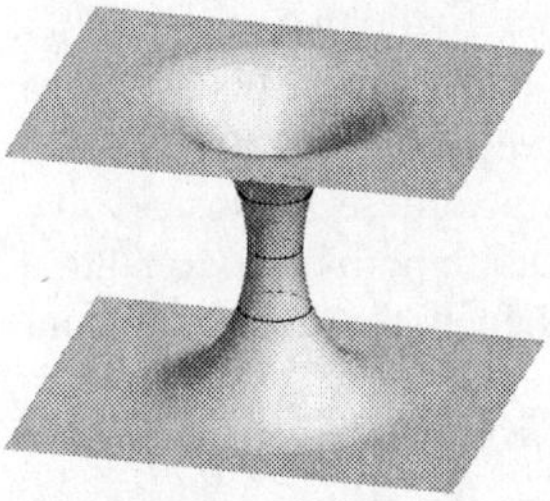

FIGURE 5.2 Foliation of a wormhole by two-spheres.

doesn't move around on some two-sphere. But it should be clear that almost all of the space is properly foliated, and this will turn out to be enough for us.

We can also have spherical symmetry without an origin to rotate things around. An example is provided by a wormhole, with topology $\mathbf{R} \times S^2$. If we suppress a dimension and draw our two-spheres as circles, such a space might look like Figure 5.2. In this case the entire manifold can be foliated by two-spheres.

Given that manifolds with SO(3) symmetry may be foliated by spheres, our second step is to show that the metric on M can be put into the form (5.24). The set of all the spheres forms a two-dimensional space (since a four-dimensional spacetime is being foliated with two-dimensional spheres). You might hope we could simply put coordinates (θ, ϕ) on each sphere, and coordinates (a, b) on the set of all spheres, for a complete set of coordinates (a, b, θ, ϕ) on M. Then each sphere is specified by $a =$ constant, $b =$ constant. We know that the metric on a round sphere is $d\Omega^2$, so this strategy would be sufficient to guarantee that the metric restricted to any fixed values $a = a_0$ and $b = b_0$ (so that $da = db = 0$) takes the form

$$ds^2(a_0, b_0, \theta, \phi) = f(a_0, b_0)\, d\Omega^2. \tag{5.27}$$

In particular, the function f must be independent of θ and ϕ, or the sphere would be lumpy rather than round. Furthermore, it's equally clear that the metric restricted to any fixed values $\theta = \theta_0$ and $\phi = \phi_0$ (so that $d\theta = d\phi = 0$) takes the form

$$ds^2(a, b, \theta_0, \phi_0) = d\tau^2(a, b). \tag{5.28}$$

Again, any dependence on θ or ϕ would destroy the symmetry; it would mean that the geometry transverse to the spheres depended on where you were on the sphere.

However, we have been too reckless by slapping down these coordinates, since we cannot rule out cross terms of the form $da d\theta + d\theta da$ and so on. In other words, we must be careful to line up our spheres appropriately, so that travel along a curve that is perpendicular to one of the spheres keeps us at constant θ and ϕ. To guarantee this we need to be more careful in setting up our coordinates. Begin by considering a single point q lying on a sphere S_q (note that q must not be a degenerate point at which all of the Killing vectors vanish). Put coordinates (θ, ϕ) on this particular sphere only, not yet through the manifold. At each point p on S_q, there will be a two dimensional orthogonal subspace O_p, consisting of points along geodesics emanating from p whose tangent vectors at p are orthogonal to S_q. Note that there will be a one-dimensional subgroup R_p of rotations that leave p fixed; indeed, these rotations keep fixed any direction perpendicular to S_q at p, and hence the entire two-surface O_p is left invariant by R_p.

Consider a point r that is not on S_q, but on some other sphere S_r in the foliation, and that lies in the two-surface O_p orthogonal to S_q at p. Since p is arbitrary, this includes any possible point r in a neighborhood of S_q. Note that O_p will be orthogonal to S_r as well as to S_q. To see this, consider the two-dimensional plane

V_r of vectors in the tangent space $T_r M$ that are orthogonal to the two-surface O_p. Since O_p is left invariant by the rotations R_p, these rotations must take V_r into itself, because they are an isometry, and hence preserve orthogonality. But R_p also takes the set of vectors tangent to S_q into itself, since these rotations leave the spheres invariant. In four dimensions, two planes that are both orthogonal to a given plane at the same point must be the same plane; hence, the vectors tangent to S_r must be orthogonal to O_p.

There will be a unique geodesic that is orthogonal to S_q and connects p to r. Traveling down such geodesics provides a map $f : S_q \to S_r$, which is both one-to-one and onto (at least in a neighborhood of the original sphere). We use this map to define coordinates on S_r (and, similarly, on any other sphere) by assigning the same values of (θ, ϕ) to $r \in S_r$ that were the coordinates at $p \in S_q$. We have therefore defined (θ, ϕ) throughout the manifold. Now to define coordinates (a, b), choose two basis vectors S, T for the subspace of $T_q M$ that generates the orthogonal space O_q. Any other sphere will be connected to q by a unique orthogonal geodesic, with tangent vector $aS + bT \in T_q M$. Assign those components (a, b) as coordinates everywhere on that sphere. This defines the full set of coordinates (a, b, θ, ϕ) throughout the manifold.

The metric in these coordinates satisfies (5.27) and (5.28); it remains to be shown that there are no cross terms between directions along the spheres and those transverse. This means, for example, that the vector field ∂_a should be orthogonal to ∂_θ, and so on; it is straightforward to verify that this is so. First, consider ∂_θ at some point $r \in S_r$; this vector is the directional derivative along a curve of the form $x^\mu(\theta) = (a_r, b_r, \theta, \phi_r)$. Since a and b are constant along the curve, the entire curve remains in the sphere S_r, so that ∂_θ is tangent to the sphere. Meanwhile, ∂_a is a derivative along $x^\mu(a) = (a, b_r, \theta_r, \phi_r)$. Since this curve remains in the orthogonal subspace O_r, ∂_a will be orthogonal to S_r, and hence to ∂_θ. Similar arguments guarantee that there will be no cross terms between (a, b) and (θ, ϕ).

We have thus succeeded in putting the metric on a spherically symmetric spacetime in the form

$$ds^2 = g_{aa}(a, b)\, da^2 + g_{ab}(a, b)(da db + db da) + g_{bb}(a, b)\, db^2 + r^2(a, b)\, d\Omega^2. \tag{5.29}$$

Here $r(a, b)$ is some as-yet-undetermined function, to which we have merely given a suggestive label. There is nothing to stop us, however, from changing coordinates from (a, b) to (a, r) by inverting $r(a, b)$, unless r were a function of a alone; in this case we could just as easily switch to (b, r), so we will not consider this situation separately. The metric is then

$$ds^2 = g_{aa}(a, r)\, da^2 + g_{ar}(a, r)(da\, dr + dr\, da) + g_{rr}(a, r)\, dr^2 + r^2\, d\Omega^2. \tag{5.30}$$

Our next step is to find a function $t(a, r)$ such that, in the (t, r) coordinate system, there are no cross terms $dt dr + dr dt$ in the metric. Notice that

$$dt = \frac{\partial t}{\partial a} da + \frac{\partial t}{\partial r} dr, \tag{5.31}$$

so

$$dt^2 = \left(\frac{\partial t}{\partial a}\right)^2 da^2 + \left(\frac{\partial t}{\partial a}\right)\left(\frac{\partial t}{\partial r}\right)(da\, dr + dr\, da) + \left(\frac{\partial t}{\partial r}\right)^2 dr^2. \quad (5.32)$$

We would like to replace the first three terms in the metric (5.30) by

$$m dt^2 + n dr^2, \quad (5.33)$$

for some functions m and n. This is equivalent to the requirements

$$m\left(\frac{\partial t}{\partial a}\right)^2 = g_{aa}, \quad (5.34)$$

$$n + m\left(\frac{\partial t}{\partial r}\right)^2 = g_{rr}, \quad (5.35)$$

and

$$m\left(\frac{\partial t}{\partial a}\right)\left(\frac{\partial t}{\partial r}\right) = g_{ar}. \quad (5.36)$$

We therefore have three equations for the three unknowns $t(a, r)$, $m(a, r)$, and $n(a, r)$, just enough to determine them precisely, up to initial conditions for t. (Of course, they are "determined" in terms of the unknown functions g_{aa}, g_{ar}, and g_{rr}, so in this sense they are still undetermined.) We can therefore put our metric in the form

$$ds^2 = m(t, r)\, dt^2 + n(t, r)\, dr^2 + r^2\, d\Omega^2. \quad (5.37)$$

To this point the only difference between the two coordinates t and r is that we have chosen r to be the one that multiplies the metric for the two-sphere. This choice was motivated by what we know about the metric for flat Minkowski space, which can be written $ds^2 = -dt^2 + dr^2 + r^2\, d\Omega^2$. We know that the spacetime under consideration is Lorentzian, so either m or n will have to be negative. Let us choose m, the coefficient of dt^2, to be negative. This is not a choice we are simply allowed to make, and in fact we will see later that it can go wrong; but we will assume it for now. The assumption is not completely unreasonable, since we know that Minkowski space is itself spherically symmetric, and will therefore be described by (5.37). With this choice we can trade in the functions m and n for new functions α and β, such that

$$ds^2 = -e^{2\alpha(t,r)}\, dt^2 + e^{2\beta(t,r)}\, dr^2 + r^2\, d\Omega^2. \quad (5.38)$$

This is the best we can do using only geometry; spherical symmetry is certainly not enough to say anything substantive about the functions $\alpha(t, r)$ and $\beta(t, r)$. Our next step is therefore to actually solve Einstein's equation; the steps follow closely along those of Section 5.1, in which we considered a metric similar to (5.38) but with the additional assumption of time-independence. Here we will see that this assumption was unnecessary, as the solution will necessarily be static.

The nonvanishing Christoffel symbols for (5.38) are

$$\begin{aligned} &\Gamma^t_{tt} = \partial_t\alpha && \Gamma^t_{tr} = \partial_r\alpha && \Gamma^t_{rr} = e^{2(\beta-\alpha)}\partial_t\beta \\ &\Gamma^r_{tt} = e^{2(\alpha-\beta)}\partial_r\alpha && \Gamma^r_{tr} = \partial_t\beta && \Gamma^r_{rr} = \partial_r\beta \\ &\Gamma^\theta_{r\theta} = \frac{1}{r} && \Gamma^r_{\theta\theta} = -re^{-2\beta} && \Gamma^\phi_{r\phi} = \frac{1}{r} \\ &\Gamma^r_{\phi\phi} = -re^{-2\beta}\sin^2\theta && \Gamma^\theta_{\phi\phi} = -\sin\theta\cos\theta && \Gamma^\phi_{\theta\phi} = \frac{\cos\theta}{\sin\theta}, \end{aligned} \tag{5.39}$$

the nonvanishing components of the Riemann tensor are

$$\begin{aligned} R^t{}_{rtr} &= e^{2(\beta-\alpha)}[\partial_t^2\beta + (\partial_t\beta)^2 - \partial_t\alpha\partial_t\beta] + [\partial_r\alpha\partial_r\beta - \partial_r^2\alpha - (\partial_r\alpha)^2] \\ R^t{}_{\theta t\theta} &= -re^{-2\beta}\partial_r\alpha \\ R^t{}_{\phi t\phi} &= -re^{-2\beta}\sin^2\theta\,\partial_r\alpha \\ R^t{}_{\theta r\theta} &= -re^{-2\alpha}\partial_t\beta \\ R^t{}_{\phi r\phi} &= -re^{-2\alpha}\sin^2\theta\,\partial_t\beta \\ R^r{}_{\theta r\theta} &= re^{-2\beta}\partial_r\beta \\ R^r{}_{\phi r\phi} &= re^{-2\beta}\sin^2\theta\,\partial_r\beta \\ R^\theta{}_{\phi\theta\phi} &= (1-e^{-2\beta})\sin^2\theta, \end{aligned} \tag{5.40}$$

and the Ricci tensor is

$$\begin{aligned} R_{tt} &= \left[\partial_t^2\beta + (\partial_t\beta)^2 - \partial_t\alpha\partial_t\beta\right] + e^{2(\alpha-\beta)}\left[\partial_r^2\alpha + (\partial_r\alpha)^2 - \partial_r\alpha\partial_r\beta + \frac{2}{r}\partial_r\alpha\right] \\ R_{rr} &= -\left[\partial_r^2\alpha + (\partial_r\alpha)^2 - \partial_r\alpha\partial_r\beta - \frac{2}{r}\partial_r\beta\right] \\ &\quad + e^{2(\beta-\alpha)}\left[\partial_t^2\beta + (\partial_t\beta)^2 - \partial_t\alpha\partial_t\beta\right] \\ R_{tr} &= \frac{2}{r}\partial_t\beta \\ R_{\theta\theta} &= e^{-2\beta}\left[r(\partial_r\beta - \partial_r\alpha) - 1\right] + 1 \\ R_{\phi\phi} &= R_{\theta\theta}\sin^2\theta. \end{aligned} \tag{5.41}$$

Our job is to solve Einstein's equation in vacuum, $R_{\mu\nu} = 0$. From $R_{tr} = 0$ we get

$$\partial_t\beta = 0. \tag{5.42}$$

If we consider taking the time derivative of $R_{\theta\theta} = 0$ and using $\partial_t \beta = 0$, we get

$$\partial_t \partial_r \alpha = 0. \tag{5.43}$$

We can therefore write

$$\begin{aligned} \beta &= \beta(r) \\ \alpha &= f(r) + g(t). \end{aligned} \tag{5.44}$$

The first term in the metric (5.38) is thus $-e^{2f(r)}e^{2g(t)}\, dt^2$. But we can always simply redefine our time coordinate by replacing $dt \to e^{-g(t)}\, dt$; in other words, we are free to choose t such that $g(t) = 0$, whence $\alpha(t, r) = f(r)$. We therefore have

$$ds^2 = -e^{2\alpha(r)}\, dt^2 + e^{2\beta(r)}\, dr^2 + r^2\, d\Omega^2. \tag{5.45}$$

All of the metric components are independent of the coordinate t. We have therefore proven a crucial result: *any spherically symmetric vacuum metric possesses a timelike Killing vector.*

This property is so interesting that it gets its own name: a metric that possesses a Killing vector that is timelike near infinity is called **stationary**. (Often, including in Schwarzschild, the Killing vector that is timelike at infinity will become spacelike somewhere in the interior.) In a stationary metric we can choose coordinates (t, x^1, x^2, x^3) in which the Killing vector is ∂_t and the metric components are independent of t; the general form of a stationary metric in these coordinates is thus

$$ds^2 = g_{00}(\vec{x})\, dt^2 + g_{0i}(\vec{x})(dt dx^i + dx^i\, dt) + g_{ij}(\vec{x})\, dx^i dx^j. \tag{5.46}$$

There is also a more restrictive property: a metric is called **static** if it possesses a timelike Killing vector that is orthogonal to a family of hypersurfaces. (For more details on hypersurfaces, see Appendix D.) In the Exercises for Chapter 4 you showed that a hypersurface-orthogonal vector field v^μ obeys

$$v_{[\mu} \nabla_\nu v_{\sigma]} = 0. \tag{5.47}$$

But there is a simpler diagnostic; if we have adapted coordinates so that the components $g_{\mu\nu}$ are all independent of t, the surfaces to which the Killing vector will be orthogonal are defined by the condition $t =$ constant. Operationally, this means that the time-space cross terms in (5.46) will be absent; the general static metric can be written

$$ds^2 = g_{00}(\vec{x})\, dt^2 + g_{ij}(\vec{x})\, dx^i dx^j. \tag{5.48}$$

We notice that only even powers of the time coordinate t appear in this form; thus, an alternative definition of "static" is "stationary, and invariant under time reversal $(t \to -t)$." The metric (5.45) is clearly static. You should think of stationary as meaning "doing exactly the same thing at every time," while static means "not doing anything at all." For example, the static spherically symmetric metric (5.45) will describe nonrotating stars or black holes, while rotating systems that keep rotating in the same way at all times will be described by metrics that are stationary but not static.

Notice that (5.45) is precisely the same as (5.11), the metric we originally used to derive the Schwarzschild solution in Section 5.1. We have therefore proven Birkhoff's theorem, that the unique spherically symmetric vacuum solution is the Schwarzschild metric,

$$ds^2 = -\left(1 - \frac{2GM}{r}\right) dt^2 + \left(1 - \frac{2GM}{r}\right)^{-1} dr^2 + r^2\, d\Omega^2, \tag{5.49}$$

as promised.

We did not say anything about the source of the Schwarzschild metric, except that it be spherically symmetric. Specifically, we did not demand that the source itself be static; it could be a collapsing star, as long as the collapse is symmetric. Therefore a process such as a supernova explosion would generate very little gravitational radiation (in comparison to the amount of energy released through other channels) if it were close to spherically symmetric, which a realistic supernova may or may not be depending on its origin. This is the same result we would have obtained in electromagnetism, where the electromagnetic fields around a spherical charge distribution do not depend on the radial distribution of the charges.

5.3 ■ SINGULARITIES

Before exploring the behavior of test particles in the Schwarzschild geometry, we should say something about singularities. From the form of (5.1), the metric coefficients become infinite at $r = 0$ and $r = 2GM$—an apparent sign that something is going wrong. The metric coefficients, of course, are coordinate-dependent quantities, and as such we should not make too much of their values; it is certainly possible to have a coordinate singularity that results from a breakdown of a specific coordinate system rather than the underlying manifold. An example occurs at the origin of polar coordinates in the plane, where the metric $ds^2 = dr^2 + r^2 d\theta^2$ becomes degenerate and the component $g^{\theta\theta} = r^{-2}$ of the inverse metric blows up, even though that point of the manifold is no different from any other.

What kind of coordinate-independent signal should we look for as a warning that something about the geometry is out of control? This turns out to be a difficult question to answer, and entire books have been written about the nature of singularities in general relativity. We won't go into this issue in detail, but rather turn to one simple criterion for when something has gone wrong—when the curvature becomes infinite. The curvature is measured by the Riemann tensor, and it is hard to say when a tensor becomes infinite, since its components are coordinate-dependent. But from the curvature we can construct various scalar quantities, and since scalars are coordinate-independent it is meaningful to say that they become infinite. The simplest such scalar is the Ricci scalar

$R = g^{\mu\nu} R_{\mu\nu}$, but we can also construct higher-order scalars such as $R^{\mu\nu} R_{\mu\nu}$, $R^{\mu\nu\rho\sigma} R_{\mu\nu\rho\sigma}$, $R_{\mu\nu\rho\sigma} R^{\rho\sigma\lambda\tau} R_{\lambda\tau}{}^{\mu\nu}$, and so on. If any of these scalars (but not necessarily all of them) goes to infinity as we approach some point, we regard that point as a singularity of the curvature. We should also check that the point is not infinitely far away; that is, that it can be reached by traveling a finite distance along a curve.

We therefore have a sufficient condition for a point to be considered a singularity. It is not a necessary condition, however, and it is generally harder to show that a given point is nonsingular; for our purposes we will simply test to see if geodesics are well-behaved at the point in question, and if so then we will consider the point nonsingular. In the case of the Schwarzschild metric (5.1), direct calculation reveals that

$$R^{\mu\nu\rho\sigma} R_{\mu\nu\rho\sigma} = \frac{48G^2M^2}{r^6}. \tag{5.50}$$

This is enough to convince us that $r = 0$ represents an honest singularity.

The other trouble spot is $r = 2GM$, the Schwarzschild radius. You could check that none of the curvature invariants blows up there. We therefore begin to think that it is actually not singular, and we have simply chosen a bad coordinate system. The best thing to do is to transform to more appropriate coordinates if possible. We will soon see that in this case it is in fact possible, and the surface $r = 2GM$ is very well-behaved (although interesting) in the Schwarzschild metric—it demarcates the event horizon of a black hole.

Having worried a little about singularities, we should point out that the behavior of the Schwarzschild metric inside the Schwarzschild radius is of little day-to-day consequence. The solution we derived is valid only in vacuum, and we expect it to hold outside a spherical body such as a star. However, in the case of the Sun we are dealing with a body that extends to a radius of

$$R_\odot = 10^6 GM_\odot. \tag{5.51}$$

Thus, $r = 2GM_\odot$ is far inside the solar interior, where we do not expect the Schwarzschild metric to apply. In fact, realistic stellar interior solutions consist of matching the exterior Schwarzschild metric to an interior metric that is perfectly smooth at the origin. Nevertheless, there are objects for which the full Schwarzschild metric is required—black holes—and therefore we will let our imaginations roam far outside the solar system in this chapter.

5.4 ■ GEODESICS OF SCHWARZSCHILD

The first step we will take to understand the Schwarzschild metric more fully is to consider the behavior of geodesics. We need the nonzero Christoffel symbols for Schwarzschild:

$$\Gamma^r_{tt} = \frac{GM}{r^3}(r - 2GM) \qquad \Gamma^r_{rr} = \frac{-GM}{r(r-2GM)} \qquad \Gamma^t_{tr} = \frac{GM}{r(r-2GM)}$$

$$\Gamma^\theta_{r\theta} = \frac{1}{r} \qquad \Gamma^r_{\theta\theta} = -(r - 2GM) \qquad \Gamma^\phi_{r\phi} = \frac{1}{r}$$

$$\Gamma^r_{\phi\phi} = -(r - 2GM)\sin^2\theta \qquad \Gamma^\theta_{\phi\phi} = -\sin\theta\cos\theta \qquad \Gamma^\phi_{\theta\phi} = \frac{\cos\theta}{\sin\theta}. \tag{5.52}$$

The geodesic equation therefore turns into the following four equations, where λ is an affine parameter:

$$\frac{d^2t}{d\lambda^2} + \frac{2GM}{r(r-2GM)}\frac{dr}{d\lambda}\frac{dt}{d\lambda} = 0,$$

$$\frac{d^2r}{d\lambda^2} + \frac{GM}{r^3}(r-2GM)\left(\frac{dt}{d\lambda}\right)^2 - \frac{GM}{r(r-2GM)}\left(\frac{dr}{d\lambda}\right)^2$$
$$-(r-2GM)\left[\left(\frac{d\theta}{d\lambda}\right)^2 + \sin^2\theta\left(\frac{d\phi}{d\lambda}\right)^2\right] = 0,$$

$$\frac{d^2\theta}{d\lambda^2} + \frac{2}{r}\frac{d\theta}{d\lambda}\frac{dr}{d\lambda} - \sin\theta\cos\theta\left(\frac{d\phi}{d\lambda}\right)^2 = 0,$$

$$\frac{d^2\phi}{d\lambda^2} + \frac{2}{r}\frac{d\phi}{d\lambda}\frac{dr}{d\lambda} + 2\frac{\cos\theta}{\sin\theta}\frac{d\theta}{d\lambda}\frac{d\phi}{d\lambda} = 0. \tag{5.53}$$

There does not seem to be much hope for simply solving this set of coupled equations by inspection. Fortunately our task is greatly simplified by the high degree of symmetry of the Schwarzschild metric. We know that there are four Killing vectors: three for the spherical symmetry, and one for time translations. Each of these will lead to a constant of the motion for a free particle. If K^μ is a Killing vector, we know that

$$K_\mu \frac{dx^\mu}{d\lambda} = \text{constant}. \tag{5.54}$$

In addition, we always have another constant of the motion for geodesics: the geodesic equation (together with metric compatibility) implies that the quantity

$$\epsilon = -g_{\mu\nu}\frac{dx^\mu}{d\lambda}\frac{dx^\nu}{d\lambda} \tag{5.55}$$

is constant along the path. (For any trajectory we can choose the parameter λ such that ϵ is a constant; we are simply noting that this is compatible with affine parameterization along a geodesic.) Of course, for a massive particle we typically choose $\lambda = \tau$, and this relation simply becomes $\epsilon = -g_{\mu\nu}U^\mu U^\nu = +1$. For massless particles, which move along null trajectories, we always have $\epsilon = 0$, and this equation does not fix the parameter λ. As discussed in Section 3.4, it is convenient to normalize λ along null geodesics such that the four-momentum and four-velocity are equal, $p^\mu = dx^\mu/d\lambda$. We might also be concerned with spacelike geodesics (even though they do not correspond to paths of particles), for which we will choose $\epsilon = -1$.

Rather than immediately writing out explicit expressions for the four conserved quantities associated with Killing vectors, let's think about what they are telling us. Notice that the symmetries they represent are also present in flat spacetime, where the conserved quantities they lead to are very familiar. Invariance under time translations leads to conservation of energy, while invariance under spatial rotations leads to conservation of the three components of angular momentum. Essentially the same applies to the Schwarzschild metric. We can think of the angular momentum as a three-vector with a magnitude (one component) and direction (two components). Conservation of the direction of angular momentum means that the particle will move in a plane. We can choose this to be the equatorial plane of our coordinate system; if the particle is not in this plane, we can rotate coordinates until it is. Thus, the two Killing vectors that lead to conservation of the direction of angular momentum imply that, for a single particle, we can choose

$$\theta = \frac{\pi}{2}. \tag{5.56}$$

The two remaining Killing vectors correspond to energy and the magnitude of angular momentum. The energy arises from the timelike Killing vector

$$K^\mu = (\partial_t)^\mu = (1, 0, 0, 0). \tag{5.57}$$

The Killing vector whose conserved quantity is the magnitude of the angular momentum is

$$R^\mu = (\partial_\phi)^\mu = (0, 0, 0, 1). \tag{5.58}$$

In both cases it is convenient to lower the index to obtain

$$K_\mu = \left(-\left(1 - \frac{2GM}{r}\right), 0, 0, 0\right) \tag{5.59}$$

and

$$R_\mu = \left(0, 0, 0, r^2 \sin^2\theta\right). \tag{5.60}$$

Since (5.56) implies that $\sin\theta = 1$ along the geodesics of interest to us, the two conserved quantities are

$$E = -K_\mu \frac{dx^\mu}{d\lambda} = \left(1 - \frac{2GM}{r}\right)\frac{dt}{d\lambda} \tag{5.61}$$

and

$$L = R_\mu \frac{dx^\mu}{d\lambda} = r^2 \frac{d\phi}{d\lambda}. \tag{5.62}$$

For massless particles, these can be thought of as the conserved energy and angular momentum, while for massive particles they are the conserved energy and angular momentum per unit mass of the particle. In the discussion of rotating black holes in the next chapter, we will use E and L to refer to the actual energy and angular momentum, not "per unit mass"; the meaning should be clear from context. Note that the constancy of (5.62) is the GR equivalent of Kepler's second law—equal areas are swept out in equal times.

Recall that in Section 3.4 we claimed that the energy of a particle with four-momentum p^μ, as measured by an observer with four-velocity U^μ, would be $-p_\mu U^\mu$. This is *not* equal, or even proportional, to (5.61), even if the observer is taken to be stationary ($U^i = 0$). Mathematically, this is because the four-velocity is normalized to $U_\mu U^\mu = -1$, which the Killing vector K^μ is not: If we tried to normalize it in that way, it would no longer solve Killing's equation. At a slightly deeper level, $-p_\mu U^\mu$ may be thought of as the inertial/kinetic energy of the particle, while $-p_\mu K^\mu$ is the total conserved energy, including the potential energy due to the gravitational field. The notion of gravitational potential energy is not always well-defined, but the total energy is well-defined in the presence of a timelike Killing vector. We will presently use E to help characterize geodesics of Schwarzschild; later we will also use $-p_\mu U^\mu$ for massless particles, where it can be thought of as the observed frequency of a photon, to describe gravitational redshift.

Together the conserved quantities E and L provide a convenient way to understand the orbits of particles in the Schwarzschild geometry. Let us expand the expression (5.55) for ϵ to obtain

$$-\left(1 - \frac{2GM}{r}\right)\left(\frac{dt}{d\lambda}\right)^2 + \left(1 - \frac{2GM}{r}\right)^{-1}\left(\frac{dr}{d\lambda}\right)^2 + r^2\left(\frac{d\phi}{d\lambda}\right)^2 = -\epsilon. \tag{5.63}$$

If we multiply this by $(1 - 2GM/r)$ and use our expressions for E and L, we obtain

$$-E^2 + \left(\frac{dr}{d\lambda}\right)^2 + \left(1 - \frac{2GM}{r}\right)\left(\frac{L^2}{r^2} + \epsilon\right) = 0. \tag{5.64}$$

This is certainly progress, since we have taken a messy system of coupled equations and obtained a single equation for $r(\lambda)$. It looks even nicer if we rewrite it as

$$\frac{1}{2}\left(\frac{dr}{d\lambda}\right)^2 + V(r) = \mathcal{E}, \tag{5.65}$$

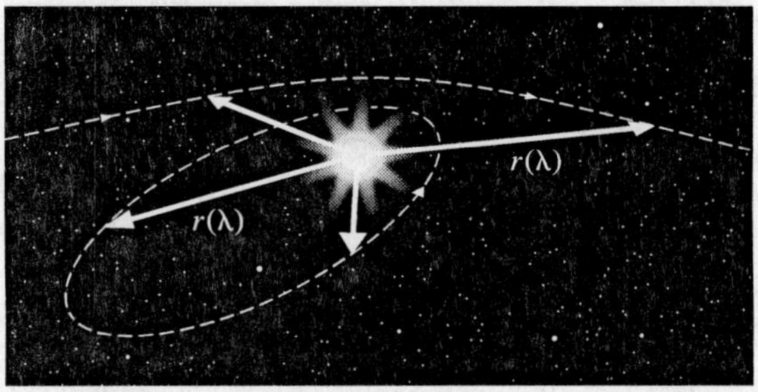

FIGURE 5.3 Orbits around a star are characterized by giving the radius r as a function of a parameter λ.

where

$$V(r) = \frac{1}{2}\epsilon - \epsilon\frac{GM}{r} + \frac{L^2}{2r^2} - \frac{GML^2}{r^3} \tag{5.66}$$

and

$$\mathcal{E} = \frac{1}{2}E^2. \tag{5.67}$$

In (5.65) we have precisely the equation for a classical particle of unit mass and "energy" $\mathcal{E}$ moving in a one-dimensional potential given by $V(r)$. It's a little confusing, but not too bad: the conserved energy per unit mass is E, but the effective potential for the coordinate r responds to $\mathcal{E} = E^2/2$.

Of course, our physical situation is quite different from a classical particle moving in one dimension; the trajectories under consideration are orbits around a star or other object, as shown in Figure 5.3. The quantities of interest to us are not only $r(\lambda)$, but also $t(\lambda)$ and $\phi(\lambda)$. Nevertheless, we can go a long way toward understanding all of the orbits by understanding their radial behavior, and it is a great help to reduce this behavior to a problem we know how to solve.

A similar analysis of orbits in Newtonian gravity would have produced a similar result; the general equation (5.65) would have been the same, but the effective potential (5.66) would not have had the last term. (Note that this equation is not a power series in $1/r$, it is exact.) In the potential (5.66) the first term is just a constant, the second term corresponds exactly to the Newtonian gravitational potential, and the third term is a contribution from angular momentum that takes the same form in Newtonian gravity and general relativity. The last term, the GR contribution, will turn out to make a great deal of difference, especially at small r.

Let us examine the effective potentials for different kinds of possible orbits, as illustrated in Figures 5.4 and 5.5. There are different curves $V(r)$ for different values of L; for any one of these curves, the behavior of the orbit can be judged by comparing $\mathcal{E}$ to $V(r)$. The general behavior of the particle will be to move in the potential until it reaches a "turning point" where $V(r) = \mathcal{E}$, when it will begin

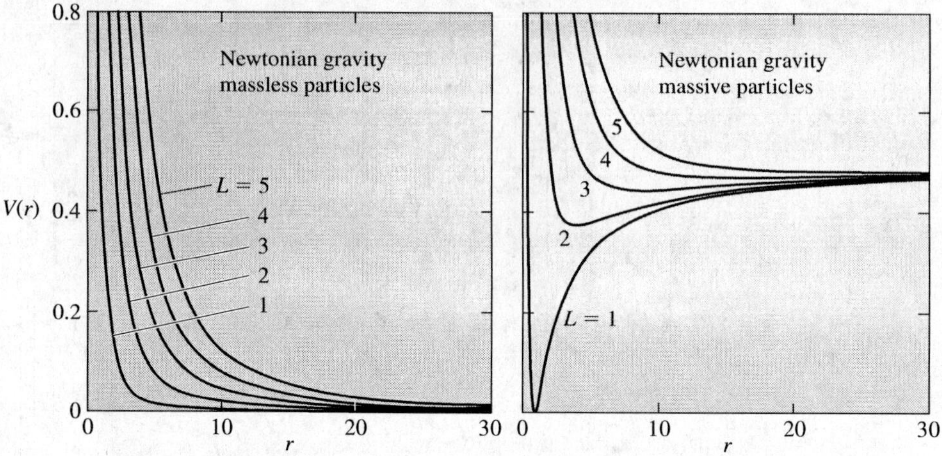

FIGURE 5.4 Effective potentials in Newtonian gravity. Five curves are shown, corresponding to the listed values of the angular momentum (per unit mass) L, and we have chosen $GM = 1$. Note that, for large enough energy, every orbit reaches a turning point and returns to infinity.

moving in the other direction. Sometimes there may be no turning point to hit, in which case the particle just keeps going. In other cases the particle may simply move in a circular orbit at radius $r_c =$ constant; this can happen at points where the potential is flat, $dV/dr = 0$. Differentiating (5.66), we find that the circular orbits occur when

$$\epsilon GMr_c^2 - L^2 r_c + 3GML^2\gamma = 0, \tag{5.68}$$

where $\gamma = 0$ in Newtonian gravity and $\gamma = 1$ in general relativity. Circular orbits will be stable if they correspond to a minimum of the potential, and unstable if they correspond to a maximum. Bound orbits that are not circular will oscillate around the radius of the stable circular orbit.

Turning to Newtonian gravity, we find that circular orbits appear at

$$r_c = \frac{L^2}{\epsilon GM}. \tag{5.69}$$

For massless particles, $\epsilon = 0$, and there are no circular orbits; this is consistent with the first plot in Figure 5.4, which illustrates that there are no bound orbits of any sort. Although it is somewhat obscured in polar coordinates, massless particles actually move in a straight line, since the Newtonian gravitational force on a massless particle is zero. Of course the standing of massless particles in Newtonian theory is somewhat problematic, so you can get different answers depending on what assumptions you make. In terms of the effective potential, a photon with a given energy E will come in from $r = \infty$ and gradually slow down (actually $dr/d\lambda$ will decrease, but the speed of light isn't changing) until it reaches the

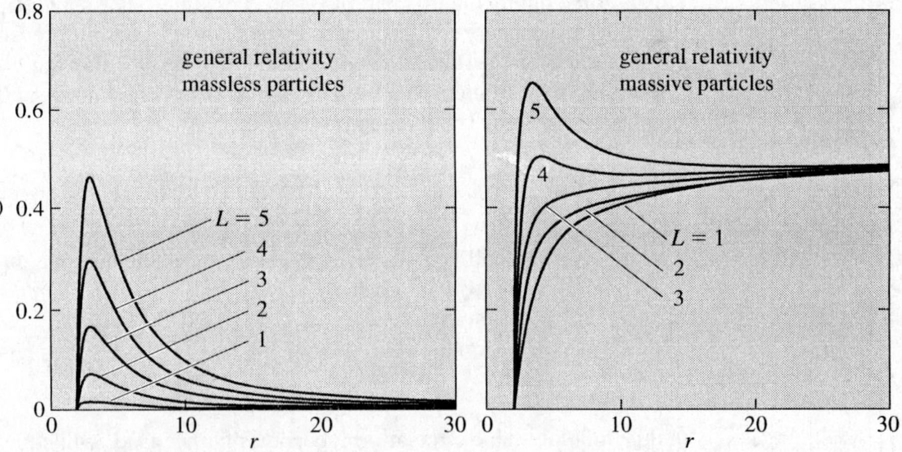

FIGURE 5.5 Effective potentials in general relativity. Again, five curves are shown, corresponding to the listed values of the angular momentum (per unit mass) L, and we have chosen $GM = 1$. In GR there is an innermost circular orbit greater than or equal to $3GM$, and any orbit that falls inside this radius continues to $r = 0$ (for particles on geodesics).

turning point, when it will start moving away back to $r = \infty$. The lower values of L, for which the photon will come closer before it starts moving away, are simply those trajectories that are initially aimed closer to the gravitating body. For massive particles there will be stable circular orbits at the radius (5.69), as well as bound orbits that oscillate around this radius. If the energy is greater than the asymptotic value $E = 1$, the orbits will be unbound, describing a particle that approaches the star and then recedes. We know that the orbits in Newton's theory are conic sections—bound orbits are either circles or ellipses, while unbound ones are either parabolas or hyperbolas—although we won't show that here.

In general relativity the situation is different, but only for r sufficiently small. Since the difference resides in the term $-GML^2/r^3$, as $r \to \infty$ the behaviors are identical in the two theories. But as $r \to 0$ the potential goes to $-\infty$ rather than $+\infty$ as in the Newtonian case. At $r = 2GM$ the potential is always zero; inside this radius is the black hole, which we will discuss more thoroughly later. For massless particles there is always a barrier (except for $L = 0$, for which the potential vanishes identically), but a sufficiently energetic photon will nevertheless go over the barrier and be dragged inexorably down to the center. Note that "sufficiently energetic" means "in comparison to its angular momentum"—in fact the frequency of the photon is immaterial, only the direction in which it is pointing. At the top of the barrier are unstable circular orbits. For $\epsilon = 0$, $\gamma = 1$, we can easily solve (5.68) to obtain

$$r_c = 3GM. \tag{5.70}$$

This is borne out by the first part of Figure 5.5, which shows a maximum of $V(r)$ at $r = 3GM$ for every L. This means that a photon can orbit forever in a circle at this radius, but any perturbation will cause it to fly away either to $r = 0$ or $r = \infty$.

For massive particles there are once again different regimes depending on the angular momentum. The circular orbits are at

$$r_c = \frac{L^2 \pm \sqrt{L^4 - 12G^2M^2L^2}}{2GM}. \tag{5.71}$$

For large L there will be two circular orbits, one stable and one unstable. In the $L \to \infty$ limit their radii are given by

$$r_c = \frac{L^2 \pm L^2(1 - 6G^2M^2/L^2)}{2GM} = \left(\frac{L^2}{GM}, 3GM\right). \tag{5.72}$$

In this limit the stable circular orbit becomes farther away, while the unstable one approaches $3GM$, behavior that parallels the massless case. As we decrease L, the two circular orbits come closer together; they coincide when the discriminant in (5.71) vanishes, which is at

$$L = \sqrt{12}GM, \tag{5.73}$$

for which

$$r_c = 6GM, \tag{5.74}$$

and they disappear entirely for smaller L. Thus $6GM$ is the smallest possible radius of a stable circular orbit in the Schwarzschild metric. There are also unbound orbits, which come in from infinity and turn around, and bound but noncircular orbits, which oscillate around the stable circular radius. Note that such orbits, which would describe exact conic sections in Newtonian gravity, will not do so in GR, although we would have to solve the equation for $d\phi/d\lambda$ to demonstrate it. Finally, there are orbits that come in from infinity and continue all the way in to $r = 0$; this can happen either if the energy is higher than the barrier, or for $L < \sqrt{12}GM$, when the barrier goes away entirely.

We have therefore found that the Schwarzschild solution possesses stable circular orbits for $r > 6GM$ and unstable circular orbits for $3GM < r < 6GM$. It's important to remember that these are only the geodesics; there is nothing to stop an accelerating particle from dipping below $r = 3GM$ and emerging, as long as it stays beyond $r = 2GM$.

5.5 ■ EXPERIMENTAL TESTS

Most experimental tests of general relativity involve the motion of test particles in the solar system, and hence geodesics of the Schwarzschild metric. Einstein

FIGURE 5.6 Orbits in general relativity describe precessing ellipses.

suggested three tests: the deflection of light, the precession of perihelia, and gravitational redshift. The deflection of light is observable in the weak-field limit, and is therefore discussed in Chapter 7. In this section we will discuss the precession of perihelia and the gravitational redshift. (The perihelion of an elliptical orbit is its point of closest approach to the Sun; orbits around the Earth or a star would have perigee or periastron, respectively.)

The precession of perihelia reflects the fact that noncircular orbits in GR are not perfect closed ellipses; to a good approximation they are ellipses that precess, describing a flower pattern as shown in Figure 5.6. Despite its conceptual simplicity, the rate of perihelion precession is somewhat cumbersome to calculate; here we follow d'Inverno (1992). The strategy is to describe the evolution of the radial coordinate r as a function of the angular coordinate ϕ; for a perfect ellipse, $r(\phi)$ would be periodic with period 2π, reflecting the fact that perihelion occurred at the same angular position each orbit. Using perturbation theory we can show how GR introduces a slight alteration of the period, giving rise to precession.

We start with our radial equation of motion of a massive particle in a Schwarzschild metric (5.65). To get an equation for $dr/d\phi$ we multiply by

$$\left(\frac{d\phi}{d\lambda}\right)^{-2} = \frac{r^4}{L^2}, \tag{5.75}$$

which yields

$$\left(\frac{dr}{d\phi}\right)^2 + \frac{1}{L^2}r^4 - \frac{2GM}{L^2}r^3 + r^2 - 2GMr = \frac{2\mathcal{E}}{L^2}r^4. \tag{5.76}$$

Two tricks are useful in solving this equation. The first trick is to define a new variable

$$x = \frac{L^2}{GMr}. \tag{5.77}$$

From (5.69) we see that $x = 1$ at a Newtonian circular orbit. Our equation of motion (5.76) becomes

$$\left(\frac{dx}{d\phi}\right)^2 + \frac{L^2}{G^2M^2} - 2x + x^2 - \frac{2G^2M^2}{L^2}x^3 = \frac{2\mathcal{E}L^2}{G^2M^2}. \tag{5.78}$$

The second trick is to differentiate this with respect to ϕ, obtaining a second-order equation for $x(\phi)$:

$$\frac{d^2x}{d\phi^2} - 1 + x = \frac{3G^2M^2}{L^2}x^2. \tag{5.79}$$

In a Newtonian calculation, the last term would be absent, and we could solve for x exactly; here, we can treat it as a perturbation.

We expand x into a Newtonian solution plus a small deviation,

$$x = x_0 + x_1. \tag{5.80}$$

The zeroth-order part of (5.79) is then

$$\frac{d^2x_0}{d\phi^2} - 1 + x_0 = 0 \tag{5.81}$$

and the first-order part is

$$\frac{d^2x_1}{d\phi^2} + x_1 = \frac{3G^2M^2}{L^2}x_0^2. \tag{5.82}$$

The solution for the zeroth-order equation can be written

$$x_0 = 1 + e\cos\phi. \tag{5.83}$$

This is the standard result of Newton or Kepler; it describes a perfect ellipse, with e the eccentricity. An ellipse is specified by the semi-major axis a, the distance from the center to the farthest point on the ellipse, and the semi-minor axis b, the distance from the center to the closest point. The eccentricity satisfies $e^2 = 1 - b^2/a^2$.

Plugging the Newtonian solution into the first-order equation (5.82), we obtain

$$\begin{aligned}\frac{d^2x_1}{d\phi^2} + x_1 &= \frac{3G^2M^2}{L^2}(1 + e\cos\phi)^2 \\ &= \frac{3G^2M^2}{L^2}\left[\left(1 + \frac{1}{2}e^2\right) + 2e\cos\phi + \frac{1}{2}e^2\cos 2\phi\right].\end{aligned} \tag{5.84}$$

To solve this equation, notice that

$$\frac{d^2}{d\phi^2}(\phi \sin\phi) + \phi \sin\phi = 2\cos\phi \tag{5.85}$$

and

$$\frac{d^2}{d\phi^2}(\cos 2\phi) + \cos 2\phi = -3\cos 2\phi. \tag{5.86}$$

Comparing these to (5.84), we see that a solution is provided by

$$x_1 = \frac{3G^2M^2}{L^2}\left[\left(1 + \frac{1}{2}e^2\right) + e\phi\sin\phi - \frac{1}{6}e^2\cos 2\phi\right], \tag{5.87}$$

as you are welcome to check. The three terms here have different characters. The first is simply a constant displacement, while the third oscillates around zero. The important effect is thus contained in the second term, which accumulates over successive orbits. We therefore combine this term with the zeroth-order solution to write

$$x = 1 + e\cos\phi + \frac{3G^2M^2e}{L^2}\phi\sin\phi. \tag{5.88}$$

This is not a full solution, even to the perturbed equation, but it encapsulates the part that we care about. In particular, this expression for x can be conveniently rewritten as the equation for an ellipse with an angular period that is not quite 2π:

$$x = 1 + e\cos\left[(1-\alpha)\phi\right], \tag{5.89}$$

where we have introduced

$$\alpha = \frac{3G^2M^2}{L^2}. \tag{5.90}$$

The equivalence of (5.88) and (5.89) can be seen by expanding $\cos[(1-\alpha)\phi]$ as a power series in the small parameter α:

$$\begin{aligned}\cos\left[(1-\alpha)\phi\right] &= \cos\phi + \alpha\frac{d}{d\alpha}\cos\left[(1-\alpha)\phi\right]_{\alpha=0}\\ &= \cos\phi + \alpha\phi\sin\phi.\end{aligned} \tag{5.91}$$

We have therefore found that, during each orbit of the planet, perihelion advances by an angle

$$\Delta\phi = 2\pi\alpha = \frac{6\pi G^2M^2}{L^2}. \tag{5.92}$$

To convert from the angular momentum L to more conventional quantities, we may use expressions valid for Newtonian orbits, since the quantity we're looking at is already a small perturbation. An ordinary ellipse satisfies

$$r = \frac{(1-e^2)a}{1+e\cos\phi}, \tag{5.93}$$

where a is the semi-major axis. Comparing to our zeroth-order solution (5.83) and the definition (5.77) of x, we see that

$$L^2 \approx GM(1-e^2)a. \tag{5.94}$$

This is an approximation, valid if the orbit were a perfect closed ellipse. Plugging this into (5.92) and restoring explicit factors of the speed of light, we obtain

$$\Delta\phi = \frac{6\pi GM}{c^2(1-e^2)a}. \tag{5.95}$$

Historically, the precession of Mercury was the first test of GR. In fact it was known before Einstein invented GR that there was an apparent discrepancy in Mercury's orbit, and a number of solutions had been proposed (including "dark matter" in the inner Solar System). Einstein knew of the discrepancy, and one of his first tasks after formulating GR was to show that it correctly accounted for Mercury's perihelion precession. For the motion of Mercury around the Sun, the relevant orbital parameters are

$$\begin{aligned} \frac{GM_\odot}{c^2} &= 1.48\times 10^5\ \text{cm}, \\ a &= 5.79\times 10^{12}\ \text{cm} \\ e &= 0.2056, \end{aligned} \tag{5.96}$$

and of course $c = 3.00\times 10^{10}$ cm/sec. This gives

$$\Delta\phi_{\text{Mercury}} = 5.01\times 10^{-7}\ \text{radians/orbit} = 0.103''/\text{orbit}, \tag{5.97}$$

where $''$ stands for arcseconds. It is more conventional to express this in terms of precession per century; Mercury orbits once every 88 days, yielding

$$\Delta\phi_{\text{Mercury}} = 43.0''/\text{century}. \tag{5.98}$$

So the major axis of Mercury's orbit precesses at a rate of 43.0 arcsecs every 100 years. The observed value is 5601 arcsecs/100 years. However, much of that is due to the precession of equinoxes in our geocentric coordinate system; 5025 arcsecs/100 years, to be precise. The gravitational perturbations of the other planets contribute an additional 532 arcsecs/100 years, leaving 43 arcsecs/100 years to be explained by GR, which it does quite well. You can imagine that Einstein must have been very pleased when he first figured this out.

In Chapter 2 we discussed the gravitational redshift of photons as a consequence of the Principle of Equivalence. The Schwarzschild metric is an exact solution of GR, and should therefore predict a redshift that reduces to the EP prediction in small regions of spacetime. Let's see how that works.

Consider an observer with four-velocity U^μ, who is stationary in the Schwarzschild coordinates ($U^i = 0$). We could allow the observer to be moving, but that would merely superimpose a conventional Doppler shift over the gravitational effect. The four-velocity satisfies $U_\mu U^\mu = -1$, which for a stationary observer in Schwarzschild implies

$$U^0 = \left(1 - \frac{2GM}{r}\right)^{-1/2}. \tag{5.99}$$

Any such observer measures the frequency of a photon following along a null geodesic $x^\mu(\lambda)$ to be

$$\omega = -g_{\mu\nu} U^\mu \frac{dx^\nu}{d\lambda}. \tag{5.100}$$

Indeed, this relation defines the normalization of λ. We therefore have

$$\omega = \left(1 - \frac{2GM}{r}\right)^{1/2} \frac{dt}{d\lambda} \tag{5.101}$$

$$= \left(1 - \frac{2GM}{r}\right)^{-1/2} E, \tag{5.102}$$

where E is defined by (5.61), applied to the photon trajectory. E is conserved, so ω will clearly take on different values when measured at different radial distances. For a photon emitted at r_1 and observed at r_2, the observed frequencies will be related by

$$\frac{\omega_2}{\omega_1} = \left(\frac{1 - 2GM/r_1}{1 - 2GM/r_2}\right)^{1/2}. \tag{5.103}$$

This is an exact result for the frequency shift; in the limit $r \gg 2GM$ we have

$$\begin{aligned}\frac{\omega_2}{\omega_1} &= 1 - \frac{GM}{r_1} + \frac{GM}{r_2} \\ &= 1 + \Phi_1 - \Phi_2,\end{aligned} \tag{5.104}$$

where $\Phi = -GM/r$ is the Newtonian potential. This tells us that the frequency goes down as Φ increases, which happens as we climb out of a gravitational field: thus, a redshift. (Photons that fall toward a gravitating body are blueshifted.) We see that the $r \gg 2GM$ result agrees with the calculation based on the Equivalence Principle.

The gravitational redshift was first detected in 1960 by Pound and Rebka, using gamma rays traveling upward a distance of only 72 feet (the height of the physics building at Harvard). Subsequent tests have become increasingly precise, often making use of artificial spacecraft or atomic clocks carried aboard airplanes. The agreement with Einstein's predictions has been excellent in all cases.

Since Einstein's proposal of the three classic tests, further tests of GR have been proposed. The most famous is of course the binary pulsar, to be discussed in Chapter 7. Another is the gravitational time delay, discovered and observed by Shapiro, also discussed in Chapter 7. In a very different context, Big-Bang nucleosynthesis provides a cosmological test of GR at an epoch when the universe was only seconds old, as discussed in Chapter 8. Modern advances have also introduced a host of new tests; for a comprehensive introduction see Will (1981).

5.6 ■ SCHWARZSCHILD BLACK HOLES

We now know something about the behavior of geodesics outside the troublesome radius $r = 2GM$, which is the regime of interest for the solar system and most other astrophysical situations. We next turn to the study of objects that are described by the Schwarzschild solution even at radii smaller than $2GM$—black holes. (We'll use the term "black hole" for the moment, even though we haven't introduced a precise meaning for such an object.)

One way to understand the geometry of a spacetime is to explore its causal structure, as defined by the light cones. We therefore consider radial null curves, those for which θ and ϕ are constant and $ds^2 = 0$:

$$ds^2 = 0 = -\left(1 - \frac{2GM}{r}\right) dt^2 + \left(1 - \frac{2GM}{r}\right)^{-1} dr^2, \tag{5.105}$$

from which we see that

$$\frac{dt}{dr} = \pm\left(1 - \frac{2GM}{r}\right)^{-1}. \tag{5.106}$$

This of course measures the slope of the light cones on a spacetime diagram of the t-r plane. For large r the slope is ± 1, as it would be in flat space, while as we approach $r = 2GM$ we get $dt/dr \to \pm\infty$, and the light cones "close up," as shown in Figure 5.7. Thus a light ray that approaches $r = 2GM$ never seems to get there, at least in this coordinate system; instead it seems to asymptote to this radius.

As we will see, the apparent inability to get to $r = 2GM$ is an illusion, and the light ray (or a massive particle) actually has no trouble reaching this radius. But an observer far away would never be able to tell. If we stayed outside while an intrepid observational general relativist dove into the black hole, sending back signals all the time, we would simply see the signals reach us more and more slowly, as portrayed in Figure 5.8. In the Exercises you are asked to look at this phenomenon more carefully. As an infalling observer approaches $r = 2GM$, any fixed interval $\Delta\tau_1$ of their proper time corresponds to a longer and longer interval $\Delta\tau_2$ from our point of view. This continues forever; we would never see

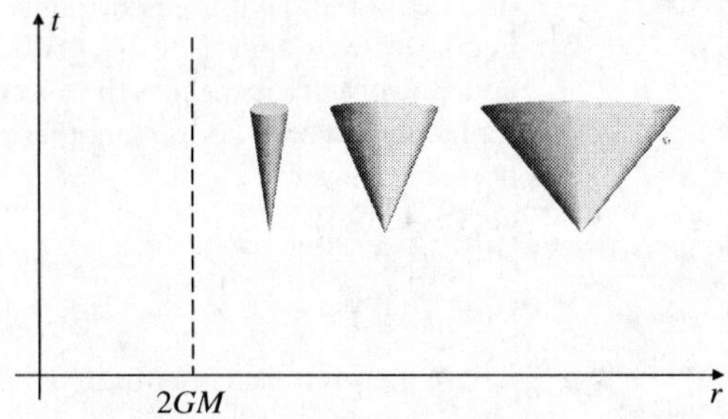

FIGURE 5.7 In Schwarzschild coordinates, light cones appear to close up as we approach $r = 2GM$.

the observer cross $r = 2GM$, we would just see them move more and more slowly (and become redder and redder, as if embarrassed to have done something as stupid as diving into a black hole).

The fact that we never see the infalling observer reach $r = 2GM$ is a meaningful statement, but the fact that their trajectory in the t-r plane never reaches there is not. It is highly dependent on our coordinate system, and we would like to ask a more coordinate-independent question (such as, "Does the observer reach this radius in a finite amount of their proper time?"). The best way to do this is to change coordinates to a system that is better behaved at $r = 2GM$. We now set out to find an appropriate set of such coordinates. There is no way to "derive" a coordinate transformation, of course, we just say what the new coordinates are and plug in the formulas. But we will develop these coordinates in several steps, in hopes of making the choices seem somewhat motivated.

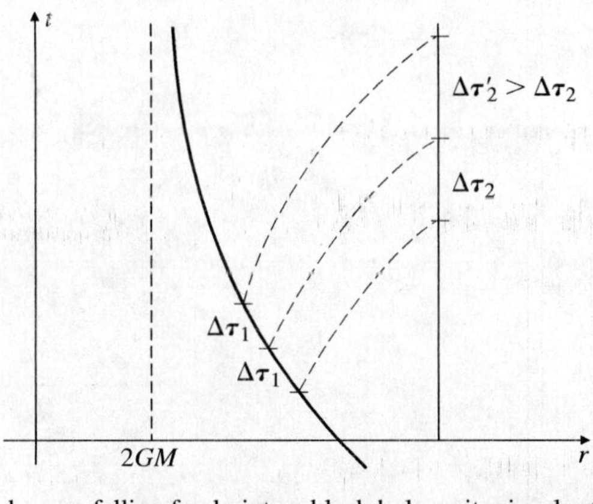

FIGURE 5.8 A beacon falling freely into a black hole emits signals at intervals of constant proper time $\Delta\tau_1$. An observer at fixed r receives the signals at successively longer time intervals $\Delta\tau_2$.

The problem with our current coordinates is that $dt/dr \to \infty$ along radial null geodesics that approach $r = 2GM$; progress in the r direction becomes slower and slower with respect to the coordinate time t. We can try to fix this problem by replacing t with a coordinate that moves more slowly along null geodesics. First notice that we can explicitly solve the condition (5.106) characterizing radial null curves to obtain

$$t = \pm r^* + \text{constant}, \tag{5.107}$$

where the **tortoise coordinate** r^* is defined by

$$r^* = r + 2GM \ln\left(\frac{r}{2GM} - 1\right). \tag{5.108}$$

(The tortoise coordinate is only sensibly related to r when $r \geq 2GM$, but beyond there our coordinates aren't very good anyway.) In terms of the tortoise coordinate the Schwarzschild metric becomes

$$ds^2 = \left(1 - \frac{2GM}{r}\right)\left(-dt^2 + dr^{*2}\right) + r^2 d\Omega^2, \tag{5.109}$$

where r is thought of as a function of r^*. This represents some progress, since the light cones now don't seem to close up, as shown in Figure 5.9; furthermore, none of the metric coefficients becomes infinite at $r = 2GM$ (although both g_{tt} and $g_{r^*r^*}$ become zero). The price we pay, however, is that the surface of interest at $r = 2GM$ has just been pushed to infinity.

Our next move is to define coordinates that are naturally adapted to the null geodesics. If we let

$$\begin{aligned} v &= t + r^* \\ u &= t - r^*, \end{aligned} \tag{5.110}$$

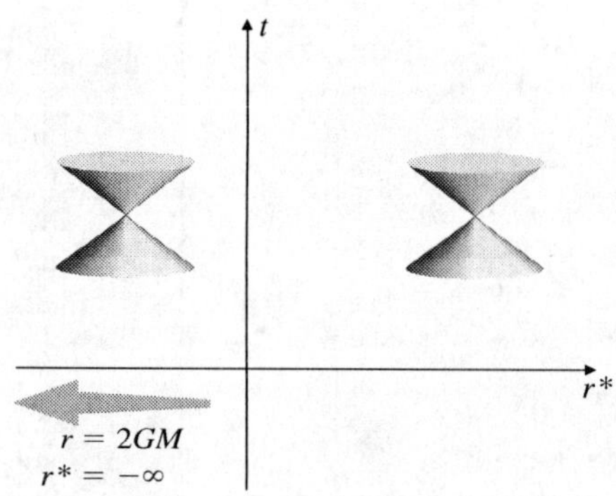

FIGURE 5.9 Schwarzschild light cones in tortoise coordinates, equation (5.109). Light cones remain nondegenerate, but the surface $r = 2GM$ has been pushed to infinity.

then infalling radial null geodesics are characterized by $v =$ constant, while the outgoing ones satisfy $u =$ constant. Now consider going back to the original radial coordinate r, but replacing the timelike coordinate t with the new coordinate v. These are known as **Eddington–Finkelstein coordinates**. In terms of these coordinates the metric is

$$ds^2 = -\left(1 - \frac{2GM}{r}\right) dv^2 + (dv\, dr + dr\, dv) + r^2 d\Omega^2. \tag{5.111}$$

Here we see our first sign of real progress. Even though the metric coefficient g_{vv} vanishes at $r = 2GM$, there is no real degeneracy; the determinant of the metric is

$$g = -r^4 \sin^2\theta, \tag{5.112}$$

which is perfectly regular at $r = 2GM$. Therefore the metric is invertible, and we see once and for all that $r = 2GM$ is simply a coordinate singularity in our original (t, r, θ, ϕ) system. In the Eddington–Finkelstein coordinates the condition for radial null curves is solved by

$$\frac{dv}{dr} = \begin{cases} 0, & \text{(infalling)} \\ 2\left(1 - \frac{2GM}{r}\right)^{-1}. & \text{(outgoing)} \end{cases} \tag{5.113}$$

We can therefore see what has happened: In this coordinate system the light cones remain well-behaved at $r = 2GM$, and this surface is at a finite coordinate value. There is no problem in tracing the paths of null or timelike particles past the surface. On the other hand, something interesting is certainly going on. Although the light cones don't close up, they do tilt over, such that for $r < 2GM$ all future-directed paths are in the direction of decreasing r, as shown in Figure 5.10.

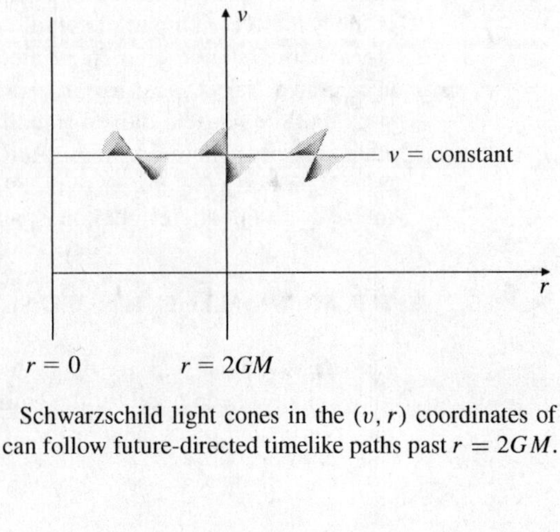

FIGURE 5.10 Schwarzschild light cones in the (v, r) coordinates of (5.111). In these coordinates we can follow future-directed timelike paths past $r = 2GM$.

The surface $r = 2GM$, while being locally perfectly regular, globally functions as a point of no return—once a test particle dips below it, it can never come back. We define an **event horizon** to be a surface past which particles can never escape to infinity; in Schwarzschild the event horizon is located at $r = 2GM$. (This is a rough definition; we will be somewhat more precise in the next chapter.) Despite being located at fixed radial coordinate, the event horizon is a null surface rather than a timelike one, so it is really the causal structure of spacetime itself that makes it impossible to cross the horizon in an outward-going direction. Since nothing can escape the event horizon, it is impossible for us to see inside—thus the name **black hole**. A black hole is simply a region of spacetime separated from infinity by an event horizon. The notion of an event horizon is a global one; the location of the horizon is a statement about the spacetime as a whole, not something you could determine just by knowing the geometry at that location. This will continue to be true in more general spacetimes.

We should mention a couple of features of black holes that sometimes get confused in the popular imagination. First, the external geometry of a black hole is the same Schwarzschild solution that we would have outside a star or planet. In particular, a black hole does not suck in everything around it any more than the Sun does; a particle well outside $r = 2GM$ behaves in exactly the same way regardless of whether the gravitating source is a black hole or not. Second, there is a misleading Newtonian analogy for black holes. The Newtonian escape velocity of a particle at distance r from a gravitating body of mass M is

$$v_{\rm esc} = \sqrt{\frac{2GM}{r}}. \tag{5.114}$$

If we naively ask where the Newtonian escape velocity equals the velocity of light, we find exactly $r = 2GM$. Despite the fact that the speed of light plays no fundamental role in Newtonian theory, it might seem provocative that light, thought of as inertial particles moving at a velocity c, is seemingly not able to escape from a body with mass M and radius less than $2GM$. But there is a profound difference between this case and what we see in GR. The escape velocity is the velocity that a particle would initially need to have in order to escape from a gravitating source on a free trajectory. But nothing stops us from considering accelerated trajectories; for example, one could imagine an acceleration chosen such that the particle moved steadily away from the massive body at some constant velocity. Therefore, a purported Newtonian black hole would not have the crucial property that *nothing* can escape; whereas in GR, arbitrary timelike paths must stay inside their light cones, and hence never escape the event horizon.

5.7 ■ THE MAXIMALLY EXTENDED SCHWARZSCHILD SOLUTION

Let's review what we have done. Acting under the suspicion that our coordinates may not have been good for the entire manifold, we have changed from our original coordinate t to the new one v, which has the nice property that if we decrease

r along a radial null curve $v =$ constant, we go right through the event horizon without any problems. Indeed, a local observer actually making the trip would not necessarily know when the event horizon had been crossed—the local geometry is no different from anywhere else. We therefore conclude that our suspicion was correct and our initial coordinate system didn't do a good job of covering the entire manifold. The region $r \leq 2GM$ should certainly be included in our spacetime, since physical particles can easily reach there and pass through. However, there is no guarantee that we are finished; perhaps we can extend our manifold in other directions.

In fact there are other directions. In the (v, r) coordinate system we can cross the event horizon on future-directed paths, but not on past-directed ones. This seems unreasonable, since we started with a time-independent solution. But we could have chosen u instead of v, in which case the metric would have been

$$ds^2 = -\left(1 - \frac{2GM}{r}\right) du^2 - (du dr + dr du) + r^2 d\Omega^2. \tag{5.115}$$

Now we can once again pass through the event horizon, but this time only along past-directed curves, as shown in Figure 5.11.

This is perhaps a surprise: we can consistently follow either future-directed or past-directed curves through $r = 2GM$, but we arrive at different places. It was actually to be expected, since from the definitions (5.110), if we keep v constant and decrease r we must have $t \to +\infty$, while if we keep u constant and decrease r we must have $t \to -\infty$. (The tortoise coordinate r^* goes to $-\infty$ as $r \to 2GM$.) So we have extended spacetime in two different directions, one to the future and one to the past.

The next step would be to follow spacelike geodesics to see if we would uncover still more regions. The answer is yes, we would reach yet another piece of the spacetime, but let's shortcut the process by defining coordinates that are good all over. A first guess might be to use both u and v at once (in place of t and r),

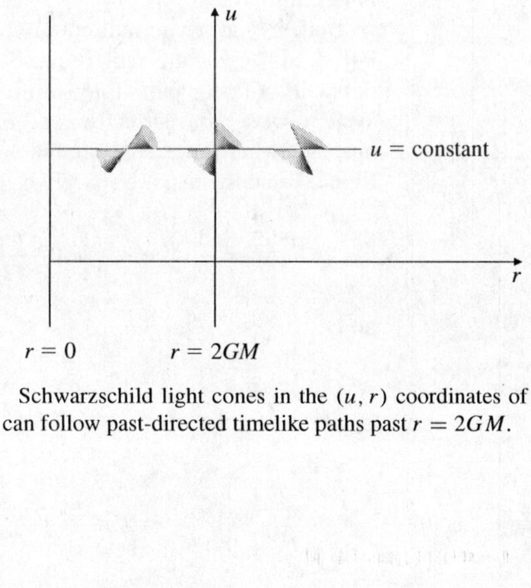

FIGURE 5.11 Schwarzschild light cones in the (u, r) coordinates of (5.115). In these coordinates we can follow past-directed timelike paths past $r = 2GM$.

which leads to

$$ds^2 = -\frac{1}{2}\left(1 - \frac{2GM}{r}\right)(dv du + du\, dv) + r^2\, d\Omega^2, \qquad (5.116)$$

with r defined implicitly in terms of v and u by

$$\frac{1}{2}(v - u) = r + 2GM \ln\left(\frac{r}{2GM} - 1\right). \qquad (5.117)$$

We have actually reintroduced the degeneracy with which we started out; in these coordinates $r = 2GM$ is "infinitely far away" (at either $v = -\infty$ or $u = +\infty$). The thing to do is to change to coordinates that pull these points into finite coordinate values; a good choice is

$$\begin{aligned} v' &= e^{v/4GM} \\ u' &= -e^{-u/4GM}, \end{aligned} \qquad (5.118)$$

which in terms of our original (t, r) system is

$$\begin{aligned} v' &= \left(\frac{r}{2GM} - 1\right)^{1/2} e^{(r+t)/4GM} \\ u' &= -\left(\frac{r}{2GM} - 1\right)^{1/2} e^{(r-t)/4GM}. \end{aligned} \qquad (5.119)$$

In the (v', u', θ, ϕ) system the Schwarzschild metric is

$$ds^2 = -\frac{16G^3M^3}{r} e^{-r/2GM}(dv' du' + du' dv') + r^2\, d\Omega^2. \qquad (5.120)$$

Finally the nonsingular nature of $r = 2GM$ becomes completely manifest; in this form none of the metric coefficients behaves in any special way at the event horizon.

Both v' and u' are null coordinates, in the sense that their partial derivatives $\partial/\partial v'$ and $\partial/\partial u'$ are null vectors. There is nothing wrong with this, since the collection of four partial derivative vectors (two null and two spacelike) in this system serve as a perfectly good basis for the tangent space. Nevertheless, we are somewhat more comfortable working in a system where one coordinate is timelike and the rest are spacelike. We therefore define

$$T = \frac{1}{2}(v' + u') = \left(\frac{r}{2GM} - 1\right)^{1/2} e^{r/4GM} \sinh\left(\frac{t}{4GM}\right) \qquad (5.121)$$

and

$$R = \frac{1}{2}(v' - u') = \left(\frac{r}{2GM} - 1\right)^{1/2} e^{r/4GM} \cosh\left(\frac{t}{4GM}\right), \qquad (5.122)$$

in terms of which the metric becomes

$$ds^2 = \frac{32G^3M^3}{r}e^{-r/2GM}(-dT^2 + dR^2) + r^2\,d\Omega^2, \tag{5.123}$$

where r is defined implicitly from

$$T^2 - R^2 = \left(1 - \frac{r}{2GM}\right)e^{r/2GM}. \tag{5.124}$$

The coordinates (T, R, θ, ϕ) are known as **Kruskal coordinates**, or sometimes Kruskal–Szekeres coordinates.

The Kruskal coordinates have a number of miraculous properties. Like the (t, r^*) coordinates, the radial null curves look like they do in flat space:.

$$T = \pm R + \text{constant}. \tag{5.125}$$

Unlike the (t, r^*) coordinates, however, the event horizon $r = 2GM$ is not infinitely far away; in fact it is defined by

$$T = \pm R, \tag{5.126}$$

consistent with it being a null surface. More generally, we can consider the surfaces $r =$ constant. From (5.124) these satisfy

$$T^2 - R^2 = \text{constant}. \tag{5.127}$$

Thus, they appear as hyperbolae in the R-T plane. Furthermore, the surfaces of constant t are given by

$$\frac{T}{R} = \tanh\left(\frac{t}{4GM}\right), \tag{5.128}$$

which defines straight lines through the origin with slope $\tanh(t/4GM)$. Note that as $t \to \pm\infty$ (5.128) becomes the same as (5.126); therefore $t = \pm\infty$ represents the same surface as $r = 2GM$.

Our coordinates (T, R) should be allowed to range over every value they can take without hitting the real singularity at $r = 0$; the allowed region is therefore

$$\begin{aligned} -\infty \le R \le \infty \\ T^2 < R^2 + 1. \end{aligned} \tag{5.129}$$

From (5.121) and (5.122), T and R seem to become imaginary for $r < 2GM$, but this is an illusion; in that region the (r, t) coordinates are no good (specifically, $|t| > \infty$). We can now draw a spacetime diagram in the T-R plane (with θ and ϕ suppressed), known as a **Kruskal diagram**, shown in Figure 5.12. Each point on the diagram is a two-sphere. This diagram represents the maximal extension

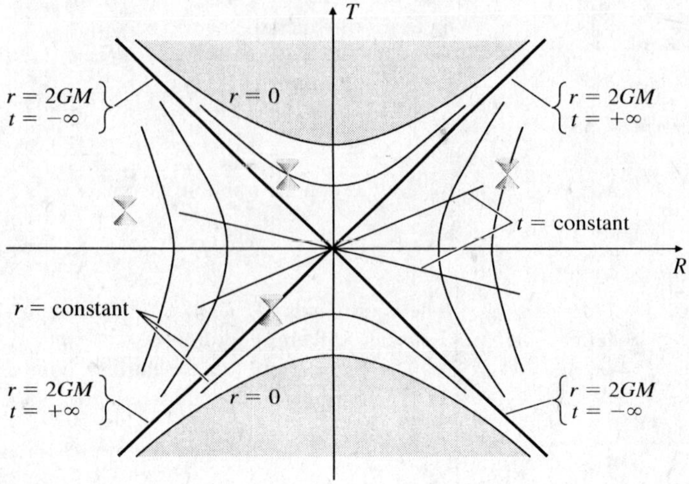

FIGURE 5.12 The Kruskal diagram—the Schwarzschild solution in Kruskal coordinates, where all light cones are at ±45°.

of the Schwarzschild geometry; the coordinates cover what we should think of as the entire manifold described by this solution.

The original Schwarzschild coordinates (t, r) were good for $r > 2GM$, which is only a part of the manifold portrayed on the Kruskal diagram. It is convenient to divide the diagram into four regions, as shown in Figure 5.13. Region I corresponds to $r > 2GM$, the patch in which our original coordinates were well-defined. By following future-directed null rays we reach region II, and by following past-directed null rays we reach region III. If we had explored spacelike geodesics, we would have been led to region IV. The definitions (5.121) and (5.122), which relate (T, R) to (t, r), are really only good in region I; in the other regions it is necessary to introduce appropriate minus signs to prevent the coordinates from becoming imaginary.

FIGURE 5.13 Regions of the Kruskal diagram.

Having extended the Schwarzschild geometry as far as it will go, we have described a remarkable spacetime. Region II, of course, is what we think of as the black hole. Once anything travels from region I into II, it can never return. In fact, every future-directed path in region II ends up hitting the singularity at $r = 0$; once you enter the event horizon, you are utterly doomed. This is worth stressing; not only can you not escape back to region I, you cannot even stop yourself from moving in the direction of decreasing r, since this is simply the timelike direction. This could have been seen in our original coordinate system; for $r < 2GM$, t becomes spacelike and r becomes timelike. Thus you can no more stop moving toward the singularity than you can stop getting older. Since proper time is maximized along a geodesic, you will live the longest if you don't struggle, but just relax as you approach the singularity. Not that you will have long to relax, nor will the voyage be very relaxing; as you approach the singularity the tidal forces become infinite. As you fall toward the singularity your feet and head will be pulled apart from each other, while your torso is squeezed to infinitesimal thinness. The grisly demise of an astrophysicist falling into a black hole is detailed in Misner, Thorne, and Wheeler (1973), Section 32.6. Note that they use orthonormal frames, as we discuss in Appendix J (not that it makes the trip any more enjoyable).

Regions III and IV might be somewhat unexpected. Region III is simply the time-reverse of region II, a part of spacetime from which things can escape to us, while we can never get there. It can be thought of as a **white hole**. There is a singularity in the past, out of which the universe appears to spring. The boundary of region III is the past event horizon, while the boundary of region II is the future event horizon. Region IV, meanwhile, cannot be reached from our region I either forward or backward in time, nor can anybody from over there reach us. It is another asymptotically flat region of spacetime, a mirror image of ours. It can be thought of as being connected to region I by a wormhole (or Einstein–Rosen bridge), a neck-like configuration joining two distinct regions. Consider slicing up the Kruskal diagram into spacelike surfaces of constant T, as shown in Figure 5.14. Now we can draw pictures of each slice, restoring one of the angular coordinates for clarity, as in Figure 5.15. In this way of slicing, the Schwarzschild geometry describes two asymptotically flat regions that reach toward each other, join together via a wormhole for a while, and then disconnect. But the wormhole closes up too quickly for any timelike observer to cross it from one region into the next.

As pleasing as the Kruskal diagram is, it is often even more useful to collapse the Schwarzschild solution into a finite region by constructing its conformal diagram. The idea of a conformal diagram is discussed in Appendix H; it is a crucial tool for analyzing spacetimes in general relativity, and you are encouraged to review that discussion now. We will not go through the manipulations necessary to construct the conformal diagram of Schwarzschild in full detail, since they parallel the Minkowski case with considerable additional algebraic complexity. We would start with the null version of the Kruskal coordinates, in which the metric

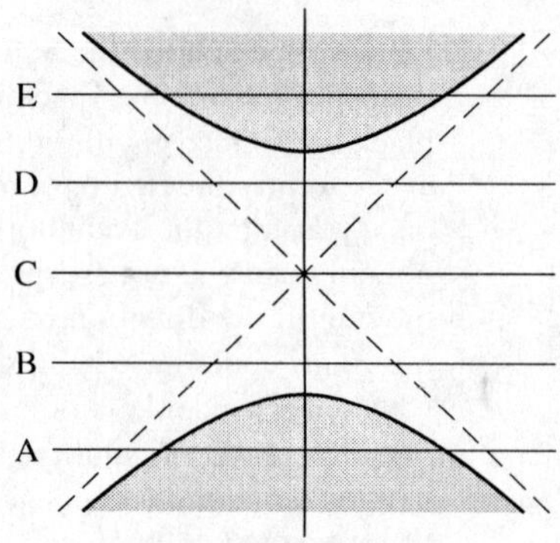

FIGURE 5.14 Spacelike slices in Kruskal coordinates.

takes the form

$$ds^2 = -\frac{16G^3M^3}{r}e^{-r/2GM}(dv'du' + du'dv') + r^2\, d\Omega^2, \tag{5.130}$$

where r is defined implicitly via

$$v'u' = -\left(\frac{r}{2GM} - 1\right)e^{r/2GM}. \tag{5.131}$$

Then essentially the same transformation used in the flat spacetime case suffices to bring infinity into finite coordinate values:

$$\begin{aligned} v'' &= \arctan\left(\frac{v'}{\sqrt{2GM}}\right) \\ u'' &= \arctan\left(\frac{u'}{\sqrt{2GM}}\right), \end{aligned} \tag{5.132}$$

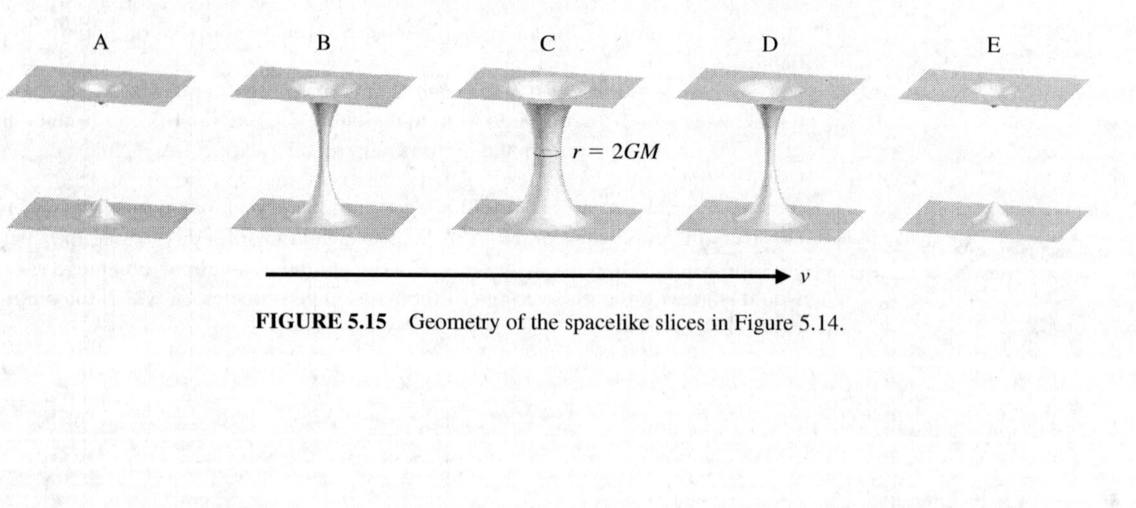

FIGURE 5.15 Geometry of the spacelike slices in Figure 5.14.

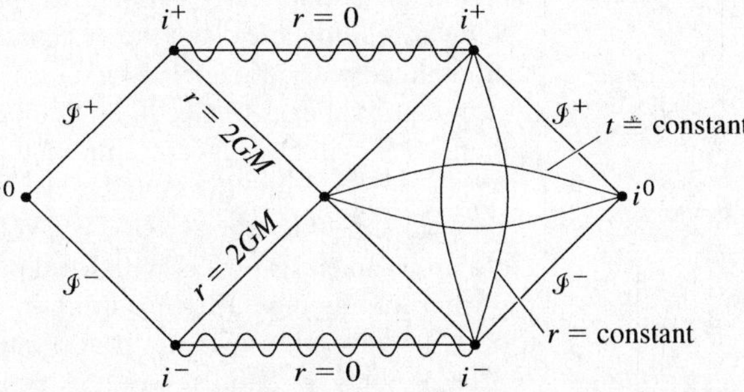

FIGURE 5.16 Conformal diagram for Schwarzschild spacetime.

with ranges

$$-\frac{\pi}{2} < v'' < +\frac{\pi}{2}$$
$$-\frac{\pi}{2} < u'' < +\frac{\pi}{2}$$
$$-\frac{\pi}{2} < v'' + u'' < \frac{\pi}{2}.$$

The $(v,'' u'')$ part of the metric (that is, at constant angular coordinates) is now conformally related to Minkowski space. In the new coordinates the singularities at $r = 0$ are straight lines that stretch from timelike infinity in one asymptotic region to timelike infinity in the other.

The conformal diagram for the maximally extended Schwarzschild solution thus looks like Figure 5.16. The only real subtlety about this diagram is the necessity to understand that i^+ and i^- (future and past infinity) are distinct from $r = 0$—there are plenty of timelike paths that do not hit the singularity. As in the Kruskal diagram, light cones in the conformal diagram are at 45°; the major difference is that the entire spacetime is represented in a finite region. Notice also that the structure of conformal infinity is just like that of Minkowski space, consistent with the claim that Schwarzschild is asymptotically flat.

5.8 ■ STARS AND BLACK HOLES

The maximally extended Schwarzschild solution we have just constructed tells a remarkable story, including not only the sought-after black hole, but also a white hole and an additional asymptotically flat region, connected to our universe by a wormhole. It would be premature, however, to imagine that such features are common in the real world. The Schwarzschild solution represents a highly idealized situation: not only spherically symmetric, but completely free of energy-momentum throughout spacetime. Birkhoff's theorem implies that any vacuum *region* of a spherically symmetric spacetime will be described by *part of* the Schwarzschild metric, but the existence of matter somewhere in the universe may dramatically alter the global picture.

A static spherical object—let's call it a star for definiteness—with radius larger than $2GM$ will be Schwarzschild in the exterior, but there won't be any singularities or horizons, and the global structure will actually be very similar to Minkowski spacetime. Of course, real stars evolve, and it may happen that a star eventually collapses under its own gravitational pull, shrinking down to below $r = 2GM$ and further into a singularity, resulting in a black hole. There is no need for a white hole, however, because the past of such a spacetime looks nothing like that of the full Schwarzschild solution. A conformal diagram describing stellar collapse would look like Figure 5.17. The interior shaded region is nonvacuum, so is not described by Schwarzschild; in particular, there is no wormhole connecting to another universe. It is asymptotically Minkowskian, except for a future region giving rise to an event horizon. We see that a realistic black hole may share the singularity and future horizon with the maximally extended Schwarzschild solution, without any white hole, past horizon, or separate asymptotic region.

We believe that gravitational collapse of this kind is by no means a necessary endpoint of stellar evolution, but will occur under certain conditions. General relativity places rigorous limits on the kind of stars that can resist gravitational collapse; for any given sort of matter, enough mass will always lead to the collapse to a black hole. Furthermore, from astrophysical observations we have excellent evidence that black holes exist in our universe.

To understand gravitational collapse to a black hole, we should first understand static configurations describing the interiors of spherically symmetric stars. We won't delve into this subject in detail, only enough to get a feeling for the basic features of interior solutions. Consider the general static, spherically symmetric

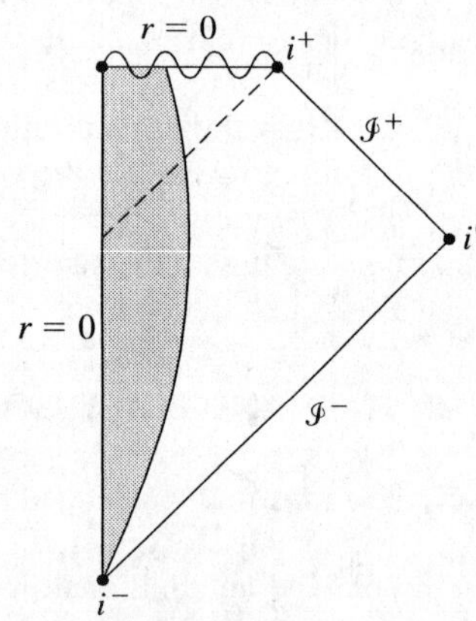

FIGURE 5.17 Conformal diagram for a black hole formed from a collapsing star. The shaded region contains matter, and will be described by an appropriate dynamical interior solution; the exterior region is Schwarzschild.

metric from (5.11):

$$ds^2 = -e^{2\alpha(r)}\,dt^2 + e^{2\beta(r)}\,dr^2 + r^2\,d\Omega^2. \tag{5.133}$$

We are now looking for nonvacuum solutions, so we turn to the full Einstein equation,

$$G_{\mu\nu} = R_{\mu\nu} - \tfrac{1}{2}Rg_{\mu\nu} = 8\pi GT_{\mu\nu}. \tag{5.134}$$

The Einstein tensor follows from the Ricci tensor (5.14) and curvature scalar (5.15),

$$\begin{aligned}
G_{tt} &= \frac{1}{r^2}e^{2(\alpha-\beta)}\left(2r\,\partial_r\beta - 1 + e^{2\beta}\right)\\
G_{rr} &= \frac{1}{r^2}\left(2r\,\partial_r\alpha + 1 - e^{2\beta}\right)\\
G_{\theta\theta} &= r^2e^{-2\beta}\left[\partial_r^2\alpha + (\partial_r\alpha)^2 - \partial_r\alpha\,\partial_r\beta + \frac{1}{r}(\partial_r\alpha - \partial_r\beta)\right]\\
G_{\phi\phi} &= \sin^2\theta\,G_{\theta\theta}.
\end{aligned} \tag{5.135}$$

We model the star itself as a perfect fluid, with energy-momentum tensor

$$T_{\mu\nu} = (\rho + p)U_\mu U_\nu + pg_{\mu\nu}. \tag{5.136}$$

The energy density ρ and pressure p will be functions of r alone. Since we seek static solutions, we can take the four-velocity to be pointing in the timelike direction. Normalized to $U^\mu U_\mu = -1$, it becomes

$$U_\mu = (e^\alpha, 0, 0, 0), \tag{5.137}$$

so that the components of the energy-momentum tensor are

$$T_{\mu\nu} = \begin{pmatrix} e^{2\alpha}\rho & & & \\ & e^{2\beta}p & & \\ & & r^2p & \\ & & & r^2(\sin^2\theta)p \end{pmatrix}. \tag{5.138}$$

We therefore have three independent components of Einstein's equation: the tt component,

$$\frac{1}{r^2}e^{-2\beta}\left(2r\,\partial_r\beta - 1 + e^{2\beta}\right) = 8\pi G\rho, \tag{5.139}$$

the rr component,

$$\frac{1}{r^2}e^{-2\beta}\left(2r\,\partial_r\alpha + 1 - e^{2\beta}\right) = 8\pi Gp, \tag{5.140}$$

and the $\theta\theta$ component,

$$e^{-2\beta}\left[\partial_r^2\alpha + (\partial_r\alpha)^2 - \partial_r\alpha\partial_r\beta + \frac{1}{r}(\partial_r\alpha - \partial_r\beta)\right] = 8\pi Gp. \tag{5.141}$$

The $\phi\phi$ equation is proportional to the $\theta\theta$ equation, so there is no need to consider it separately.

We notice that the tt equation (5.139) involves only β and ρ. It is convenient to replace $\beta(r)$ with a new function $m(r)$, given by

$$m(r) = \frac{1}{2G}(r - re^{-2\beta}), \tag{5.142}$$

or equivalently

$$e^{2\beta} = \left[1 - \frac{2Gm(r)}{r}\right]^{-1}, \tag{5.143}$$

so that

$$ds^2 = -e^{2\alpha(r)}\,dt^2 + \left[1 - \frac{2Gm(r)}{r}\right]^{-1} dr^2 + r^2\,d\Omega^2. \tag{5.144}$$

The metric component g_{rr} is an obvious generalization of the Schwarzschild case, but this will not be true for g_{tt}. The tt equation (5.139) becomes

$$\frac{dm}{dr} = 4\pi r^2\rho, \tag{5.145}$$

which can be integrated to obtain

$$m(r) = 4\pi\int_0^r \rho(r')r'^2 dr'. \tag{5.146}$$

Let's imagine that our star extends to a radius R, after which we are in vacuum and described by Schwarzschild. In order that the metrics match at this radius, the Schwarzschild mass M must be given by

$$M = m(R) = 4\pi\int_0^R \rho(r)r^2 dr. \tag{5.147}$$

It looks like $m(r)$ is simply the integral of the energy density over the stellar interior, and can be interpreted as the mass within a radius r.

There is one subtlety with interpreting $m(r)$ as the integrated energy density; in a proper spatial integral, the volume element should be

$$\sqrt{\gamma}\,d^3x = e^{\beta}r^2\sin\theta\,drd\theta d\phi, \tag{5.148}$$

where

$$\gamma_{ij}\,dx^i dx^j = e^{2\beta}\,dr^2 + r^2 d\theta^2 + r^2\sin^2\theta\,d\phi^2 \tag{5.149}$$

is the spatial metric. The true integrated energy density is therefore

$$\begin{aligned}\bar{M} &= 4\pi \int_0^R \rho(r) r^2 e^{\beta(r)}\, dr \\ &= 4\pi \int_0^R \frac{\rho(r) r^2}{\left[1 - \frac{2Gm(r)}{r}\right]^{1/2}}\, dr.\end{aligned} \tag{5.150}$$

The difference, of course, arises because there is a binding energy due to the mutual gravitational attraction of the fluid elements in the star, which is given by

$$E_{\rm B} = \bar{M} - M > 0. \tag{5.151}$$

The binding energy is the amount of energy that would be required to disperse the matter in the star to infinity. It is not always a well-defined notion in general relativity, but makes sense for spherical stars.

In terms of $m(r)$, the rr equation (5.140) can be written

$$\frac{d\alpha}{dr} = \frac{Gm(r) + 4\pi G r^3 p}{r[r - 2Gm(r)]}. \tag{5.152}$$

It is convenient not to use the $\theta\theta$ equation directly, but instead appeal to energy-momentum conservation, $\nabla_\mu T^{\mu\nu} = 0$. For our metric (5.144), it is straightforward to derive that $\nu = r$ is the only nontrivial component, and it gives

$$(\rho + p)\frac{d\alpha}{dr} = -\frac{dp}{dr}. \tag{5.153}$$

Combining this with (5.152) allows us to eliminate $\alpha(r)$ to obtain

$$\frac{dp}{dr} = -\frac{(\rho + p)[Gm(r) + 4\pi G r^3 p]}{r[r - 2Gm(r)]}. \tag{5.154}$$

This is the **Tolman–Oppenheimer–Volkoff equation**, or simply the equation of hydrostatic equilibrium. Since $m(r)$ is related to $\rho(r)$ via (5.146), this equation relates $p(r)$ to $\rho(r)$. To get a closed system of equations, we need one more relation: the equation of state. In general this will give the pressure in terms of the energy density and specific entropy, $p = p(\rho, S)$. Often we care about situations in which the entropy is very small, and can be neglected; the equation of state then takes the form

$$p = p(\rho). \tag{5.155}$$

Astrophysical systems often obey a polytropic equation of state, $p = K\rho^\gamma$ for some constants K and γ.

A simple and semi-realistic model of a star comes from assuming that the fluid is incompressible: the density is a constant ρ_* out to the surface of the star, after which it vanishes,

$$\rho(r) = \begin{cases} \rho_*, & r < R \\ 0, & r > R. \end{cases} \tag{5.156}$$

Specifying $\rho(r)$ explicitly takes the place of an equation of state, since $p(r)$ can be determined from hydrostatic equilibrium. It is then straightforward to integrate (5.146) to get

$$m(r) = \begin{cases} \frac{4}{3}\pi r^3 \rho_*, & r < R \\ \frac{4}{3}\pi R^3 \rho_* = M, & r > R. \end{cases} \tag{5.157}$$

Integrating the equation of hydrostatic equilibrium yields

$$p(r) = \rho_* \left[\frac{R\sqrt{R - 2GM} - \sqrt{R^3 - 2GMr^2}}{\sqrt{R^3 - 2GMr^2} - 3R\sqrt{R - 2GM}} \right]. \tag{5.158}$$

Finally we can get the metric component $g_{tt} = -e^{2\alpha(r)}$ from (5.152); we find that

$$e^{\alpha(r)} = \frac{3}{2}\left(1 - \frac{2GM}{R}\right)^{1/2} - \frac{1}{2}\left(1 - \frac{2GMr^2}{R^3}\right)^{1/2}, \quad r < R. \tag{5.159}$$

The pressure increases near the core of the star, as one would expect. Indeed, for a star of fixed radius R, the central pressure $p(0)$ will need to be greater than infinity if the mass exceeds

$$M_{\text{max}} = \frac{4}{9G}R. \tag{5.160}$$

Thus, if we try to squeeze a greater mass than this inside a radius R, general relativity admits no static solutions; a star that shrinks to such a size must inevitably keep shrinking, eventually forming a black hole. We derived this result from the rather strong assumption that the density is constant, but it continues to hold when that assumption considerably weakened; **Buchdahl's theorem** states that any reasonable static, spherically symmetric interior solution has $M < 4R/9G$. Although a careful proof requires more work, this result makes sense; if we imagine that there is some maximum sustainable density in nature, the most massive object we could in principle make would have that density everywhere, which is the specific case we considered.

Of course, this still doesn't mean that realistic astrophysical objects will always ultimately collapse to black holes. An ordinary planet, supported by material pressures, will persist essentially forever (apart from some fantastically unlikely quantum tunneling from a planet to something very different, or the possibility of eventual proton decay). But massive stars are a different story. The pressure supporting a star comes from the heat produced by fusion of light nuclei into heavier ones. When the nuclear fuel is used up, the temperature declines and the star begins to shrink under the influence of gravity. The collapse may eventually be halted by Fermi degeneracy pressure: Electrons are pushed so close together that they resist further compression simply on the basis of the Pauli exclusion principle (no two fermions can be in the same state). A stellar remnant supported by electron degeneracy pressure is called a **white dwarf**; a typical white dwarf is comparable in size to the Earth. Lower-mass particles become degenerate at lower number densities than high-mass particles, so nucleons do not contribute appreciably to the pressure in a white dwarf. White dwarfs are the end state for most stars, and are extremely common throughout the universe.

If the total mass is sufficiently high, however, the star will reach the **Chandrasekhar limit**, where even the electron degeneracy pressure is not enough to resist the pull of gravity. Calculations put the Chandrasekhar limit at about $1.4\ M_\odot$, where $M_\odot = 2 \times 10^{33}$ g is the mass of the Sun. When it is reached, the star is forced to collapse to an even smaller radius. At this point electrons combine with protons to make neutrons and neutrinos (inverse beta decay), and the neutrinos simply fly away. The result is a **neutron star**, with a typical radius of about 10 km. Neutron stars have a low total luminosity, but often are rapidly spinning and possess strong magnetic fields. This combination gives rise to **pulsars**, which accelerate particles in jets emanating from the magnetic poles, appearing to rapidly flash as the neutron star spins. Pulsars were discovered by Bell in 1967; after a brief speculation that they might represent signals from an extraterrestrial civilization, the more prosaic astrophysical explanation was settled on.

Since the conditions at the center of a neutron star are very different from those on Earth, we do not have a perfect understanding of the equation of state. Nevertheless, we believe that a sufficiently massive neutron star will itself be unable to resist the pull of gravity, and will continue to collapse; current estimates of the maximum possible neutron-star mass are around 3–4 $M_\odot$, the **Oppenheimer-Volkoff limit**. Since a fluid of neutrons is the densest material we know about (apart from some very speculative suggestions), it is believed that the outcome of such a collapse is a black hole.

How would we know if there were a black hole? The fundamental obstacle to direct detection is, of course, blackness: a black hole will not itself give off any radiation (neglecting Hawking radiation, which is a very small effect to be discussed in Chapter 9). But black holes will feature extremely strong gravitational fields, so we can hope to detect them indirectly by observing matter being influenced by these fields. As matter falls into a black hole, it will heat up and emit X-rays, which we can detect with satellite observatories. A large number of black-hole candidates have been detected by this method, and the case for real black holes in our universe is extremely strong.[1] The large majority of candidates fall into one of two classes. There are black holes with masses of order a solar mass or somewhat higher; these are thought to be the endpoints of evolution for very massive stars. The other category describes supermassive black holes, be-

[1] For a review on astrophysical evidence for black holes, see A. Celotti, J.C. Miller, and D.W. Sciama (1999), *Class. Quant. Grav.* **16**, A3; `http://arxiv.org/abs/astro-ph/9912186`.

tween 10^6 and 10^9 solar masses. These are found at the centers of galaxies, and are thought to be the engines that powered quasars in the early era of galaxy formation. Our own Milky Way galaxy contains an object (Sgr A*) that is believed to be a black hole of at least $2 \times 10^6 M_\odot$. The precise history of the formation of these supermassive holes is not well understood. Other possibilities include very small primordial black holes produced in the very early universe, and so-called "middleweight" black holes of order a thousand solar masses.

As matter falls into a black hole, it tends to settle into a rotating accretion disk, and both energy and angular momentum are gradually fed into the hole. As a result, the black holes we expect to see in astrophysical situations should be spinning, and indeed observations are consistent with very high spin rates for observed black holes. In this chapter we have excluded the possibility of black hole spin by focusing on the spherically symmetric Schwarzschild solution; in the next chapter we turn to more general types of black holes.

5.9 ■ EXERCISES

1. A space monkey is happily orbiting a Schwarzschild black hole in a circular geodesic orbit. An evil baboon, far from the black hole, tries to send the monkey to its death inside the black hole by dropping a carefully timed coconut radially toward the black hole, knowing that the monkey can't resist catching the falling coconut. Given the monkey's mass and initial orbital radius and the mass of the coconut, explain how you would go about solving the problem (but do not do the calculation). What are the possible fates for our intrepid space monkey?

2. Consider a perfect fluid in a static, circularly symmetric $(2+1)$-dimensional spacetime, equivalently, a cylindrical configuration in $(3+1)$ dimensions with perfect rotational symmetry.

(a) Derive the analogue of the Tolman–Oppenhiemer–Volkov (TOV) equation for $(2+1)$ dimensions.

(b) Show that the vacuum solution can be written as

$$ds^2 = -dt^2 + \frac{1}{1-8GM}dr^2 + r^2 d\theta^2$$

Here M is a constant.

(c) Show that another way to write the same solution is

$$ds^2 = -d\tau^2 + d\xi^2 + \xi^2 d\phi^2$$

where $\phi \in [0, 2\pi(1-8GM)^{1/2}]$.

(d) Solve the $(2+1)$ TOV equation for a constant density star. Find $p(r)$ and solve for the metric.

(e) Solve the $(2+1)$ TOV equation for a star with equation of state $p = \kappa\rho^{3/2}$. Find $p(r)$ and solve for the metric.

(f) Find the mass $M(R) = \int_0^{2\pi}\int_0^R \rho\, dr\, d\theta$ and the proper mass $\bar{M}(R) = \int_0^{2\pi}\int_0^R \rho\sqrt{-g}\, dr\, d\theta$ for the solutions in parts (d) and (e).

3. Consider a particle (not necessarily on a geodesic) that has fallen inside the event horizon, $r < 2GM$. Use the ordinary Schwarzschild coordinates (t, r, θ, ϕ). Show that the radial coordinate must decrease at a minimum rate given by

$$\left|\frac{dr}{d\tau}\right| \geq \sqrt{\frac{2GM}{r} - 1}.$$

Calculate the maximum lifetime for a particle along a trajectory from $r = 2GM$ to $r = 0$. Express this in seconds for a black hole with mass measured in solar masses. Show that this maximum proper time is achieved by falling freely with $E \to 0$.

4. Consider Einstein's equations in vacuum, but with a cosmological constant, $G_{\mu\nu} + \Lambda g_{\mu\nu} = 0$.
   - **(a)** Solve for the most general spherically symmetric metric, in coordinates (t, r) that reduce to the ordinary Schwarzschild coordinates when $\Lambda = 0$.
   - **(b)** Write down the equation of motion for radial geodesics in terms of an effective potential, as in (5.66). Sketch the effective potential for massive particles.

5. Consider a comoving observer sitting at constant spatial coordinates (r_*, θ_*, ϕ_*), around a Schwarzschild black hole of mass M. The observer drops a beacon into the black hole (straight down, along a radial trajectory). The beacon emits radiation at a constant wavelength $\lambda_{\rm em}$ (in the beacon rest frame).
   - **(a)** Calculate the coordinate speed dr/dt of the beacon, as a function of r.
   - **(b)** Calculate the proper speed of the beacon. That is, imagine there is a comoving observer at fixed r, with a locally inertial coordinate system set up as the beacon passes by, and calculate the speed as measured by the comoving observer. What is it at $r = 2GM$?
   - **(c)** Calculate the wavelength $\lambda_{\rm obs}$, measured by the observer at r_*, as a function of the radius $r_{\rm em}$ at which the radiation was emitted.
   - **(d)** Calculate the time $t_{\rm obs}$ at which a beam emitted by the beacon at radius $r_{\rm em}$ will be observed at r_*.
   - **(e)** Show that at late times, the redshift grows exponentially: $\lambda_{\rm obs}/\lambda_{\rm em} \propto e^{t_{\rm obs}/T}$. Give an expression for the time constant T in terms of the black hole mass M.

CHAPTER 6

More General Black Holes

6.1 ■ THE BLACK HOLE ZOO

Birkhoff's theorem ensures that the Schwarzschild metric is the only spherically symmetric vacuum solution to general relativity. This shouldn't be too surprising, as it is reminiscent of the situation in electromagnetism, where the only spherically symmetric field configuration in a region free of charges will be a Coulomb field. Moving beyond spherical symmetry, there is an unlimited variety of possible gravitational fields. For a planet like the Earth, for example, the external field will depend on the density and profile of all the various mountain ranges and valleys on the surface. We could imagine decomposing the metric into multipole moments, and an infinite number of coefficients would have to be specified to describe the field exactly.

It might therefore come as something of a surprise that black holes do not share this property. Only a small number of stationary black-hole solutions exist, described by a small number of parameters. The specific set of parameters will depend on what matter fields we include in our theory; if electromagnetism is the only long-range nongravitational field, we have a **no-hair theorem**:

> Stationary, asymptotically flat black hole solutions to general relativity coupled to electromagnetism that are nonsingular outside the event horizon are fully characterized by the parameters of mass, electric and magnetic charge, and angular momentum.

Stationary solutions are of special interest because we expect them to be the end states of gravitational collapse. The alternative might be some sort of oscillating configuration, but oscillations will ultimately be damped as energy is lost through the emission of gravitational radiation; in fact, typical evolutions will evolve quite rapidly to a stationary configuration.

We speak of "a" no-hair theorem, rather than "the" no-hair theorem, because the result depends not only on general relativity, but also on the matter content of our theory. In the Standard Model of particle physics, electromagnetism is the only long-range field, and the above theorem applies; but for different kinds of fields there might be other sorts of hair.[1] Examples have even been found of static (nonrotating) black holes that are axisymmetric but not completely spherically

[1]For a discussion see M. Heusler, "Stationary Black Holes: Uniqueness and Beyond," Living Rev. Relativity **1**, (1998), 6; `http://www.livingreviews.org/Articles/Volume1/1998-6heusler/`.

symmetric. The central point, however, remains unaltered: black hole solutions are characterized by a very small number of parameters, rather than the potentially infinite set of parameters characterizing, say, a planet.

As we will discuss at the end of this chapter and again in Chapter 9, the no-hair property leads to a puzzling situation. In most physical theories, we hope to have a well-defined initial value problem, so that information about a state at any one moment of time can be used to predict (or retrodict) the state at any other moment of time. As a consequence, any two states that are connected by a solution to the equations of motion should require the same amount of information to be specified. But in GR, it seems, we can take a very complicated collection of matter, collapse it into a black hole, and end up with a configuration described completely by mass, charge, and spin. In classical GR this might not bother us so much, since the information can be thought of as hidden behind the event horizon rather than truly being lost. But when quantum field theory is taken into account, we find that black holes evaporate and eventually disappear, and the information seems to be truly lost. Conceivably, the outgoing Hawking radiation responsible for the evaporation somehow encodes information about what state was originally used to make a black hole, but how that could happen is completely unclear. Understanding this "information loss paradox" is considered by many to be a crucial step in building a sensible theory of quantum gravity.

In this chapter, however, we will stick to considerations of classical GR. We begin with some general discussion of black hole properties, especially those of event horizons and Killing horizons. This subject can be subtle and technical, and our philosophy here will be to try to convey the main ideas without being rigorous about definitions or proofs of theorems. We then discuss the specific solutions corresponding to charged (Reissner–Nordström) and spinning (Kerr) black holes; consistent with our approach, we will not carefully go through the coordinate redefinitions necessary to construct the maximally extended spacetimes, but instead simply draw the associated conformal diagrams. The reader interested in further details should consult the review article by Townsend,[2] or the books by Hawking and Ellis (1973) and Wald (1984), all of which we draw on heavily in this chapter.

6.2 ■ EVENT HORIZONS

Black holes are characterized by the fact that you can enter them, but never exit. Thus, their most important feature is actually not the singularity at the center, but the event horizon at the boundary. An event horizon is a hypersurface separating those spacetime points that are connected to infinity by a timelike path from those that are not. To understand what this means in practice, we should think a little more carefully about what we mean by "infinity." In general relativity, the global structure of spacetime can take many different forms, with correspondingly different notions of infinity. But to think about black holes in the real universe, we

[2] P.K. Townsend, "Black Holes: Lecture Notes," `http://arxiv.org/abs/gr-qc/9707012`.

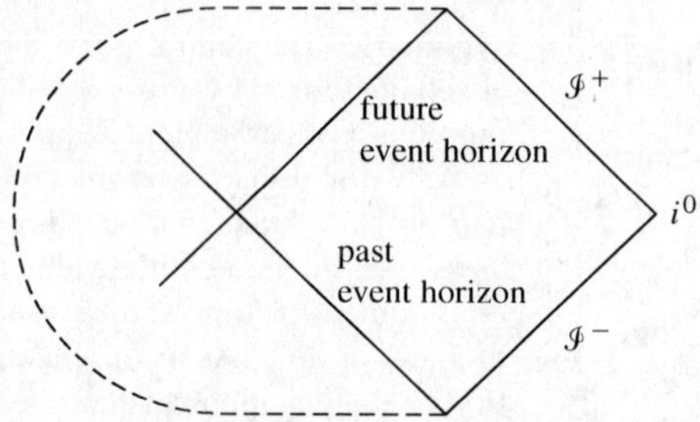

FIGURE 6.1 An asymptotically flat spacetime is one for which infinity in a conformal diagram matches that of Minkowski spacetime, with future null infinity $\mathcal{I}^+$, spacelike infinity i^0, and past null infinity $\mathcal{I}^-$. The future event horizon is the boundary of the past of $\mathcal{I}^+$. The dashed region represents the rest of the spacetime, which may take a number of different forms in different examples.

aren't actually concerned with what happens infinitely far away; we use infinity as a proxy for "well outside the black hole," and imagine that spacetime sufficiently far away from the hole can be approximated by Minkowski space.

As mentioned in the Chapter 5, a spacetime that looks Minkowskian at infinity is referred to as asymptotically flat. The meaning of this concept is made clear in a conformal diagram such as in Figure 6.1. From our discussion in Appendix H of the conformal diagram for Minkowski, we know that conformal infinity comes in five pieces: future and past timelike infinity $i^\pm$, future and past null infinity $\mathcal{I}^\pm$, and spatial infinity i^0. An asymptotically flat spacetime (or region of spacetime) is one for which $\mathcal{I}^\pm$ and i^0 have the same structure as for Minkowski; timelike infinity is not necessary. Such spacetimes will have the general form shown in Figure 6.1.

With this picture, it is clear how we should think of the future event horizon: it is the surface beyond which timelike curves cannot escape to infinity. Recalling that the causal past J^- of a region is the set of all points we can reach from that region by moving along past-directed timelike paths, the event horizon can be equivalently defined as the boundary of $J^-(\mathcal{I}^+)$, the causal past of future null infinity. (The event horizon is really the boundary of the *closure* of this set, but we're not being rigorous.) Analogous definitions hold for the past horizon. As we have seen in the case of maximally extended Schwarzschild, there may be more than one asymptotically flat region in a spacetime, and correspondingly more than one event horizon.

From the definition, it is clear that the event horizon is a null hypersurface. Properties of null hypersurfaces are discussed in Appendix D; here we can recall the major features. A hypersurface Σ can be defined by $f(x) =$ constant for some function $f(x)$. The gradient $\partial_\mu f$ is normal to Σ; if the normal vector is null, the hypersurface is said to be null, and the normal vector is also tangent to Σ. Null hypersurfaces can be thought of as a collection of null geodesics $x^\mu(\lambda)$, called the generators of the hypersurface. The tangent vectors ξ^μ to these geodesics are proportional to the normal vectors,

$$\xi^{\mu} = \frac{dx^{\mu}}{d\lambda} = h(x) g^{\mu\nu} \partial_{\nu} f, \tag{6.1}$$

and therefore also serve as normal vectors to the hypersurface. We may choose the function $h(x)$ so that the geodesics are affinely parameterized, so the tangent vectors will obey

$$\xi_{\mu}\xi^{\mu} = 0, \quad \xi^{\mu}\nabla_{\mu}\xi^{\nu} = 0. \tag{6.2}$$

For future event horizons, the generators may end in the past (for example, when a black hole is formed by stellar collapse) but will always continue indefinitely into the future (and similarly with future and past interchanged).

Because the event horizon is a global concept, it might be difficult to actually locate one when you are handed a metric in an arbitrary set of coordinates. Fortunately, in this chapter we will be concerned with quite special metrics—stationary, asymptotically flat, and containing event horizons with spherical topology. In such spacetimes, there are convenient coordinate systems in which there is a simple way to identify the event horizon. For the Schwarzschild solution, the event horizon is a place where the light cones "tilt over" so that $r = 2GM$ is a null surface rather than a timelike surface, as $r =$ constant would be for large r. Light-cone tilting is clearly a coordinate-dependent notion (it doesn't happen, for example, in Kruskal coordinates), but the metrics of concern to us will allow for analogous constructions. A stationary metric has a Killing vector ∂_t that is asymptotically timelike, and we can adapt the metric components to be time independent ($\partial_t g_{\mu\nu} = 0$). On hypersurfaces $t =$ constant, we can choose coordinates (r, θ, ϕ) in which the metric at infinity looks like Minkowski space in spherical polar coordinates. Hypersurfaces $r =$ constant will be timelike cylinders with topology $S^2 \times \mathbf{R}$ at $r \to \infty$. Now imagine we have chosen our coordinates cleverly, so that as we decrease r from infinity the $r =$ constant hypersurfaces remain timelike until some fixed $r = r_{\rm H}$, for which the surface is everywhere null. (In nonclever coordinates, $r =$ constant hypersurfaces will become null or spacelike for some values of θ and ϕ but remain timelike for others.) This will clearly represent an event horizon, since timelike paths crossing into the region $r < r_{\rm H}$ will never be able to escape back to infinity. Determining the point at which $r =$ constant hypersurfaces become null is easy; $\partial_{\mu} r$ is a one-form normal to such hypersurfaces, with norm

$$g^{\mu\nu}(\partial_{\mu} r)(\partial_{\nu} r) = g^{rr}. \tag{6.3}$$

We are looking for the place where the norm of our one-form vanishes; hence, in the coordinates we have described, the event horizon $r = r_{\rm H}$ will simply be the hypersurface at which g^{rr} switches from being positive to negative,

$$g^{rr}(r_{\rm H}) = 0. \tag{6.4}$$

This criterion clearly works for Schwarzschild, for which $g^{rr} = 1 - 2GM/r$. We will present the Reissner–Nordström and Kerr solutions in coordinates that are similarly adapted to the horizons.

The reason why we make such a big deal about event horizons is that they are nearly inevitable in general relativity. This conclusion is reached by concatenating two interesting results: Singularities are nearly inevitable, and singularities are hidden behind event horizons. Of course both results hold under appropriate sets of assumptions; it is not that hard to come up with spacetimes that have no singularities (Minkowski would be an example), nor is it even that hard to find singularities without horizons (as we will see below in our discussion of charged black holes). But we believe that "generic" solutions will have singularities hidden behind horizons.

The ubiquity of singularities is guaranteed by the **singularity theorems** of Hawking and Penrose. Before these theorems were proven, it was possible to hope that collapse to a Schwarzschild singularity was an artifact of spherical symmetry, and typical geometries would remain nonsingular (as happens, for example, in Newtonian gravity). But the Hawking–Penrose theorems demonstrate that once collapse reaches a certain point, evolution to a singularity is inevitable. The way we know there is a singularity is through geodesic incompleteness—there exists some geodesic that cannot be extended within the manifold, but nevertheless ends at a finite value of the affine parameter. The way we know collapse has reached a point of no return is the appearance of a **trapped surface**. To understand what a trapped surface is, first picture a two-sphere in Minkowski space, taken as a set of points some fixed radial distance from the origin, embedded in a constant-time slice. If we follow null rays emanating into spacetime from this spatial sphere, one set (pointed inward) will describe a shrinking set of spheres, while the other (pointed outward) will describe a growing set of spheres. But this would not be the case for a sphere of fixed radius $r < 2GM$ in the Schwarzschild geometry; inside the event horizon, both sets of null rays emanating from such a sphere would evolve to smaller values of r (since r is a timelike coordinate), and thus to smaller areas $4\pi r^2$. This is what is meant by a trapped surface: a compact, spacelike, two-dimensional submanifold with the property that outgoing future-directed light rays *converge* in both directions everywhere on the submanifold. (The formal definition of "converge" is that the expansion θ, as described in the discussion of geodesic congruences in Appendix F, is negative.)

With these definitions in hand, we can present an example of a singularity theorem.

> Let M be a manifold with a generic metric $g_{\mu\nu}$, satisfying Einstein's equation with the strong energy condition imposed. If there is a trapped surface in M, there must be either a closed timelike curve or a singularity (as manifested by an incomplete timelike or null geodesic).

In this case, by "a generic metric" we mean that the **generic condition** is satisfied for both timelike and null geodesics. For timelike geodesics, the generic condition states that every geodesic with tangent vector U^μ must have at least one point on which $R_{\alpha\beta\gamma\delta}U^\alpha U^\delta \neq 0$; for null geodesics, the generic condition states that every geodesic with tangent vector k^μ must have at least one point on which $k_{[\alpha}R_{\beta]\gamma\delta[\epsilon}k_{\zeta]}k^\gamma k^\delta \neq 0$. These fancy conditions simply serve to exclude very special metrics for which the curvature consistently vanishes in some directions.

Singularity theorems exist in many forms, proceeding from various different sets of assumptions. The moral of the story seems to be that typical time-dependent solutions in general relativity usually end in singularities. (Or begin in them; some theorems imply the existence of cosmological singularities, such as the Big Bang.) This represents something of a problem for GR, in the sense that the theory doesn't really apply to the singularities themselves, whose existence therefore represents an incompleteness of description. The traditional attitude toward this issue is to hope that a sought-after quantum theory of gravity will somehow resolve the singularities of classical GR.

In the meantime, we can take solace in the idea that singularities are hidden behind event horizons. This belief is encompassed in the **cosmic censorship conjecture**:

> Naked singularities cannot form in gravitational collapse from generic, initially nonsingular states in an asymptotically flat spacetime obeying the dominant energy condition.

A **naked singularity** is one from which signals can reach $\mathscr{I}^+$; that is, one that is not hidden behind an event horizon. Notice that the conjecture refers to the formation of naked singularities, not their existence; there are certainly solutions in which spacelike naked singularities exist in the past (such as the Schwarzschild white hole) or timelike singularities exist for all times (such as in super-extremal charged black holes, discussed below). The cosmic censorship conjecture has not been proven, although a great deal of effort has gone into finding convincing counterexamples, without success. The requirement that the initial data be in some sense "generic" is important, as numerical experiments have shown that finely-tuned initial conditions are able to give rise to naked singularities. A precise proof of some form of the cosmic censorship conjecture remains one of the outstanding problems of classical general relativity.[3]

A consequence of cosmic censorship (or of certain equivalent assumptions) is that classical black holes never shrink, they only grow bigger. The size of a black hole is measured by the area of the event horizon, by which we mean the spatial area of the intersection of the event horizon with a spacelike slice. We then have Hawking's **area theorem**:

> Assuming the weak energy condition and cosmic censorship, the area of a future event horizon in an asymptotically flat spacetime is nondecreasing.

[3] For a review of cosmic censorship see R.M. Wald, "Gravitational Collapse and Cosmic Censorship," `http://arxiv.org/abs/gr-qc/9710068`.

For a Schwarzschild black hole, the area depends monotonically on the mass, so this theorem implies that Schwarzschild black holes can only increase in mass. But for spinning black holes this is no longer the case; the area depends on a combination of mass and angular momentum, and we will see below that we can actually extract energy from a black hole by decreasing its spin. We can also decrease the mass of a black hole through quantum-mechanical Hawking radiation; this can be traced to the fact that quantum field theory in curved spacetime can violate the weak energy condition.

6.3 ■ KILLING HORIZONS

In the Schwarzschild metric, the Killing vector $K = \partial_t$ goes from being timelike to spacelike at the event horizon. In general, if a Killing vector field χ^μ is null along some null hypersurface Σ, we say that Σ is a **Killing horizon** of χ^μ. Note that the vector field χ^μ will be normal to Σ, since a null surface cannot have two linearly independent null tangent vectors.

The notion of a Killing horizon is logically independent from that of an event horizon, but in spacetimes with time-translation symmetry the two are closely related. Under certain reasonable conditions (made explicit below), we have the following classification:

> Every event horizon Σ in a stationary, asymptotically flat spacetime is a Killing horizon for some Killing vector field χ^μ.
>
> If the spacetime is static, χ^μ will be the Killing vector field $K^\mu = (\partial_t)^\mu$ representing time translations at infinity.
>
> If the spacetime is stationary but not static, it will be axisymmetric with a rotational Killing vector field $R^\mu = (\partial_\phi)^\mu$, and χ^μ will be a linear combination $K^\mu + \Omega_H R^\mu$ for some constant Ω_H.

For example, below we will examine the Kerr metric for spinning black holes, in which the event horizon is a Killing horizon for a linear combination of the Killing vectors for rotations and time translations. In Kerr, the hypersurface on which ∂_t becomes null is actually timelike, so is not a Killing horizon.

Let's be precise about the conditions under which this classification scheme actually holds.[4] Carter has shown that, for static black holes, the event horizon is a Killing horizon for K^μ; this is a purely geometric fact, which holds even without invoking Einstein's equation. In the stationary case, if we assume the existence of a rotational Killing field R^μ with the property that 2-planes spanned by K^μ and R^μ are orthogonal to a family of two-dimensional surfaces, then the event horizon will be a Killing horizon for a linear combination of the two Killing fields, again from purely geometric considerations. If on the other hand we only assume that the black hole is stationary, we cannot prove in general that the event horizon

[4] For a discussion see R. M. Wald, "The thermodynamics of black holes," *Living Rev. Rel.* **4**, 6 (2001), http://arxiv.org/gr-qc/9912119.

is axisymmetric. Given Einstein's equation and some conditions on the matter fields, Hawking was able to show that the event horizon of any stationary black hole must be a Killing horizon for some vector field, and furthermore that such horizons must either be stationary or axisymmetric. For the rest of this chapter we will speak as if the above classification holds; however, making assumptions about matter fields is notoriously tricky, and we should keep in mind the possibility in principle of finding black holes that are not static or axisymmetric, for which the event horizon might not be a Killing horizon.

It's important to point out that, while event horizons for stationary asymptotically flat spacetimes will typically be Killing horizons, it's easy to have Killing horizons that have nothing to do with event horizons. Consider Minkowski space in inertial coordinates, $ds^2 = -dt^2 + dx^2 + dy^2 + dz^2$; clearly there are no event horizons in this spacetime. The Killing vector that generates boosts in the x-direction is

$$\chi = x\partial_t + t\partial_x, \tag{6.5}$$

with norm

$$\chi_\mu \chi^\mu = -x^2 + t^2. \tag{6.6}$$

This goes null at the null surfaces

$$x = \pm t, \tag{6.7}$$

which are therefore Killing horizons. By combining the boost Killing vector with translational and rotational Killing vectors, we can move these horizons through the manifold; there are Killing horizons all over. In more interesting spacetimes, of course, there will be fewer Killing vector fields, and the associated horizons (if any) will have greater physical significance.

To every Killing horizon we can associate a quantity called the **surface gravity**. Consider a Killing vector χ^μ with Killing horizon Σ. Because χ^μ is a normal vector to Σ, along the Killing horizon it obeys the geodesic equation,

$$\chi^\mu \nabla_\mu \chi^\nu = -\kappa \chi^\nu, \tag{6.8}$$

where the right-hand side arises because the integral curves of χ^μ may not be affinely parameterized. The parameter κ is the surface gravity; it will be constant over the horizon, except for a "bifurcation two-sphere" where the Killing vector vanishes and κ can change sign. (This happens, for example, at the center of the Kruskal diagram in the Schwarzschild solution.) Using Killing's equation $\nabla_{(\mu}\chi_{\nu)} = 0$ and the fact that $\chi_{[\mu}\nabla_\nu \chi_{\sigma]} = 0$ (since χ^μ is normal to Σ), it is straightforward to derive a nice formula for the surface gravity:

$$\kappa^2 = -\tfrac{1}{2}(\nabla_\mu \chi_\nu)(\nabla^\mu \chi^\nu). \tag{6.9}$$

The expression on the right-hand side is to be evaluated at the horizon Σ. You are encouraged to check this formula yourself.

The surface gravity associated with a Killing horizon is in principle arbitrary, since we can always scale a Killing field by a real constant and obtain another Killing field. In a static, asymptotically flat spacetime, the time-translation Killing vector $K = \partial_t$ can be normalized by setting

$$K_\mu K^\mu (r \to \infty) = -1. \tag{6.10}$$

This in turn fixes the surface gravity of any associated Killing horizon. If we are in a stationary spacetime, where the Killing horizon is associated with a linear combination of time translations and rotations, fixing the normalization of $K = \partial_t$ also fixes this linear combination, so the surface gravity remains unique.

The reason why κ is called the "surface gravity" becomes clear only when the spacetime is static. In that case we have the following interpretation:

> In a static, asymptotically flat spacetime, the surface gravity is the acceleration of a static observer near the horizon, as measured by a static observer at infinity.

To make sense of such a statement, let's first consider static observers. By a static observer we mean one whose four-velocity U^μ is proportional to the time-translation Killing field K^μ:

$$K^\mu = V(x)U^\mu. \tag{6.11}$$

Since the four-velocity is normalized to $U_\mu U^\mu = -1$, the function V is simply the magnitude of the Killing field,

$$V = \sqrt{-K_\mu K^\mu}, \tag{6.12}$$

and hence ranges from zero at the Killing horizon to unity at infinity. V is sometimes called the "redshift factor," since it relates the emitted and observed frequencies of a photon as measured by static observers. Recall that the conserved energy of a photon with four-momentum p^μ is $E = -p_\mu K^\mu$, while the frequency measured by an observer with four-velocity U^μ will be $\omega = -p_\mu U^\mu$. Therefore

$$\omega = \frac{E}{V}, \tag{6.13}$$

and a photon emitted by static observer 1 will be observed by static observer 2 to have wavelength $\lambda = 2\pi/\omega$ given by

$$\lambda_2 = \frac{V_2}{V_1}\lambda_1. \tag{6.14}$$

In particular, at infinity where $V = 1$, we will observe a wavelength $\lambda_\infty = \lambda_1/V_1$.

Now we turn to the idea of "acceleration as viewed from infinity." A static observer will not typically be moving on a geodesic; for example, particles tend to fall into black holes rather than hovering next to them at fixed spatial coordinates.

We can express the four-acceleration $a^\mu = U^\sigma \nabla_\sigma U^\mu$ in terms of the redshift factor as

$$a_\mu = \nabla_\mu \ln V, \tag{6.15}$$

as you can easily check. The magnitude of the acceleration,

$$a = \sqrt{a_\mu a^\mu} = V^{-1}\sqrt{\nabla_\mu V \nabla^\mu V}, \tag{6.16}$$

will go to infinity at the Killing horizon—it will take an infinite acceleration to keep an object on a static trajectory. But an observer at infinity will detect the acceleration to be "redshifted" by a factor V; this turns out to be the surface gravity. Thus, we claim that

$$\kappa = Va = \sqrt{\nabla_\mu V \nabla^\mu V}, \tag{6.17}$$

evaluated at the horizon Σ. You can check that this expression agrees with (6.9). The surface gravity is the product of zero (V) and infinity (a), but will typically be finite. When we say that the observed acceleration is redshifted, we have in mind stretching a test string from a static object at the horizon to an observer at infinity, and measuring the acceleration on the end of the string at infinity. (It is worth taking the time to see if you can promote this hand-waving argument to something more rigorous.)

What goes wrong with the above considerations if the spacetime is stationary but not static? We still have an asymptotically time-translation Killing vector $K = \partial_t$, and we can define stationary observers as ones whose four-velocities are parallel to K^μ, as in (6.11); the redshift will continue to be given by (6.14). The problem is that K^μ won't become null at a Killing horizon, but generally at some timelike surface outside the horizon. This place where $K^\mu K_\mu = 0$ is called the **stationary limit surface** (or sometimes "ergosurface"), since inside this surface K^μ is spacelike, and consequently no observer can remain stationary, even if it is still outside the event horizon. Such an observer has to move with respect to the Killing field, but need not move in the direction of the black hole. From (6.12) and (6.14), the redshift of a stationary observer diverges as we approach the stationary limit surface, which is therefore also called the **infinite redshift surface**. As we will see in our discussion of the Kerr metric, the region between the stationary limit surface and the event horizon, the ergosphere, is a place where timelike paths are inevitably dragged along with the rotation of the black hole. We will continue to use "surface gravity" as a label in stationary spacetimes, which we will calculate using the Killing vector χ^μ, which actually does go null on the event horizon, even if the resulting quantity cannot be interpreted as the gravitational acceleration of a stationary observer as seen at infinity.

Let's apply these notions to Schwarzschild to see how they work. For the metric

$$ds^2 = -\left(1 - \frac{2GM}{r}\right)dt^2 + \left(1 - \frac{2GM}{r}\right)^{-1} dr^2 + r^2 d\Omega^2, \tag{6.18}$$

the Killing vector and static four-velocity are

$$K^\mu = (1, 0, 0, 0), \quad U^\mu = \left[\left(1 - \frac{2GM}{r}\right)^{-1/2}, 0, 0, 0\right], \tag{6.19}$$

so the redshift factor is

$$V = \sqrt{1 - \frac{2GM}{r}}. \tag{6.20}$$

(Note the agreement with our calculation of the redshift in the previous chapter.) From (6.15), the acceleration is

$$a_\mu = \frac{GM}{r^2\left(1 - \frac{2GM}{r}\right)} \nabla_\mu r, \tag{6.21}$$

where of course $\nabla_\mu r = \delta^r_\mu$. The magnitude of the acceleration is thus

$$a = \frac{GM}{r^2\left(1 - \frac{2GM}{r}\right)^{1/2}}. \tag{6.22}$$

The surface gravity is $\kappa = Va$ evaluated at the event horizon $r = 2GM$, and

$$Va = \frac{GM}{r^2}, \tag{6.23}$$

so the surface gravity of a Schwarzschild black hole is

$$\kappa = \frac{1}{4GM}. \tag{6.24}$$

It might seem surprising that the surface gravity decreases as the mass increases, but a glance at (6.23) reveals what is going on; at fixed radius increasing M acts to increase the combination Va, but increasing the mass also increases the Schwarzschild radius, and that effect wins out. Thus, the surface gravity of a big black hole is actually weaker than that of a small black hole; this is consistent with an examination of the tidal forces, which are also smaller for bigger black holes.

6.4 ■ MASS, CHARGE, AND SPIN

Since we have claimed above that the most general stationary black-hole solution to general relativity is characterized by mass, charge, and spin, we should consider how these quantities might be defined in GR. Charge is the easiest to consider, so we start there; more details are found in our discussion of Stokes's theorem in Appendix E. We'll look specifically at electric charge, although magnetic charge could be examined in the same way.

Maxwell's equations relate the electromagnetic field strength tensor $F_{\mu\nu}$ to the electric current four-vector J_e^μ,

$$\nabla_\nu F^{\mu\nu} = J_e^\mu. \tag{6.25}$$

The charge passing through a spacelike hypersurface Σ is given by an integral over coordinates x^i on the hypersurface,

$$\begin{aligned} Q &= -\int_\Sigma d^3x\sqrt{\gamma}\, n_\mu J_e^\mu \\ &= -\int_\Sigma d^3x\sqrt{\gamma}\, n_\mu \nabla_\nu F^{\mu\nu}, \end{aligned} \tag{6.26}$$

where γ_{ij} is the induced metric, and n^μ is the unit normal vector, associated with Σ. The minus sign ensures that a positive charge density and a future-pointing normal vector will give a positive total charge. Stokes's theorem can then be used to express the charge as a boundary integral,

$$Q = -\int_{\partial\Sigma} d^2x\sqrt{\gamma^{(2)}}\, n_\mu \sigma_\nu F^{\mu\nu}, \tag{6.27}$$

where the boundary $\partial\Sigma$, typically a two-sphere at spatial infinity, has metric $\gamma_{ij}^{(2)}$ and outward-pointing normal vector σ^μ. The magnetic charge could be determined by replacing $F^{\mu\nu}$ with the dual tensor $*F^{\mu\nu} = \frac{1}{2}\epsilon^{\mu\nu\rho\sigma} F_{\rho\sigma}$. Thus, to calculate the total charge, we need know only the behavior of the electromagnetic field at spatial infinity. In Appendix E we do an explicit calculation for a point charge in Minkowski space, which yields a predictable result but serves as a good check that our conventions work out correctly.

We turn now to the concept of the total energy (or mass) of an asymptotically flat spacetime. This is a much trickier notion than that of the charge; for one thing, energy-momentum is a tensor rather than a vector in general relativity, and for another, the energy-momentum tensor $T_{\mu\nu}$ only describes the properties of matter, not of the gravitational field. But recall that in Chapter 3 we discussed how we could nevertheless define a conserved total energy if spacetime were stationary, with a timelike Killing vector field K^μ. We first construct a current

$$J_T^\mu = K_\nu T^{\mu\nu}, \tag{6.28}$$

where $T^{\mu\nu}$ is the energy-momentum tensor. Because this current is divergenceless (from Killing's equation and conservation of $T^{\mu\nu}$), we can find a conserved energy by integrating over a spacelike surface Σ,

$$E_T = \int_\Sigma d^3x\sqrt{\gamma}\, n_\mu J_T^\mu, \tag{6.29}$$

just as for the charge. As interesting as this expression is, there are clearly some inadequacies with it. For example, consider the Schwarzschild metric. It has a

Killing vector, but $T^{\mu\nu}$ vanishes everywhere. Is the energy of a Schwarzschild black hole therefore zero? On both physical and mathematical grounds, there is reason to suspect not; there is a singularity, after all, which renders the integral difficult to evaluate. Furthermore, a Schwarzschild black hole can evolve from a massive star with a definite nonzero energy, and we might like that energy to be conserved. It is worth searching for an alternative definition of energy that better captures our intuitive picture for black hole spacetimes.

Sticking for the moment to spacetimes with a timelike Killing vector K^μ, consider a new current

$$J_R^\mu = K_\nu R^{\mu\nu}. \tag{6.30}$$

Using Einstein's equation, we can equivalently write this as

$$J_R^\mu = 8\pi G K_\nu \left(T^{\mu\nu} - \tfrac{1}{2}T g^{\mu\nu}\right). \tag{6.31}$$

The Ricci tensor is not divergenceless; instead we have the contracted Bianchi identity,

$$\nabla_\mu R^{\mu\nu} = \tfrac{1}{2}\nabla^\nu R. \tag{6.32}$$

But this and Killing's equation suffice to guarantee that our new current is conserved. To see this, we simply compute

$$\nabla_\mu J_R^\mu = (\nabla_\mu K_\nu) R^{\mu\nu} + K_\nu(\nabla_\mu R^{\mu\nu}). \tag{6.33}$$

The first term vanishes because $R^{\mu\nu}$ is symmetric and $\nabla_\mu K_\nu$ is antisymmetric (from Killing's equation). Using (6.32) we therefore have

$$\nabla_\mu J_R^\mu = \tfrac{1}{2} K_\nu \nabla^\nu R = 0, \tag{6.34}$$

which we know vanishes because the directional derivative of R vanishes along a Killing vector, (3.178).

As before, we can define a conserved energy associated with this current,

$$E_R = \frac{1}{4\pi G}\int_\Sigma d^3x \sqrt{\gamma}\, n_\mu J_R^\mu, \tag{6.35}$$

where the normalization is chosen for future convenience. The energy E_R will be independent of the spacelike hypersurface Σ, and hence conserved. This notion of energy has a significant advantage over E_T, arising from the fact that E_R can be rewritten as a surface integral over a two-sphere at spatial infinity. To see this, recall from (3.177) that any Killing vector satisfies $\nabla_\mu \nabla_\nu K^\mu = K^\mu R_{\mu\nu}$; the current itself can thus be written as a total derivative,

$$J_R^\mu = \nabla_\nu(\nabla^\mu K^\nu), \tag{6.36}$$

so that

$$E_R = \frac{1}{4\pi G}\int_\Sigma d^3x\sqrt{\gamma}\, n_\mu \nabla_\nu(\nabla^\mu K^\nu). \tag{6.37}$$

Note that, from raising indices on Killing's equation, $\nabla^\mu K^\nu = -\nabla^\nu K^\mu$. We can therefore again use Stokes's theorem just as we did for electric charge, to write E_R as an integral at spatial infinity,

$$E_R = \frac{1}{4\pi G}\int_{\partial\Sigma} d^2x\sqrt{\gamma^{(2)}}\, n_\mu \sigma_\nu \nabla^\mu K^\nu. \tag{6.38}$$

This expression is the **Komar integral** associated with the timelike Killing vector K^μ; it can be interpreted as the total energy of a stationary spacetime.

To convince ourselves that we're on the right track, let's calculate the Komar integral for Schwarzschild, with metric (6.18). The normal vectors, normalized to $n_\mu n^\mu = -1$ and $\sigma_\mu \sigma^\mu = +1$, have nonzero components

$$n_0 = -\left(1 - \frac{2GM}{r}\right)^{1/2}, \quad \sigma_1 = \left(1 - \frac{2GM}{r}\right)^{-1/2}, \tag{6.39}$$

with other components vanishing. We therefore have

$$n_\mu \sigma_\nu \nabla^\mu K^\nu = -\nabla^0 K^1. \tag{6.40}$$

The Killing vector is $K^\mu = (1, 0, 0, 0)$, so we can readily calculate

$$\begin{aligned}
\nabla^0 K^1 &= g^{00}\nabla_0 K^1 \\
&= g^{00}\left(\partial_0 K^1 + \Gamma^1_{0\lambda} K^\lambda\right) \\
&= g^{00}\Gamma^1_{00} K^0 \\
&= -\left(1 - \frac{2GM}{r}\right)^{-1} \frac{GM}{r^2}\left(1 - \frac{2GM}{r}\right) \\
&= -\frac{GM}{r^2}.
\end{aligned} \tag{6.41}$$

The metric on the two-sphere at infinity is

$$\gamma^{(2)}_{ij}\,\mathrm{d}x^i \mathrm{d}x^j = r^2(\mathrm{d}\theta^2 + \sin^2\theta\,\mathrm{d}\phi^2), \tag{6.42}$$

so that

$$\sqrt{\gamma^{(2)}} = r^2 \sin\theta. \tag{6.43}$$

Putting it all together, the energy of a Schwarzschild black hole is

$$E_R = \frac{1}{4\pi G}\int d\theta d\phi\, r^2 \sin\theta \left(\frac{GM}{r^2}\right)$$
$$= M. \tag{6.44}$$

This is of course the desired result, explaining the normalization chosen in (6.35).

Despite getting the right answer, we should think about what just happened. In particular, we obtained this energy by integrating the current $J^\mu_R = K_\nu R^{\mu\nu}$ over a spacelike slice, finding that the result could be written as an integral at spatial infinity. But for Schwarzschild, the metric solves the vacuum Einstein equation, $R_{\mu\nu} = 0$; it therefore seems difficult to get a nonzero answer from integrating J^μ_R, just as it did for (6.29). If we think about the structure of the maximally extended Schwarzschild solution, we realize that we could draw two kinds of spacelike slices: those that extend through the wormhole to the second asymptotic region, and those that end on the singularity. If the slice extends through the wormhole, the other asymptotic region provides another component to $\partial\Sigma$, and thus another contribution to (6.38); this contribution would exactly cancel, so the total energy would indeed be zero. If the slice intersected the singularity, we wouldn't know quite how to deal with it. Nevertheless, in either case it is sensible to treat our result (6.44) as the correct answer. The point is that, since (6.38) involves contributions only at spatial infinity, it should be a valid expression for the energy no matter what happens in the interior. We could even imagine time-dependent behavior in the interior; so long as K^μ was *asymptotically* a timelike Killing vector, the Komar energy will be well-defined. We could, for example, consider spherically symmetric gravitational collapse from an initially static star. Evaluating the integral (6.35) directly over Σ would give a sensible answer for the total mass, which should not change as the star collapsed to a black hole (we are imagining spherical symmetry, so that gravitational radiation cannot carry away energy to infinity). So the Komar integral (6.38), which would be valid before the collapse, may be safely interpreted as the energy even after collapse to a black hole. Of course for some purposes we might want to allow for energy loss through gravitational radiation, in which case we need to be careful about how we extend our slice to infinity; one can define a "Bondi mass" at future null infinity which allows us to keep track of energy loss through radiation.

Another worry about the Komar formula is whether it is really what we should think of as the "energy," which is typically the conserved quantity associated with time translation invariance. The best argument in favor of this interpretation is simply that E_R is certainly a conserved quantity of some sort, and it agrees with what we think should be the energy of Schwarzschild (and of a collection of masses in the Newtonian limit, as you could check), so what else could it be? Alternatively, one could think about a Hamiltonian formulation of general relativity, and carefully define the generator of time translations in an asymptotically flat spacetime, and then identify that with the total energy. This was first done by Arnowitt, Deser, and Misner, and their result is known as the **ADM energy**. In an asymptotically flat spacetime, we can write the metric just as we do in perturbation theory,

$$g_{\mu\nu} = \eta_{\mu\nu} + h_{\mu\nu}, \tag{6.45}$$

except that here we only ask that the components $h_{\mu\nu}$ be small at spatial infinity, not necessarily everywhere. The ADM energy can then be written as an integral over a two-sphere at spatial infinity, as

$$E_{\rm ADM} = \frac{1}{16\pi G}\int_{\partial\Sigma} d^2x\sqrt{\gamma^{(2)}}\,\sigma^i\left(\partial_j h^j{}_i - \partial_i h^j{}_j\right), \tag{6.46}$$

where spatial indices are raised with δ^{ij} (the spatial metric at infinity). This formula looks coordinate-dependent, but is actually well-defined under our assumptions. If $h_{\mu\nu}$ is time-independent at infinity, it can be verified that the ADM energy and the Komar energy actually agree. This gives us even more confidence that the Komar integral really represents the energy. However, there is a sense in which the ADM energy is more respectable; for example, the Komar integral can run into trouble if we have long-range scalar fields nonminimally coupled to gravity. But for our immediate purposes the Komar energy is quite acceptable.

One quality that we would like something called "energy" to have is that it be positive for any physical configuration; otherwise a zero-energy state could decay into pieces of positive energy and negative energy. The energy conditions discussed in Chapter 4 give a notion of positive energy for matter fields, but we might worry about a negative gravitational contribution leading to problems. Happily, in GR we have the **positive energy theorem**, first proven by Shoen and Yau:

> The ADM energy of a nonsingular, asymptotically flat spacetime obeying Einstein's equation and the dominant energy condition is nonnegative. Furthermore, Minkowski is the only such spacetime with vanishing ADM energy.

If we allow for singularities, there are clearly counterexamples, such as Schwarzschild with $M < 0$. However, if a spacetime with a singularity (such as Schwarzschild with $M > 0$) is reached as the evolution of nonsingular initial data, the theorem will apply. Thus we seem to be safe from negative-energy isolated systems in general relativity.

Finally, we may turn to spin (angular momentum), which is perfectly straightforward after our discussion of energy. Imagine that we have a rotational Killing vector $R = \partial_\phi$. In exact analogy with the time-translation case, we can define a conserved current

$$J^\mu_\phi = R_\nu R^{\mu\nu}, \tag{6.47}$$

which will lead to an expression for the conserved angular momentum J as an integral over spatial infinity,

$$J = -\frac{1}{8\pi G}\int_{\partial\Sigma} d^2x\sqrt{\gamma^{(2)}}\, n_\mu \sigma_\nu \nabla^\mu R^\nu. \tag{6.48}$$

(It is too bad that "J" is used for both the current and the angular momentum, just as it is too bad that "R" is both the rotational Killing vector and the Ricci tensor. But there are only so many letters to go around.) Just as with the energy, this expression will still be valid even if R^μ is only asymptotically a Killing vector. Note that the normalization is different than in the energy integral; it could be justified, for example, by evaluating the expression for slowly-moving masses with weak gravitational fields.

6.5 ■ CHARGED (REISSNER–NORDSTRÖM) BLACK HOLES

We turn now to the exact solutions representing electrically charged black holes. Such solutions are not extremely relevant to realistic astrophysical situations; in the real world, a highly-charged black hole would be quickly neutralized by interactions with matter in the vicinity of the hole. But charged holes nevertheless illustrate a number of important features of more general situations. In this case the full spherical symmetry of the problem is still present; we know therefore that we can write the metric as

$$ds^2 = -e^{2\alpha(r,t)}dt^2 + e^{2\beta(r,t)}dr^2 + r^2 d\Omega^2. \tag{6.49}$$

Now, however, we are no longer in vacuum, since the hole will have a nonzero electromagnetic field, which in turn acts as a source of energy-momentum. The energy-momentum tensor for electromagnetism is given by

$$T_{\mu\nu} = F_{\mu\rho}F_\nu{}^\rho - \tfrac{1}{4}g_{\mu\nu}F_{\rho\sigma}F^{\rho\sigma}, \tag{6.50}$$

where $F_{\mu\nu}$ is the electromagnetic field strength tensor. Since we have spherical symmetry, the most general field strength tensor will have components

$$\begin{aligned} F_{tr} &= f(r,t) = -F_{rt} \\ F_{\theta\phi} &= g(r,t)\sin\theta = -F_{\phi\theta}, \end{aligned} \tag{6.51}$$

where $f(r,t)$ and $g(r,t)$ are some functions to be determined by the field equations, and components not written are zero. F_{tr} corresponds to a radial electric field, while $F_{\theta\phi}$ corresponds to a radial magnetic field. For those of you wondering about the $\sin\theta$, recall that the thing that should be independent of θ and ϕ is the radial component of the magnetic field, $B^r = \epsilon^{01\mu\nu}F_{\mu\nu}$. For a spherically symmetric metric,

$$\epsilon^{\rho\sigma\mu\nu} = \frac{1}{\sqrt{-g}}\tilde{\epsilon}^{\rho\sigma\mu\nu}$$

is proportional to $(\sin\theta)^{-1}$, so we want a factor of $\sin\theta$ in $F_{\theta\phi}$.

The field equations in this case are both Einstein's equation and Maxwell's equations:

$$g^{\mu\nu}\nabla_\mu F_{\nu\sigma} = 0$$
$$\nabla_{[\mu} F_{\nu\rho]} = 0. \tag{6.52}$$

The two sets are coupled together, since the electromagnetic field strength tensor enters Einstein's equation through the energy-momentum tensor, while the metric enters explicitly into Maxwell's equations.

The difficulties are not insurmountable, however, and a procedure similar to the one we followed for the vacuum case leads to a solution for the charged case as well. We will not go through the steps explicitly, but merely quote the final answer. The solution is known as the **Reissner–Nordström (RN) metric**, and is given by

$$\boxed{ds^2 = -\Delta dt^2 + \Delta^{-1} dr^2 + r^2 d\Omega^2,} \tag{6.53}$$

where

$$\Delta = 1 - \frac{2GM}{r} + \frac{G(Q^2 + P^2)}{r^2}. \tag{6.54}$$

In this expression, M is once again interpreted as the mass of the hole; Q is the total electric charge, and P is the total magnetic charge. Isolated magnetic charges (monopoles) have never been observed in nature, but that doesn't stop us from writing down the metric that they would produce if they did exist.[5] There are good theoretical reasons to think that monopoles may exist if forces are "grand unified" at very high energies, but they must be very heavy and extremely rare. Of course, a black hole could possibly have magnetic charge even if there aren't any monopoles. In fact, the electric and magnetic charges enter the metric in the same way, so we are not introducing any additional complications by keeping P in our expressions. Conservatives are welcome to set $P = 0$ if they like. The electromagnetic fields associated with this solution are given by

$$E_r = F_{rt} = \frac{Q}{r^2}$$
$$B_r = \frac{F_{\theta\phi}}{r^2 \sin\theta} = \frac{P}{r^2}. \tag{6.55}$$

The $1/r^2$ dependence of these fields is just what we are used to in flat space; of course, here we know that this depends on our precise choice of radial coordinate.

The RN metric has a true curvature singularity at $r = 0$, as could be checked by computing the curvature invariant scalar $R_{\mu\nu\rho\sigma}R^{\mu\nu\rho\sigma}$. The horizon structure,

[5] In this chapter we are using units in which there is no factor of 4π in Coloumb's law. To compare with other chapters, divide each appearance of Q or P by 4π.

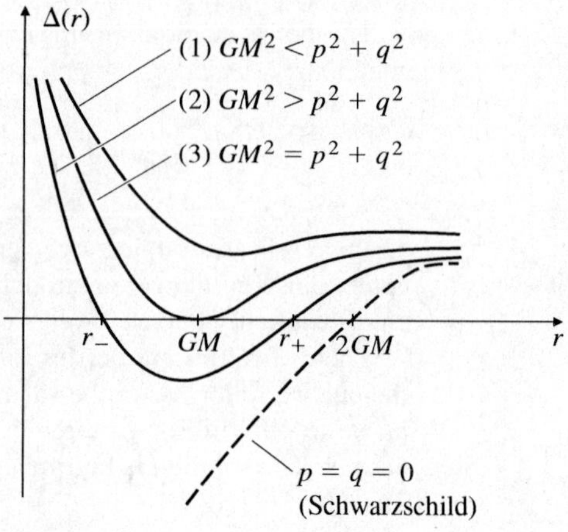

FIGURE 6.2 The function $\Delta(r) = 1 - 2GM/r + G(Q^2 + P^2)/r^2$ for the Reissner–Nordström solutions; zeroes indicate the location of an event horizon.

however, is more complicated than in Schwarzschild. In the discussion of event horizons above, we suggested that $g^{rr} = 0$ would be a useful diagnostic for locating event horizons, if we had cleverly chosen coordinates so that this condition is satisfied at some fixed value of r. Fortunately the coordinates of (6.53) have this property, and the event horizon will be located at

$$g^{rr}(r) = \Delta(r) = 1 - \frac{2GM}{r} + \frac{G(Q^2 + P^2)}{r^2} = 0. \tag{6.56}$$

This will occur at

$$r_\pm = GM \pm \sqrt{G^2M^2 - G(Q^2 + P^2)}. \tag{6.57}$$

As shown in Figure 6.2, this might constitute two, one, or zero solutions, depending on the relative values of GM^2 and $Q^2 + P^2$. We therefore consider each case separately.

Case One: $GM^2 < Q^2 + P^2$

In this case the coefficient Δ is always positive (never zero), and the metric is completely regular in the (t, r, θ, ϕ) coordinates all the way down to $r = 0$. The coordinate t is always timelike, and r is always spacelike. But still there is the singularity at $r = 0$, which is now a timelike line. Since there is no event horizon, there is no obstruction to an observer traveling to the singularity and returning to report on what was observed. This is a naked singularity, as discussed earlier. A careful analysis of the geodesics reveals that the singularity is repulsive—timelike geodesics never intersect $r = 0$; instead they approach and then reverse course and move away. (Null geodesics can reach the singularity, as can nongeodesic timelike curves.)

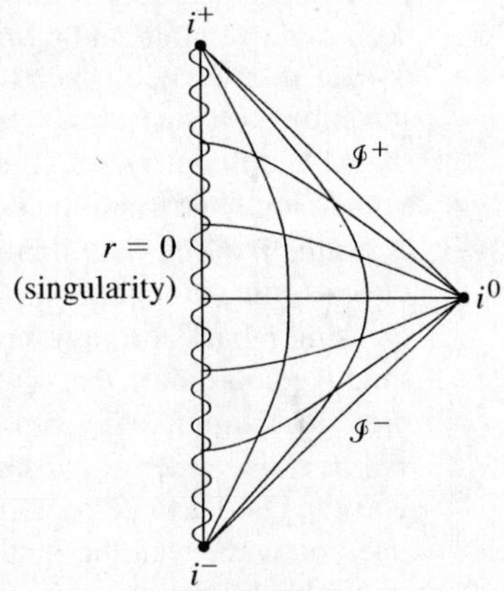

FIGURE 6.3 Conformal diagram for Reissner–Nordström solution with $Q^2 + P^2 > GM^2$. There is a naked singularity at the origin.

As $r \to \infty$ the solution approaches flat spacetime, and as we have just seen the causal structure seems normal everywhere. The conformal diagram will therefore be just like that of Minkowski space, except that now $r = 0$ is a singularity, as shown in Figure 6.3.

The nakedness of the singularity offends our sense of decency, as well as the cosmic censorship conjecture. In fact, we should never expect to find a black hole with $GM^2 < Q^2 + P^2$ as the result of gravitational collapse. Roughly speaking, this condition states that the total energy of the hole is less than the contribution to the energy from the electromagnetic fields alone—that is, the mass of the matter that carried the charge would have had to be negative. This solution is therefore generally considered to be unphysical. Notice also that there are no Cauchy surfaces in this spacetime, since timelike lines can begin and end at the singularity.

Case Two: $GM^2 > Q^2 + P^2$

We expect this situation to apply in realistic gravitational collapse; the energy in the electromagnetic field is less than the total energy. In this case the metric coefficient $\Delta(r)$ is positive at large r and small r, and negative inside the two vanishing points $r_\pm = GM \pm \sqrt{G^2M^2 - G(Q^2 + P^2)}$. The metric has coordinate singularities at both r_+ and r_-; in both cases these could be removed by a change of coordinates as we did with Schwarzschild.

The surfaces defined by $r = r_\pm$ are both null, and they are both event horizons. The singularity at $r = 0$ is a timelike line, not a spacelike surface as in Schwarzschild. If you are an observer falling into the black hole from far away, r_+ is just like $2GM$ in the Schwarzschild metric; at this radius r switches from being a spacelike coordinate to a timelike coordinate, and you necessarily move in the direction of decreasing r. Witnesses outside the black hole also see the same phenomena that they would outside an uncharged hole—the infalling observer is seen to move more and more slowly, and is increasingly redshifted.

But the inevitable fall from r_+ to ever-decreasing radii only lasts until you reach the null surface $r = r_-$, where r switches back to being a spacelike coordinate and the motion in the direction of decreasing r can be arrested. Therefore you do not have to hit the singularity at $r = 0$; this is to be expected, since $r = 0$ is a timelike line (and therefore not necessarily in your future). In fact you can choose either to continue on to $r = 0$, or begin to move in the direction of increasing r back through the null surface at $r = r_-$. Then r will once again be a timelike coordinate, but with reversed orientation; you are forced to move in the direction of *increasing* r. You will eventually be spit out past $r = r_+$ once more, which is like emerging from a white hole into the rest of the universe. From here you can choose to go back into the black hole—this time, a different hole than the one you entered in the first place—and repeat the voyage as many times as you like. This little story corresponds to the conformal diagram in Figure 6.4, which of course can be derived more rigorously by choosing appropriate coordinates and analytically extending the Reissner–Nordström metric as far as it will go.

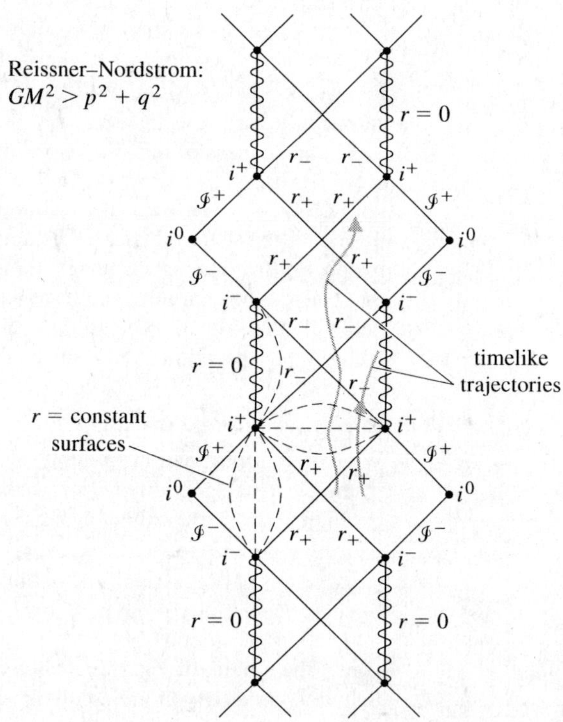

FIGURE 6.4 Conformal diagram for Reissner–Nordström solution with $GM^2 > Q^2 + P^2$. There are an infinite number of copies of the region outside the black hole.

How much of this is science, as opposed to science fiction? Probably not much. If you think about the world as seen from an observer inside the black hole who is about to cross the event horizon at r_-, you notice that the observer can look back in time to see the entire history of the external (asymptotically flat) universe, at least as seen from the black hole. But they see this (infinitely long) history in a finite amount of their proper time—thus, any signal that gets to them as they approach r_- is infinitely blueshifted. Therefore it is likely that any nonspherically-symmetric perturbation that comes into an RN black hole will violently disturb the geometry we have described. It's hard to say what the actual geometry will look like, but there is no very good reason to believe that it must contain an infinite number of asymptotically flat regions connecting to each other via various wormholes.[5]

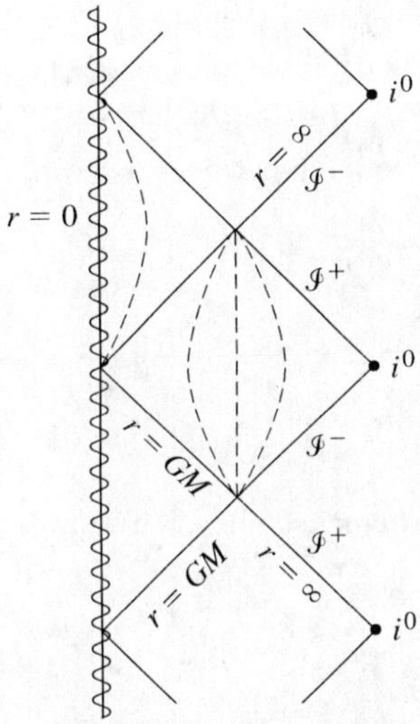

FIGURE 6.5 Conformal diagram for the extremal Reissner–Nordström solution, $GM^2 = Q^2 + P^2$. There is a naked singularity at the origin, and an infinite number of external regions.

Case Three: $GM^2 = Q^2 + P^2$

This case is known as the **extreme** Reissner–Nordström solution. On the one hand the extremal hole is an amusing theoretical toy; this solution is often examined in studies of the role of black holes in quantum gravity. In supersymmetric theories, extremal black holes can leave certain symmetries unbroken, which is a considerable aid in calculations. On the other hand it appears unstable, since adding just a little bit of matter will bring it to Case Two.

The extremal black holes have $\Delta(r) = 0$ at a single radius, $r = GM$. This represents an event horizon, but the r coordinate is never timelike; it becomes null at $r = GM$, but is spacelike on either side. The singularity at $r = 0$ is a timelike line, as in the other cases. So for this black hole you can again avoid the singularity and continue to move to the future to extra copies of the asymptotically flat region, but the singularity is always "to the left." The conformal diagram is shown in Figure 6.5.

A fascinating property of extremal black holes is that the mass is in some sense balanced by the charge. More specifically, two extremal holes with same-sign charges will attract each other gravitationally, but repel each other electromagnetically, and it turns out that these effects precisely cancel. Indeed, we can find *exact* solutions to the coupled Einstein–Maxwell equations representing any number of such black holes in a stationary configuration. To see this, turn first to the Reissner–Nordström metric itself, and let's stick with electric charges rather than magnetic charges, just for simplicity. At extremality, $GM^2 = Q^2$, and the metric takes the form

$$ds^2 = -\left(1 - \frac{GM}{r}\right)^2 dt^2 + \left(1 - \frac{GM}{r}\right)^{-2} dr^2 + r^2 d\Omega^2. \tag{6.58}$$

By defining a shifted radial coordinate

$$\rho = r - GM, \tag{6.59}$$

[5]For some work on this issue, see E. Poisson and W. Israel, *Phys. Rev. D* **41**, 1796 (1990).

the metric takes the isotropic form

$$ds^2 = -H^{-2}(\rho)\mathrm{d}t^2 + H^2(\rho)[\mathrm{d}\rho^2 + \rho^2 d\Omega^2], \tag{6.60}$$

where

$$H(\rho) = 1 + \frac{GM}{\rho}. \tag{6.61}$$

Because $\mathrm{d}\rho^2 + \rho^2 d\Omega^2$ is just the flat metric in three spatial dimensions, we can write (6.60) equally well as

$$ds^2 = -H^{-2}(\vec{x})\mathrm{d}t^2 + H^2(\vec{x})[\mathrm{d}x^2 + \mathrm{d}y^2 + \mathrm{d}z^2], \tag{6.62}$$

where H can be written

$$H = 1 + \frac{GM}{|\vec{x}|}. \tag{6.63}$$

In the original r coordinate, the electric field of the extremal solution can be expressed in terms of a vector potential A_μ as

$$E_r = F_{rt} = \frac{Q}{r^2} = \partial_r A_0, \tag{6.64}$$

where the timelike component of the vector potential is

$$A_0 = -\frac{Q}{r}, \tag{6.65}$$

and we imagine the spatial components vanish (having set the magnetic field to zero). In our new ρ coordinate, and with the extremality condition $Q^2 = GM^2$, this becomes

$$A_0 = -\frac{\sqrt{G}M}{\rho + GM}, \tag{6.66}$$

or equivalently

$$\sqrt{G}A_0 = H^{-1} - 1. \tag{6.67}$$

But now let's forget that we know that H obeys (6.61), and simply plug the metric (6.62) and the electrostatic potential (6.67) into Einstein's equation and Maxwell's equations, imagining that H is time-independent ($\partial_0 H = 0$) but otherwise unconstrained. We can straightforwardly show (see the Exercises) that they can be simultaneously satisfied by any time-independent function $H(\vec{x})$ that obeys

$$\nabla^2 H = 0, \tag{6.68}$$

where $\nabla^2 = \partial_x^2 + \partial_y^2 + \partial_z^2$. This is simply Laplace's equation, and it is straightforward to write down all of the solutions that are well-behaved at infinity; they take the form

$$H = 1 + \sum_{a=1}^{N} \frac{GM_a}{|\vec{x} - \vec{x}_a|}, \tag{6.69}$$

for some set of N spatial points defined by $\vec{x}_a$. These points describe the locations of N extremal RN black holes with masses M_a and charges $Q_a = \sqrt{G}M_a$. This multi-extremal-black hole metric is undoubtedly one of the most remarkable exact solutions to Einstein's equation.

6.6 ■ ROTATING (KERR) BLACK HOLES

We could go into a good deal more detail about the charged solutions, but let's instead move on to rotating black holes. To find the exact solution for the metric in this case is much more difficult, since we have given up on spherical symmetry. Instead we look for solutions with axial symmetry around the axis of rotation that are also stationary (a timelike Killing vector). Although the Schwarzschild and Reissner–Nordström solutions were discovered soon after general relativity was invented, the solution for a rotating black hole was found by Kerr only in 1963. His result, the **Kerr metric**, is given by the following mess:

$$\boxed{\begin{aligned} ds^2 = &-\left(1 - \frac{2GMr}{\rho^2}\right) dt^2 - \frac{2GMar\sin^2\theta}{\rho^2}(dt\,d\phi + d\phi\,dt) \\ &+ \frac{\rho^2}{\Delta}dr^2 + \rho^2 d\theta^2 + \frac{\sin^2\theta}{\rho^2}\left[(r^2+a^2)^2 - a^2\Delta\sin^2\theta\right]d\phi^2, \end{aligned}} \tag{6.70}$$

where

$$\boxed{\Delta(r) = r^2 - 2GMr + a^2} \tag{6.71}$$

and

$$\boxed{\rho^2(r,\theta) = r^2 + a^2\cos^2\theta.} \tag{6.72}$$

The two constants M and a parameterize the possible solutions. To verify that the mass M is equal to the Komar energy (6.38) is straightforward but tedious, while a is the angular momentum per unit mass,

$$a = J/M, \tag{6.73}$$

where J is the Komar angular momentum (6.48). It is easy to include electric and magnetic charges Q and P, simply by replacing $2GMr$ with $2GMr - G(Q^2 +$

P^2); the result is the **Kerr–Newman metric**. The associated one-form potential has nonvanishing components

$$A_t = \frac{Qr - Pa\cos\theta}{\rho^2}, \quad A_\phi = \frac{-Qar\sin^2\theta + P(r^2+a^2)\cos\theta}{\rho^2}. \tag{6.74}$$

All of the essential phenomena persist in the absence of charges, so we will set $Q = P = 0$ from now on.

The coordinates (t, r, θ, ϕ) are known as **Boyer–Lindquist coordinates**. It is straightforward to check that as $a \to 0$ they reduce to Schwarzschild coordinates. If we keep a fixed and let $M \to 0$, however, we recover flat spacetime but not in ordinary polar coordinates. The metric becomes

$$ds^2 = -dt^2 + \frac{(r^2 + a^2\cos^2\theta)}{(r^2+a^2)}dr^2 + (r^2+a^2\cos^2\theta)^2 d\theta^2 + (r^2+a^2)\sin^2\theta\, d\phi^2, \tag{6.75}$$

and we recognize the spatial part of this as flat space in ellipsoidal coordinates, as shown in Figure 6.6. They are related to Cartesian coordinates in Euclidean 3-space by

$$\begin{aligned} x &= (r^2+a^2)^{1/2}\sin\theta\,\cos\phi \\ y &= (r^2+a^2)^{1/2}\sin\theta\,\sin\phi \\ z &= r\cos\theta. \end{aligned} \tag{6.76}$$

There are two Killing vectors of the metric (6.70), both of which are manifest; since the metric coefficients are independent of t and ϕ, both $K = \partial_t$ and $R = \partial_\phi$ are Killing vectors. Of course R^μ expresses the axial symmetry of the solution. The vector K^μ is not orthogonal to t = constant hypersurfaces, and in fact is not orthogonal to any hypersurfaces at all; hence this metric is stationary, but not

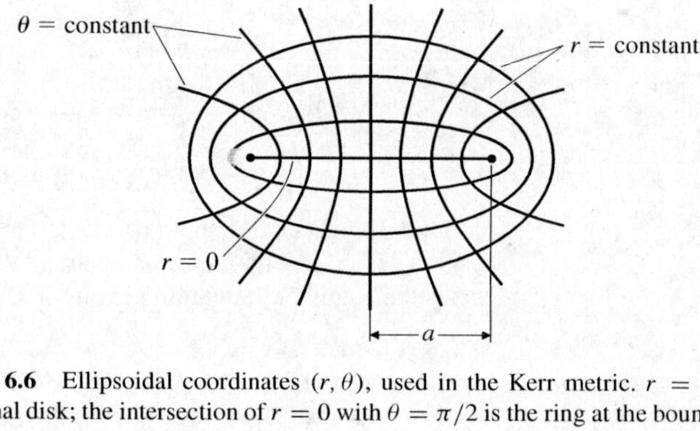

FIGURE 6.6 Ellipsoidal coordinates (r, θ), used in the Kerr metric. $r = 0$ is a two-dimensional disk; the intersection of $r = 0$ with $\theta = \pi/2$ is the ring at the boundary of this disk.

static. This makes sense; the black hole is spinning, so it's not static, but it is spinning in exactly the same way at all times, so it's stationary. Alternatively, the metric can't be static because it's not time-reversal invariant, since that would reverse the angular momentum of the hole.

The Kerr metric also possesses a Killing tensor. These were defined in Chapter 3 as any symmetric $(0, n)$ tensor $\sigma_{\mu_1\cdots\mu_n}$ satisfying

$$\nabla_{(\lambda}\sigma_{\mu_1\cdots\mu_n)} = 0. \tag{6.77}$$

In the Kerr geometry we can define the $(0, 2)$ tensor

$$\sigma_{\mu\nu} = 2\rho^2 l_{(\mu}n_{\nu)} + r^2 g_{\mu\nu}. \tag{6.78}$$

In this expression the two vectors l and n are given (with indices raised) by

$$\begin{aligned} l^\mu &= \frac{1}{\Delta}\left(r^2 + a^2, \Delta, 0, a\right) \\ n^\mu &= \frac{1}{2\rho^2}\left(r^2 + a^2, -\Delta, 0, a\right). \end{aligned} \tag{6.79}$$

Both vectors are null and satisfy

$$l^\mu l_\mu = 0, \quad n^\mu n_\mu = 0, \quad l^\mu n_\mu = -1. \tag{6.80}$$

With these definitions, you can check for yourself that $\sigma_{\mu\nu}$ is a Killing tensor.

We have chosen coordinates for Kerr such that the event horizons occur at those fixed values of r for which $g^{rr} = 0$. Since $g^{rr} = \Delta/\rho^2$, and $\rho^2 \geq 0$, this occurs when

$$\Delta(r) = r^2 - 2GMr + a^2 = 0. \tag{6.81}$$

As in the Reissner–Nordström solution, there are three possibilities: $GM > a$, $GM = a$, and $GM < a$. The last case features a naked singularity, and the extremal case $GM = a$ is unstable, just as in Reissner–Nordström. Since these cases are of less physical interest, we will concentrate on $GM > a$. Then there are two radii at which Δ vanishes, given by

$$r_\pm = GM \pm \sqrt{G^2M^2 - a^2}. \tag{6.82}$$

Both radii are null surfaces that will turn out to be event horizons; a side view of a Kerr black hole is portrayed in Figure 6.7. The analysis of these surfaces proceeds in close analogy with the Reissner–Nordström case; it is straightforward to find coordinates that extend through the horizons.

Because Kerr is stationary but not static, the event horizons at $r_\pm$ are not Killing horizons for the asymptotic time-translation Killing vector $K = \partial_t$. The norm of K^μ is given by

$$K^\mu K_\mu = -\frac{1}{\rho^2}(\Delta - a^2\sin^2\theta). \tag{6.83}$$

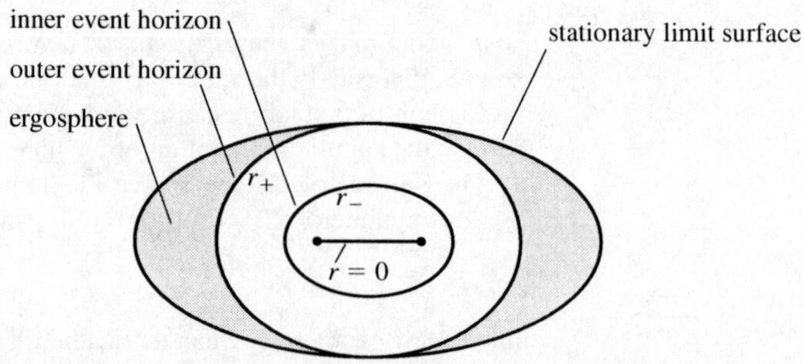

FIGURE 6.7 Horizon structure around the Kerr solution (side view). The event horizons are null surfaces that demarcate points past which it becomes impossible to return to a certain region of space. The stationary limit surface, in contrast, is timelike except where it is tangent to the event horizon (at the poles); it represents the place past which it is impossible to be a stationary observer. The ergosphere between the stationary limit surface and the outer event horizon is a region in which it is possible to enter and leave again, but not to remain stationary.

This does not vanish at the outer event horizon; in fact, at $r = r_+$ (where $\Delta = 0$), we have

$$K^\mu K_\mu = \frac{a^2}{\rho^2}\sin^2\theta \geq 0. \tag{6.84}$$

So the Killing vector is already spacelike at the outer horizon, except at the north and south poles ($\theta = 0, \pi$) where it is null. The locus of points where $K^\mu K_\mu = 0$ is of course the stationary limit surface, and is given by

$$(r - GM)^2 = G^2M^2 - a^2\cos^2\theta, \tag{6.85}$$

while the outer event horizon is given by

$$(r_+ - GM)^2 = G^2M^2 - a^2. \tag{6.86}$$

There is thus a region between these two surfaces, known as the **ergosphere**. Inside the ergosphere, you must move in the direction of the rotation of the black hole (the ϕ direction); however, you can still move toward or away from the event horizon (and have no trouble exiting the ergosphere). The ergosphere is evidently a place where interesting things can happen even before you cross the horizon; more details on this later.

Before rushing to draw conformal diagrams, we need to understand the nature of the true curvature singularity; this does not occur at $r = 0$ in this spacetime, but rather at $\rho = 0$ (where the curvature invariant $R_{\rho\sigma\mu\nu}R^{\rho\sigma\mu\nu}$ diverges). Since $\rho^2 = r^2 + a^2\cos^2\theta$ is the sum of two manifestly nonnegative quantities, it can only vanish when both quantities are zero, or

$$r = 0, \quad \theta = \frac{\pi}{2}. \tag{6.87}$$

This seems like a funny result, but remember that $r = 0$ is not a point in space, but a disk; the set of points $r = 0, \theta = \pi/2$ is actually the *ring* at the edge of this disk. The rotation has "softened" the Schwarzschild singularity, spreading it out over a ring.

What happens if you go inside the ring? A careful analytic continuation (which we will not perform) would reveal that you exit to another asymptotically flat spacetime, but not an identical copy of the one you came from. The new spacetime is described by the Kerr metric with $r < 0$. As a result, Δ never vanishes and there are no horizons. The conformal diagram, Figure 6.8, is much like that for Reissner–Nordström, except now you can pass through the singularity. Because the Kerr metric is not spherically symmetric, the conformal diagram is not quite

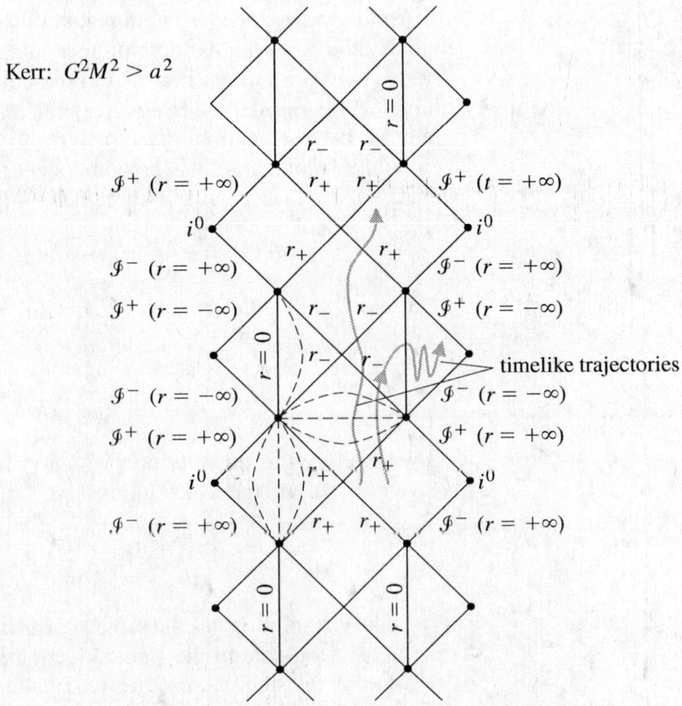

FIGURE 6.8 Conformal diagram for the Kerr solution with $G^2M^2 > a^2$. As with the analogous charged solution, there are an infinite number of copies of the region outside the black hole.

as faithful as in the previous cases; a single point on the diagram represents fixed values of t and r, and will have a different geometry for different values of θ.

Not only do we have the usual strangeness of these distinct asymptotically flat regions connected to ours through the black hole, but the region near the ring singularity has additional pathologies: closed timelike curves. If you consider trajectories that wind around in ϕ while keeping θ and t constant and r a small negative value, the line element along such a path is

$$ds^2 \approx a^2 \left(1 + \frac{2GM}{r}\right) d\phi^2, \tag{6.88}$$

which is negative for small negative r. Since these paths are closed, they are obviously CTCs. You can therefore meet yourself in the past, with all that entails.

Of course, everything we say about the analytic extension of Kerr is subject to the same caveats we mentioned for Schwarzschild and Reissner–Nordström; it is unlikely that realistic gravitational collapse leads to these bizarre spacetimes. It is nevertheless always useful to have exact solutions. Furthermore, for the Kerr metric strange things are happening even if we stay outside the event horizon, to which we now turn.

We begin by considering more carefully the angular velocity of the hole. Obviously the conventional definition of angular velocity will have to be modified somewhat before we can apply it to something as abstract as the metric of spacetime. Let us consider the fate of a photon that is emitted in the ϕ direction at some radius r in the equatorial plane ($\theta = \pi/2$) of a Kerr black hole. The instant it is emitted its momentum has no components in the r or θ direction, and therefore the condition that the trajectory be null is

$$ds^2 = 0 = g_{tt}dt^2 + g_{t\phi}(dt d\phi + d\phi dt) + g_{\phi\phi}d\phi^2. \tag{6.89}$$

This can be immediately solved to obtain

$$\frac{d\phi}{dt} = -\frac{g_{t\phi}}{g_{\phi\phi}} \pm \sqrt{\left(\frac{g_{t\phi}}{g_{\phi\phi}}\right)^2 - \frac{g_{tt}}{g_{\phi\phi}}}. \tag{6.90}$$

If we evaluate this quantity on the stationary limit surface of the Kerr metric, we have $g_{tt} = 0$, and the two solutions are

$$\frac{d\phi}{dt} = 0, \quad \frac{d\phi}{dt} = \frac{a}{2G^2M^2 + a^2}. \tag{6.91}$$

The nonzero solution has the same sign as a; we interpret this as the photon moving around the hole in the same direction as the hole's rotation. The zero solution means that the photon directed against the hole's rotation doesn't move at all in this coordinate system. Note that we haven't given a full solution to the photon's trajectory, only shown that its instantaneous velocity is zero. This is an example of a phenomenon known as the "dragging of inertial frames"; it is explored more in one of the exercises to Chapter 7. Massive particles, which must move more slowly than photons, are necessarily dragged along with the hole's rotation once they are inside the stationary limit surface. This dragging continues as we approach the outer event horizon at r_+; we can define the angular velocity of the event horizon itself, Ω_H, to be the minimum angular velocity of a particle at the horizon. Directly from (6.90) we find that

$$\Omega_H = \left(\frac{d\phi}{dt}\right)_-(r_+) = \frac{a}{r_+^2 + a^2}. \tag{6.92}$$

6.7 ■ THE PENROSE PROCESS AND BLACK-HOLE THERMODYNAMICS

Black hole thermodynamics is one of the most fascinating and mysterious subjects in general relativity. To get there, however, let us begin with something apparently very straightforward: motion along geodesics in the Kerr metric. We know that such a discussion will be simplified by considering the conserved quantities associated with the Killing vectors $K = \partial_t$ and $R = \partial_\phi$. For the purposes at hand we can restrict our attention to massive particles, for which we can work with the four-momentum

$$p^\mu = m\frac{dx^\mu}{d\tau}, \tag{6.93}$$

where m is the rest mass of the particle. Then we can take as our two conserved quantities the actual energy and angular momentum of the particle,

$$E = -K_\mu p^\mu = m\left(1 - \frac{2GMr}{\rho^2}\right)\frac{dt}{d\tau} + \frac{2mGMar}{\rho^2}\sin^2\theta\,\frac{d\phi}{d\tau} \tag{6.94}$$

and

$$L = R_\mu p^\mu = -\frac{2mGMar}{\rho^2}\sin^2\theta\,\frac{dt}{d\tau} + \frac{m(r^2+a^2)^2 - m\Delta a^2\sin^2\theta}{\rho^2}\sin^2\theta\,\frac{d\phi}{d\tau}. \tag{6.95}$$

These differ from the definitions for the conserved quantities used in the last chapter, where E and L were taken to be the energy and angular momentum *per unit mass*. They are conserved either way, of course.

The minus sign in the definition of E is there because at infinity both K^μ and p^μ are timelike, so their inner product is negative, but we want the energy to be positive. Inside the ergosphere, however, K^μ becomes spacelike; we can therefore imagine particles for which

$$E = -K_\mu p^\mu < 0. \tag{6.96}$$

The extent to which this bothers us is ameliorated somewhat by the realization that *all* particles must have positive energies if they are outside the stationary limit surface; therefore a particle inside the ergosphere with negative energy must either remain in the ergosphere, or be accelerated until its energy is positive if it is to escape.

Still, this realization leads to a way to extract energy from a rotating black hole; the method is known as the **Penrose process**. The idea is simple; starting from outside the ergosphere, you arm yourself with a large rock and leap toward the black hole. If we call the four-momentum of the (you + rock) system $p^{(0)\mu}$, then the energy $E^{(0)} = -K_\mu p^{(0)\mu}$ is certainly positive, and conserved as you move along your geodesic. Once you enter the ergosphere, you hurl the rock with all your might, in a very specific way. If we call your momentum $p^{(1)\mu}$ and that of the rock $p^{(2)\mu}$, then at the instant you throw it we have conservation of momentum just as in special relativity:

$$p^{(0)\mu} = p^{(1)\mu} + p^{(2)\mu}. \tag{6.97}$$

Contracting with the Killing vector K_μ gives

$$E^{(0)} = E^{(1)} + E^{(2)}. \tag{6.98}$$

But, if we imagine that you are arbitrarily strong (and accurate), you can arrange your throw such that $E^{(2)} < 0$, as per (6.96). Furthermore, Penrose was able to show that you can arrange the initial trajectory and the throw as shown in Figure 6.9, such that afterward you follow a geodesic trajectory back outside the stationary limit surface into the external universe. Since your energy is conserved

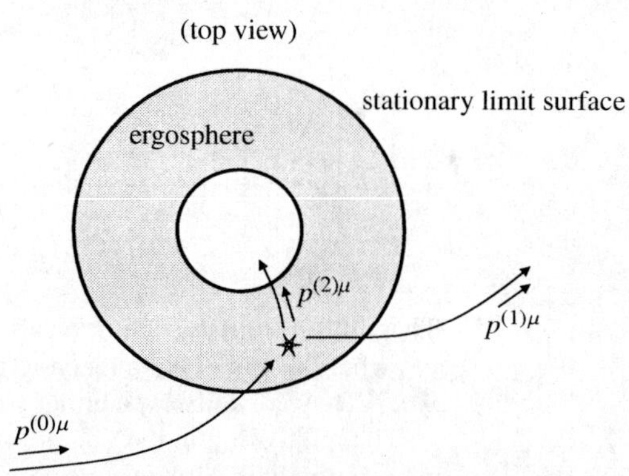

FIGURE 6.9 The Penrose process (top view). An object falls toward a Kerr black hole and splits in two while in the ergosphere (within the stationary limit surface, but outside the outer event horizon). One piece falls into the horizon with a negative energy $E^{(2)}$, while the other escapes to infinity with a larger energy than that of the original infalling object.

along the way, at the end we will have

$$E^{(1)} > E^{(0)}. \tag{6.99}$$

Thus, you have emerged with *more* energy than you entered with.

There is no such thing as a free lunch; the energy you gained came from somewhere, and that somewhere is the black hole. In fact, the Penrose process extracts energy from the rotating black hole by decreasing its angular momentum; you have to throw the rock against the hole's rotation to get the trick to work. To see this more precisely, recall that we claimed earlier in this chapter that any event horizon in a stationary spacetime would be a Killing horizon for some Killing vector. For Kerr this is a linear combination of the time-translation and rotational Killing vectors,

$$\chi^\mu = K^\mu + \Omega_H R^\mu, \tag{6.100}$$

where Ω_H is precisely the angular velocity of the horizon as defined in (6.92). Using $K = \partial_t$ and $R = \partial_\phi$, it is straightforward to verify that χ^μ becomes null at the outer event horizon. The statement that the particle with momentum $p^{(2)\mu}$ crosses the event horizon "moving forward in time" is simply

$$p^{(2)\mu}\chi_\mu < 0. \tag{6.101}$$

Plugging in the definitions of E and L, we see that this condition is equivalent to

$$L^{(2)} < \frac{E^{(2)}}{\Omega_H}. \tag{6.102}$$

Since we have arranged $E^{(2)}$ to be negative, and Ω_H positive, we see that the particle must have a negative angular momentum—it is moving against the hole's rotation. Once you have escaped the ergosphere and the rock has fallen inside the event horizon, the mass and angular momentum of the hole are what they used to be plus the negative contributions of the rock:

$$\begin{aligned} \delta M &= E^{(2)} \\ \delta J &= L^{(2)}, \end{aligned} \tag{6.103}$$

where $J = Ma$ is the angular momentum of the black hole. Then (6.102) becomes a limit on how much you can decrease the angular momentum:

$$\delta J < \frac{\delta M}{\Omega_H}. \tag{6.104}$$

If we exactly reach this limit, as the rock we throw in becomes more and more null, we have the "ideal" process, in which $\delta J = \delta M/\Omega_H$.

We will now use these ideas to verify that, although you can use the Penrose process to extract energy from the black hole (thereby decreasing M), you cannot violate the area theorem: The area of the event horizon is nondecreasing. Although the mass decreases, the angular momentum must also decrease, in a combination which only allows the area to increase. To see this, let's calculate the area of the outer event horizon, which is located at

$$r_+ = GM + \sqrt{G^2M^2 - a^2}. \tag{6.105}$$

The induced metric γ_{ij} on the horizon (where i and j run over $\{\theta, \phi\}$) can be found straightforwardly by setting $r = r_+$ (so $\Delta = 0$), $\mathrm{d}t = 0$ and $\mathrm{d}r = 0$ in (6.70):

$$\begin{aligned}\gamma_{ij}\mathrm{d}x^i\mathrm{d}x^j &= ds^2(\mathrm{d}t = 0, \mathrm{d}r = 0, r = r_+)\\ &= (r_+^2 + a^2\cos^2\theta)\mathrm{d}\theta^2 + \left[\frac{(r_+^2 + a^2)^2\sin^2\theta}{r_+^2 + a^2\cos^2\theta}\right]\mathrm{d}\phi^2. \end{aligned}\tag{6.106}$$

The horizon area is then the integral of the induced volume element,

$$A = \int \sqrt{|\gamma|}\, d\theta d\phi. \tag{6.107}$$

The determinant is

$$|\gamma| = (r_+^2 + a^2)^2\sin^2\theta, \tag{6.108}$$

so the horizon area is simply

$$A = 4\pi(r_+^2 + a^2). \tag{6.109}$$

To show that the area doesn't decrease, it is convenient to work instead in terms of the **irreducible mass** of the black hole, defined by

$$\begin{aligned}M_{\mathrm{irr}}^2 &= \frac{A}{16\pi G^2}\\ &= \frac{1}{4G^2}(r_+^2 + a^2)\\ &= \frac{1}{2}\left(M^2 + \sqrt{M^4 - (Ma/G)^2}\right)\\ &= \frac{1}{2}\left(M^2 + \sqrt{M^4 - (J/G)^2}\right). \end{aligned}\tag{6.110}$$

We can differentiate to obtain, after a bit of work, how M_{irr} is affected by changes in the mass or angular momentum,

$$\delta M_{\mathrm{irr}} = \frac{a}{4GM_{\mathrm{irr}}\sqrt{G^2M^2 - a^2}}(\Omega_{\mathrm{H}}^{-1}\delta M - \delta J). \tag{6.111}$$

Then our limit (6.104) becomes

$$\delta M_{\rm irr} > 0. \tag{6.112}$$

The irreducible mass can never be reduced; hence the name. It follows that the maximum amount of energy we can extract from a black hole before we slow its rotation to zero is

$$M - M_{\rm irr} = M - \frac{1}{\sqrt{2}}\left(M^2 + \sqrt{M^4 - (J/G)^2}\right)^{1/2}. \tag{6.113}$$

The result of this complete extraction is a Schwarzschild black hole of mass $M_{\rm irr}$. It turns out that the best we can do is to start with an extreme Kerr black hole; then we can get out approximately 29% of its total energy.

The irreducibility of $M_{\rm irr}$ leads immediately to the fact that the area A can never decrease. From (6.110) and (6.111) we have

$$\delta A = 8\pi G \frac{a}{\Omega_{\rm H}\sqrt{G^2M^2 - a^2}}(\delta M - \Omega_{\rm H}\delta J), \tag{6.114}$$

which can be recast as

$$\delta M = \frac{\kappa}{8\pi G}\delta A + \Omega_{\rm H}\delta J, \tag{6.115}$$

where we have introduced

$$\kappa = \frac{\sqrt{G^2M^2 - a^2}}{2GM(GM + \sqrt{G^2M^2 - a^2})}. \tag{6.116}$$

The quantity κ is of course just the surface gravity of the Kerr solution, as you could verify by plugging (6.100) into (6.9).

Equations like (6.115) first started people thinking about a correspondence between black holes and thermodynamics. Consider the first law of thermodynamics,

$$dE = T\,dS - p\,dV, \tag{6.117}$$

where T is the temperature, S is the entropy, p is the pressure, and V is the volume, so the $p dV$ term represents work we do to the system. It is natural to think of the term $\Omega_{\rm H}\delta J$ in (6.115) as work that we do on the black hole by throwing rocks into it. Then the correspondence begins to take shape if we think of identifying the thermodynamic quantities energy, entropy, and temperature with the black-hole mass, area, and surface gravity:

$$\begin{aligned} E &\leftrightarrow M \\ S &\leftrightarrow A/4G \\ T &\leftrightarrow \kappa/2\pi. \end{aligned} \tag{6.118}$$

(Remember we are using units in which $\hbar = c = k = 1$.) In the context of classical general relativity the analogy is essentially perfect, with each law of thermodynamics corresponding to a law of black hole mechanics. A system in thermal equilibrium will have settled to a stationary state, corresponding to a stationary black hole. The zeroth law of thermodynamics states that in thermal equilibrium the temperature is constant throughout the system; the analogous statement for black holes is that stationary black holes have constant surface gravity on the entire horizon. This will be true, at least under the same reasonable assumptions under which the event horizon is a Killing horizon. As we have seen, the first law (6.117) is equivalent to (6.115). The second law, that entropy never decreases, is simply the statement that the area of the horizon never decreases. Finally, the usual statement of the third law is that it is impossible to achieve $T = 0$ in any physical process, or that the entropy must go to zero as the temperature goes to zero. For black holes this doesn't quite work; it turns out that $\kappa = 0$ corresponds to extremal black holes, which don't necessarily have a vanishing area. But the thermodynamic third law doesn't really work either, in the sense that there are ordinary physical systems that violate it; the third law applies to some situations but is not truly fundamental.

We have cheated a little in proposing the correspondence (6.118); you will notice that by equating TdS with $\kappa dA/8\pi G$ we do not know how to separately normalize S/A or T/κ, only their combination. As we will discuss in Chapter 9, however, Hawking showed that quantum fields in a black-hole background allow the hole to radiate at a temperature $T = \kappa/2\pi$. Once this is known, we can interpret $A/4G$ as an actual entropy of the black hole. Bekenstein has proposed a **generalized second law**, that the combined entropy of matter and black holes never decreases:

$$\delta\left(S + \frac{A}{4G}\right) \geq 0. \tag{6.119}$$

The generalized second law can actually be proven under a variety of assumptions. Usually, however, we like to associate the entropy of a system with the logarithm of the number of accessible quantum states. There is therefore some tension between this concept and the no-hair theorem, which indicates that there are very few possible states for a black hole of fixed charge, mass, and spin (only one, in fact). It seems likely that this behavior is an indication of a profound feature of the interaction between quantum mechanics and gravitation.

6.8 ■ EXERCISES

1. Show that the coupled Einstein–Maxwell equations can be simultaneously solved by the metric (6.62) and the electrostatic potential (6.67) if $H(\vec{x})$ obeys Laplace's equation,

$$\nabla^2 H = 0. \tag{6.120}$$

2. Consider the orbits of massless particles, with affine parameter λ, in the equatorial plane of a Kerr black hole.

(a) Show that

$$\left(\frac{dr}{d\lambda}\right)^2 = \frac{\Sigma^2}{\rho^4}(E - LW_+(r))(E - LW_-(r)), \tag{6.121}$$

where $\Sigma^2 = (r^2 + a^2)^2 - a^2\Delta(r)\sin^2\theta$, E and L are the conserved energy and angular momentum, and you have to find expressions for $W_\pm(r)$.

(b) Using this result, and assuming that $\Sigma^2 > 0$ everywhere, show that the orbit of a photon in the equatorial plane cannot have a turning point inside the outer event horizon r_+. This means that ingoing light rays cannot escape once they cross r_+, so it really is an event horizon.

3. In the presence of an electromagnetic field, a particle of charge e and mass m obeys

$$\frac{d^2x^\mu}{d\tau^2} + \Gamma^\mu_{\rho\sigma}\frac{dx^\rho}{d\tau}\frac{dx^\sigma}{d\tau} = \frac{e}{m}F^\mu{}_\nu\frac{dx^\nu}{d\tau}. \tag{6.122}$$

Imagine that such a particle is moving in the field of a Reissner–Nordström black hole with charge Q and mass M.

(a) Show that the energy

$$E = m\left(1 - \frac{2GM}{r} + \frac{GQ^2}{r^2}\right)\frac{dt}{d\tau} + \frac{eQ}{r} \tag{6.123}$$

is conserved.

(b) Will a Penrose-type process work for a charged black hole? What is the change in the black hole mass, δM, for the maximum physical process?

4. Consider de Sitter space in static coordinates:

$$ds^2 = -\left(1 - \frac{\Lambda}{3}r^2\right)dt^2 + \frac{dr^2}{1 - \frac{\Lambda}{3}r^2} + r^2 d\Omega^2.$$

This space has a Killing vector ∂_t that is timelike near $r = 0$ and null on a Killing horizon. Locate the radial position of the Killing horizon, r_K. What is the surface gravity, κ, of the horizon? Consider the Euclidean signature version of de Sitter space obtained by making the replacement $t \to i\tau$. Show that a coordinate transformation can be made to make the Euclidean metric regular at the horizon, so long as τ is made periodic.

5. What is the magnetic field seen by an observer orbiting a Riessner–Nordström black hole of electric charge Q and mass M in a circular orbit with circumference $2\pi R$?

6. Consider a Kerr black hole with an accretion disk of negligible mass in the equatorial plane. Assume that particles in the disk follow geodesics (that is, ignore any pressure support). Now suppose the disk contains some iron atoms that are being excited by a source of radiation. When the iron atoms de-excite they emit radiation with a known frequency ν_0, as measured in their rest frame. Suppose we detect this radiation far from the black hole (we also lie in the equatorial plane). What is the observed frequency of photons emitted from either edge of the disk, and from the center of the disk? Consider cases where the disk and the black hole are rotating in the same and opposite directions. Can we use these measurements to determine the mass and angular momentum of the black hole?

CHAPTER

7 Perturbation Theory and Gravitational Radiation

7.1 ■ LINEARIZED GRAVITY AND GAUGE TRANSFORMATIONS

When we first derived Einstein's equation, we checked that we were on the right track by considering the Newtonian limit. We took this to mean not only that the gravitational field was weak, but also that it was static (no time derivatives), and that test particles were moving slowly. The weak-field limit described in this chapter is less restrictive, assuming that the field is still weak but it can vary with time, and without any restrictions on the motion of test particles. This will allow us to discuss phenomena that are absent or ambiguous in the Newtonian theory, such as gravitational radiation (where the field varies with time) and the deflection of light (which involves fast-moving particles).

The weakness of the gravitational field is once again expressed as our ability to decompose the metric into the flat Minkowski metric plus a small perturbation,

$$g_{\mu\nu} = \eta_{\mu\nu} + h_{\mu\nu}, \qquad |h_{\mu\nu}| \ll 1. \tag{7.1}$$

We will restrict ourselves to coordinates in which $\eta_{\mu\nu}$ takes its canonical form, $\eta_{\mu\nu} = \mathrm{diag}(-1, +1, +1, +1)$. The assumption that $h_{\mu\nu}$ is small allows us to ignore anything that is higher than first order in this quantity, from which we immediately obtain

$$g^{\mu\nu} = \eta^{\mu\nu} - h^{\mu\nu}, \tag{7.2}$$

where $h^{\mu\nu} = \eta^{\mu\rho}\eta^{\nu\sigma}h_{\rho\sigma}$. As before, we can raise and lower indices using $\eta^{\mu\nu}$ and $\eta_{\mu\nu}$, since the corrections would be of higher order in the perturbation. In fact, we can think of the linearized version of general relativity (where effects of higher than first order in $h_{\mu\nu}$ are neglected) as describing a theory of a symmetric tensor field $h_{\mu\nu}$ propagating on a flat background spacetime. This theory is Lorentz invariant in the sense of special relativity; under a Lorentz transformation $x^{\mu'} = \Lambda^{\mu'}{}_{\mu}x^{\mu}$, the flat metric $\eta_{\mu\nu}$ is invariant, while the perturbation transforms as

$$h_{\mu'\nu'} = \Lambda^{\mu'}{}_{\mu}\Lambda^{\nu'}{}_{\nu}h_{\mu\nu}. \tag{7.3}$$

Note that we could have considered small perturbations about some other background spacetime besides Minkowski space. In that case the metric would have been written $g_{\mu\nu} = g^{(0)}_{\mu\nu} + h_{\mu\nu}$, and we would have derived a theory of a symmetric tensor propagating on the curved space with metric $g^{(0)}_{\mu\nu}$. Such an approach is necessary, for example, in cosmology.

We want to find the equations of motion obeyed by the perturbations $h_{\mu\nu}$, which come by examining Einstein's equation to first order. We begin with the Christoffel symbols, which are given by

$$\begin{aligned}\Gamma^{\rho}_{\mu\nu} &= \tfrac{1}{2}g^{\rho\lambda}(\partial_\mu g_{\nu\lambda} + \partial_\nu g_{\lambda\mu} - \partial_\lambda g_{\mu\nu}) \\ &= \tfrac{1}{2}\eta^{\rho\lambda}(\partial_\mu h_{\nu\lambda} + \partial_\nu h_{\lambda\mu} - \partial_\lambda h_{\mu\nu}).\end{aligned} \tag{7.4}$$

Since the connection coefficients are first-order quantities, the only contribution to the Riemann tensor will come from the derivatives of the Γ's, not the Γ^2 terms. Lowering an index for convenience, we obtain

$$\begin{aligned}R_{\mu\nu\rho\sigma} &= \eta_{\mu\lambda}\partial_\rho\Gamma^{\lambda}_{\nu\sigma} - \eta_{\mu\lambda}\partial_\sigma\Gamma^{\lambda}_{\nu\rho} \\ &= \tfrac{1}{2}(\partial_\rho\partial_\nu h_{\mu\sigma} + \partial_\sigma\partial_\mu h_{\nu\rho} - \partial_\sigma\partial_\nu h_{\mu\rho} - \partial_\rho\partial_\mu h_{\nu\sigma}).\end{aligned} \tag{7.5}$$

The Ricci tensor comes from contracting over μ and ρ, giving

$$R_{\mu\nu} = \tfrac{1}{2}(\partial_\sigma\partial_\nu h^{\sigma}{}_{\mu} + \partial_\sigma\partial_\mu h^{\sigma}{}_{\nu} - \partial_\mu\partial_\nu h - \Box h_{\mu\nu}), \tag{7.6}$$

which is manifestly symmetric in μ and ν. In this expression we have defined the trace of the perturbation as $h = \eta^{\mu\nu}h_{\mu\nu} = h^{\mu}{}_{\mu}$, and the d'Alembertian is simply the one from flat space, $\Box = -\partial_t^2 + \partial_x^2 + \partial_y^2 + \partial_z^2$. Contracting again to obtain the Ricci scalar yields

$$R = \partial_\mu\partial_\nu h^{\mu\nu} - \Box h. \tag{7.7}$$

Putting it all together we obtain the Einstein tensor:

$$\begin{aligned}G_{\mu\nu} &= R_{\mu\nu} - \tfrac{1}{2}\eta_{\mu\nu}R \\ &= \tfrac{1}{2}(\partial_\sigma\partial_\nu h^{\sigma}{}_{\mu} + \partial_\sigma\partial_\mu h^{\sigma}{}_{\nu} - \partial_\mu\partial_\nu h - \Box h_{\mu\nu} - \eta_{\mu\nu}\partial_\rho\partial_\lambda h^{\rho\lambda} + \eta_{\mu\nu}\Box h).\end{aligned} \tag{7.8}$$

Consistent with our interpretation of the linearized theory as one describing a symmetric tensor on a flat background, the linearized Einstein tensor (7.8) can be derived by varying the following Lagrangian with respect to $h_{\mu\nu}$:

$$\begin{aligned}\mathcal{L} = \tfrac{1}{2}\big[&(\partial_\mu h^{\mu\nu})(\partial_\nu h) - (\partial_\mu h^{\rho\sigma})(\partial_\rho h^{\mu}{}_{\sigma}) + \tfrac{1}{2}\eta^{\mu\nu}(\partial_\mu h^{\rho\sigma})(\partial_\nu h_{\rho\sigma}) \\ &- \tfrac{1}{2}\eta^{\mu\nu}(\partial_\mu h)(\partial_\nu h)\big].\end{aligned} \tag{7.9}$$

You are asked to verify the appropriateness of the Lagrangian in the exercises.

The linearized field equation is of course $G_{\mu\nu} = 8\pi G T_{\mu\nu}$, where $G_{\mu\nu}$ is given by (7.8) and $T_{\mu\nu}$ is the energy-momentum tensor, calculated to zeroth order in $h_{\mu\nu}$. We do not include higher-order corrections to the energy-momentum tensor because the amount of energy and momentum must itself be small for the weak-field limit to apply. In other words, the lowest nonvanishing order in $T_{\mu\nu}$ is automatically of the same order of magnitude as the perturbation. Notice that the conservation law to lowest order is simply $\partial_\mu T^{\mu\nu} = 0$. We will often be concerned with the vacuum equation, which as usual is just $R_{\mu\nu} = 0$, where $R_{\mu\nu}$ is given by (7.6).

With the linearized field equation in hand, we are almost prepared to set about solving it. First, however, we should deal with the thorny issue of gauge invariance. This issue arises because the demand that $g_{\mu\nu} = \eta_{\mu\nu} + h_{\mu\nu}$ does not completely specify the coordinate system on spacetime; there may be other coordinate systems in which the metric can still be written as the Minkowski metric plus a small perturbation, but the perturbation will be different. Thus, the decomposition of the metric into a flat background plus a perturbation is not unique. To examine this issue, we will draw upon ideas about diffeomorphisms discussed in Appendices A and B; readers who have not yet read those sections can skip to equation (7.14) and the two paragraphs following, which contain the essential ideas.

Let's think about gauge invariance from a highbrow point of view. The notion that the linearized theory can be thought of as one governing the behavior of tensor fields on a flat background can be formalized in terms of a *background spacetime* M_b, a *physical spacetime* M_p, and a diffeomorphism $\phi : M_b \rightarrow M_p$. As manifolds M_b and M_p are the same (since they are diffeomorphic), but we imagine that they possess some different tensor fields; on M_b we have defined the flat Minkowski metric $\eta_{\mu\nu}$, while on M_p we have some metric $g_{\alpha\beta}$ that obeys Einstein's equation. (We imagine that M_b is equipped with coordinates x^μ and M_p is equipped with coordinates y^α, although these will not play a prominent role.) The diffeomorphism ϕ allows us to move tensors back and forth between the background and physical spacetimes, as in Figure 7.1. Since we would like to construct our linearized theory as one taking place on the flat background spacetime, we are interested in the pullback $(\phi^* g)_{\mu\nu}$ of the physical metric. We can define the perturbation as the difference between the pulled-back physical metric

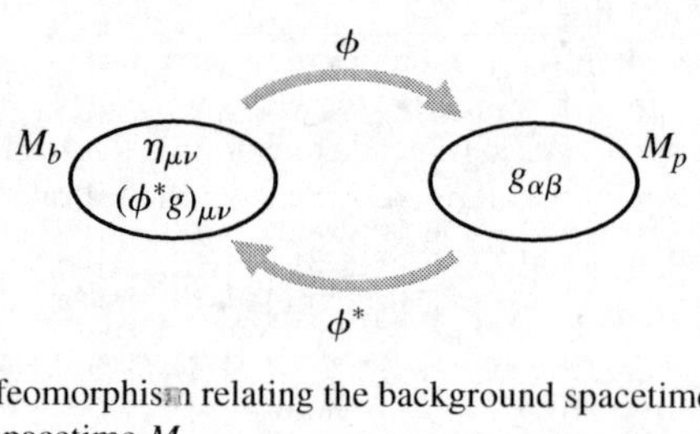

FIGURE 7.1 A diffeomorphism relating the background spacetime M_b (with flat metric $\eta_{\mu\nu}$) to the physical spacetime M_p.

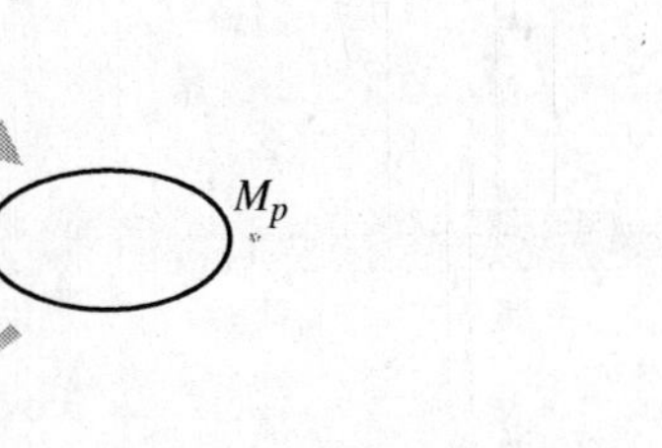

FIGURE 7.2 A one-parameter family of diffeomorphisms ψ_ϵ, generated by the vector field ξ^μ on the background spacetime M_b.

and the flat one:

$$h_{\mu\nu} = (\phi^* g)_{\mu\nu} - \eta_{\mu\nu}. \tag{7.10}$$

From this definition, there is no reason for the components of $h_{\mu\nu}$ to be small; however, if the gravitational fields on M_p are weak, then for *some* diffeomorphisms ϕ we will have $|h_{\mu\nu}| \ll 1$. We therefore limit our attention only to those diffeomorphisms for which this is true. Then the fact that $g_{\alpha\beta}$ obeys Einstein's equation on the physical spacetime means that $h_{\mu\nu}$ will obey the linearized equation on the background spacetime (since ϕ, as a diffeomorphism, can be used to pull back Einstein's equation themselves).

In this language, the issue of gauge invariance is simply that there are a large number of permissible diffeomorphisms between M_b and M_p (where "permissible" means that the perturbation is small). Consider a vector field $\xi^\mu(x)$ on the background spacetime. This vector field generates a one-parameter family of diffeomorphisms $\psi_\epsilon : M_b \to M_b$, as shown in Figure 7.2. For ϵ sufficiently small, if ϕ is a diffeomorphism for which the perturbation defined by (7.10) is small, then so will $(\phi \circ \psi_\epsilon)$ be, although the perturbation will have a different value. Specifically, we can define a family of perturbations parameterized by ϵ:

$$\begin{aligned} h^{(\epsilon)}_{\mu\nu} &= [(\phi \circ \psi_\epsilon)^* g]_{\mu\nu} - \eta_{\mu\nu} \\ &= [\psi^*_\epsilon(\phi^* g)]_{\mu\nu} - \eta_{\mu\nu}. \end{aligned} \tag{7.11}$$

The second equality is based on the fact that the pullback under a composition is given by the composition of the pullbacks in the opposite order, which follows from the fact that the pullback itself moves things in the opposite direction from the original map. Plugging in the relation (7.10), we find

$$\begin{aligned} h^{(\epsilon)}_{\mu\nu} &= \psi^*_\epsilon(h + \eta)_{\mu\nu} - \eta_{\mu\nu} \\ &= \psi^*_\epsilon(h_{\mu\nu}) + \psi^*_\epsilon(\eta_{\mu\nu}) - \eta_{\mu\nu}, \end{aligned} \tag{7.12}$$

since the pullback of the sum of two tensors is the sum of the pullbacks. Now we use our assumption that ϵ is small; in this case $\psi^*_\epsilon(h_{\mu\nu})$ will be equal to $h_{\mu\nu}$ to lowest order, while the other two terms give us a Lie derivative:

$$h^{(\epsilon)}_{\mu\nu} = \psi^*_\epsilon(h_{\mu\nu}) + \epsilon\left[\frac{\psi^*_\epsilon(\eta_{\mu\nu}) - \eta_{\mu\nu}}{\epsilon}\right]$$
$$= h_{\mu\nu} + \epsilon\mathcal{L}_\xi \eta_{\mu\nu}. \tag{7.13}$$

In Appendix B we show that the Lie derivative of the metric along a vector field ξ_μ is $\mathcal{L}_\xi g_{\mu\nu} = 2\nabla_{(\mu}\xi_{\nu)}$. In the current context the background metric is flat, and covariant derivatives become partial derivatives; we therefore have

$$\boxed{h^{(\epsilon)}_{\mu\nu} = h_{\mu\nu} + 2\epsilon\,\partial_{(\mu}\xi_{\nu)}.} \tag{7.14}$$

This formula represents the change of the metric perturbation under an infinitesimal diffeomorphism along the vector field $\epsilon\xi^\mu$: we will call this a **gauge transformation** in linearized theory.

The diffeomorphisms ψ_ϵ provide a different representation of the same physical situation, while maintaining our requirement that the perturbation be small. Therefore, the result (7.12) tells us what kind of metric perturbations denote physically equivalent spacetimes—those related to each other by $2\epsilon\,\partial_{(\mu}\xi_{\nu)}$, for some vector ξ^μ. The invariance of our theory under such transformations is analogous to traditional gauge invariance of electromagnetism under $A_\mu \to A_\mu + \partial_\mu\lambda$. (The analogy is different from another analogy we draw with electromagnetism in Appendix J, relating local Lorentz transformations in the orthonormal-frame formalism to changes of basis in an internal vector bundle.) In electromagnetism the invariance comes about because the field strength $F_{\mu\nu} = \partial_\mu A_\nu - \partial_\nu A_\mu$ is left unchanged by gauge transformations; similarly, we find that the transformation (7.14) changes the linearized Riemann tensor by

$$\delta R_{\mu\nu\rho\sigma} = \tfrac{1}{2}(\partial_\rho\partial_\nu\partial_\mu\xi_\sigma + \partial_\rho\partial_\nu\partial_\sigma\xi_\mu + \partial_\sigma\partial_\mu\partial_\nu\xi_\rho + \partial_\sigma\partial_\mu\partial_\rho\xi_\nu$$
$$- \partial_\sigma\partial_\nu\partial_\mu\xi_\rho - \partial_\sigma\partial_\nu\partial_\rho\xi_\mu - \partial_\rho\partial_\mu\partial_\nu\xi_\sigma - \partial_\rho\partial_\mu\partial_\sigma\xi_\nu)$$
$$= 0. \tag{7.15}$$

Our abstract derivation of the appropriate gauge transformation for the metric perturbation is verified by the fact that it leaves the curvature (and hence the physical spacetime) unchanged.

Gauge invariance can also be understood from the slightly more lowbrow but considerably more direct route of infinitesimal coordinate transformations. Our diffeomorphism ψ_ϵ can be thought of as changing coordinates from x^μ to $x^\mu - \epsilon\xi^\mu$. (The minus sign, which is unconventional, comes from the fact that the "new" metric is pulled back from a small distance forward along the integral curves, which is equivalent to replacing the coordinates by those a small distance backward along the curves.) Following through the usual rules for transforming tensors under coordinate transformations, you can derive precisely (7.14)—although you have to cheat somewhat by equating components of tensors in two different coordinate systems.

7.2 ■ DEGREES OF FREEDOM

With the expression (7.8) for the linearized Einstein tensor, and the expression (7.14) for the effect of gauge transformations, we could immediately set about choosing a gauge and solving Einstein's equation. However, we can accumulate some additional physical insight by first choosing a fixed inertial coordinate system in the Minkowski background spacetime, and decomposing the components of the metric perturbation according to their transformation properties under spatial rotations. You might worry that such a decomposition is contrary to the coordinate-independent spirit of general relativity, but it is really no different than decomposing the electromagnetic field strength tensor into electric and magnetic fields. Even though both **E** and **B** are components of a (0, 2) tensor, it is nevertheless sometimes convenient to assume the role of some fixed observer and think of them as three-vectors.[1]

The metric perturbation is a (0, 2) tensor, but symmetric rather than antisymmetric. Under spatial rotations, the 00 component is a scalar, the $0i$ components (equal to the $i0$ components) form a three-vector, and the ij components form a two-index symmetric spatial tensor. This spatial tensor can be further decomposed into a trace and a trace-free part. (In group theory language, we are looking for "irreducible representations" of the rotation group. In other words, we decompose the tensor into individual pieces, which transform only into themselves under spatial rotations.) We therefore write $h_{\mu\nu}$ as

$$\begin{aligned} h_{00} &= -2\Phi \\ h_{0i} &= w_i \\ h_{ij} &= 2s_{ij} - 2\Psi\delta_{ij}, \end{aligned} \tag{7.16}$$

where Ψ encodes the trace of h_{ij}, and s_{ij} is traceless:

$$\begin{aligned} \Psi &= -\tfrac{1}{6}\delta^{ij}h_{ij} \\ s_{ij} &= \tfrac{1}{2}\left(h_{ij} - \tfrac{1}{3}\delta^{kl}h_{kl}\delta_{ij}\right). \end{aligned} \tag{7.17}$$

The entire metric is thus written as

$$ds^2 = -(1+2\Phi)dt^2 + w_i(dt\,dx^i + dx^i dt) + [(1-2\Psi)\delta_{ij} + 2s_{ij}]dx^i dx^j. \tag{7.18}$$

[1] The discussion here follows that in E. Bertschinger, "Cosmological Dynamics," a talk given at Summer School on Cosmology and Large Scale Structure (Session 60), Les Houches, France, 1–28 Aug 1993; `http://arXiv.org/abs/astro-ph/9503125`. Bertschinger focuses on cosmological perturbation theory, in which spacelike hypersurfaces are expanding with time, but it is simple enough to specialize to the case of a nonexpanding universe.

We have not chosen a gauge or solved any equations, just defined some convenient notation. The traceless tensor s_{ij} is known as the *strain*, and will turn out to contain gravitational radiation. Sometimes the decomposition of the spatial components into trace and trace-free parts is not helpful, and we can just stick with h_{ij}; we will use whichever notation is appropriate in individual cases. Note that, just as in Chapter 1, the spatial metric is now simply δ_{ij}, and we can freely raise and lower spatial indices without changing the components.

To get a feeling for the physical interpretation of the different fields in the metric perturbation, we consider the motion of test particles as described by the geodesic equation. The Christoffel symbols for (7.18) are

$$\begin{aligned}
\Gamma^0_{00} &= \partial_0 \Phi \\
\Gamma^i_{00} &= \partial_i \Phi + \partial_0 w_i \\
\Gamma^0_{j0} &= \partial_j \Phi \\
\Gamma^i_{j0} &= \partial_{[j} w_{i]} + \tfrac{1}{2}\partial_0 h_{ij} \\
\Gamma^0_{jk} &= -\partial_{(j} w_{k)} + \tfrac{1}{2}\partial_0 h_{jk} \\
\Gamma^i_{jk} &= \partial_{(j} h_{k)i} - \tfrac{1}{2}\partial_i h_{jk}.
\end{aligned} \tag{7.19}$$

In these expressions we have stuck with h_{ij} rather than s_{ij} and Ψ, since they enter only in the combination $h_{ij} = 2s_{ij} - 2\Psi\delta_{ij}$. The distinction will become important once we start taking traces to get to the Ricci tensor and Einstein's equation. Since we have fixed an inertial frame, it is convenient to express the four-momentum $p^\mu = dx^\mu/d\lambda$ (where $\lambda = \tau/m$ if the particle is massive) in terms of the energy E and three-velocity $v^i = dx^i/dt$, as

$$p^0 = \frac{dt}{d\lambda} = E, \quad p^i = Ev^i. \tag{7.20}$$

Then we can take the geodesic equation

$$\frac{dp^\mu}{d\lambda} + \Gamma^\mu_{\rho\sigma} p^\rho p^\sigma = 0, \tag{7.21}$$

move the second term to the right-hand side so that it takes on the appearance of a force term, and divide both sides by E to obtain

$$\frac{dp^\mu}{dt} = -\Gamma^\mu_{\rho\sigma}\frac{p^\rho p^\sigma}{E}. \tag{7.22}$$

The $\mu = 0$ component describes the evolution of the energy,

$$\frac{dE}{dt} = -E\left[\partial_0\Phi + 2(\partial_k\Phi)v^k - \left(\partial_{(j} w_{k)} - \frac{1}{2}\partial_0 h_{jk}\right)v^j v^k\right]. \tag{7.23}$$

You might think that the energy should be conserved, but $E = p^0 = m\gamma$ only includes the "inertial" energy of the particle—in the slowly-moving limit, the rest energy and the kinetic energy—and not the energy from interactions with the gravitational field.

The spatial components $\mu = i$ of the geodesic equation become

$$\frac{dp^i}{dt} = -E\left[\partial_i\Phi + \partial_0 w_i + 2(\partial_{[i}w_{j]} + \partial_0 h_{ij})v^j + \left(\partial_{(j}h_{k)i} - \frac{1}{2}\partial_i h_{jk}\right)v^j v^k\right]. \tag{7.24}$$

To interpret this physically, it is convenient to define the "gravito-electric" and "gravito-magnetic" three-vector fields,

$$\begin{aligned} G^i &\equiv -\partial_i\Phi - \partial_0 w_i \\ H^i &\equiv (\nabla\times\vec{w})^i = \epsilon^{ijk}\partial_j w_k, \end{aligned} \tag{7.25}$$

which bear an obvious resemblance to the definitions of the ordinary electric and magnetic field in terms of a scalar and vector potential. Then (7.24) becomes

$$\frac{dp^i}{dt} = E\left[G^i + (\vec{v}\times H)^i - 2(\partial_0 h_{ij})v^j - \left(\partial_{(j}h_{k)i} - \frac{1}{2}\partial_i h_{jk}\right)v^j v^k\right]. \tag{7.26}$$

The first two terms on the right-hand side describe how the test particle, moving along a geodesic, responds to the scalar and vector perturbations Φ and w_i in a way reminiscent of the Lorentz force law in electromagnetism. We also find couplings to the spatial perturbations h_{ij}, of linear and quadratic order in the three-velocity. The relative importance of the different perturbations will of course depend on the physical situation under consideration, as we will soon demonstrate.

In addition to the motion of test particles, we should examine the field equations for the metric perturbations, which are of course the linearized Einstein equations. The Riemann tensor in our variables is

$$\begin{aligned} R_{0j0l} &= \partial_j\partial_l\Phi + \partial_0\partial_{(j}w_{l)} - \tfrac{1}{2}\partial_0\partial_0 h_{jl} \\ R_{0jkl} &= \partial_j\partial_{[k}w_{l]} - \partial_0\partial_{[k}h_{l]j} \\ R_{ijkl} &= \partial_j\partial_{[k}h_{l]i} - \partial_i\partial_{[k}h_{l]j}, \end{aligned} \tag{7.27}$$

with other components related by symmetries. We contract using $\eta^{\mu\nu}$ to obtain the Ricci tensor,

$$\begin{aligned} R_{00} &= \nabla^2\Phi + \partial_0\partial_k w^k + 3\partial_0^2\Psi \\ R_{0j} &= -\tfrac{1}{2}\nabla^2 w_j + \tfrac{1}{2}\partial_j\partial_k w^k + 2\partial_0\partial_j\Psi + \partial_0\partial_k s_j{}^k \\ R_{ij} &= -\partial_i\partial_j(\Phi-\Psi) - \partial_0\partial_{(i}w_{j)} + \Box\Psi\delta_{ij} - \Box s_{ij} + 2\partial_k\partial_{(i}s_{j)}{}^k, \end{aligned} \tag{7.28}$$

where $\nabla^2 = \delta^{ij}\partial_i\partial_j$ is the three-dimensional flat Laplacian. Since the Ricci tensor involves contractions, the trace-free and trace parts of the spatial perturbations now enter in different ways. Finally, we can calculate the Einstein tensor,

$$\begin{aligned} G_{00} &= 2\nabla^2\Psi + \partial_k\partial_l s^{kl} \\ G_{0j} &= -\tfrac{1}{2}\nabla^2 w_j + \tfrac{1}{2}\partial_j\partial_k w^k + 2\partial_0\partial_j\Psi + \partial_0\partial_k s_j{}^k \\ G_{ij} &= (\delta_{ij}\nabla^2 - \partial_i\partial_j)(\Phi - \Psi) + \delta_{ij}\partial_0\partial_k w^k - \partial_0\partial_{(i}w_{j)} \\ &\quad + 2\delta_{ij}\partial_0^2\Psi - \Box s_{ij} + 2\partial_k\partial_{(i}s_{j)}{}^k - \delta_{ij}\partial_k\partial_l s^{kl}. \end{aligned} \tag{7.29}$$

Using this expression in Einstein's equation $G_{\mu\nu} = 8\pi G T_{\mu\nu}$ reveals that only a small fraction of the metric components are true degrees of freedom of the gravitational field; the rest obey constraints that determine them in terms of the other fields. To see this, start with $G_{00} = 8\pi G T_{00}$, which we write using (7.29) as

$$\nabla^2\Psi = 4\pi G T_{00} - \tfrac{1}{2}\partial_k\partial_l s^{kl}. \tag{7.30}$$

This is an equation for Ψ with no time derivatives; if we know what T_{00} and s_{ij} are doing at any time, we can determine what Ψ must be (up to boundary conditions at spatial infinity). Thus, Ψ is not by itself a propagating degree of freedom; it is determined by the energy-momentum tensor and the gravitational strain s_{ij}. Next turn to the $0j$ equation, which we write as

$$(\delta_{jk}\nabla^2 - \partial_j\partial_k)w^k = -16\pi G T_{0j} + 4\partial_0\partial_j\Psi + 2\partial_0\partial_k s_j{}^k. \tag{7.31}$$

This is an equation for w^i with no time derivatives; once again, if we know the energy-momentum tensor and the strain (from which we can find Ψ), the vector w^i will be determined. Finally, the ij equation is

$$\begin{aligned} (\delta_{ij}\nabla^2 - \partial_i\partial_j)\Phi &= 8\pi G T_{ij} + (\delta_{ij}\nabla^2 - \partial_i\partial_j - 2\delta_{ij}\partial_0^2)\Psi \\ &\quad - \delta_{ij}\partial_0\partial_k w^k + \partial_0\partial_{(i}w_{j)} + \Box s_{ij} - 2\partial_k\partial_{(i}s_{j)}{}^k - \delta_{ij}\partial_k\partial_l s^{jl}. \end{aligned} \tag{7.32}$$

Once again, we see that there are no time derivatives acting on Φ, which is therefore determined as a function of the other fields.

Thus, the only propagating degrees of freedom in Einstein's equations are those in the strain tensor s_{ij}; as we will see, these are used to describe gravitational waves. The other components of $h_{\mu\nu}$ are determined in terms of s_{ij} and the matter fields—they do not require separate initial data. In alternative theories, such as those discussed in Section 4.8 with either additional fields or higher-order terms in the action, the other components of the metric may become dynamical variables. As we discuss briefly at the end of Section 7.4, propagating tensor fields give rise upon quantization to particles of different spins, depending on the behavior of the field under spatial rotations. Thus, the scalars Φ and Ψ would be spin-0, the vector w_i would be spin-1, and the tensor s_{ij} is spin-2. Only the spin-2 piece is a true particle excitation in ordinary GR.

In the previous section we showed how gauge transformations $h_{\mu\nu} \rightarrow h_{\mu\nu} + \partial_\mu \xi_\nu + \partial_\nu \xi_\mu$ are generated by a vector field ξ^μ. Henceforth we set the parameter ϵ of (7.14) equal to unity, and think of the vector field ξ^μ itself as being small. Under such a transformation, the different metric perturbation fields change by

$$\begin{aligned}
\Phi &\rightarrow \Phi + \partial_0 \xi^0 \\
w_i &\rightarrow w_i + \partial_0 \xi^i - \partial_i \xi^0 \\
\Psi &\rightarrow \Psi - \tfrac{1}{3} \partial_i \xi^i \\
s_{ij} &\rightarrow s_{ij} + \partial_{(i} \xi_{j)} - \tfrac{1}{3} \partial_k \xi^k \delta_{ij},
\end{aligned} \tag{7.33}$$

as you can easily check. Just as in electromagnetism and other gauge theories, different gauges can be appropriate to different circumstances; here we list some popular choices.

Consider first the **transverse gauge** (a generalization of the conformal Newtonian or Poisson gauge sometimes used in cosmology.) The transverse gauge is closely related to the Coulomb gauge of electromagnetism, $\partial_i A^i = 0$. We begin by fixing the strain to be spatially transverse,

$$\partial_i s^{ij} = 0, \tag{7.34}$$

by choosing ξ^j to satisfy

$$\nabla^2 \xi^j + \tfrac{1}{3} \partial_j \partial_i \xi^i = -2 \partial_i s^{ij}. \tag{7.35}$$

The value of ξ^0 is still undetermined, so we can use this remaining freedom to render the vector perturbation transverse,

$$\partial_i w^i = 0, \tag{7.36}$$

by choosing ξ^0 to satisfy

$$\nabla^2 \xi^0 = \partial_i w^i + \partial_0 \partial_i \xi^i. \tag{7.37}$$

The meaning of transverse becomes clear upon taking the Fourier transform, after which a vanishing divergence implies that a tensor is orthogonal to the wave vector. Neither (7.35) nor (7.37) completely fixes the value of ξ^μ; they are both second-order differential equations in spatial derivatives, which require boundary conditions to specify a solution. For our present purposes, it suffices that solutions will always exist. The conditions (7.34) and (7.36) together define the transverse gauge. In this gauge, Einstein's equation becomes

$$\boxed{G_{00} = 2\nabla^2 \Psi = 8\pi G T_{00},} \tag{7.38}$$

$$\boxed{G_{0j} = -\frac{1}{2}\nabla^2 w_j + 2\partial_0\partial_j\Psi = 8\pi G T_{0j},} \tag{7.39}$$

and

$$\boxed{G_{ij} = (\delta_{ij}\nabla^2 - \partial_i\partial_j)(\Phi - \Psi) - \partial_0\partial_{(i}w_{j)} + 2\delta_{ij}\partial_0^2\Psi - \Box s_{ij} = 8\pi G T_{ij}.} \tag{7.40}$$

In the remainder of this chapter, we will use these equations to find weak-field solutions in different situations.

Another popular gauge is known as the **synchronous gauge**. It is equivalent to the choice of Gaussian normal coordinates, discussed in Appendix D. It may be thought of as the gravitational analogue of the temporal gauge of electromagnetism, $A^0 = 0$, since it kills off the nonspatial components of the perturbation. We begin by setting the scalar potential Φ to vanish,

$$\Phi = 0, \tag{7.41}$$

by choosing ξ^0 to satisfy

$$\partial_0\xi^0 = -\Phi. \tag{7.42}$$

This leaves us the ability to choose ξ^i. We can set the vector components to zero,

$$w^i = 0, \tag{7.43}$$

by choosing ξ^i to satisfy

$$\partial_0\xi^i = -w^i + \partial_i\xi^0. \tag{7.44}$$

The metric in synchronous gauge therefore takes on the attractive form

$$ds^2 = -dt^2 + (\delta_{ij} + h_{ij})dx^i dx^j. \tag{7.45}$$

This is just a matter of gauge choice, and is applicable to any spacetime slightly perturbed away from Minkowski. It is straightforward to write down Einstein's equation in synchronous gauge, but we won't bother as we won't actually be using it in the rest of this chapter.

In addition to transverse and synchronous gauges, in calculating the production of gravitational waves it is convenient to use yet a third choice, the Lorenz/harmonic gauge. As we will discuss below, it is equivalent to setting

$$\partial_\mu h^\mu{}_\nu - \frac{1}{2}\partial_\nu h = 0, \tag{7.46}$$

where $h = \eta^{\mu\nu} h_{\mu\nu}$. This gauge does not have any especially simple expression in terms of our decomposed perturbation fields, but it does make the linearized Einstein equation take on a particularly simple form.

Before moving on to applications of the weak-field limit, we conclude our discussion of degrees of freedom by drawing attention to the distinction between our *algebraic* decomposition of the metric perturbation components in (7.16), and an additional decomposition that becomes possible if we consider tensor *fields* rather than tensors defined at a point. This additional decomposition helps to bring out the physical degrees of freedom more directly, and is crucial in cosmological perturbation theory. Its basis is the standard observation that a vector field can be decomposed into a transverse part $w^i_\perp$ and a longitudinal part $w^i_\parallel$:

$$w^i = w^i_\perp + w^i_\parallel, \tag{7.47}$$

where a transverse vector is divergenceless and a longitudinal vector is curl-free,

$$\partial_i w^i_\perp = 0, \quad \epsilon^{ijk} \partial_j w_{\parallel k} = 0. \tag{7.48}$$

Notice that these are differential equations, so clearly they only make sense when applied to tensor fields. A transverse vector can be represented as the curl of some other vector ξ^i, although the choice of ξ^i is not unique unless we impose a subsidiary condition such as $\partial_i \xi^i = 0$. A longitudinal vector is the divergence of a scalar λ,

$$w^i_\perp = \epsilon^{ijk} \partial_j \xi_k, \quad w_{\parallel i} = \partial_i \lambda. \tag{7.49}$$

Just like our original decomposition of the metric perturbation into scalar, vector and tensor pieces, this decomposition of a vector field into parts depending on a scalar and a transverse vector is invariant under spatial rotations. The scalar λ clearly represents one degree of freedom; the vector ξ^i looks like three degrees of freedom, but one of these is illusory due to the nonuniqueness of the choice of ξ^i (which you will notice is equivalent to the freedom to make gauge transformations $\xi_i \to \xi_i + \partial_i \omega$). There are thus three degrees of freedom in total, as there should be to describe the original vector field w^i.

A similar procedure applies to the traceless symmetric tensor s^{ij}, which can be decomposed into a transverse part $s^{ij}_\perp$, a solenoidal part $s^{ij}_{\rm S}$, and a longitudinal part $s^{ij}_\parallel$,

$$s^{ij} = s^{ij}_\perp + s^{ij}_{\rm S} + s^{ij}_\parallel. \tag{7.50}$$

The transverse part is divergenceless, while the divergence of the solenoidal part is a transverse (divergenceless) vector, and the divergence of the longitudinal part is a longitudinal (curl-free) vector:

$$\partial_i s^{ij}_\perp = 0$$

$$\partial_i \partial_j s_{\mathrm{S}}^{ij} = 0$$
$$\epsilon^{jkl} \partial_k \partial_i s_{\|}{}^{i}{}_{j} = 0. \tag{7.51}$$

This means that the longitudinal part can be derived from a scalar field θ, and the solenoidal part can be derived from a transverse vector ζ^i,

$$s_{\| ij} = \left(\partial_i \partial_j - \tfrac{1}{3}\delta_{ij}\nabla^2\right)\theta$$
$$s_{\mathrm{S}ij} = \partial_{(i}\zeta_{j)}, \tag{7.52}$$

where

$$\partial_i \zeta^i = 0. \tag{7.53}$$

Thus, the longitudinal part describes a single degree of freedom, while the solenoidal part describes two degrees of freedom. The transverse part cannot be further decomposed; it describes the remaining two degrees of freedom of the symmetric traceless 3×3 tensor s_{ij}. Later in this chapter we will introduce the transverse-traceless gauge for describing gravitational waves propagating in vacuum; in this gauge, the only nonvanishing metric perturbation is the transverse tensor perturbation $s_{\perp}^{ij}$.

With this decomposition of tensor fields, we have succeeded in writing the original ten-component metric perturbation $h_{\mu\nu}$ in terms of four scalars (Φ, Ψ, λ, and θ) with one degree of freedom each, two transverse vectors (ξ^i and ζ^i) with two degrees of freedom each, and one transverse-traceless tensor ($s_{\perp}^{ij}$) with two degrees of freedom. People refer to this set of fields when they speak of "scalar," "vector," and "tensor" modes. We can then decompose the energy-momentum tensor in a similar way, write Einstein's equation in terms of these variables, and isolate the physical (gauge-invariant) degrees of freedom. We won't use this decomposition in this book, but you should be aware of its existence when referring to the literature.

7.3 ■ NEWTONIAN FIELDS AND PHOTON TRAJECTORIES

We previously defined the "Newtonian limit" as describing weak fields for which sources were static and test particles were slowly moving. In this section we will extend this definition somewhat, still restricting ourselves to static sources but allowing the test particles to move at any velocity. There is clearly an important difference, as we previously only needed to consider effects of the g_{00} component of the metric, but we will find that relativistic particles respond to spatial components of the metric as well.

We can model our static gravitating sources by dust, a perfect fluid for which the pressure vanishes. (Most of the matter in the universe is well approximated by dust, including stars, planets, galaxies, and even dark matter.) We work in the rest frame of the dust, where the energy-momentum tensor takes the form

$$T_{\mu\nu} = \rho U_\mu U_\nu = \begin{pmatrix} \rho & & & \\ & 0 & & \\ & & 0 & \\ & & & 0 \end{pmatrix}. \tag{7.54}$$

Since our background is flat Minkowski space, it is straightforward to accommodate moving sources by simply Lorentz-transforming into their rest frame; what we are unable to deal with in this limit is multiple sources with large relative velocities.

Turn to Einstein's equation in the transverse gauge, (7.38)–(7.40). For static sources we drop all time-derivative terms, and simultaneously plug in the energy-momentum tensor (7.54), to obtain

$$\begin{aligned} \nabla^2 \Psi &= 4\pi G\rho \\ \nabla^2 w_j &= 0 \\ (\delta_{ij}\nabla^2 - \partial_i\partial_j)(\Phi - \Psi) - \nabla^2 s_{ij} &= 0. \end{aligned} \tag{7.55}$$

We will look for solutions that are both nonsingular and well-behaved at infinity; consequently, only those fields that are sourced by the right-hand side will be nonvanishing. For example, the second equation in (7.55) immediately implies $w^i = 0$. We next take the trace of the third equation (summing over δ^{ij}):

$$2\nabla^2(\Phi - \Psi) = 0. \tag{7.56}$$

This enforces equality of the two scalar potentials,

$$\Phi = \Psi. \tag{7.57}$$

Recall that in our initial discussion of the Newtonian limit in Chapter 4, we argued that the 00 component Φ of the perturbation (which is responsible for the motion of nonrelativistic particles) obeyed the Poisson equation; from (7.55) it appears as if it is actually the scalar perturbation Ψ to the spatial components that obeys this equation. The implicit connection is provided by (7.56), which sets the two potentials equal when the trace of T_{ij} (the sum of the three principle pressures) vanishes. Finally we can plug $\Phi = \Psi$ into the last equation of (7.55) to get

$$\nabla^2 s_{ij} = 0, \tag{7.58}$$

which implies $s_{ij} = 0$ for a well-behaved solution.

The perturbed metric for static Newtonian sources is therefore

$$\boxed{ds^2 = -(1 + 2\Phi)dt^2 + (1 - 2\Phi)(dx^2 + dy^2 + dz^2),} \tag{7.59}$$

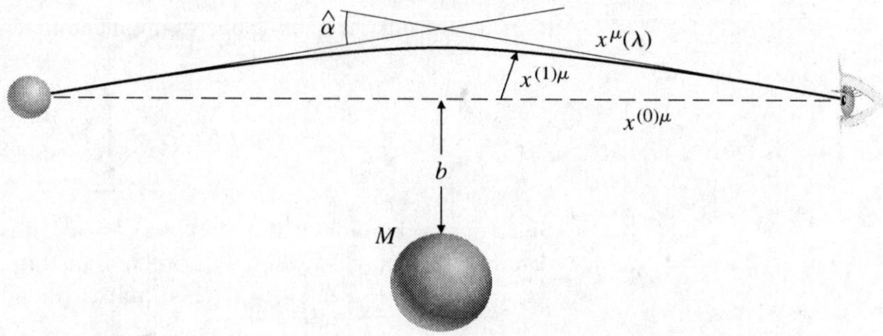

FIGURE 7.3 A deflected geodesic $x^{\mu}(\lambda)$, decomposed into a background geodesic $x^{(0)\mu}$ and a perturbation $x^{(1)\mu}$. The deflection angle $\widehat{\alpha}$ represents (minus) the amount by which the wave vector rotates along the path. A single mass M with impact parameter b is depicted, although the setup is more general.

or equivalently

$$h_{\mu\nu} = \begin{pmatrix} -2\Phi & & & \\ & -2\Phi & & \\ & & -2\Phi & \\ & & & -2\Phi \end{pmatrix}, \tag{7.60}$$

where the potential obeys the conventional Poisson equation,

$$\nabla^2\Phi = 4\pi G\rho. \tag{7.61}$$

This is an important extension of our result from Chapter 4, since we now know the perturbation of the spatial metric as well as h_{00}.

Now let us consider the path of a photon (or other massless particle) through this geometry; in other words, solve the perturbed geodesic equation for a null trajectory $x^{\mu}(\lambda)$.[2] The geometry we consider is portrayed in Figure 7.3. Recall that our philosophy is to consider the metric perturbation as a field defined on a flat background spacetime. Similarly, we can decompose the geodesic into a background path plus a perturbation,

$$x^{\mu}(\lambda) = x^{(0)\mu}(\lambda) + x^{(1)\mu}(\lambda), \tag{7.62}$$

where $x^{(0)\mu}$ solves the geodesic equation in the background (in other words, is just a straight null path). We then evaluate all quantities along the background path, to solve for $x^{(1)\mu}(\lambda)$. For this procedure to make sense, we need to assume that the potential Φ is not appreciably different along the background and true geodesics; this condition amounts to requiring that $x^{(1)i}\partial_i\Phi \ll \Phi$. If this condition is not true, however, all is not lost. If we consider only very short paths, the deviation

[2]The approach we use is outlined in T. Pyne and M. Birkinshaw, *Astrophys. Journ.* **458**, 46 (1996), `http://arxiv.org/abs/astro-ph/9504060`.

$x^{(1)\mu}$ will necessarily be small, and our approximation will be valid. But then we can assemble larger paths out of such short segments. As a result, we will derive true equations, but the paths over which we integrate will be the *actual* path $x^{\mu}(\lambda)$, rather than the background path $x^{(0)\mu}(\lambda)$. As long as this is understood, our results will be valid for any trajectories in the perturbed spacetime.

For convenience we denote the wave vector of the background path as k^{μ}, and the derivative of the deviation vector as ℓ^{μ}:

$$k^{\mu} \equiv \frac{dx^{(0)\mu}}{d\lambda}, \qquad \ell^{\mu} \equiv \frac{dx^{(1)\mu}}{d\lambda}. \tag{7.63}$$

The condition that a path be null is of course

$$g_{\mu\nu}\frac{dx^{\mu}}{d\lambda}\frac{dx^{\nu}}{d\lambda} = 0, \tag{7.64}$$

which we must solve order-by-order. At zeroth order we simply have $\eta_{\mu\nu}k^{\mu}k^{\nu} = 0$, or

$$(k^0)^2 = (\vec{k})^2 \equiv k^2, \tag{7.65}$$

where $\vec{k}$ is the three-vector with components k^i. This equation serves as the definition of the constant k. Then at first order we obtain

$$2\eta_{\mu\nu}k^{\mu}\ell^{\nu} + h_{\mu\nu}k^{\mu}k^{\nu} = 0, \tag{7.66}$$

or

$$-k\ell^0 + \vec{\ell}\cdot\vec{k} = 2k^2\Phi. \tag{7.67}$$

We now turn to the perturbed geodesic equation,

$$\frac{d^2x^{\mu}}{d\lambda^2} + \Gamma^{\mu}_{\rho\sigma}\frac{dx^{\rho}}{d\lambda}\frac{dx^{\sigma}}{d\lambda} = 0. \tag{7.68}$$

The Christoffel symbols can be found by setting $w^i = 0$ and $h_{ij} = -2\Phi\delta_{ij}$ in (7.19):

$$\begin{aligned}\Gamma^0_{0i} &= \Gamma^i_{00} = \partial_i\Phi,\\ \Gamma^i_{jk} &= \delta_{jk}\partial_i\Phi - \delta_{ik}\partial_j\Phi - \delta_{ij}\partial_k\Phi.\end{aligned} \tag{7.69}$$

The zeroth-order geodesic equation simply tells us that $x^{(0)\mu}$ is a straight trajectory, while at first order we have

$$\frac{d\ell^{\mu}}{d\lambda} = -\Gamma^{\mu}_{\rho\sigma}k^{\rho}k^{\sigma}. \tag{7.70}$$

There are no factors of ℓ^μ on the right-hand side, since the Christoffel symbols are already first-order in the perturbation. The $\mu = 0$ component of (7.70) is

$$\frac{d\ell^0}{d\lambda} = -2k(\vec{k} \cdot \vec{\nabla}\Phi), \tag{7.71}$$

while the spatial components are

$$\frac{d\vec{\ell}}{d\lambda} = -2k^2\vec{\nabla}_\perp\Phi. \tag{7.72}$$

Here we have introduced the gradient transverse to the path, defined as the total gradient minus the gradient along the path,

$$\begin{aligned}\vec{\nabla}_\perp\Phi &\equiv \vec{\nabla}\Phi - \vec{\nabla}_\parallel\Phi \\ &= \vec{\nabla}\Phi - k^{-2}(\vec{k} \cdot \vec{\nabla}\Phi)\vec{k}.\end{aligned} \tag{7.73}$$

In all of these expressions, the path means the background path.

Note that, to first order in Φ, the spatial wave vector perturbation $\vec{\ell}$ is orthogonal to the original spatial wave vector $\vec{k}$. To see this, we can get an expression for ℓ^0 by integrating (7.71) to get

$$\begin{aligned}\ell^0 &= \int \frac{d\ell^0}{d\lambda} d\lambda \\ &= -2k \int (\vec{k} \cdot \vec{\nabla}\Phi) d\lambda \\ &= -2k \int \left(\frac{d\vec{x}}{d\lambda} \cdot \vec{\nabla}\Phi\right) d\lambda \\ &= -2k \int \vec{\nabla}\Phi \cdot d\vec{x} \\ &= -2k\Phi.\end{aligned} \tag{7.74}$$

The constant of integration is fixed by demanding that $\ell^0 = 0$ when $\Phi = 0$. Plugging this into (7.67) reveals

$$\vec{\ell} \cdot \vec{k} = k\ell^0 + 2k^2\Phi = 0, \tag{7.75}$$

verifying that $\vec{\ell}$ and $\vec{k}$ are orthogonal to first order.

The **deflection angle** $\widehat{\alpha}$ is the amount by which the original spatial wave vector is deflected as it travels from a source to the observer; it is a two-dimensional vector in the plane perpendicular to $\vec{k}$. (We use the notation $\widehat{\alpha}$ rather than $\vec{\alpha}$, as the latter is used for the reduced deflection angle introduced in Chapter 8.) From the geometry portrayed in Figure 7.3, the deflection angle can be expressed as

$$\widehat{\alpha} = -\frac{\Delta\vec{\ell}}{k}, \tag{7.76}$$

where the minus sign simply accounts for the fact that the deflection angle is measured by an observer looking backward along the photon path. The rotation of the wave vector can be calculated from (7.72) as

$$\begin{aligned} \Delta\vec{\ell} &= \int \frac{d\vec{\ell}}{d\lambda} d\lambda \\ &= -2k^2 \int \vec{\nabla}_\perp \Phi \, d\lambda. \end{aligned} \tag{7.77}$$

The deflection angle can therefore be expressed as an integral over the physical spatial distance traversed, $s = k\lambda$, as

$$\boxed{\widehat{\alpha} = 2 \int \vec{\nabla}_\perp \Phi \, ds.} \tag{7.78}$$

We can evaluate the deflection angle in the case of a point mass, where we imagine the background path to be along the x-direction with an impact parameter defined by a transverse vector $\vec{b}$ pointing from the path to the mass at the point of closest approach. Setting $b = |\vec{b}|$, the potential is

$$\Phi = -\frac{GM}{r} = -\frac{GM}{(b^2 + x^2)^{1/2}}, \tag{7.79}$$

and its transverse gradient is therefore

$$\vec{\nabla}_\perp \Phi = \frac{GM}{(b^2 + x^2)^{3/2}} \vec{b}. \tag{7.80}$$

The deflection angle is thus

$$\begin{aligned} \widehat{\alpha} &= 2GMb \int \frac{dx}{(b^2 + x^2)^{3/2}} \\ &= \frac{4GM}{b}, \end{aligned} \tag{7.81}$$

where the integral has been taken from $-\infty$ to ∞, presuming that both source and observer are very far from the deflecting mass. Note that $c = 1$ in our units; a factor of c^2 should be inserted in the denominator of (7.81) in other systems.

Deflection of light by the Sun was historically a crucial test of general relativity. Einstein proposed three such tests: precession of the perihelion of Mercury, gravitational redshift, and deflection of light. The precession of Mercury's perihelion was successfully explained by GR, but this explained a discrepancy that had already been observed; gravitational redshift was not observed until much later, so deflection of light was the first time that Einstein's theory correctly predicted a phenomenon that had not yet been detected. A famous expedition led by Eddington observed the positions of stars near the Sun during a 1919 total eclipse; the observations were in agreement with the GR prediction, leading to front-page stories in newspapers around the world. The predicted effect is quite small: for the Sun we have $GM_\odot/c^2 = 1.48 \times 10^5$ cm, and the solar radius is $R_\odot = 6.96 \times 10^{10}$ cm, leading to a maximum deflection angle of $\widehat{\alpha} = 1.75$ arcsecs. Later re-evaluation of Eddington's results has cast doubt upon whether he actually obtained the precision that was originally claimed; contemporary measurements use high-precision interferometric observations of quasars passing behind the Sun to obtain very accurate tests of GR (which it has so far passed). Meanwhile, observation of light deflection by astrophysical sources such as galaxies and stars has become a vibrant area of research, under the name of "gravitational lensing." Of course in these circumstances we rarely know the mass of the lens well enough to provide precision tests of GR; instead, it is more common to use the observed deflection angle as a way to measure the mass. We will discuss lensing more in Chapter 8.

In addition to the deflection of light, in 1964 Shapiro pointed out another observable consequence of weak-field general relativity on photon trajectories: gravitational time delay. The total coordinate time elapsed along a null path is

$$t = \int \frac{dx^0}{d\lambda}\, d\lambda. \tag{7.82}$$

We are putting ourselves in the position of an observer far from any sources, at rest in the background inertial frame, so coordinate time is our proper time. In the presence of a Newtonian potential, the photons appear to "slow down" with respect to the background light cones, leading to an additional time delay of

$$\begin{aligned}\Delta t &= \int \frac{dx^{(1)0}}{d\lambda}\, d\lambda \\ &= \int \ell^0\, d\lambda \\ &= -2k \int \Phi\, d\lambda,\end{aligned} \tag{7.83}$$

or

$$\Delta t = -2 \int \Phi\, ds. \tag{7.84}$$

According to our rules, the integral is performed over the background path. In addition to this Shapiro delay, there can be an additional "geometric" time delay because the spatial distance traversed by the real path is longer than that of the background path. For deflection of light by the Sun the geometric delay effect is negligible, but in cosmological applications it can be comparable to the Shapiro effect. The time delay has been observed, most precisely by making use of spacecraft rather than naturally-occurring objects; for details see Will (1981).

The motion of photons through a Newtonian potential, leading to both the deflection of light and the gravitational time delay, could equivalently be derived by imagining that the photons are propagating in a medium with refractive index

$$n = 1 - 2\Phi, \tag{7.85}$$

to first order. Indeed, we could have found the equations of motion for the photon by using Fermat's principle of least time; you are asked to demonstrate this in the exercises.

7.4 ■ GRAVITATIONAL WAVE SOLUTIONS

An even more exciting application of the weak-field limit is to gravitational radiation. Here we are studying the freely-propagating degrees of freedom of the gravitational field, requiring no local sources for their existence (although they can of course be generated by such sources). We therefore turn once again to the weak-field equations in transverse gauge, (7.38)–(7.40), this time keeping time derivatives but completely turning off the energy-momentum tensor, $T_{\mu\nu} = 0$. The 00 equation is then

$$\nabla^2 \Psi = 0, \tag{7.86}$$

which with well-behaved boundary conditions implies $\Psi = 0$. Then the $0j$ equation is

$$\nabla^2 w_j = 0, \tag{7.87}$$

which again implies $w_j = 0$.

We turn next to the trace of the ij equation, which (plugging in the above results) yields

$$\nabla^2 \Phi = 0, \tag{7.88}$$

which implies $\Phi = 0$.

We are therefore left with the trace-free part of the ij equation, which becomes a wave equation for the traceless strain tensor:

$$\Box s_{ij} = 0. \tag{7.89}$$

Although it has been convenient thus far to work with s_{ij}, it is far more common in the literature to find expressions written in terms of the entire metric perturbation $h_{\mu\nu}$, but in an ansatz where all of the other degrees of freedom (Φ, Ψ, w_i) are set to zero (and s_{ij} is transverse). This is commonly known as the **transverse**

traceless gauge, in which we have

$$h_{\mu\nu}^{\rm TT} = \begin{pmatrix} 0 & 0 & 0 & 0 \\ 0 & & & \\ 0 & & 2s_{ij} & \\ 0 & & & \end{pmatrix}. \tag{7.90}$$

The equation of motion is then

$$\Box h_{\mu\nu}^{\rm TT} = 0. \tag{7.91}$$

To make it easier to compare with other resources, in our discussion of gravitational waves we will use $h_{\mu\nu}^{\rm TT}$ rather than s_{ij}, keeping in mind that $h_{\mu\nu}^{\rm TT}$ is purely spatial, traceless and transverse:

$$\begin{aligned} h_{0\nu}^{\rm TT} &= 0 \\ \eta^{\mu\nu} h_{\mu\nu}^{\rm TT} &= 0 \\ \partial_\mu h_{\rm TT}^{\mu\nu} &= 0. \end{aligned} \tag{7.92}$$

From the wave equation (7.91) we begin finding solutions. Those familiar with the analogous problem in electromagnetism will notice that the procedure is almost precisely the same. A particularly useful set of solutions to this wave equation are the plane waves, given by

$$h_{\mu\nu}^{\rm TT} = C_{\mu\nu} e^{ik_\sigma x^\sigma}, \tag{7.93}$$

where $C_{\mu\nu}$ is a constant, symmetric, (0, 2) tensor, which is obviously traceless and purely spatial:

$$\begin{aligned} C_{0\nu} &= 0 \\ \eta^{\mu\nu} C_{\mu\nu} &= 0. \end{aligned} \tag{7.94}$$

Of course $e^{ik_\sigma x^\sigma}$ is complex, while $h_{\mu\nu}^{\rm TT}$ is real; we carry both real and imaginary parts through the calculation, and take the real part at the end. The constant vector k^σ is the wave vector. To check that we have a solution, we plug in:

$$\begin{aligned} 0 &= \Box h_{\mu\nu}^{\rm TT} \\ &= \eta^{\rho\sigma} \partial_\rho \partial_\sigma h_{\mu\nu}^{\rm TT} \\ &= \eta^{\rho\sigma} \partial_\rho (i k_\sigma h_{\mu\nu}^{\rm TT}) \\ &= -\eta^{\rho\sigma} k_\rho k_\sigma h_{\mu\nu}^{\rm TT} \\ &= -k_\sigma k^\sigma h_{\mu\nu}^{\rm TT}. \end{aligned} \tag{7.95}$$

Since, for an interesting solution, not all of the components of $h^{\rm TT}_{\mu\nu}$ will be zero everywhere, we must have

$$k_\sigma k^\sigma = 0. \tag{7.96}$$

The plane wave (7.93) is therefore a solution to the linearized equation if the wave vector is null; this is loosely translated into the statement that gravitational waves propagate at the speed of light. The timelike component of the wave vector is the frequency of the wave, and we write $k^\sigma = (\omega, k^1, k^2, k^3)$. (More generally, an observer moving with four-velocity U^μ would observe the wave to have a frequency $\omega = -k_\mu U^\mu$.) Then the condition that the wave vector be null becomes

$$\omega^2 = \delta_{ij} k^i k^j. \tag{7.97}$$

Of course our wave is far from the most general solution; any (possibly infinite) number of distinct plane waves can be added together and will still solve the linear equation (7.91). Indeed, any solution can be written as such a superposition.

We still need to ensure that the perturbation is transverse. This means that

$$\begin{aligned} 0 &= \partial_\mu h^{\mu\nu}_{\rm TT} \\ &= i C^{\mu\nu} k_\mu e^{i k_\sigma x^\sigma}, \end{aligned} \tag{7.98}$$

which is only true if

$$k_\mu C^{\mu\nu} = 0. \tag{7.99}$$

We say that the wave vector is orthogonal to $C^{\mu\nu}$.

Our solution can be made more explicit by choosing spatial coordinates such that the wave is traveling in the x^3 direction; that is,

$$k^\mu = (\omega, 0, 0, k^3) = (\omega, 0, 0, \omega), \tag{7.100}$$

where we know that $k^3 = \omega$ because the wave vector is null. In this case, $k^\mu C_{\mu\nu} = 0$ and $C_{0\nu} = 0$ together imply

$$C_{3\nu} = 0. \tag{7.101}$$

The only nonzero components of $C_{\mu\nu}$ are therefore C_{11}, C_{12}, C_{21}, and C_{22}. But $C_{\mu\nu}$ is traceless and symmetric, so in general we can write

$$C_{\mu\nu} = \begin{pmatrix} 0 & 0 & 0 & 0 \\ 0 & C_{11} & C_{12} & 0 \\ 0 & C_{12} & -C_{11} & 0 \\ 0 & 0 & 0 & 0 \end{pmatrix}. \tag{7.102}$$

Thus, for a plane wave in this gauge traveling in the x^3 direction, the two components C_{11} and C_{12} (along with the frequency ω) completely characterize the wave.

To get a feeling for the physical effect of a passing gravitational wave, consider the motion of test particles in the presence of a wave. It is certainly insufficient to solve for the trajectory of a single particle, since that would only tell us about the values of the coordinates along the world line. In fact, for any single particle we can find transverse traceless coordinates in which the particle appears stationary to first order in $h^{\rm TT}_{\mu\nu}$. To obtain a coordinate-independent measure of the wave's effects, we consider the relative motion of nearby particles, as described by the geodesic deviation equation. If we consider some nearby particles with four-velocities described by a single vector field $U^\mu(x)$ and separation vector S^μ, we have

$$\frac{D^2}{d\tau^2}S^\mu = R^\mu{}_{\nu\rho\sigma}U^\nu U^\rho S^\sigma. \tag{7.103}$$

We would like to compute the right-hand side to first order in $h^{\rm TT}_{\mu\nu}$. If we take our test particles to be moving slowly, we can express the four-velocity as a unit vector in the time direction plus corrections of order $h^{\rm TT}_{\mu\nu}$ and higher; but we know that the Riemann tensor is already first order, so the corrections to U^ν may be ignored, and we write

$$U^\nu = (1, 0, 0, 0). \tag{7.104}$$

Therefore we only need to compute $R^\mu{}_{00\sigma}$, or equivalently $R_{\mu 00\sigma}$. From (7.5) we have

$$R_{\mu 00\sigma} = \tfrac{1}{2}(\partial_0\partial_0 h^{\rm TT}_{\mu\sigma} + \partial_\sigma\partial_\mu h^{\rm TT}_{00} - \partial_\sigma\partial_0 h^{\rm TT}_{\mu 0} - \partial_\mu\partial_0 h^{\rm TT}_{\sigma 0}). \tag{7.105}$$

But $h^{\rm TT}_{\mu 0} = 0$, so

$$R_{\mu 00\sigma} = \tfrac{1}{2}\partial_0\partial_0 h^{\rm TT}_{\mu\sigma}. \tag{7.106}$$

Meanwhile, for our slowly-moving particles we have $\tau = x^0 = t$ to lowest order, so the geodesic deviation equation becomes

$$\frac{\partial^2}{\partial t^2}S^\mu = \frac{1}{2}S^\sigma\frac{\partial^2}{\partial t^2}h^{{\rm TT}\mu}{}_\sigma. \tag{7.107}$$

For our wave traveling in the x^3 direction, this implies that only S^1 and S^2 will be affected—the test particles are only disturbed in directions perpendicular to the wave vector. This is of course familiar from electromagnetism, where the electric and magnetic fields in a plane wave are perpendicular to the wave vector.

Our wave is characterized by the two numbers, which for future convenience we will rename as follows:

$$\begin{aligned} h_+ &= C_{11} \\ h_\times &= C_{12}, \end{aligned} \tag{7.108}$$

so that

$$C_{\mu\nu} = \begin{pmatrix} 0 & 0 & 0 & 0 \\ 0 & h_+ & h_\times & 0 \\ 0 & h_\times & -h_+ & 0 \\ 0 & 0 & 0 & 0 \end{pmatrix}. \tag{7.109}$$

Let's consider their effects separately, beginning with the case $h_\times = 0$. Then we have

$$\frac{\partial^2}{\partial t^2} S^1 = \frac{1}{2} S^1 \frac{\partial^2}{\partial t^2} (h_+ e^{ik_\sigma x^\sigma}) \tag{7.110}$$

and

$$\frac{\partial^2}{\partial t^2} S^2 = -\frac{1}{2} S^2 \frac{\partial^2}{\partial t^2} (h_+ e^{ik_\sigma x^\sigma}). \tag{7.111}$$

These can be immediately solved to yield, to lowest order,

$$S^1 = \left(1 + \tfrac{1}{2} h_+ e^{ik_\sigma x^\sigma}\right) S^1(0) \tag{7.112}$$

and

$$S^2 = \left(1 - \tfrac{1}{2} h_+ e^{ik_\sigma x^\sigma}\right) S^2(0). \tag{7.113}$$

Thus, particles initially separated in the x^1 direction will oscillate in the x^1 direction, and likewise for those with an initial x^2 separation. That is, if we start with a ring of stationary particles in the x-y plane, as the wave passes they will bounce back and forth in the shape of a "+," as shown in Figure 7.4. On the other hand, the equivalent analysis for the case where $h_+ = 0$ but $h_\times \neq 0$ would yield the solution

$$S^1 = S^1(0) + \tfrac{1}{2} h_\times e^{ik_\sigma x^\sigma} S^2(0) \tag{7.114}$$

and

$$S^2 = S^2(0) + \tfrac{1}{2} h_\times e^{ik_\sigma x^\sigma} S^1(0). \tag{7.115}$$

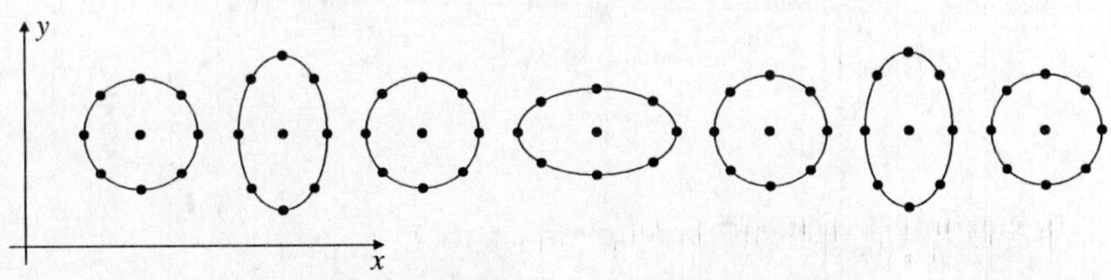

FIGURE 7.4 The effect of a gravitational wave with + polarization is to distort a circle of test particles into ellipses oscillating in a + pattern.

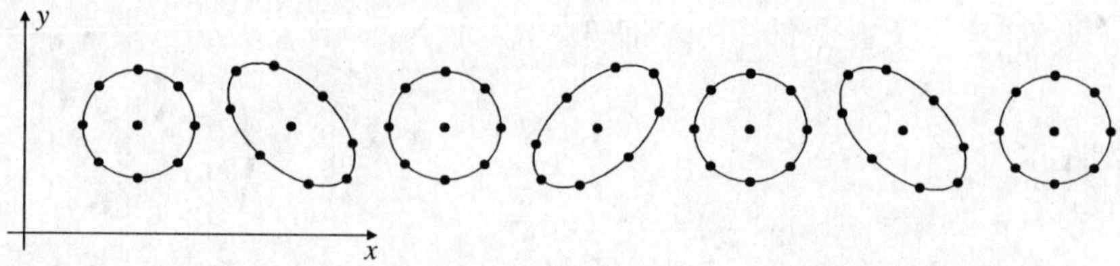

FIGURE 7.5 The effect of a gravitational wave with $\times$ polarization is to distort a circle of test particles into ellipses oscillating in a $\times$ pattern.

In this case the circle of particles would bounce back and forth in the shape of a "$\times$," as shown in Figure 7.5. The notation h_+ and $h_\times$ should therefore be clear. These two quantities measure the two independent modes of linear polarization of the gravitational wave, known as the "plus" and "cross" polarizations. If we liked we could consider right- and left-handed circularly polarized modes by defining

$$\begin{aligned} h_R &= \frac{1}{\sqrt{2}}(h_+ + ih_\times), \\ h_L &= \frac{1}{\sqrt{2}}(h_+ - ih_\times). \end{aligned} \tag{7.116}$$

The effect of a pure h_R wave would be to rotate the particles in a right-handed sense, as shown in Figure 7.6, and similarly for the left-handed mode h_L. Note that the individual particles do not travel around the ring; they just move in little epicycles.

We can relate the polarization states of classical gravitational waves to the kinds of particles we would expect to find upon quantization. The spin of a quantized field is directly related to the transformation properties of that field under spatial rotations. The electromagnetic field has two independent polarization states described by vectors in the x-y plane; equivalently, a single polarization mode is invariant under a rotation by 360° in this plane. Upon quantization this theory yields the photon, a massless spin-1 particle. The neutrino, on the other hand, is also a massless particle, described by a field that picks up a minus sign under rotations by 360°; it is invariant under rotations of 720°, and we say it has

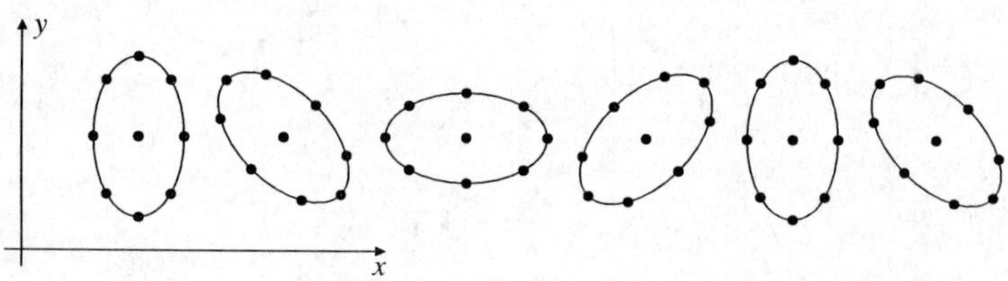

FIGURE 7.6 The effect of a gravitational wave with R polarization is to distort a circle of test particles into an ellipse that rotates in a right-handed sense.

spin-$\frac{1}{2}$. The general rule is that the spin S is related to the angle θ under which the polarization modes are invariant by $S = 360°/\theta$. The gravitational field, whose waves propagate at the speed of light, should lead to massless particles in the quantum theory. Noticing that the polarization modes we have described are invariant under rotations of 180° in the x-y plane, we expect the associated particles—gravitons—to be spin-2. We are a long way from detecting such particles (and it would not be a surprise if we never detected them directly), but any respectable quantum theory of gravity should predict their existence.

In fact, starting with a theory of spin-2 gravitons and requiring some simple properties provides a nice way to *derive* the full Einstein's equation of general relativity. Imagine starting with the Lagrangian (7.9) for the symmetric tensor $h_{\mu\nu}$, but now imagining that this "really is" a physical field propagating in Minkowski spacetime rather than a perturbation to a dynamical metric. (This Lagrangian doesn't include couplings to matter, but it is straightforward to do so.) Now make the additional demand that $h_{\mu\nu}$ couple to its own energy-momentum tensor (discussed below), as well as to the matter energy-momentum tensor. This induces higher-order nonlinear terms in the action, and consequently induces additional "energy-momentum" terms of even higher order. By repeating this process, an infinite series of terms is introduced, but the series can be summed to a simple expression, perhaps because you already know the answer—the Einstein–Hilbert action (possibly with some higher-order terms). In the process, we find that matter couples to the unique combination $g_{\mu\nu} = \eta_{\mu\nu} + h_{\mu\nu}$. In other words, by asking for a theory of a spin-2 field coupling to the energy-momentum tensor, we end up with the fully nonlinear glory of general relativity. The background metric $\eta_{\mu\nu}$ becomes completely unobservable. Of course, some of the global geometric aspects of GR are obscured by this procedure, which ultimately is just another way of justifying Einstein's equation.

While we are noting amusing things, let's point out that the behavior of gravitational waves yields a clue as to why string theory gives rise to a quantum theory of gravity. Consider the fundamental vibrational modes of a loop of string, as shown in Figure 7.7. There are three lowest-energy modes for a loop of string: an over-

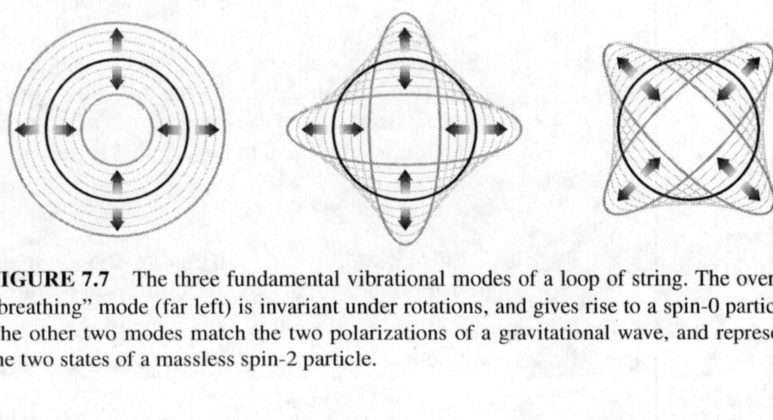

FIGURE 7.7 The three fundamental vibrational modes of a loop of string. The overall "breathing" mode (far left) is invariant under rotations, and gives rise to a spin-0 particle. The other two modes match the two polarizations of a gravitational wave, and represent the two states of a massless spin-2 particle.

all oscillation of its size, plus two independent ways it can oscillate into ellipses. These give rise to three massless degrees of freedom: a spin-0 particle (the dilaton) and a massless spin-2 particle (the graviton). Notice the obvious similarity between the string oscillations and the motion of test particles under the influence of a gravitational wave; this is no accident, and is the reason why quantized strings inevitably give rise to gravity. (String theory was originally investigated as a theory of the strong interactions, but different models would inevitably predict an unnecessary massless spin-2 particle; eventually it was realized that this flaw could be a virtue, if the theory came to be thought of as a quantum theory of gravity.) The extra unwanted spin-0 (scalar) mode reflects the fact that string theory actually predicts a scalar-tensor theory of gravity (as discussed in Section 4.8) rather than ordinary GR. Since a massless scalar of this sort is not observed in nature, some mechanism must work to give a mass to the scalar at low energies.

7.5 ■ PRODUCTION OF GRAVITATIONAL WAVES

With plane-wave solutions to the linearized vacuum equation in our possession, it remains to discuss the generation of gravitational radiation by sources. For this purpose it is necessary to consider Einstein's equation coupled to matter, $G_{\mu\nu} = 8\pi G T_{\mu\nu}$. Because $T_{\mu\nu}$ doesn't vanish, the metric perturbation will include nonzero scalar and vector components as well as the strain tensor representing gravitational waves; we cannot assume that our solution takes the transverse-traceless form (7.90). Instead, we will keep the entire perturbation $h_{\mu\nu}$ and solve for the produced gravitational wave far from the source, where we can then impose transverse-traceless gauge.

There are still some convenient simplifications we can introduce, even in the presence of sources. We first define the trace-reversed perturbation,

$$\bar{h}_{\mu\nu} = h_{\mu\nu} - \tfrac{1}{2} h \eta_{\mu\nu}. \tag{7.117}$$

The name of the trace-reversed perturbation makes sense, since

$$\bar{h} = \eta^{\mu\nu} \bar{h}_{\mu\nu} = -h. \tag{7.118}$$

Obviously we can reconstruct the original perturbation from the trace-reversed form, so no information has been lost. Note also that, if we are in vacuum far away from any sources and can go to transverse-traceless gauge, the trace-reversed perturbation will be equal to the original perturbation,

$$\bar{h}^{\rm TT}_{\mu\nu} = h^{\rm TT}_{\mu\nu}. \tag{7.119}$$

Meanwhile, we are still free to choose some sort of gauge. Under a gauge transformation (7.14), the trace-reversed perturbation transforms as

$$\bar{h}_{\mu\nu} \to \bar{h}_{\mu\nu} + 2\partial_{(\mu}\xi_{\nu)} - \partial_\lambda \xi^\lambda \eta_{\mu\nu}. \tag{7.120}$$

By choosing a gauge parameter ξ_μ satisfying

$$\Box\xi_\mu = -\partial_\lambda \bar{h}^\lambda{}_\mu, \tag{7.121}$$

we can therefore set

$$\partial_\mu \bar{h}^{\mu\nu} = 0. \tag{7.122}$$

This condition is known as the **Lorenz gauge**, analogous with the similar condition $\partial_\mu A^\mu = 0$ often used in electromagnetism.[3] Note that the original perturbation is not transverse in this gauge; rather, we have

$$\partial_\mu h^{\mu\nu} = \tfrac{1}{2}\partial^\nu h. \tag{7.123}$$

Plugging the definition of the trace-reversed perturbation into our expression for the Einstein tensor (7.8), and using the Lorenz gauge condition, yields the very concise expression

$$G_{\mu\nu} = -\tfrac{1}{2}\Box\bar{h}_{\mu\nu}. \tag{7.124}$$

The analogous expression in terms of the original perturbation $h_{\mu\nu}$ is slightly messier; this is the reason for introducing $\bar{h}_{\mu\nu}$. The linearized Einstein equation in this gauge is therefore simply a wave equation for each component,

$$\Box\bar{h}_{\mu\nu} = -16\pi G T_{\mu\nu}. \tag{7.125}$$

The solution to such an equation can be obtained using a Green function, in precisely the same way as the analogous problem in electromagnetism. Here we will review the outline of the method, following Wald (1984).

The Green function $G(x^\sigma - y^\sigma)$ for the d'Alembertian operator $\Box$ is the solution of the wave equation in the presence of a delta-function source:

$$\Box_x G(x^\sigma - y^\sigma) = \delta^{(4)}(x^\sigma - y^\sigma), \tag{7.126}$$

where $\Box_x$ denotes the d'Alembertian with respect to the coordinates x^σ. The usefulness of such a function resides in the fact that the general solution to an equation such as (7.125) can be written

$$\bar{h}_{\mu\nu}(x^\sigma) = -16\pi G \int G(x^\sigma - y^\sigma) T_{\mu\nu}(y^\sigma)\, d^4y, \tag{7.127}$$

as can be verified immediately. (Notice that no factors of $\sqrt{-g}$ are necessary, since our background is simply flat spacetime.) The solutions to (7.126) have of course been worked out long ago, and they can be thought of as either "retarded" or "advanced," depending on whether they represent waves traveling forward or

[3]Note the spelling. The "gauge" was originated by Ludwig Lorenz (1829–1891), while the more famous "transformation" was invented by Hendrick Antoon Lorentz (1853–1928). See J.D. Jackson and L.B. Okun, *Rev. Mod. Phys.* **73**, 663 (2001).

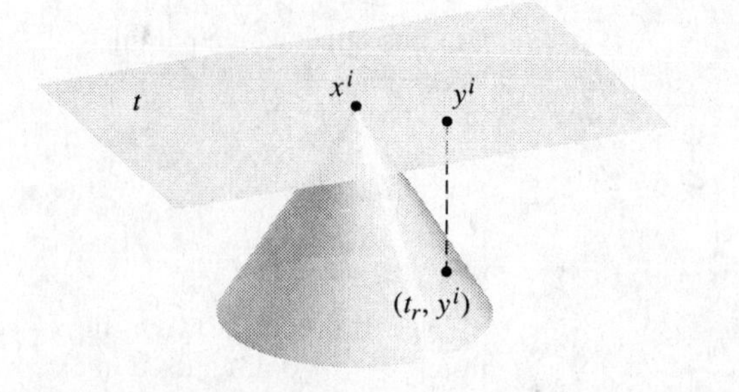

FIGURE 7.8 Disturbances in the gravitational field at (t, x^i) are calculated in terms of events on the past light cone.

backward in time. Our interest is in the retarded Green function, which represents the accumulated effect of signals to the past of the points under consideration. It is given by

$$G(x^\sigma - y^\sigma) = -\frac{1}{4\pi|\mathbf{x}-\mathbf{y}|}\delta[|\mathbf{x}-\mathbf{y}| - (x^0 - y^0)]\,\theta(x^0 - y^0). \tag{7.128}$$

Here we have used boldface to denote the spatial vectors $\mathbf{x} = (x^1, x^2, x^3)$ and $\mathbf{y} = (y^1, y^2, y^3)$, with norm $|\mathbf{x}-\mathbf{y}| = [\delta_{ij}(x^i - y^i)(x^j - y^j)]^{1/2}$. The theta function $\theta(x^0 - y^0)$ equals 1 when $x^0 > y^0$, and zero otherwise. The derivation of (7.128) would take us too far afield, but it can be found in any standard text on electrodynamics or partial differential equations in physics.

Upon plugging (7.128) into (7.127), we can use the delta function to perform the integral over y^0, leaving us with

$$\bar{h}_{\mu\nu}(t, \mathbf{x}) = 4G\int \frac{1}{|\mathbf{x}-\mathbf{y}|}T_{\mu\nu}(t - |\mathbf{x}-\mathbf{y}|, \mathbf{y})\, d^3y, \tag{7.129}$$

where $t = x^0$. The term "retarded time" is used to refer to the quantity

$$t_r = t - |\mathbf{x}-\mathbf{y}|. \tag{7.130}$$

The interpretation of (7.129) should be clear: the disturbance in the gravitational field at $(t, \mathbf{x})$ is a sum of the influences from the energy and momentum sources at the point $(t_r, \mathbf{x}-\mathbf{y})$ on the past light cone, as depicted in Figure 7.8.

Let us take this general solution and consider the case where the gravitational radiation is emitted by an isolated source, fairly far away, comprised of nonrelativistic matter; these approximations will be made more precise as we go on. First we need to set up some conventions for Fourier transforms, which always make life easier when dealing with oscillatory phenomena. Given a function of spacetime $\phi(t, \mathbf{x})$, we are interested in its Fourier transform (and inverse) with respect to time alone,

$$\tilde{\phi}(\omega, \mathbf{x}) = \frac{1}{\sqrt{2\pi}} \int dt\, e^{i\omega t} \phi(t, \mathbf{x}),$$
$$\phi(t, \mathbf{x}) = \frac{1}{\sqrt{2\pi}} \int d\omega\, e^{-i\omega t} \tilde{\phi}(\omega, \mathbf{x}). \tag{7.131}$$

Taking the transform of the metric perturbation, we obtain

$$\begin{aligned}
\tilde{\bar{h}}_{\mu\nu}(\omega, \mathbf{x}) &= \frac{1}{\sqrt{2\pi}} \int dt\, e^{i\omega t} \bar{h}_{\mu\nu}(t, \mathbf{x}) \\
&= \frac{4G}{\sqrt{2\pi}} \int dt\, d^3y\, e^{i\omega t} \frac{T_{\mu\nu}(t - |\mathbf{x}-\mathbf{y}|, \mathbf{y})}{|\mathbf{x}-\mathbf{y}|} \\
&= \frac{4G}{\sqrt{2\pi}} \int dt_r\, d^3y\, e^{i\omega t_r} e^{i\omega|\mathbf{x}-\mathbf{y}|} \frac{T_{\mu\nu}(t_r, \mathbf{y})}{|\mathbf{x}-\mathbf{y}|} \\
&= 4G \int d^3y\, e^{i\omega|\mathbf{x}-\mathbf{y}|} \frac{\tilde{T}_{\mu\nu}(\omega, \mathbf{y})}{|\mathbf{x}-\mathbf{y}|}.
\end{aligned} \tag{7.132}$$

In this sequence, the first equation is simply the definition of the Fourier transform, the second line comes from the solution (7.129), the third line is a change of variables from t to t_r, and the fourth line is once again the definition of the Fourier transform.

We now make the approximations that our source is isolated, far away, and slowly moving. This means that we can consider the source to be centered at a (spatial) distance r, with the different parts of the source at distances $r + \delta r$ such that $\delta r \ll r$, as shown in Figure 7.9. Since it is slowly moving, most of the radiation emitted will be at frequencies ω sufficiently low that $\delta r \ll \omega^{-1}$. (Essentially, light traverses the source much faster than the components of the source itself do.) Under these approximations, the term $e^{i\omega|\mathbf{x}-\mathbf{y}|}/|\mathbf{x}-\mathbf{y}|$ can be replaced by $e^{i\omega r}/r$ and brought outside the integral. This leaves us with

$$\tilde{\bar{h}}_{\mu\nu}(\omega, \mathbf{x}) = 4G \frac{e^{i\omega r}}{r} \int d^3y\, \tilde{T}_{\mu\nu}(\omega, \mathbf{y}). \tag{7.133}$$

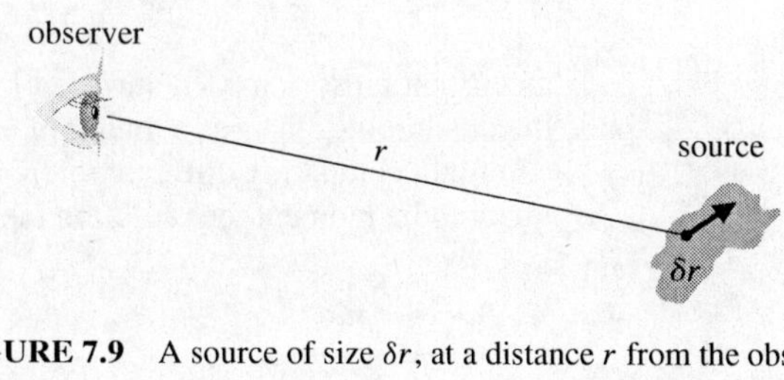

FIGURE 7.9 A source of size δr, at a distance r from the observer.

In fact there is no need to compute all of the components of $\tilde{\bar{h}}_{\mu\nu}(\omega, \mathbf{x})$, since the Lorenz gauge condition $\partial_\mu \bar{h}^{\mu\nu}(t, \mathbf{x}) = 0$ in Fourier space implies

$$\tilde{\bar{h}}^{0\nu} = \frac{i}{\omega}\partial_i \tilde{\bar{h}}^{i\nu}. \tag{7.134}$$

We therefore only need to concern ourselves with the spacelike components of $\tilde{\bar{h}}_{\mu\nu}(\omega, \mathbf{x})$, and recover $\tilde{\bar{h}}^{0\nu}$ from (7.134). The first thing to do is to set $\nu = j$ to find $\tilde{\bar{h}}^{0j}$ from $\tilde{\bar{h}}^{ij}$, which we would then use to find $\tilde{\bar{h}}^{00}$ from $\tilde{\bar{h}}^{i0}$. From (7.133) we therefore want to take the integral of the spacelike components of $\widetilde{T}_{\mu\nu}(\omega, \mathbf{y})$. We begin by integrating by parts in reverse:

$$\int d^3y\, \widetilde{T}^{ij}(\omega, \mathbf{y}) = \int \partial_k(y^i \widetilde{T}^{kj})\, d^3y - \int y^i(\partial_k \widetilde{T}^{kj})\, d^3y. \tag{7.135}$$

The first term is a surface integral which will vanish since the source is isolated, while the second can be related to $\widetilde{T}^{0j}$ by the Fourier-space version of $\partial_\mu T^{\mu\nu} = 0$:

$$-\partial_k \widetilde{T}^{k\mu} = i\omega \widetilde{T}^{0\mu}. \tag{7.136}$$

Thus,

$$\begin{aligned}
\int d^3y\, \widetilde{T}^{ij}(\omega, \mathbf{y}) &= i\omega \int y^i \widetilde{T}^{0j}\, d^3y \\
&= \frac{i\omega}{2}\int (y^i \widetilde{T}^{0j} + y^j \widetilde{T}^{0i})\, d^3y \\
&= \frac{i\omega}{2}\int \left[\partial_l(y^i y^j \widetilde{T}^{0l}) - y^i y^j(\partial_l \widetilde{T}^{0l})\right] d^3y \\
&= -\frac{\omega^2}{2}\int y^i y^j \widetilde{T}^{00}\, d^3y.
\end{aligned} \tag{7.137}$$

The second line is justified since we know that the left-hand side is symmetric in i and j, while the third and fourth lines are simply repetitions of reverse integration by parts and conservation of $T^{\mu\nu}$. It is conventional to define the **quadrupole moment tensor** of the energy density of the source,

$$I_{ij}(t) = \int y^i y^j T^{00}(t, \mathbf{y})\, d^3y, \tag{7.138}$$

a constant tensor on each surface of constant time. The overall normalization of the quadrupole tensor is a matter of convention, and by no means universal, so be careful in comparing different references. In terms of the Fourier transform of the quadrupole moment, our solution takes on the compact form

$$\tilde{\bar{h}}_{ij}(\omega, \mathbf{x}) = -2G\omega^2 \frac{e^{i\omega r}}{r}\tilde{I}_{ij}(\omega). \tag{7.139}$$

We can transform this back to t to obtain the **quadrupole formula,**

$$\bar{h}_{ij}(t, \mathbf{x}) = \frac{2G}{r}\frac{d^2 I_{ij}}{dt^2}(t_r), \tag{7.140}$$

where as before $t_r = t - r$.

The gravitational wave produced by an isolated nonrelativistic object is therefore proportional to the second derivative of the quadrupole moment of the energy density at the point where the past light cone of the observer intersects the source. In contrast, the leading contribution to electromagnetic radiation comes from the changing *dipole* moment of the charge density. The difference can be traced back to the universal nature of gravitation. A changing dipole moment corresponds to motion of the center of density—charge density in the case of electromagnetism, energy density in the case of gravitation. While there is nothing to stop the center of charge of an object from oscillating, oscillation of the center of mass of an isolated system violates conservation of momentum. (You can shake a body up and down, but you and the earth shake ever so slightly in the opposite direction to compensate.) The quadrupole moment, which measures the shape of the system, is generally smaller than the dipole moment, and for this reason, as well as the weak coupling of matter to gravity, gravitational radiation is typically much weaker than electromagnetic radiation.

One case of special interest is the gravitational radiation emitted by a binary star (two stars in orbit around each other). For simplicity let us consider two stars of mass M in a circular orbit in the x^1-x^2 plane, at distance R from their common center of mass, as shown in Figure 7.10. We will treat the motion of the stars in the Newtonian approximation, where we can discuss their orbit just as Kepler would have. Circular orbits are most easily characterized by equating the force due to gravity to the outward "centrifugal" force:

$$\frac{GM^2}{(2R)^2} = \frac{Mv^2}{R}, \tag{7.141}$$

which gives us

$$v = \left(\frac{GM}{4R}\right)^{1/2}. \tag{7.142}$$

The time it takes to complete a single orbit is simply

$$T = \frac{2\pi R}{v}, \tag{7.143}$$

but more useful to us is the angular frequency of the orbit,

$$\Omega = \frac{2\pi}{T} = \left(\frac{GM}{4R^3}\right)^{1/2}. \tag{7.144}$$

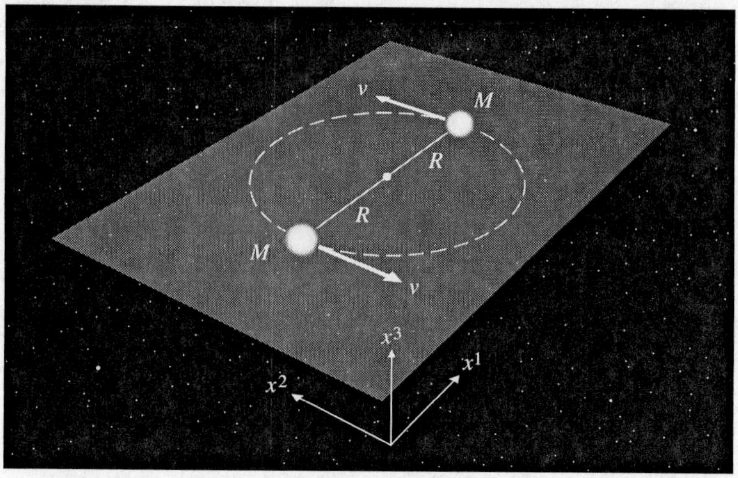

FIGURE 7.10 A binary star system. Two stars of mass M orbit in the x^1-x^2 plane with an orbital radius R.

In terms of Ω we can write down the explicit path of star a,

$$x_a^1 = R\cos\Omega t, \qquad x_a^2 = R\sin\Omega t, \tag{7.145}$$

and star b,

$$x_b^1 = -R\cos\Omega t, \qquad x_b^2 = -R\sin\Omega t. \tag{7.146}$$

The corresponding energy density is

$$\begin{aligned} T^{00}(t,\mathbf{x}) = M\delta(x^3)\big[&\delta(x^1 - R\cos\Omega t)\delta(x^2 - R\sin\Omega t) \\ &+ \delta(x^1 + R\cos\Omega t)\delta(x^2 + R\sin\Omega t)\big]. \end{aligned} \tag{7.147}$$

The profusion of delta functions allows us to integrate this straightforwardly to obtain the quadrupole moment from (7.138):

$$\begin{aligned} I_{11} &= 2MR^2\cos^2\Omega t = MR^2(1+\cos 2\Omega t) \\ I_{22} &= 2MR^2\sin^2\Omega t = MR^2(1-\cos 2\Omega t) \\ I_{12} = I_{21} &= 2MR^2(\cos\Omega t)(\sin\Omega t) = MR^2\sin 2\Omega t \\ I_{i3} &= 0. \end{aligned} \tag{7.148}$$

From this in turn it is easy to get the components of the metric perturbation from (7.140):

$$\bar{h}_{ij}(t,\mathbf{x}) = \frac{8GM}{r}\Omega^2 R^2 \begin{pmatrix} -\cos 2\Omega t_r & -\sin 2\Omega t_r & 0 \\ -\sin 2\Omega t_r & \cos 2\Omega t_r & 0 \\ 0 & 0 & 0 \end{pmatrix}. \tag{7.149}$$

The remaining components of $\bar{h}_{\mu\nu}$ could be derived from demanding that the Lorenz gauge condition be satisfied.

7.6 ■ ENERGY LOSS DUE TO GRAVITATIONAL RADIATION

It is natural at this point to talk about the energy emitted via gravitational radiation. Such a discussion, however, is immediately beset by problems, both technical and philosophical. As we have mentioned before, there is no true local measure of the energy in the gravitational field. Of course, in the weak field limit, where we think of gravitation as being described by a symmetric tensor propagating on a fixed background metric, we might hope to derive an energy-momentum tensor for the fluctuations $h_{\mu\nu}$, just as we would for electromagnetism or any other field theory. To some extent this is possible, but still difficult. As a result of these difficulties there are a number of different proposals in the literature for what we should use as the energy-momentum tensor for gravitation in the weak field limit; all of them are different, but for the most part they give the same answers for physically well-posed questions such as the rate of energy emitted by a binary system.

At a technical level, the difficulties begin to arise when we consider what form the energy-momentum tensor should take. We have previously mentioned the energy-momentum tensors for electromagnetism and scalar field theory, both of which share an important feature—they are quadratic in the relevant fields. By hypothesis, our approach to the weak field limit has been to keep only terms that are linear in the metric perturbation. Hence, in order to keep track of the energy carried by the gravitational waves, we will have to extend our calculations to at least second order in $h_{\mu\nu}$. In fact we have been cheating slightly all along. In discussing the effects of gravitational waves on test particles, and the generation of waves by a binary system, we have been using the fact that test particles move along geodesics. But as we know, this is derived from the covariant conservation of energy-momentum, $\nabla_\mu T^{\mu\nu} = 0$. In the order to which we have been working, however, we actually have $\partial_\mu T^{\mu\nu} = 0$, which would imply that test particles move on straight lines in the flat background metric. This is a symptom of the inability of the weak field limit to describe self-gravitating systems. In practice, the best that can be done is to solve the weak field equation to some appropriate order, and then justify after the fact the validity of the solution. We will follow the procedure outlined in Chapters 35 and 36 of Misner, Thorne, and Wheeler (1973), where additional discussion of subtleties may be found. See also Wald (1984) and Schutz (1985).

Let us now examine Einstein's vacuum equation $R_{\mu\nu} = 0$ to second order, and see how the result can be interpreted in terms of an energy-momentum tensor for the gravitational field. We expand both the metric and the Ricci tensor,

$$g_{\mu\nu} = \eta_{\mu\nu} + h^{(1)}_{\mu\nu} + h^{(2)}_{\mu\nu}$$
$$R_{\mu\nu} = R^{(0)}_{\mu\nu} + R^{(1)}_{\mu\nu} + R^{(2)}_{\mu\nu}, \tag{7.150}$$

where $R^{(1)}_{\mu\nu}$ is taken to be of the same order as $h^{(1)}_{\mu\nu}$, while $R^{(2)}_{\mu\nu}$ and $h^{(2)}_{\mu\nu}$ are of order $(h^{(1)}_{\mu\nu})^2$. Because we work in a flat background, the zeroth-order equation $R^{(0)}_{\mu\nu} = 0$ is automatically solved. The first-order vacuum equation is simply

$$R^{(1)}_{\mu\nu}[h^{(1)}] = 0, \tag{7.151}$$

which determines the first-order perturbation $h^{(1)}_{\mu\nu}$ (up to gauge transformations). The second-order perturbation $h^{(2)}_{\mu\nu}$ will be determined by the second-order equation

$$R^{(1)}_{\mu\nu}[h^{(2)}] + R^{(2)}_{\mu\nu}[h^{(1)}] = 0. \tag{7.152}$$

The notation $R^{(1)}_{\mu\nu}[h^{(2)}]$ indicates the parts of the expanded Ricci tensor that are linear in the metric perturbation, as given by (7.6), applied to the second-order perturbation $h^{(2)}_{\mu\nu}$; meanwhile, $R^{(2)}_{\mu\nu}[h^{(1)}]$ stands for the quadratic part of the expanded Ricci tensor,

$$\begin{aligned} R^{(2)}_{\mu\nu} = {} & \tfrac{1}{2}h^{\rho\sigma}\partial_\mu\partial_\nu h_{\rho\sigma} + \tfrac{1}{4}(\partial_\mu h_{\rho\sigma})\partial_\nu h^{\rho\sigma} + (\partial^\sigma h^\rho{}_\nu)\partial_{[\sigma}h_{\rho]\mu} - h^{\rho\sigma}\partial_\rho\partial_{(\mu}h_{\nu)\sigma} \\ & + \tfrac{1}{2}\partial_\sigma(h^{\rho\sigma}\partial_\rho h_{\mu\nu}) - \tfrac{1}{4}(\partial_\rho h_{\mu\nu})\partial^\rho h - (\partial_\sigma h^{\rho\sigma} - \tfrac{1}{2}\partial^\rho h)\partial_{(\mu}h_{\nu)\rho}, \end{aligned} \tag{7.153}$$

applied to the first-order perturbation $h^{(1)}_{\mu\nu}$. There are no cross terms, as they would necessarily be higher order.

Now let's write the vacuum equation as $G_{\mu\nu} = 0$; this is of course equivalent to $R_{\mu\nu} = 0$, but will enable us to express the result in a suggestive form. At second order we have

$$R^{(1)}_{\mu\nu}[h^{(2)}] - \tfrac{1}{2}\eta^{\rho\sigma}R^{(1)}_{\rho\sigma}[h^{(2)}]\eta_{\mu\nu} = 8\pi G t_{\mu\nu}, \tag{7.154}$$

where we have defined

$$t_{\mu\nu} \equiv -\frac{1}{8\pi G}\left\{R^{(2)}_{\mu\nu}[h^{(1)}] - \frac{1}{2}\eta^{\rho\sigma}R^{(2)}_{\rho\sigma}[h^{(1)}]\eta_{\mu\nu}\right\}. \tag{7.155}$$

Notice a couple of things about this expression. First, we have not included terms of the form $h^{(1)\rho\sigma}R^{(1)}_{\rho\sigma}[h^{(1)}]$, since $R^{(1)}_{\mu\nu}[h^{(1)}] = 0$. Second, the left-hand side of (7.154) is not the full second-order Einstein tensor, as we have moved terms involving $R^{(2)}_{\mu\nu}[h^{(1)}]$ to the right-hand side and provocatively relabeled them as an energy-momentum tensor for the first-order perturbations, $t_{\mu\nu}$. Such an identification seems eminently reasonable; $t_{\mu\nu}$ is a symmetric tensor, quadratic in $h_{\mu\nu}$, which represents how the perturbations affect the spacetime metric in just the way that a matter energy-momentum tensor would. (Linear terms in $h_{\mu\nu}$ have no effect, since $G^{(1)}_{\mu\nu}[h^{(1)}]$ is simply set to zero by the first-order equation.) Notice that $t_{\mu\nu}$ is also conserved, in the background flat-space sense,

$$\partial_\mu t^{\mu\nu} = 0, \tag{7.156}$$

which we know from the Bianchi identity $\partial_\mu G^{\mu\nu} = 0$.

Unfortunately there are some limitations on our interpretation of $t_{\mu\nu}$ as an energy-momentum tensor. Of course it is not a tensor at all in the full theory, but we are leaving that aside by hypothesis. More importantly, it is not invariant under gauge transformations (infinitesimal diffeomorphisms), as you could check by direct calculation. One way of circumventing this difficulty is to average the energy-momentum tensor over several wavelengths, an operation we denote by angle brackets $\langle \cdots \rangle$. This procedure has both philosophical and practical advantages. From a philosophical viewpoint, we know that our ability to choose Riemann normal coordinates at any one point makes it impossible to define a reliable measure of the gravitational energy-momentum that is purely local (defined at each point in terms of the metric and its first derivatives at precisely that point). If we average over several wavelengths, however, we may hope to capture enough of the physical curvature in a small region to describe a gauge-invariant measure. From a practical standpoint, any terms that are derivatives (as opposed to products of derivatives) will average to zero,

$$\langle \partial_\mu (X) \rangle = 0. \tag{7.157}$$

We are therefore empowered to integrate by parts under the averaging brackets,

$$\langle A(\partial_\mu B) \rangle = -\langle (\partial_\mu A) B \rangle, \tag{7.158}$$

which will greatly simplify our expressions.

With this in mind, let us calculate $t_{\mu\nu}$ as defined in (7.155), using the expression (7.153) for the second-order Ricci tensor. (Henceforth we will no longer use superscripts on the metric perturbation, as we will only be interested in the first-order perturbation.) Although part of the motivation for averaging is to obtain a gauge-invariant answer, the actual calculation is a mess, so for illustrative purposes we will carry it out in transverse-traceless gauge,

$$\partial^\mu h^{\rm TT}_{\mu\nu} = 0, \quad h^{\rm TT} = 0. \tag{7.159}$$

Don't forget that we are only allowed to choose this gauge in vacuum. The non-vanishing parts of $R^{(2)\rm TT}_{\mu\nu}$ in this gauge can be written as

$$\begin{aligned} R^{(2)\rm TT}_{\mu\nu} = {} & \tfrac{1}{2} h^{\rho\sigma}_{\rm TT} \partial_\mu \partial_\nu h^{\rm TT}_{\rho\sigma} + \tfrac{1}{4} (\partial_\mu h^{\rm TT}_{\rho\sigma}) \partial_\nu h^{\rho\sigma}_{\rm TT} + \tfrac{1}{2} \eta^{\rho\lambda} (\partial^\sigma h^{\rm TT}_{\rho\nu}) \partial_\sigma h^{\rm TT}_{\lambda\mu} \\ & - \tfrac{1}{2} (\partial^\sigma h^{\rm TT}_{\rho\nu}) \partial^\rho h^{\rm TT}_{\sigma\mu} - h^{\rho\sigma}_{\rm TT} \partial_\rho \partial_{(\mu} h^{\rm TT}_{\nu)\sigma} + \tfrac{1}{2} h^{\rho\sigma}_{\rm TT} \partial_\sigma \partial_\rho h^{\rm TT}_{\mu\nu}. \end{aligned} \tag{7.160}$$

Now let's apply the averaging brackets, and integrate by parts where convenient. The last three terms in (7.160) all go away, as integration by parts leads to divergences that vanish. We are left with

$$\left\langle R^{(2)\mathrm{TT}}_{\mu\nu}\right\rangle = -\tfrac{1}{4}\left\langle(\partial_\mu h^{\mathrm{TT}}_{\rho\sigma})(\partial_\nu h^{\rho\sigma}_{\mathrm{TT}}) + 2\eta^{\rho\lambda}(\Box h^{\mathrm{TT}}_{\rho\nu})h^{\mathrm{TT}}_{\lambda\mu}\right\rangle. \tag{7.161}$$

But the perturbation obeys the first-order equation of motion, which sets $\Box h^{\mathrm{TT}}_{\mu\nu} = 0$. So we are finally left with

$$\left\langle R^{(2)\mathrm{TT}}_{\mu\nu}\right\rangle = -\tfrac{1}{4}\left\langle(\partial_\mu h^{\mathrm{TT}}_{\rho\sigma})(\partial_\nu h^{\rho\sigma}_{\mathrm{TT}})\right\rangle. \tag{7.162}$$

We can take the trace to get the curvature scalar; after integration by parts we again find a $\Box h^{\mathrm{TT}}_{\mu\nu}$ term which we set to zero, so

$$\left\langle \eta^{\mu\nu} R^{(2)\mathrm{TT}}_{\mu\nu}\right\rangle = 0. \tag{7.163}$$

These expressions can be inserted into (7.155) to obtain a simple expression for the gravitational-wave energy-momentum tensor in transverse-traceless gauge:

$$\boxed{t_{\mu\nu} = \frac{1}{32\pi G}\left\langle(\partial_\mu h^{\mathrm{TT}}_{\rho\sigma})(\partial_\nu h^{\rho\sigma}_{\mathrm{TT}})\right\rangle.} \tag{7.164}$$

Remember that, in this gauge, nonspatial components vanish, $h^{\mathrm{TT}}_{0\nu} = 0$. You will therefore sometimes see the above expression written with spatial indices ij instead of spacetime indices $\rho\sigma$; the two versions are clearly equivalent. If we had been strong enough to do the corresponding calculation without first choosing a gauge, we would have found

$$t_{\mu\nu} = \frac{1}{32\pi G}\Big\langle(\partial_\mu h_{\rho\sigma})(\partial_\nu h^{\rho\sigma}) - \frac{1}{2}(\partial_\mu h)(\partial_\nu h) - (\partial_\rho h^{\rho\sigma})(\partial_\mu h_{\nu\sigma}) - (\partial_\rho h^{\rho\sigma})(\partial_\nu h_{\mu\sigma})\Big\rangle. \tag{7.165}$$

A bit of straightforward manipulation suffices to check that this expression is actually gauge invariant, as you are asked to show in the exercises.

Let's calculate the transverse-traceless expression (7.164) for a single plane wave,

$$h^{\mathrm{TT}}_{\mu\nu} = C_{\mu\nu}\sin(k_\lambda x^\lambda). \tag{7.166}$$

We have taken the real part and set the phase arbitrarily so that the wave is a sine rather than cosine. The energy-momentum tensor is then

$$t_{\mu\nu} = \frac{1}{32\pi G}k_\mu k_\nu C_{\rho\sigma}C^{\rho\sigma}\left\langle\cos^2(k_\lambda x^\lambda)\right\rangle. \tag{7.167}$$

Averaging the $\cos^2$ term over several wavelengths yields

$$\left\langle \cos^2(k_\lambda x^\lambda) \right\rangle = \tfrac{1}{2}. \tag{7.168}$$

For simplicity we can take the wave to be moving along the z-axis, so that

$$k_\lambda = (-\omega, 0, 0, \omega) \tag{7.169}$$

the minus sign coming from lowering an index on k^λ, and from (7.109),

$$C_{\rho\sigma} C^{\rho\sigma} = 2(h_+^2 + h_\times^2). \tag{7.170}$$

It is more common in the gravitational-wave literature to express observables in terms of the ordinary frequency $f = \omega/2\pi$, rather than the angular frequency ω. Putting it all together reveals

$$t_{\mu\nu} = \frac{\pi}{8G} f^2 (h_+^2 + h_\times^2) \begin{pmatrix} 1 & 0 & 0 & -1 \\ 0 & 0 & 0 & 0 \\ 0 & 0 & 0 & 0 \\ -1 & 0 & 0 & 1 \end{pmatrix}. \tag{7.171}$$

As we will discuss in the next section, typical gravitational-wave sources we might expect to observe at Earth will have frequencies between 10^{-4} and 10^4 Hz, and amplitudes $h \sim 10^{-22}$. It is therefore useful to express the energy flux in the z direction, $-T_{0z}$, at an order-of-magnitude level as

$$-T_{0z} \sim 10^{-4} \left(\frac{f}{\text{Hz}}\right)^2 \frac{(h_+^2 + h_\times^2)}{(10^{-21})^2} \frac{\text{erg}}{\text{cm}^2 \cdot \text{s}}. \tag{7.172}$$

This is the amount of energy that could in principle be deposited in each square centimeter of a detector every second. As pointed out by Thorne,[4] this is actually a substantial energy flux, especially at the upper end of the frequency range. For comparison purposes, a supernova at cosmological distances is characterized by a peak electromagnetic flux of approximately 10^{-9} erg/cm^2/s; the gravitational-wave signal, however, only lasts for milliseconds, while the visible electromagnetic signal extends for months.

Now let's use our formula for the gravitational-wave energy-momentum tensor to calculate the rate of energy loss from a system emitting gravitational radiation according to the quadrupole formula (7.140). The total energy contained in gravitational radiation on a surface Σ of constant time is defined as

$$E = \int_\Sigma t_{00}\, d^3x, \tag{7.173}$$

while the total energy radiated through to infinity may be expressed as

[4]K.S. Thorne, in *Three Hundred Years of Gravitation*, Cambridge: Cambridge University Press, 1987.

$$\Delta E = \int P\, dt, \tag{7.174}$$

where the power P is

$$P = \int_{S^2_\infty} t_{0\mu} n^\mu r^2 d\Omega. \tag{7.175}$$

Here, the integral is taken over a two-sphere at spatial infinity S^2_∞, and n^μ is a unit spacelike vector normal to S^2_∞. In polar coordinates $\{t, r, \theta, \phi\}$, the components of the normal vector are

$$n^\mu = (0, 1, 0, 0). \tag{7.176}$$

We would like to calculate the power P using our expression for $t_{\mu\nu}$, (7.164). The first issue we face is that this expression is written in terms of the transverse-traceless perturbation, while the quadrupole formula (7.140) is written in terms of the spatial components $\bar{h}_{ij}$ of the Lorenz-gauge trace-reversed perturbation. The simplest procedure (although it's not that simple) is to first convert $\bar{h}_{ij}$ into transverse-traceless gauge, which is permissible because we are interested in the behavior of the waves in vacuum, far from the source from which they are emitted, plug into the formula for $t_{\mu\nu}$, then convert back into nontransverse-traceless form. Let's see how this works.

We begin by introducing the (spatial) projection tensor

$$P_{ij} = \delta_{ij} - n_i n_j, \tag{7.177}$$

which projects tensor components into a surface orthogonal to the unit vector n^i. (See Appendix D for more discussion.) In our case, we choose n^i to point along the direction of propagation of the wave, so that P_{ij} will project onto the two-sphere at spatial infinity. We can use the projection tensor to construct the transverse-traceless version of a symmetric spatial tensor X_{ij} via

$$X^{\rm TT}_{ij} = \left(P_i{}^k P_j{}^l - \tfrac{1}{2} P_{ij} P^{kl}\right) X_{kl}. \tag{7.178}$$

You can check for yourself that $X^{\rm TT}_{ij}$ is indeed transverse and traceless. Because it is traceless, $\bar{h}^{\rm TT}_{ij}$ is equal to the original perturbation $h^{\rm TT}_{ij}$; plugging into the quadrupole formula (7.140), we get

$$h^{\rm TT}_{ij} = \bar{h}^{\rm TT}_{ij} = \frac{2G}{r} \frac{d^2 I^{\rm TT}_{ij}}{dt^2}(t - r), \tag{7.179}$$

where the transverse-traceless part of the quadrupole moment is also constructed via (7.178). In fact the quadrupole moment defined by (7.138) is not the most convenient quantity to use in expressing the generated wave, as it involves an integral over the energy density that might be difficult to determine. Instead we can use the **reduced quadrupole moment,**

$$J_{ij} = I_{ij} - \tfrac{1}{3}\delta_{ij}\delta^{kl}I_{kl}, \tag{7.180}$$

which is just the traceless part of I_{ij}. The reduced quadrupole moment has the nice property of being the coefficient of the r^{-3} term in the multipole expansion of the Newtonian potential,

$$\Phi = -\frac{GM}{r} - \frac{G}{r^3}D_i x^i - \frac{3G}{2r^5}J_{ij}x^i x^j + \cdots, \tag{7.181}$$

and is therefore more readily approximated for realistic sources. (Here D_i is the dipole moment, $D_i = \int T^{00}x^i\, d^3x$.) Of course, the transverse-traceless part of the quadrupole moment is the same as the transverse-traceless part of the reduced (that is, traceless) quadrupole moment, so (7.179) becomes

$$h_{ij}^{\rm TT} = \frac{2G}{r}\frac{d^2 J_{ij}^{\rm TT}}{dt^2}(t-r). \tag{7.182}$$

To calculate the power, we are interested in $t_{0\mu}n^\mu = t_{0r}$. Because the quadrupole moment depends only on the retarded time $t_r = t - r$, we have

$$\begin{aligned}
\partial_0 h_{ij}^{\rm TT} &= \frac{2G}{r}\frac{d^3 J_{ij}^{\rm TT}}{dt^3}, \\
\partial_r h_{ij}^{\rm TT} &= -\frac{2G}{r}\frac{d^3 J_{ij}^{\rm TT}}{dt^3} - \frac{2G}{r^2}\frac{d^2 J_{ij}^{\rm TT}}{dt^2} \\
&\approx -\frac{2G}{r}\frac{d^3 J_{ij}^{\rm TT}}{dt^3},
\end{aligned} \tag{7.183}$$

where we have dropped the r^{-2} term because we are interested in the $r \to \infty$ limit. The important component of the energy-momentum tensor is therefore

$$t_{0r} = -\frac{G}{8\pi r^2}\left\langle\left(\frac{d^3 J_{ij}^{\rm TT}}{dt^3}\right)\left(\frac{d^3 J^{ij}_{\rm TT}}{dt^3}\right)\right\rangle. \tag{7.184}$$

The next step is to convert back to J_{ij} from the transverse-traceless part. Applying (7.178) and some messy algebra, it is straightforward to show that

$$X_{ij}^{\rm TT}X^{ij}_{\rm TT} = X_{ij}X^{ij} - 2X_i{}^j X^{ik}n_j n_k + \tfrac{1}{2}X^{ij}X^{kl}n_i n_j n_k n_l - \tfrac{1}{2}X^2 + XX^{ij}n_i n_j, \tag{7.185}$$

where $X = \delta^{ij}X_{ij}$. Because J_{ij} is traceless, we have

$$J_{ij}^{\rm TT}J^{ij}_{\rm TT} = J_{ij}J^{ij} - 2J_i{}^j J^{ik}n_j n_k + \tfrac{1}{2}J^{ij}J^{kl}n_i n_j n_k n_l, \tag{7.186}$$

and the power is

$$P = -\frac{G}{8\pi}\int_{S^2_\infty}\left\langle \frac{d^3 J_{ij}}{dt^3}\frac{d^3 J^{ij}}{dt^3} - 2\frac{d^3 J_i{}^j}{dt^3}\frac{d^3 J^{ik}}{dt^3}n_j n_k + \frac{1}{2}\frac{d^3 J^{ij}}{dt^3}\frac{d^3 J^{kl}}{dt^3}n_i n_j n_k n_l\right\rangle d\Omega. \tag{7.187}$$

To evaluate this expression, it is best to switch back to Cartesian coordinates in space, where $n^i = x^i/r$. The quadrupole tensors are independent of the angular coordinates, since they are defined by integrals over all of space. We may therefore pull them outside the integral, and use the identities

$$\int d\Omega = 4\pi$$
$$\int n_i n_j \, d\Omega = \frac{4\pi}{3}\delta_{ij}$$
$$\int n_i n_j n_k n_l \, d\Omega = \frac{4\pi}{15}(\delta_{ij}\delta_{kl} + \delta_{ik}\delta_{jl} + \delta_{il}\delta_{jk}). \tag{7.188}$$

When all is said and done, the expression for the power collapses to

$$P = -\frac{G}{5}\left\langle \frac{d^3 J_{ij}}{dt^3}\frac{d^3 J^{ij}}{dt^3}\right\rangle, \tag{7.189}$$

where we should remember that the quadrupole moment is evaluated at the retarded time $t_r = t - r$. Our formula has a minus sign because it represents the rate at which the energy is changing, and radiating sources will be losing energy.

For the binary system represented by (7.148), the reduced quadrupole moment is

$$J_{ij} = \frac{MR^2}{3}\begin{pmatrix} (1+3\cos 2\Omega t) & 3\sin 2\Omega t & 0 \\ 3\sin 2\Omega t & (1-3\cos 2\Omega t) & 0 \\ 0 & 0 & -2 \end{pmatrix}, \tag{7.190}$$

and its third time derivative is therefore

$$\frac{d^3 J_{ij}}{dt^3} = 8MR^2\Omega^3\begin{pmatrix} \sin 2\Omega t & -\cos 2\Omega t & 0 \\ -\cos 2\Omega t & -\sin 2\Omega t & 0 \\ 0 & 0 & 0 \end{pmatrix}. \tag{7.191}$$

The power radiated by the binary is thus

$$P = -\frac{128}{5}GM^2R^4\Omega^6, \tag{7.192}$$

or, using expression (7.144) for the frequency,

$$P = -\frac{2}{5}\frac{G^4M^5}{R^5}. \tag{7.193}$$

Of course, energy loss through the emission of gravitational radiation has been observed. In 1974 Hulse and Taylor discovered a binary system, PSR1913+16, in which both stars are very small, so classical effects are negligible, or at least under control, and one is a pulsar. The period of the orbit is eight hours, extremely small by astrophysical standards. The fact that one of the stars is a pulsar provides a very accurate clock, with respect to which the change in the period as the system loses energy can be measured. The result is consistent with the prediction of general relativity for energy loss through gravitational radiation.

7.7 ■ DETECTION OF GRAVITATIONAL WAVES

One of the highest-priority goals of contemporary gravitational physics and astrophysics is to detect gravitational radiation directly. (By direct we mean "by observing the influence of the gravitational wave on test bodies," in contrast to observing the indirect effect of energy loss, as in the binary pulsar.) There is every reason to believe that such a detection will happen soon, either in already-existing gravitational-wave observatories or those being planned for the near future. Once we detect gravitational radiation, of course, the goal will immediately become to extract useful astrophysical information from the observations. Our current understanding of the universe outside the Solar System comes almost exclusively from observations of electromagnetic radiation, with some additional input from neutrinos and cosmic rays; the advent of gravitational-wave astrophysics will open an entirely new window onto energetic phenomena in the distant universe.[5]

Before discussing how we might go about detecting astrophysical gravitational waves, we should think about what sources are likely to be most readily observable. The first important realization is that the necessary conditions for the generation of appreciable gravitational radiation are very different from those for electromagnetic radiation. The difference can be traced to the fact that gravitational waves are produced by the bulk motion of large masses, while electromagnetic waves are produced (typically) by incoherent excitations of individual particles. Electromagnetic radiation can therefore be produced by a source that is static in bulk, such as a star, which is a substantial advantage to the astronomer. However, gravitational waves are produced coherently by large moving masses (every particle in the mass contributes in the same sense to the wave), which can partially compensate for the impossibility of emission from static sources.

We therefore need massive sources with substantial bulk motions. As a simple example, consider the binary system of Section 7.5, in which both stars have mass M and the orbital radius is R. We will cheat somewhat by applying the Newtonian formulae for the orbital parameters in a regime where GR has begun to become important, but this will suffice for an order-of-magnitude estimate. The relevant parameters can be distilled down to the Schwarzschild radius $R_S = 2GM/c^2$,

[5]For an overview of gravitational-wave astrophysics, see S.A. Hughes, S. Márka, P.L. Bender, and C.J. Hogan, "New physics and astronomy with the new gravitational-wave observatories," `http://arxiv.org/astro-ph/0110349`.

the orbital radius R, and the distance r between us and the binary. (We will now restore explicit factors of c, to facilitate comparison with experiment.) In terms of these, the frequency of the orbit and thus of the produced gravitational waves is approximately

$$f = \frac{\Omega}{2\pi} \sim \frac{cR_S^{1/2}}{10R^{3/2}}. \tag{7.194}$$

From the formula (7.149) for the resulting perturbation, we can estimate the gravitational-wave amplitude received as

$$h \sim \frac{R_S^2}{rR}. \tag{7.195}$$

Let's see what this implies for the kind of source we might hope to observe. A paradigmatic example is the coalescence of a black-hole/black-hole binary. For typical parameters we can take both black holes to be 10 solar masses, the binary to be at cosmological distances $\sim$ 100 Mpc, and the components to be separated by ten times their Schwarzschild radii:

$$\begin{aligned} R_S &\sim 10^6 \text{ cm} \\ R &\sim 10^7 \text{ cm} \\ r &\sim 10^{26} \text{ cm}. \end{aligned} \tag{7.196}$$

Such a source is thus characterized by

$$f \sim 10^2 \text{ s}^{-1}, \qquad h \sim 10^{-21}. \tag{7.197}$$

If we are to have any hope of detecting the coalescence of a binary with these parameters, we need to be sensitive to frequencies near 100 Hz and strains of order 10^{-21} or less.

Fortunately, these parameters are within the reach of our experimental capabilities (with the heroic efforts of many scientists). The most promising technique for gravitational-wave detection currently under consideration is interferometry, and here we will stick exclusively to a discussion of interferometers, although it is certainly conceivable that a new technology could be invented that would have better sensitivity.

Recall that the physical effect of a passing gravitational wave is to slightly perturb the relative positions of freely-falling masses. If two test masses are separated by a distance L, the change in their distance will be roughly

$$\frac{\delta L}{L} \sim h. \tag{7.198}$$

Imagine that we contemplate building an observatory with test bodies separated by some distance of order kilometers. Then to detect a wave with amplitude of order $h \sim 10^{-21}$ would require a sensitivity to changes of

$$\delta L \sim 10^{-16} \left(\frac{h}{10^{-21}}\right)\left(\frac{L}{\text{km}}\right) \text{ cm.} \tag{7.199}$$

Compare this to the size of a typical atom, set by the Bohr radius,

$$a_0 \sim 5 \times 10^{-9} \text{ cm,} \tag{7.200}$$

or for that matter the size of a typical nucleus, of approximately a Fermi,

$$1 \text{ fm} = 10^{-13} \text{ cm.} \tag{7.201}$$

The point we are belaboring here is that a feasible terrestrial gravitational-wave observatory will have to be sensitive to changes in distance much smaller than the size of the constituent atoms out of which any conceivable test masses would have to be made.

Laser interferometers provide a way to overcome the difficulty of measuring such miniscule perturbations. Consider the schematic set-up portrayed in Figure 7.11. A laser (typically with characteristic wavelength $\lambda \sim 10^{-4}$ cm) is directed at a beamsplitter, which sends the photons down two evacuated tubes of length L. At the ends of the cavities are test masses, represented by mirrors suspended from pendulums. The light actually bounces off partially-reflective mirrors near the beamsplitter, so that a typical photon travels up and down the cavity

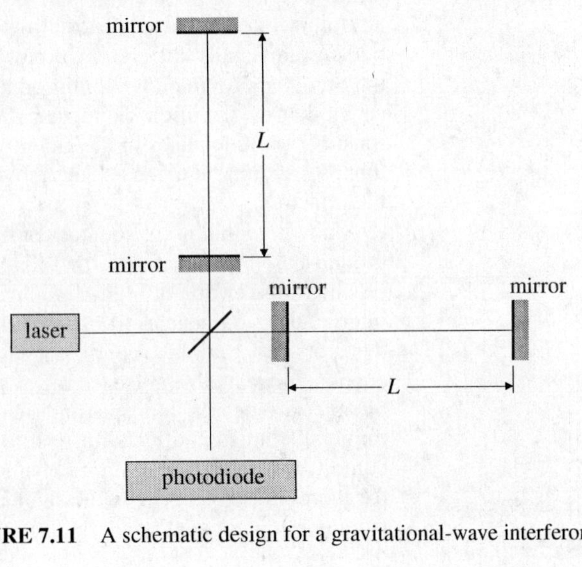

FIGURE 7.11 A schematic design for a gravitational-wave interferometer.

of order 100 times before returning to the beamsplitter and being directed into a photodiode. The system is arranged such that, if the test masses are perfectly stationary, the returning beams destructively interfere, sending no signal to the photodiode. As we have seen, the effect of a passing gravitational wave will be to perturb orthogonal lengths in opposite senses, leading to a phase shift in the laser pulse that will disturb the destructive interference. During 100 round trips through the cavity arms, the accumulated phase shift will be

$$\delta\phi \sim 200\left(\frac{2\pi}{\lambda}\right)\delta L \sim 10^{-9}, \tag{7.202}$$

where 200 rather than 100 represents the fact that the shifts in the two arms add together. Such a tiny shift can be measured if the number of photons N is sufficiently large to overcome the "shot noise"; in particular, if $\sqrt{N} > \delta\phi$.

The technological challenges associated with building sufficiently quiet and sensitive gravitational-wave observatories are being tackled in a number of different locations, including the United States (LIGO), Italy (Virgo), Germany (GEO), Japan (TAMA), and Australia (ACIGA). LIGO (Laser Interferometric Gravitational-Wave Observatory) is presently the most advanced detector; it consists of two facilities (one in Washington state and one in Louisiana), each with four-kilometer arms. A single gravitational-wave observatory will be unable to localize a source's position on the sky; multiple detectors will be crucial for this task (as well as for verifying that an apparent signal is actually real).

Fundamental noise sources limit the ability of terrestrial observatories to detect low-frequency gravitational waves. Figure 7.12 shows the sensitivity regions, as a function of frequency, for two dramatically different designs: a terrestrial observatory such as LIGO, and a space-based mission such as LISA (Laser Interferometer Space Antenna). The general principle behind LISA is the same as any other interferometer, but the implementation is (or will be, if it is actually built) dramatically different. Current designs envision three spacecraft orbiting the Sun at approximately 30 million kilometers behind the Earth, separated from each other by 5 million kilometers. Due to the much larger separations, LISA is sensitive to frequencies in the vicinity of 10^{-2} Hz. The sensitivities portrayed in this plot should be taken as suggestive, as they depend on integration times and other factors.

Many potential noise sources confront the gravitational-wave astronomer. For ground-based observatories, the dominant effect at low frequencies is typically seismic noise, while at high frequencies it comes from photon shot noise and at intermediate frequencies from thermal noise. Advanced versions of ground-based detectors may be able to compensate for seismic noise at low frequencies, but will encounter irreducible noise from gravity gradients due to atmospheric phenomena or objects (such as cars) passing nearby. Satellite observatories, of course, are immune from such effects. Instead, the fundamental limitations are expected to come from errors in measuring changes in the distances between the spacecraft (or more properly, between the shielded proof masses within the spacecraft) and from nongravitational accelerations of the spacecraft.

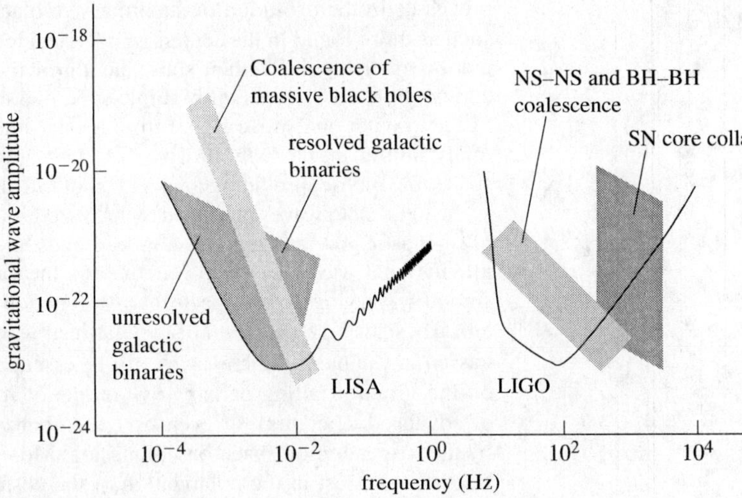

FIGURE 7.12 Sensitivities as a function of frequency for representative ground-based (LIGO) and space-based (LISA) gravitational-wave observatories, along with the expected signals from possible sources. Figure from the LISA collaboration home page (`http://lisa.jpl.nasa.gov/`).

We can conclude with a very brief overview of possible sources for gravitational-wave observatories. We have already mentioned the possibility of compact binaries of various sorts. For ground-based observatories, such sources will not become visible until they are very close to coalescence, and then only if the components are sufficiently massive (neutron stars or black holes). Extrapolating from what we know about such systems suggests that there may be several coalescences per year within a distance of a few hundred Mpc. Another promising possibility is core collapse in massive stars, giving rise to supernovae. Although a perfectly spherically-symmetric collapse would not generate any gravitational waves, realistic events are expected to be subject to instabilities that would break this symmetry. An exciting prospect is the coordinated observation of supernovae by ordinary telescopes and gravitational-wave observatories. Lastly, among possible sources for ground-based observatories are periodic sources such as (not-completely-axially-symmetric) rotating neutron stars. The amplitudes from such sources are expected to be small, but not necessarily completely out of reach of advanced detectors.

The interesting sources for space-based detectors are somewhat different. Most importantly, the known population of binaries in our galaxy will certainly provide a gravitational-wave signal of detectable magnitude. Indeed, unresolved binaries represent a source of confusion noise for the detector, as it will be impossible to pick out individual low-intensity sources from the background. Nevertheless, numerous higher-intensity sources should be easily observable. In addition, various processes in the evolution of supermassive black holes (greater than 1000 $M_\odot$, such as those found in the centers of galaxies) lead to interesting sources: the formation of such objects, their subsequent growth via accretion of smaller objects, and possible coalescence of multiple supermassive holes. Tracking the evolution of the gravitational-wave signal from a solar-mass black hole orbiting and eventually falling into a supermassive hole will allow for precision mapping of the spacetime metric, providing a novel test of GR.

In addition to waves produced by localized sources, we also face the possibility of stochastic gravitational-wave backgrounds. By this we mean an isotropic set of gravitational waves, perhaps generated in the early universe, characterized by a smoothly-varying power spectrum as a function of frequency. One possibility is a nearly scale-free spectrum of gravitational waves produced by inflation, as discussed in Chapter 8. Such waves will be essentially impossible to detect directly on the ground (falling perhaps five orders of magnitude below the capabilities of advanced detectors), or even by LISA, but could conceivably be observable by a next-generation space-based mission. More likely, any such waves will first become manifest in the polarization of the cosmic microwave background. Another possibility, however, is generation of primordial gravitational waves from a violent (first-order) phase transition. Such waves will have a spectrum with a well-defined peak frequency, related to the temperature T of the phase transition by

$$f_{\text{peak}} \sim 10^{-3} \left(\frac{T}{1000 \text{ GeV}} \right) \text{Hz}. \tag{7.203}$$

Thus, a first-order electroweak phase transition ($T \sim 200$ GeV) falls within the band potentially observable by LISA. This is especially intriguing, as some models of baryogenesis require a strong phase transition at this scale; it is provocative to think that we could learn something significant about electroweak physics through a gravitational experiment.

7.8 ■ EXERCISES

1. Show that the Lagrangian (7.9) gives rise to the linearized version of Einstein's equation.

2. Consider a thin spherical shell of matter, with mass M and radius R, slowly rotating with an angular velocity Ω.

   (a) Show that the gravito-electric field $\vec{G}$ vanishes, and calculate the gravito-magnetic field $\vec{H}$ in terms of M, R, and Ω.

   (b) The nonzero gravito-magnetic field caused by the shell leads to dragging of inertial frames, known as the **Lense–Thirring effect**. Calculate the rotation (relative to the inertial frame defined by the background Minkowski metric) of a freely-falling observer sitting at the center of the shell. In other words, calculate the precession of the spatial components of a parallel-transported vector located at the center.

3. Fermat's principle states that a light ray moves along a path of least time. For a medium with refractive index $n(\mathbf{x})$, this is equivalent to extremizing the time

$$t = \int n(\mathbf{x})[\delta_{ij} dx^i dx^j]^{1/2} \tag{7.204}$$

along the path. Show that Fermat's principle, with the refractive index given by $n = 1 - 2\Phi$, leads to the correct equation of motion for a photon in a spacetime perturbed by a Newtonian potential.

4. Show that the Lorenz gauge condition $\partial_\mu \bar{h}^{\mu\nu} = 0$ is equivalent to the **harmonic gauge** condition. This gauge is defined by

$$\Box x^\mu = 0, \tag{7.205}$$

where each coordinate x^μ is thought of as a scalar function on spacetime. (Any function satisfying $\Box f = 0$ is known as an "harmonic function.")

5. In the exercises for Chapter 3, we introduced the metric

$$ds^2 = -(du\,dv + dv\,du) + a^2(u)dx^2 + b^2(u)dy^2, \tag{7.206}$$

where a and b are unspecified functions of u. For appropriate functions a and b, this represents an *exact* gravitational plane wave.

(a) Calculate the Christoffel symbols and Riemann tensor for this metric.

(b) Use Einstein's equation in vacuum to derive equations obeyed by $a(u)$ and $b(u)$.

(c) Show that an exact solution can be found, in which both a and b are determined in terms of an *arbitrary* function $f(u)$.

6. Two objects of mass M have a head-on collision at event $(0, 0, 0, 0)$. In the distant past, $t \to -\infty$, the masses started at $x \to \pm\infty$ with zero velocity.

(a) Using Newtonian theory, show that $x(t) = \pm(9Mt^2/8)^{1/3}$.

(b) For what separations is the Newtonian approximation reasonable?

(c) Calculate $h^{TT}_{xx}(t)$ at $(x, y, z) = (0, R, 0)$.

7. Gravitational waves can be detected by monitoring the distance between two free flying masses. If one of the masses is equipped with a laser and an accurate clock, and the other with a good mirror, the distance between the masses can be measured by timing how long it takes for a pulse of laser light to make the round-trip journey. How would you want your detector oriented to register the largest response from a plane wave of the form

$$ds^2 = -dt^2 + [1 + A\cos(\omega(t - z))]\,dx^2 + [1 - A\cos(\omega(t - z))]\,dy^2 + dz^2?$$

If the masses have a mean separation L, what is the largest change in the arrival time of the pulses caused by the wave? What frequencies ω would go undetected?

8. The gravitational analog of *bremsstrahlung* radiation is produced when two masses scatter off each other. Consider what happens when a small mass m scatters off a large mass M with impact parameter b and total energy $E = 0$. Take $M \gg m$ and $M/b \ll 1$. The motion of the small mass can be described by Newtonian physics, since $M/b \ll 1$. If the orbit lies in the (x, y) plane and if the large mass sits at

$(x, y, z) = (0, 0, 0)$, calculate the gravitational wave amplitude for both polarizations at $(x, y, z) = (0, 0, r)$. Since the motion is not periodic, the gravitational waves will be burst-like and composed of many different frequencies. On physical grounds, what do you expect the dominant frequency to be? Estimate the total energy radiated by the system. How does this compare to the peak kinetic energy of the small mass?

Hint: The solution for the orbit can be found in Goldstein (2002). The solution is:

$$r = \frac{2b}{1 + \cos\theta},$$

$$t = \sqrt{\frac{2b^3}{M}}\left(\tan\frac{\theta}{2} + \frac{1}{3}\tan^3\frac{\theta}{2}\right).$$

Time runs from $t = (-\infty, \infty)$. Rather than using the above implicit solution for $\theta(t)$ you might want to use

$$\dot{\theta} = \sqrt{\frac{M}{8b^3}}\,(1 + \cos\theta)^2\,.$$

9. Verify that the expression (7.165) for the gravitational-wave energy-momentum tensor is invariant under gauge transformations $h_{\mu\nu} \rightarrow h_{\mu\nu} + 2\partial_{(\mu}\xi_{\nu)}$.

10. Show that the integral expression (7.173) for the total energy in gravitational perturbations is independent of the spatial hypersurface Σ.

CHAPTER

8 Cosmology

8.1 ■ MAXIMALLY SYMMETRIC UNIVERSES

Contemporary cosmological models are based on the idea that the universe is pretty much the same everywhere—a stance sometimes known as the **Copernican principle**. On the face of it, such a claim seems crazy; the center of the sun, for example, bears little resemblance to the desolate cold of interstellar space. But we take the Copernican principle to apply only on the very largest scales, where local variations in density are averaged over. Its validity on such scales is manifested in a number of different observations, such as number counts of galaxies and observations of diffuse X-ray and γ-ray backgrounds, but is most clear in the 3K cosmic microwave background (CMB). Although we now know that the microwave background radiation is not perfectly smooth (and nobody ever expected that it was), the deviations from regularity are on the order of 10^{-5} or less, certainly an adequate basis for an approximate description of spacetime on large scales.

The Copernican principle is related to two more mathematically precise properties that a manifold might have: isotropy and homogeneity. **Isotropy** applies at some specific point in the manifold, and states that the space looks the same no matter in what direction you look. More formally, a manifold M is isotropic around a point p if, for any two vectors V and W in T_pM, there is an isometry of M such that the pushforward of W under the isometry is parallel with V (not pushed forward). It is isotropy of space that is indicated by the observations of the microwave background.

Homogeneity is the statement that the metric is the same throughout the manifold. In other words, given any two points p and q in M, there is an isometry that takes p into q. Note that there is no necessary relationship between homogeneity and isotropy; a manifold can be homogeneous but nowhere isotropic (such as $\mathbf{R} \times S^2$ in the usual metric), or it can be isotropic around a point without being homogeneous (such as a cone, which is isotropic around its vertex but certainly not homogeneous). On the other hand, if a space is isotropic *everywhere*, then it is homogeneous. Likewise if it is isotropic around one point and also homogeneous, it will be isotropic around every point. Since there is ample observational evidence for isotropy, and the Copernican principle would have us believe that we are not the center of the universe and therefore observers elsewhere should also observe isotropy, we will henceforth assume both homogeneity and isotropy.

The usefulness of homogeneity and isotropy is that they imply that a space is maximally symmetric. Think of isotropy as invariance under rotations, and homogeneity as invariance under translations, suitably generalized. Then homogeneity and isotropy together imply that a space has its maximum possible number of Killing vectors. An extreme application of the Copernican principle would be to insist that spacetime itself is maximally symmetric. In fact this will turn out not to be true; observationally we know that the universe is homogeneous and isotropic in *space*, but not in all of *spacetime*. However, it is interesting to begin by considering spacetimes that are maximally symmetric (which are, after all, special cases of the more general situation in which only space is maximally symmetric). As we shall see, there is a sense in which such universes are "ground states" of general relativity. This discussion is less relevant to the observed universe than subsequent parts of this chapter, and empirically-minded readers are welcome to skip ahead to the next section.

We mentioned in Chapter 3 that the Riemann tensor for a maximally symmetric n-dimensional manifold with metric $g_{\mu\nu}$ can be written

$$R_{\rho\sigma\mu\nu} = \kappa(g_{\rho\mu}g_{\sigma\nu} - g_{\rho\nu}g_{\sigma\mu}), \tag{8.1}$$

where κ is a normalized measure of the Ricci curvature,

$$\kappa = \frac{R}{n(n-1)}, \tag{8.2}$$

and the Ricci scalar R will be a constant over the manifold. Since at any single point we can always put the metric into its canonical form ($g_{\mu\nu} = \eta_{\mu\nu}$), the kinds of maximally symmetric manifolds are characterized locally by the signature of the metric and the sign of the constant κ. The modifier "locally" is necessary to account for possible global differences, such as between the plane and the torus. We are interested in metrics of signature $(-+++)$. For vanishing curvature ($\kappa = 0$) the maximally symmetric spacetime is well known; it is simply Minkowski space, with metric

$$ds^2 = -dt^2 + dx^2 + dy^2 + dz^2. \tag{8.3}$$

The conformal diagram for Minkowski space is derived in Appendix H.

The maximally symmetric spacetime with positive curvature ($\kappa > 0$) is called **de Sitter space.** Consider a five-dimensional Minkowski space with metric $ds_5^2 = -du^2 + dx^2 + dy^2 + dz^2 + dw^2$, and embed a hyperboloid given by

$$-u^2 + x^2 + y^2 + z^2 + w^2 = \alpha^2. \tag{8.4}$$

Now induce coordinates $\{t, \chi, \theta, \phi\}$ on the hyperboloid via

$$u = \alpha \sinh(t/\alpha)$$
$$w = \alpha \cosh(t/\alpha) \cos\chi$$

$$
\begin{aligned}
x &= \alpha \cosh(t/\alpha) \sin\chi \cos\theta \\
y &= \alpha \cosh(t/\alpha) \sin\chi \sin\theta \cos\phi \\
z &= \alpha \cosh(t/\alpha) \sin\chi \sin\theta \sin\phi .
\end{aligned} \tag{8.5}
$$

The metric on the hyperboloid is then

$$
ds^2 = -dt^2 + \alpha^2 \cosh^2(t/\alpha)\left[d\chi^2 + \sin^2\chi\,(d\theta^2 + \sin^2\theta\, d\phi^2)\right]. \tag{8.6}
$$

We recognize the expression in round parentheses as the metric on a two-sphere, $d\Omega_2^2$, and the expression in square brackets as the metric on a three-sphere, $d\Omega_3^2$. Thus, de Sitter space describes a spatial three-sphere that initially shrinks, reaching a minimum size at $t = 0$, and then re-expands. Of course this particular description is inherited from a certain coordinate system; we will see that there are equally valid alternative descriptions.

These coordinates cover the entire manifold. You can generally check this by, for example, following the behavior of geodesics near the edges of the coordinate system; if the coordinates were incomplete, geodesics would appear to terminate in finite affine parameter. The topology of de Sitter is thus $\mathbf{R} \times S^3$. This makes it very simple to derive the conformal diagram, since the important step in constructing conformal diagrams is to write the metric in a form in which it is conformally related to the Einstein static universe (a spacetime with topology $\mathbf{R} \times S^3$, describing a spatial three-sphere of constant radius through time). Consider the coordinate transformation from t to t' via

$$
\cosh(t/\alpha) = \frac{1}{\cos(t')}. \tag{8.7}
$$

The metric (8.6) now becomes

$$
ds^2 = \frac{\alpha^2}{\cos^2(t')} d\bar{s}^2, \tag{8.8}
$$

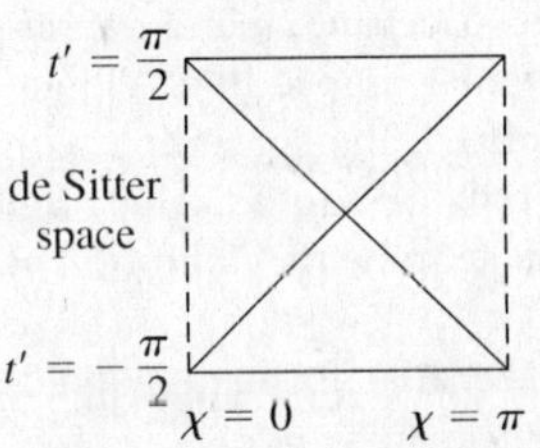

FIGURE 8.1 Conformal diagram for de Sitter spacetime. Spacelike slices are three-spheres, so that points on the diagram represent two-spheres except for those at left and right edges, which are points.

where $d\bar{s}^2$ represents the metric on the Einstein static universe,

$$
d\bar{s}^2 = -(dt')^2 + d\chi^2 + \sin^2\chi\, d\Omega_2^2. \tag{8.9}
$$

The range of the new time coordinate is

$$
-\pi/2 < t' < \pi/2. \tag{8.10}
$$

The conformal diagram of de Sitter space will simply be a representation of the patch of the Einstein static universe to which de Sitter is conformally related. It looks like a square, as shown in Figure 8.1. A spacelike slice of constant t' represents a three-sphere; the dashed lines at the left and right edges are the north and south poles of this sphere. The diagonal lines represent null rays; a photon released at past infinity will get to precisely the antipodal point on the sphere at future infinity. Keep in mind that the spacetime "ends" to the past and the future only through the magic of conformal transformations; the actual de Sitter space extends indefinitely into the future and past. Note also that two points can have future (or past) light cones that are completely disconnected; this reflects the fact that the spherical spatial sections are expanding so rapidly that light from one point can never come into contact with light from the other.

A similar hyperboloid construction reveals the $\kappa < 0$ spacetime of maximal symmetry, known as **anti-de Sitter space**. Begin with a fictitious five-dimensional flat manifold with metric $ds_5^2 = -du^2 - dv^2 + dx^2 + dy^2 + dz^2$, and embed a hyperboloid given by

$$-u^2 - v^2 + x^2 + y^2 + z^2 = -\alpha^2. \tag{8.11}$$

Note all the minus signs. Then we can induce coordinates $\{t', \rho, \theta, \phi\}$ on the hyperboloid via

$$\begin{aligned} u &= \alpha \sin(t') \cosh(\rho) \\ v &= \alpha \cos(t') \cosh(\rho) \\ x &= \alpha \sinh(\rho) \cos\theta \\ y &= \alpha \sinh(\rho) \sin\theta \cos\phi \\ z &= \alpha \sinh(\rho) \sin\theta \sin\phi, \end{aligned} \tag{8.12}$$

yielding a metric on this hyperboloid of the form

$$ds^2 = \alpha^2\big(-\cosh^2(\rho)\, dt'^2 + d\rho^2 + \sinh^2(\rho)\, d\Omega_2^2\big). \tag{8.13}$$

These coordinates have a strange feature, namely that t' is periodic. From (8.12), t' and $t'+2\pi$ represent the same place on the hyperboloid. Since $\partial_{t'}$ is everywhere timelike, a curve with constant $\{\rho, \theta, \phi\}$ as t' increases will be a closed timelike curve. However, this is not an intrinsic property of the spacetime, merely an artifact of how we have derived the metric from a particular embedding. We are welcome to consider the "covering space" of this manifold, the spacetime with metric given by (8.13) in which we allow t' to range from $-\infty$ to ∞. There are no closed timelike curves in this space, which we will take to be the definition of anti-de Sitter space.

To derive the conformal diagram, perform a coordinate transformation analogous to that used for de Sitter, but now on the radial coordinate:

$$\cosh(\rho) = \frac{1}{\cos\chi}, \tag{8.14}$$

so that

$$ds^2 = \frac{\alpha^2}{\cos^2\chi} d\bar{s}^2, \tag{8.15}$$

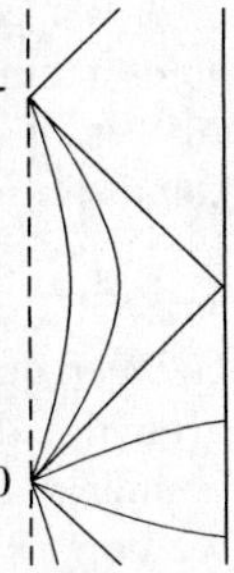

FIGURE 8.2 Conformal diagram for anti-de Sitter spacetime. Spacelike slices have the topology of $\mathbf{R}^3$, which we have represented in polar coordinates, so that points on the diagram stand for two-spheres except those at the left side, which stand for single points at the spatial origin. Infinity is a timelike surface at the right side.

where $d\bar{s}^2$ represents the metric on the Einstein static universe (8.9). Unlike in de Sitter, the radial coordinate now appears in the conformal factor. In addition, for anti-de Sitter, the t' coordinate goes from minus infinity to plus infinity, while the range of the radial coordinate is

$$0 \leq \chi < \frac{\pi}{2}. \tag{8.16}$$

Thus, anti-de Sitter space is conformally related to half of the Einstein static universe. The conformal diagram is shown in Figure 8.2, which illustrates a few representative timelike and spacelike geodesics passing through the point $t' = 0$, $\chi = 0$. Since χ only goes to $\pi/2$ rather than all the way to π, a spacelike slice of this spacetime has the topology of the interior of a hemisphere of S^3; that is, it is topologically $\mathbf{R}^3$ (and the entire spacetime therefore has the topology $\mathbf{R}^4$). Note that we have drawn the diagram in polar coordinates, such that a point on the left side represents a point at the spatial origin, while one on the right side represents a two-sphere at spatial infinity. Another popular representation is to draw the spacetime in cross-section, so that the spatial origin lies in the middle and the right and left sides together comprise spatial infinity.

An interesting feature of anti-de Sitter is that infinity takes the form of a timelike hypersurface, defined by $\chi = \pi/2$. Because infinity is timelike, the space is not globally hyperbolic, we do not have a well-posed initial value problem in terms of information specified on a spacelike slice, since information can always "flow in from infinity." Another interesting feature is that the exponential map is not onto the entire spacetime; geodesics, such as those drawn on the figure, which leave from a specified point do not cover the whole manifold. The future-pointing timelike geodesics, as indicated, can initially move radially outward from $t' = 0$, $\chi = 0$, but eventually refocus to the point $t' = \pi$, $\chi = 0$ and will then move radially outward once again.

As an aside, it is irresistible to point out that the timelike nature of infinity enables a remarkable feature of string theory, the "AdS/CFT correspondence." Here, AdS is of course the anti-de Sitter space we have been discussing, while CFT stands for a conformally-invariant field theory defined on the boundary [which is, for an n-dimensional AdS, an $(n-1)$-dimensional spacetime in its own right]. The AdS/CFT correspondence suggests that, in a certain limit, there is an equivalence between quantum gravity (or a supersymmetric version thereof) on an AdS background and a conformally-invariant nongravitational field theory defined on the boundary. Since we know a lot about nongravitational quantum field theory that we don't know about quantum gravity, this correspondence (if it is true, which seems likely but remains unproven) reveals a great deal about what can happen in quantum gravity.[1]

So we have three spacetimes of maximal symmetry: Minkowski ($\kappa = 0$), de Sitter ($\kappa > 0$), and anti-de Sitter ($\kappa < 0$). Are any one of these useful models for the real world? For that matter, are they solutions to Einstein's equation? Start by taking the trace of the Riemann tensor as given by (8.1), specifying to four dimensions:

$$R_{\mu\nu} = 3\kappa g_{\mu\nu}, \qquad R = 12\kappa. \tag{8.17}$$

So the Ricci tensor is proportional to the metric in a maximally symmetric space. A spacetime with this property is sometimes called an Einstein space; the Einstein static universe is *not* an example of an Einstein space, which can sometimes be confusing. What is worse, we will later encounter the Einstein–de Sitter cosmology, which is not related to Einstein spaces, the Einstein static universe, or to de Sitter space. The Einstein tensor is

$$G_{\mu\nu} = R_{\mu\nu} - \tfrac{1}{2}Rg_{\mu\nu} = -3\kappa g_{\mu\nu}. \tag{8.18}$$

Therefore, Einstein's equation $G_{\mu\nu} = 8\pi G T_{\mu\nu}$ implies (in a maximally symmetric spacetime, not in general) that the energy-momentum tensor is proportional to the metric:

$$T_{\mu\nu} = -\frac{3\kappa}{8\pi G} g_{\mu\nu}. \tag{8.19}$$

Such an energy-momentum tensor corresponds to a vacuum energy or cosmological constant, as discussed in Chapter 4. The energy density and pressure are given by

$$\rho = -p = \frac{3\kappa}{8\pi G}. \tag{8.20}$$

If ρ is positive, we get a de Sitter solution; if ρ is negative, we get anti-de Sitter.

But in our universe, we have ordinary matter and radiation, as well as a possible vacuum energy. Our maximally symmetric spacetimes are not compatible

[1] For a comprehensive review article, see O. Aharony, S.S. Gubser, J.M. Maldacena, H. Ooguri, and Y. Oz, *Phys. Rept.* **323**, 183 (2000), `http://arxiv.org/hep-th/9905111`.

with a dynamically interesting amount of matter and/or radiation. Furthermore, since we observe the visible matter in the universe to be moving apart (the universe is expanding, as discussed below), the density of matter was higher in the past; so even if the matter contribution to the total energy were negligible today, it would have been appreciable in the earlier universe. The maximally symmetric spacetimes are therefore not reasonable models of the real world. They do, however, represent the (locally) unique solutions to Einstein's equation in the absence of any ordinary matter or gravitational radiation; it is in this sense that they may be thought of as ground states of general relativity.

8.2 ■ ROBERTSON–WALKER METRICS

To describe the real world, we are forced to give up the "perfect" Copernican principle, which implies symmetry throughout space and time, and postulate something more forgiving. It turns out to be straightforward, and consistent with observation, to posit that the universe is *spatially* homogeneous and isotropic, but evolving in time. In general relativity this translates into the statement that the universe can be foliated into spacelike slices such that each three-dimensional slice is maximally symmetric. We therefore consider our spacetime to be $\mathbf{R} \times \Sigma$, where $\mathbf{R}$ represents the time direction and Σ is a maximally symmetric three-manifold. The spacetime metric thus takes the form

$$ds^2 = -dt^2 + R^2(t)d\sigma^2, \tag{8.21}$$

where t is the timelike coordinate, $R(t)$ is a function known as the **scale factor**, and $d\sigma^2$ is the metric on Σ, which can be expressed as

$$d\sigma^2 = \gamma_{ij}(u)\, du^i du^j, \tag{8.22}$$

where (u^1, u^2, u^3) are coordinates on Σ and γ_{ij} is a maximally symmetric three-dimensional metric. The scale factor tells us how big the spacelike slice Σ is at the moment t. (Don't confuse it with the curvature scalar.) The coordinates used here, in which the metric is free of cross terms $dt\, du^i$ and coefficient of dt^2 is independent of the u^i, are known as **comoving coordinates**, a special case of the Gaussian normal coordinates discussed in Appendix D. An observer who stays at constant u^i is also called "comoving." Only a comoving observer will think that the universe looks isotropic; in fact on Earth we are not quite comoving, and as a result we see a dipole anisotropy in the cosmic microwave background as a result of the conventional Doppler effect.

Our interest is therefore in maximally symmetric Euclidean three-metrics γ_{ij}. We know that maximally symmetric metrics obey

$${}^{(3)}R_{ijkl} = k(\gamma_{ik}\gamma_{jl} - \gamma_{il}\gamma_{jk}), \tag{8.23}$$

where for future convenience we have introduced

$$k = {}^{(3)}R/6, \tag{8.24}$$

and we put a superscript $^{(3)}$ on the Riemann tensor to remind us that it is associated with the three-metric γ_{ij}, not the metric of the entire spacetime. The Ricci tensor is then

$$^{(3)}R_{jl} = 2k\gamma_{jl}. \tag{8.25}$$

If the space is to be maximally symmetric, then it will certainly be spherically symmetric. We already know something about spherically symmetric spaces from our exploration of the Schwarzschild solution; the metric can be put in the form

$$d\sigma^2 = \gamma_{ij}\, du^i\, du^j = e^{2\beta(\bar{r})}\, d\bar{r}^2 + \bar{r}^2 d\Omega^2, \tag{8.26}$$

where $\bar{r}$ is the radial coordinate and the metric on the two-sphere is $d\Omega^2 = d\theta^2 + \sin^2\theta\, d\phi^2$ as usual. The components of the Ricci tensor for such a metric can be obtained from (5.14), the Ricci tensor for a static, spherically symmetric spacetime, by setting $\alpha = 0$ and $r = \bar{r}$, which gives

$$\begin{aligned} ^{(3)}R_{11} &= \frac{2}{\bar{r}}\partial_1\beta \\ ^{(3)}R_{22} &= e^{-2\beta}(\bar{r}\partial_1\beta - 1) + 1 \\ ^{(3)}R_{33} &= [e^{-2\beta}(\bar{r}\partial_1\beta - 1) + 1]\sin^2\theta\,. \end{aligned} \tag{8.27}$$

We set these proportional to the metric using (8.25), and can solve for $\beta(\bar{r})$:

$$\beta = -\tfrac{1}{2}\ln(1 - k\bar{r}^2), \tag{8.28}$$

which yields the metric on the three-surface Σ,

$$d\sigma^2 = \frac{d\bar{r}^2}{1 - k\bar{r}^2} + \bar{r}^2 d\Omega^2. \tag{8.29}$$

Notice from (8.24) that the value of k sets the curvature, and therefore the size, of the spatial surfaces. It is common to normalize this so that

$$k \in \{+1, 0, -1\}, \tag{8.30}$$

and absorb the physical size of the manifold into the scale factor $R(t)$.

The $k = -1$ case corresponds to constant negative curvature on Σ, and is sometimes called **open**; the $k = 0$ case corresponds to no curvature on Σ, and is called **flat**; the $k = +1$ case corresponds to positive curvature on Σ, and is sometimes called **closed**. The physical interpretation of these cases is made more clear using an alternative form of the metric, obtained by introducing a new radial coordinate χ defined by

$$d\chi = \frac{d\bar{r}}{\sqrt{1 - k\bar{r}^2}}. \tag{8.31}$$

This can be integrated to obtain

$$\bar{r} = S_k(\chi), \tag{8.32}$$

where

$$S_k(\chi) \equiv \begin{cases} \sin(\chi), & k = +1 \\ \chi, & k = 0 \\ \sinh(\chi), & k = -1, \end{cases} \tag{8.33}$$

so that

$$d\sigma^2 = \mathrm{d}\chi^2 + S_k{}^2(\chi)d\Omega^2. \tag{8.34}$$

For the flat case $k = 0$, the metric on Σ becomes

$$\begin{aligned} d\sigma^2 &= \mathrm{d}\chi^2 + \chi^2 d\Omega^2 \\ &= \mathrm{d}x^2 + \mathrm{d}y^2 + \mathrm{d}z^2, \end{aligned} \tag{8.35}$$

which is simply flat Euclidean space. Globally, it could describe $\mathbf{R}^3$ or a more complicated manifold, such as the three-torus $S^1 \times S^1 \times S^1$. For the closed case $k = +1$ we have

$$d\sigma^2 = \mathrm{d}\chi^2 + \sin^2\chi\, d\Omega^2, \tag{8.36}$$

which is the metric of a three-sphere. In this case the only possible global structure is the complete three-sphere (except for the nonorientable manifold $\mathbf{RP}^3$, obtained by identifying antipodal points on S^3). Finally in the open $k = -1$ case we obtain

$$d\sigma^2 = \mathrm{d}\chi^2 + \sinh^2\chi\, d\Omega^2. \tag{8.37}$$

This is the metric for a three-dimensional space of constant negative curvature, a generalization of the hyperboloid discussed in Section 3.9. Globally such a space could extend forever (which is the origin of the word "open"), but it could also describe a nonsimply-connected compact space (so "open" is really not the most accurate description).

The metric on spacetime describes one of these maximally-symmetric hypersurfaces evolving in size, and can be written

$$ds^2 = -\mathrm{d}t^2 + R^2(t)\left[\frac{\mathrm{d}\bar{r}^2}{1 - k\bar{r}^2} + \bar{r}^2 d\Omega^2\right]. \tag{8.38}$$

This is the **Robertson–Walker (RW) metric**. We have not yet made use of Einstein's equation; that will determine the behavior of the scale factor $R(t)$. Note that the substitutions

$$
\begin{aligned}
R &\to \lambda^{-1} R \\
\bar{r} &\to \lambda \bar{r} \\
k &\to \lambda^{-2} k
\end{aligned}
\tag{8.39}
$$

leave (8.38) invariant. Therefore we can choose a convenient normalization. In the variables where the curvature k is normalized to $\{+1, 0, -1\}$, the scale factor has units of distance and the radial coordinate $\bar{r}$ (or χ) is actually dimensionless; this is the most popular choice. We will flout the conventional wisdom and instead work with a dimensionless scale factor

$$
a(t) = \frac{R(t)}{R_0}, \tag{8.40}
$$

a coordinate with dimensions of distance

$$
r = R_0 \bar{r}, \tag{8.41}
$$

and a curvature parameter with dimensions of (length)$^{-2}$,

$$
\kappa = \frac{k}{R_0^2}. \tag{8.42}
$$

Note that κ can take on any value, not just $\{+1, 0, -1\}$. In these variables the Robertson–Walker metric is

$$
\boxed{ds^2 = -dt^2 + a^2(t)\left[\frac{dr^2}{1 - \kappa r^2} + r^2\, d\Omega^2\right].} \tag{8.43}
$$

To convert to the more common notation, just plug in the relations (8.40), (8.41), and (8.42).

With the metric in hand, we can set about computing the connection coefficients and curvature tensor. Setting $\dot{a} \equiv da/dt$, the Christoffel symbols are given by

$$
\begin{aligned}
\Gamma^0_{11} &= \frac{a\dot{a}}{1 - \kappa r^2} & \Gamma^1_{11} &= \frac{\kappa r}{1 - \kappa r^2} \\
\Gamma^0_{22} &= a\dot{a} r^2 & \Gamma^0_{33} &= a\dot{a} r^2 \sin^2\theta \\
\Gamma^1_{01} &= \Gamma^2_{02} & \Gamma^3_{03} &= \frac{\dot{a}}{a} \\
\Gamma^1_{22} &= -r(1 - \kappa r^2) & \Gamma^1_{33} &= -r(1 - \kappa r^2)\sin^2\theta \\
\Gamma^2_{12} &= \Gamma^3_{13} = \frac{1}{r} & & \\
\Gamma^2_{33} &= -\sin\theta\,\cos\theta & \Gamma^3_{23} &= \cot\theta,
\end{aligned}
\tag{8.44}
$$

or related to these by symmetry. The nonzero components of the Ricci tensor are

$$\begin{aligned} R_{00} &= -3\frac{\ddot{a}}{a} \\ R_{11} &= \frac{a\ddot{a} + 2\dot{a}^2 + 2\kappa}{1 - \kappa r^2} \\ R_{22} &= r^2(a\ddot{a} + 2\dot{a}^2 + 2\kappa) \\ R_{33} &= r^2(a\ddot{a} + 2\dot{a}^2 + 2\kappa)\sin^2\theta, \end{aligned} \tag{8.45}$$

and the Ricci scalar is then

$$R = 6\left[\frac{\ddot{a}}{a} + \left(\frac{\dot{a}}{a}\right)^2 + \frac{\kappa}{a^2}\right]. \tag{8.46}$$

8.3 ■ THE FRIEDMANN EQUATION

The RW metric is defined for any behavior of the scale factor $a(t)$; our next step will be to plug it into Einstein's equation to derive the Friedmann equation(s) relating the scale factor to the energy-momentum of the universe. We will choose to model matter and energy by a perfect fluid. It is clear that, if a fluid that is isotropic in some frame leads to a metric that is isotropic in some frame, the two frames will coincide; that is, the fluid will be at rest in comoving coordinates. The four-velocity is then

$$U^\mu = (1, 0, 0, 0), \tag{8.47}$$

and the energy-momentum tensor

$$T_{\mu\nu} = (\rho + p)U_\mu U_\nu + p g_{\mu\nu} \tag{8.48}$$

becomes

$$T_{\mu\nu} = \begin{pmatrix} \rho & 0 & 0 & 0 \\ 0 & & & \\ 0 & & g_{ij}p & \\ 0 & & & \end{pmatrix}. \tag{8.49}$$

With one index raised this takes the convenient form

$$T^\mu{}_\nu = \mathrm{diag}(-\rho, p, p, p). \tag{8.50}$$

Note that the trace is given by

$$T = T^\mu{}_\mu = -\rho + 3p. \tag{8.51}$$

Before plugging in to Einstein's equation, it is educational to consider the zero component of the conservation of energy equation:

$$\begin{aligned}0 &= \nabla_\mu T^\mu{}_0 \\ &= \partial_\mu T^\mu{}_0 + \Gamma^\mu_{\mu\lambda} T^\lambda{}_0 - \Gamma^\lambda_{\mu 0} T^\mu{}_\lambda \\ &= -\partial_0 \rho - 3\frac{\dot{a}}{a}(\rho + p).\end{aligned} \tag{8.52}$$

To make progress we can choose an **equation of state**, a relationship between ρ and p. Often the perfect fluids relevant to cosmology obey the simple equation of state

$$p = w\rho, \tag{8.53}$$

where w is a constant independent of time. Of course we are free to define the parameter $w = p/\rho$ whether or not it remains constant; if w varies, however, it is not really legitimate to call $p = w\rho$ the "equation of state." The conservation of energy equation becomes

$$\boxed{\frac{\dot{\rho}}{\rho} = -3(1 + w)\frac{\dot{a}}{a}.} \tag{8.54}$$

If w is a constant, this can be integrated to obtain

$$\rho \propto a^{-3(1+w)}. \tag{8.55}$$

To get an idea about what values of w are allowed, refer to the discussion of energy conditions in Chapter 4. The Null Dominant Energy Condition, which allows for a vacuum energy of either sign but otherwise requires matter that cannot destabilize the vacuum, implies

$$|w| \leq 1. \tag{8.56}$$

While this requirement is by no means set in stone, it seems like a sensibly conservative starting point for investigations of what might happen in the real world.

The two most popular examples of cosmological fluids are known as **matter** and **radiation**. Matter is any set of collisionless, nonrelativistic particles, which will have essentially zero pressure:

$$p_{\mathrm{M}} = 0. \tag{8.57}$$

Examples include ordinary stars and galaxies, for which the pressure is negligible in comparison with the energy density. Matter is also known as *dust*, and universes whose energy density is mostly due to matter are known as **matter-dominated**. The energy density in matter falls off as

$$\rho_{\mathrm{M}} \propto a^{-3}. \tag{8.58}$$

This is simply interpreted as the decrease in the number density of particles as the universe expands. For matter the energy density is dominated by the rest energy, which is proportional to the number density. Radiation may be used to describe either actual electromagnetic radiation, or massive particles moving at relative velocities sufficiently close to the speed of light that they become indistinguishable from photons (at least as far as their equation of state is concerned). Although an isotropic gas of relativistic particles is a perfect fluid and thus has an energy-momentum tensor given by (8.48), we also know that $T_{\mu\nu}$ for electromagnetism can be expressed in terms of the field strength as

$$T^{\mu\nu} = F^{\mu\lambda}F^{\nu}{}_{\lambda} - \tfrac{1}{4}g^{\mu\nu}F^{\lambda\sigma}F_{\lambda\sigma}. \tag{8.59}$$

The trace of this is given by

$$T^{\mu}{}_{\mu} = F^{\mu\lambda}F_{\mu\lambda} - \tfrac{1}{4}(4)F^{\lambda\sigma}F_{\lambda\sigma} = 0. \tag{8.60}$$

But this must also equal (8.51), so the equation of state is

$$p_{\rm R} = \tfrac{1}{3}\rho_{\rm R}. \tag{8.61}$$

A universe in which most of the energy density is in the form of radiation is known as **radiation-dominated**. The energy density in radiation falls off as

$$\rho_{\rm R} \propto a^{-4}. \tag{8.62}$$

Thus, the energy density in radiation falls off slightly faster than that in matter; this is because the number density of photons decreases in the same way as the number density of nonrelativistic particles, but individual photons also lose energy as a^{-1} as they redshift, which we will see later. Likewise, massive but relativistic particles will lose energy as they "slow down" in comoving coordinates. We believe that today the radiation energy density is much less than that of matter, with $\rho_{\rm M}/\rho_{\rm R} \sim 10^3$. However, in the past the universe was much smaller, and the energy density in radiation would have dominated at very early times.

As we have discussed, vacuum energy also takes the form of a perfect fluid, with an equation of state $p_\Lambda = -\rho_\Lambda$. The energy density is constant,

$$\rho_\Lambda \propto a^0. \tag{8.63}$$

Since the energy density in matter and radiation decreases as the universe expands, if there is a nonzero vacuum energy it tends to win out over the long term, as long as the universe doesn't start contracting. If this happens, we say that the universe becomes **vacuum-dominated**. de Sitter and anti-de Sitter are vacuum-dominated solutions.

We now turn to Einstein's equation. Recall that it can be written in the form (4.45):

$$R_{\mu\nu} = 8\pi G\left(T_{\mu\nu} - \tfrac{1}{2}g_{\mu\nu}T\right). \tag{8.64}$$

The $\mu\nu = 00$ equation is

$$-3\frac{\ddot{a}}{a} = 4\pi G(\rho + 3p), \tag{8.65}$$

and the $\mu\nu = ij$ equations give

$$\frac{\ddot{a}}{a} + 2\left(\frac{\dot{a}}{a}\right)^2 + 2\frac{\kappa}{a^2} = 4\pi G(\rho - p). \tag{8.66}$$

There is only one distinct equation from $\mu\nu = ij$, due to isotropy. We can use (8.65) to eliminate second derivatives in (8.66), and do a little cleaning up to obtain

$$\boxed{\left(\frac{\dot{a}}{a}\right)^2 = \frac{8\pi G}{3}\rho - \frac{\kappa}{a^2},} \tag{8.67}$$

and

$$\boxed{\frac{\ddot{a}}{a} = -\frac{4\pi G}{3}(\rho + 3p).} \tag{8.68}$$

Together these are known as the **Friedmann equations**, and metrics of the form (8.43) obey these equations define Friedmann–Robertson–Walker (FRW) universes. In fact, if we know the dependence of ρ on a, the first of these (8.67) is enough to solve for $a(t)$; when you hear people refer to *the* Friedmann equation, this is the one to which they are referring, whereas (8.68) is sometimes called the *second* Friedmann equation.

A bunch of terminology is associated with the cosmological parameters, and we will just introduce the basics here. The rate of expansion is characterized by the **Hubble parameter**,

$$\boxed{H = \frac{\dot{a}}{a}.} \tag{8.69}$$

The value of the Hubble parameter at the present epoch is the Hubble constant, H_0. Current measurements lead us to believe that the Hubble constant is 70 ± 10 km/sec/Mpc. (Mpc stands for megaparsec, which is 3.09×10^{24} cm.) Since there is still some uncertainty in this value, we often parameterize the Hubble constant as

$$H_0 = 100h \text{ km/sec/Mpc}, \tag{8.70}$$

so that $h \approx 0.7$. Typical cosmological scales are set by the **Hubble length**

$$\begin{aligned} d_H &= H_0^{-1} c \\ &= 9.25 \times 10^{27} h^{-1} \text{ cm} \\ &= 3.00 \times 10^3 h^{-1} \text{ Mpc}, \end{aligned} \tag{8.71}$$

and the **Hubble time**

$$\begin{aligned} t_H &= H_0^{-1} \\ &= 3.09 \times 10^{17} h^{-1} \text{ sec} \\ &= 9.78 \times 10^9 h^{-1} \text{ yr}. \end{aligned} \tag{8.72}$$

Of course since we usually set $c = 1$, you will see H_0^{-1} referred to as both the Hubble length and the Hubble time. There is also the **deceleration parameter**,

$$q = -\frac{a\ddot{a}}{\dot{a}^2}, \tag{8.73}$$

which measures the rate of change of the rate of expansion.

Another useful quantity is the **density parameter**,

$$\boxed{\Omega = \frac{8\pi G}{3H^2}\rho = \frac{\rho}{\rho_{\rm crit}},} \tag{8.74}$$

where the **critical density** is defined by

$$\rho_{\rm crit} = \frac{3H^2}{8\pi G}. \tag{8.75}$$

This quantity, which will generally change with time, is called the *critical* density because the Friedmann equation (8.67) can be written

$$\Omega - 1 = \frac{\kappa}{H^2 a^2}. \tag{8.76}$$

The sign of κ is therefore determined by whether Ω is greater than, equal to, or less than, one. We have

$$\begin{array}{lllllll} \rho < \rho_{\rm crit} & \leftrightarrow & \Omega < 1 & \leftrightarrow & \kappa < 0 & \leftrightarrow & \text{open} \\ \rho = \rho_{\rm crit} & \leftrightarrow & \Omega = 1 & \leftrightarrow & \kappa = 0 & \leftrightarrow & \text{flat} \\ \rho > \rho_{\rm crit} & \leftrightarrow & \Omega > 1 & \leftrightarrow & \kappa > 0 & \leftrightarrow & \text{closed.} \end{array}$$

The density parameter, then, tells us which of the three Robertson–Walker geometries describes our universe. Determining it observationally is of crucial importance; recent measurements of the cosmic microwave background anisotropy lead us to believe that Ω is very close to unity.

8.4 ■ EVOLUTION OF THE SCALE FACTOR

Given a specification of the amounts of energy density ρ_i in different species i, along with their equations of state $p_i = p_i(\rho_i)$, and the amount of spatial curvature κ, one can solve the Friedmann equation (8.67) to obtain a complete history of the evolution of the scale factor, $a(t)$. In general we simply numerically integrate the Friedmann equation (which is just a first-order differential equation), but it is useful to get a feeling for the types of solutions appropriate to different cosmological parameters.

To simplify our task, let us imagine that all of the different components of energy density evolve as power laws,

$$\rho_i = \rho_{i0} a^{-n_i}. \tag{8.77}$$

Comparing to (8.55), this is equivalent to positing that each equation-of-state parameter $w_i = p_i/\rho_i$ is a constant equal to

$$w_i = \tfrac{1}{3}n_i - 1. \tag{8.78}$$

We can further streamline our expressions by treating the contribution of spatial curvature as a fictitious energy density

$$\rho_c \equiv -\frac{3\kappa}{8\pi G a^2}, \tag{8.79}$$

with a corresponding density parameter

$$\Omega_c = -\frac{\kappa}{H^2 a^2}. \tag{8.80}$$

It's *not* an energy density, of course, so don't forget that this is just notational sleight-of-hand. The behaviors of our favorite sources are summarized in the following table.

	w_i	n_i
matter	0	3
radiation	$\frac{1}{3}$	4
curvature	$-\frac{1}{3}$	2
vacuum	-1	0

(8.81)

In these variables, the Friedmann equation (8.67) can be written

$$H^2 = \frac{8\pi G}{3}\sum_{i(c)} \rho_i, \tag{8.82}$$

where the notation $\sum_{i(c)}$ indicates that we sum not only over all the actual components of energy density ρ_i, but also over the contribution of spatial curvature

ρ_c. Note that if we divide both sides by H^2, we obtain

$$1 = \sum_{i(c)} \Omega_i. \tag{8.83}$$

The right-hand side is *not* the total density parameter Ω, which only gets contributions from actual energy density (not curvature); we therefore have

$$\Omega_c = 1 - \Omega. \tag{8.84}$$

Let's begin by asking what can happen if all of the ρ_i's (including ρ_c) are nonnegative. Because H^2 is proportional to $\sum_{i(c)} \rho_i$, the universe will never undergo a transition from expanding to contracting so long as $\sum_{i(c)} \rho_i \neq 0$. We can also take the time derivative of the Hubble parameter,

$$\dot{H} = \frac{\ddot{a}}{a} - \left(\frac{\dot{a}}{a}\right)^2, \tag{8.85}$$

and plug in the two Friedmann equations (8.67) and (8.68) to obtain

$$\dot{H} = -4\pi G \sum_{i(c)} (1 + w_i)\rho_i. \tag{8.86}$$

Since we are imagining that $|w_i| \leq 1$, when all the ρ_i's are nonnegative we will always have $\dot{H} \leq 0$. In other words, the universe keeps expanding, but the expansion rate continually decreases (which suggests the excellent question, what made it so large in the first place?).

From (8.85) we see that $\ddot{a}$ can be positive and $\dot{H}$ be negative at the same time—the scale factor can be "accelerating" even though the expansion rate as measured by the Hubble parameter is decreasing (for example, if $a \propto t^2$). This is an unavoidable subtlety of non-Euclidean geometry. The Hubble parameter and the derivative of the scale factor are the answers to two different questions. If we set two test particles at a fixed initial distance, and ask by how much they have separated a short time thereafter, the answer is given by the Hubble parameter. If, on the other hand, we pick some fixed source, and ask how it appears to move away from us with time, the answer is given by the change in the scale factor. There are consequently two very different and equally legitimate senses of "accelerating" (or "decelerating"). In practice, "accelerating" usually refers to a situation in which $\ddot{a} > 0$, even if $\dot{H} < 0$. This discussion is not completely academic; as we will see below, our current real universe seems to be of this type.

It is by no means necessary that each ρ_i should be nonnegative. Matter and radiation arise from dynamical particles and fields, and we consequently expect that their energy densities will never be negative; if they could be, empty space could decay into a collection of positive- and negative-energy fields. But vacuum and curvature are different stories. Vacuum energy is nondynamical, so a negative value cannot induce any instabilities, while curvature is simply a property of the spatial geometry, and can have either sign. If we therefore have either a negative vacuum energy or a positive spatial curvature (remember $\rho_c \propto -\kappa$), the Hubble parameter can vanish and even change sign. An example is provided by the de Sitter metric (8.6), which has a positive vacuum energy but also a positive spatial curvature; it describes a universe that initially collapses, reaches a turning point, and thereafter begins to expand.

The real world is an untidy place, consisting of numerous different kinds of energy density. Because different sources evolve at different rates, however, for long periods the energy density will be clearly dominated by one kind of source. It is therefore very useful to examine solutions to the Friedmann equation when there is only one kind of energy density $\rho \propto a^{-n}$. Because we are including spatial curvature as an effective energy source, this means we are considering either flat universes dominated by a single source, or completely empty universes with spatial curvature. The Friedmann equation then implies

$$\dot{a} \propto a^{1-n/2}. \tag{8.87}$$

This can be immediately integrated to obtain

$$\boxed{a \propto t^{2/n} \qquad (\text{for } \rho \propto a^{-n}).} \tag{8.88}$$

Consider for example a flat universe dominated by matter, $\Omega = \Omega_M = 1$; this is known as the Einstein–de Sitter model, and for a long time was the favorite (at least among theorists) to describe the real world. In an Einstein–de Sitter universe, the scale factor evolves as $a \propto t^{2/3}$. A flat radiation-dominated universe, meanwhile, evolves as $a \propto t^{1/2}$. The conformal diagram for any such universe with $n > 2$ is derived in Appendix H. Even though we believe there are nonzero amounts of matter, radiation, and vacuum energy in the real universe, these solutions are still very useful; as we discuss later, the universe was radiation-dominated at early times, and was matter dominated as the universe expanded from $a \sim 1/3000$ to $a \sim 1/2$.

These solutions all feature a singularity at $a = 0$, known as the **Big Bang**. It represents the creation of the universe from a singular state, not an explosion of matter into a pre-existing spacetime. It might be hoped that the perfect symmetry of our FRW universes is responsible for this singularity, but in fact that's not true; cosmological singularity theorems show that any universe with $\rho > 0$ and $p \geq 0$ must have begun at a singularity. Of course the energy density becomes arbitrarily high as $a \to 0$, and we don't expect classical general relativity to be an accurate description of nature in this regime; presumably quantum gravity becomes important, although it is unclear how at present.

Looking at (8.88), we see that a universe dominated by vacuum energy ($n = 0$) is clearly a special case. The scale factor then expands as an exponential rather than a power law; the entire metric is

$$ds^2 = -dt^2 + e^{Ht}[dx^2 + dy^2 + dz^2], \tag{8.89}$$

where the Hubble parameter H is a constant. Of course, in Section 8.1 we already described a cosmological spacetime with a positive cosmological constant: de Sitter space, which featured $\kappa > 0$ and $a \propto \cosh(t/\alpha)$. What is the relationship between that solution and the one here, with $\kappa = 0$ and $a \propto \exp(Ht)$? They are the same spacetime, represented in different coordinates. One way to verify this is to calculate the Riemann tensor for (8.89) and check that it has the characteristic form of a maximally symmetric spacetime, (8.1). Since maximally symmetric spacetimes with positive curvature are locally unique, the metrics (8.6) and (8.89) must describe the same manifold, or parts thereof. In fact, the coordinates of (8.89) only cover part of de Sitter; they are incomplete in the past. In the exercises you are asked to show that comoving geodesics in these coordinates reach $t = -\infty$ in finite affine parameter; they run into the edge of the coordinates. In the conformal diagram of Figure 8.1, these coordinates cover the upper-right triangular portion of the square. See Hawking and Ellis (1973) for a more complete description of different coordinate systems on de Sitter and anti-de Sitter.

Another interesting special case is the completely empty universe, with $\rho = 0$, but with spatial curvature. The Friedmann equation becomes

$$H^2 = -\frac{\kappa}{a^2}, \tag{8.90}$$

so the curvature κ must be negative. Thinking of curvature as a fictitious energy density $\rho_c \propto a^{-2}$, from (8.88) we know that such a universe will expand linearly, $a \propto t$. This spacetime is known as the **Milne universe**. However, just as with de Sitter, we know of another cosmological spacetime with $\rho = 0$—in this case, flat Minkowski space. Once again, the Milne spacetime is just a patch of Minkowski in a certain incomplete coordinate system. It can be thought of as the interior of the future light cone of some fixed point in Minkowski, foliated by negatively-curved hyperboloids. To check, it would suffice to calculate all of the components of the Riemann tensor, which turn out to vanish; any spacetime with vanishing Riemann curvature is locally Minkowski.

In contrast to these idealized solutions, a realistic cosmology will feature several forms of energy-momentum. In the current universe, we feel confident that the radiation density is significantly lower than the matter density, but that vacuum and matter are both dynamically important. It is therefore convenient to parameterize universes like ours by Ω_M and Ω_Λ, with the curvature fixed by $\Omega_c = 1 - \Omega_M - \Omega_\Lambda$. The expansion history of some particular examples of such universes is shown in Figure 8.3. As these universes expand, the relative influences of matter, curvature, and vacuum are altered, since the corresponding densities evolve at different rates:

$$\Omega_\Lambda \propto \Omega_c a^2 \propto \Omega_M a^3. \tag{8.91}$$

As $a \to 0$ in the past, curvature and vacuum will be negligible, and the universe will behave as Einstein–de Sitter. As $a \to \infty$ in the future, curvature and matter will be negligible, and the universe will asymptote to de Sitter; unless the scale

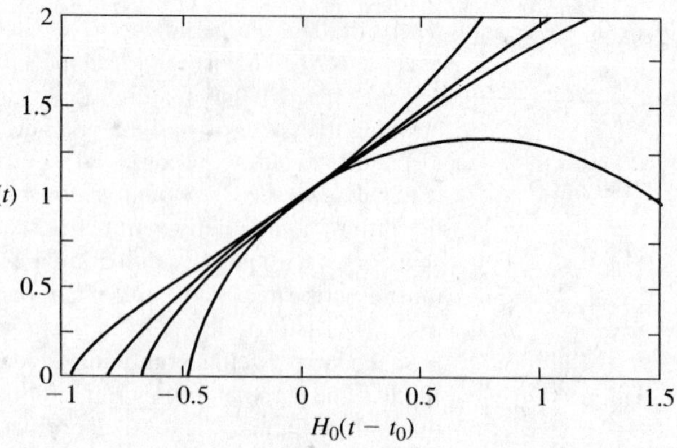

FIGURE 8.3 Expansion histories for different values of Ω_M and Ω_Λ. From top to bottom, the curves describe $(\Omega_M, \Omega_\Lambda) = (0.3, 0.7)$, $(0.3, 0.0)$, $(1.0, 0.0)$, and $(4.0, 0.0)$.

factor never reaches infinity, because the universe begins to recollapse at some finite time.

Recollapse will *always* occur if the vacuum energy is negative; as the universe expands, the vacuum energy eventually dominates, and the effect of $\Omega_\Lambda < 0$ is to cause deceleration and recollapse (just as the effect of $\Omega_\Lambda > 0$ is to push the universe apart). Recollapse is also possible with $\Omega_\Lambda \geq 0$, if Ω_M is sufficiently large that it halts the universal expansion before Ω_Λ has a chance to take over. The possibilities are expressed as different regions of the Ω_M/Ω_Λ parameter space in Figure 8.4. The diagonal line represents $\Omega_{\rm total} = 1$, implying $\kappa = 0$.

To determine the dividing line between perpetual expansion and eventual recollapse, note that collapse requires the Hubble parameter to pass through zero as it changes from positive to negative. The scale factor a_* at which this turnaround occurs can be found by setting $H = 0$ in the Friedmann equation,

$$H^2 = 0 = \frac{8\pi G}{3}\left(\rho_{M0}a_*^{-3} + \rho_{\Lambda 0} + \rho_{c0}a_*^{-2}\right). \tag{8.92}$$

We can divide this by H_0^2, use $\Omega_{c0} = 1 - \Omega_{M0} - \Omega_{\Lambda 0}$, and rearrange a bit to obtain

$$\Omega_{\Lambda 0}a_*^3 + (1 - \Omega_{M0} - \Omega_\Lambda)a_* + \Omega_{M0} = 0. \tag{8.93}$$

This is a cubic equation for a_*, the scale factor at turnaround. Of course we don't actually care very much about a_*; what we care about are the values of $\Omega_{\Lambda 0}$, given Ω_{M0}, for which a real solution to (8.93) exists. Solving the cubic equation and doing some math, we find that the value of $\Omega_{\Lambda 0}$ for which the universe will expand forever is given by

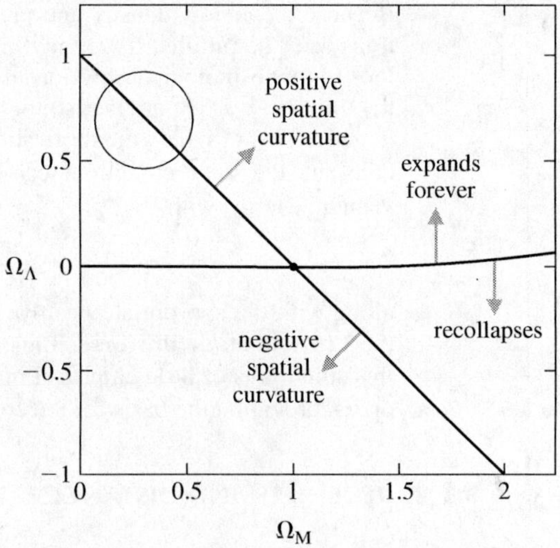

FIGURE 8.4 Properties of universes dominated by matter and vacuum energy, as a function of the density parameters Ω_M and Ω_Λ. The circular region in the upper-left corner represents roughly those values favored by experimental data (as of 2003).

$$\Omega_{\Lambda 0} \geq \begin{cases} 0 & 0 \leq \Omega_{M0} \leq 1 \\ 4\Omega_{M0}\cos^3\left[\frac{1}{3}\cos^{-1}\left(\frac{1-\Omega_{M0}}{\Omega_{M0}}\right) + \frac{4\pi}{3}\right] & \Omega_{M0} > 1. \end{cases} \tag{8.94}$$

Note that, when $\Omega_{\Lambda 0} = 0$, open and flat universes ($\Omega_0 = \Omega_{M0} \leq 1$) will expand forever, while closed universes ($\Omega_0 = \Omega_{M0} > 1$) will recollapse. Traditional disdain for the cosmological constant has led to a folk belief that this is a necessary correspondence; once the possibility of vacuum energy is admitted, however, any combination of spatial geometry and eventual fate is possible.

In the upper-left corner of Figure 8.4, we have indicated the currently favored values of the cosmological parameters: $\Omega_{M0} \sim 0.3$, $\Omega_{\Lambda 0} \sim 0.7$, as we will discuss in Section 8.7. This is well into the regime of perpetual expansion; if the vacuum energy remains truly constant (which it might not), our universe is fated to continue its expansion for all time.

We end this section by noting the difficulty of finding static solutions to the Friedmann equations. To be static, we must have not only $\dot{a} = 0$, but also $\ddot{a} = 0$. From (8.68), this can only happen if the pressure is

$$p = -\tfrac{1}{3}\rho, \tag{8.95}$$

and from (8.67), there must be a nonvanishing spatial curvature

$$\frac{\kappa}{a^2} = \frac{8\pi G}{3}\rho. \tag{8.96}$$

Because the energy density and pressure must be of opposite sign, these conditions can't be fulfilled if we only invoke matter or radiation. When Einstein first looked for cosmological solutions in GR, astronomers had not yet discovered that the universe was expanding, so the lack of static solutions was considered problematic. This provided the motivation for Einstein to introduce the cosmological constant; the static conditions can be satisfied by a combination of matter and vacuum energy, with

$$\rho_\Lambda = \tfrac{1}{2}\rho_M, \tag{8.97}$$

along with the appropriate positive spatial curvature. These parameters describe the **Einstein static universe**. Today we know that the universe is expanding, so this solution is of little empirical interest; it is, however, extremely useful to theorists, providing the basis for the construction of conformal diagrams.

8.5 ■ REDSHIFTS AND DISTANCES

It is clear that we would like to determine a number of quantities observationally to decide which of the FRW models corresponds to our universe. Obviously we would like to determine H_0, since that is related to the age of the universe. We would also like to know Ω, which determines κ through (8.76). To understand how these quantities might conceivably be measured, let's consider geodesic motion in an FRW universe. There are a number of spacelike Killing vectors, but no timelike Killing vector to give us a notion of conserved energy. There is, however, a Killing tensor. If $U^\mu = (1, 0, 0, 0)$ is the four-velocity of comoving observers, then the tensor

$$K_{\mu\nu} = a^2(g_{\mu\nu} + U_\mu U_\nu) \tag{8.98}$$

satisfies $\nabla_{(\sigma}K_{\mu\nu)} = 0$ (as you can check), and is therefore a Killing tensor. This means that if a particle has four-velocity $V^\mu = dx^\mu/d\lambda$, the quantity

$$K^2 = K_{\mu\nu}V^\mu V^\nu = a^2[V_\mu V^\mu + (U_\mu V^\mu)^2] \tag{8.99}$$

will be a constant along geodesics. Let's think about this, first for massive particles. Then we will have $V_\mu V^\mu = -1$, so

$$(V^0)^2 = 1 + |\vec{V}|^2, \tag{8.100}$$

where $|\vec{V}|^2 = g_{ij}V^iV^j$. We also have $U_\mu V^\mu = -V^0$, so (8.99) implies

$$|\vec{V}| = \frac{K}{a}. \tag{8.101}$$

The particle therefore "slows down" with respect to the comoving coordinates as the universe expands. In fact this is an actual slowing down, in the sense that a gas of particles with initially high relative velocities will cool down as the universe expands.

A similar thing happens to null geodesics. In this case $V_\mu V^\mu = 0$, and (8.99) implies

$$U_\mu V^\mu = \frac{K}{a}. \tag{8.102}$$

But the frequency of the photon as measured by a comoving observer is $\omega = -U_\mu V^\mu$. The frequency of the photon emitted with frequency $\omega_{\rm em}$ will therefore be observed with a lower frequency $\omega_{\rm obs}$ as the universe expands:

$$\frac{\omega_{\rm obs}}{\omega_{\rm em}} = \frac{a_{\rm em}}{a_{\rm obs}}. \tag{8.103}$$

Cosmologists like to speak of this in terms of the **redshift** z between the two events, defined by the fractional change in wavelength:

$$z_{\rm em} = \frac{\lambda_{\rm obs} - \lambda_{\rm em}}{\lambda_{\rm em}}. \tag{8.104}$$

If the observation takes place today ($a_{\rm obs} = a_0 = 1$), this implies

$$\boxed{a_{\rm em} = \frac{1}{1 + z_{\rm em}}.} \tag{8.105}$$

So the redshift of an object tells us the scale factor when the photon was emitted.

Notice that this redshift is not the same as the conventional Doppler effect; it is the expansion of space, not the relative velocities of the observer and emitter, which leads to the redshift. Nevertheless, if we observe galaxies over distances that are small compared to the Hubble radius H_0^{-1} and the radius of spatial curvature $\kappa^{-1/2}$, the expansion of the universe looks very much like a set of galaxies moving apart from each other and the redshift looks very much like the Doppler effect. Consequently, astronomers often think of the redshift in terms of a "velocity" $v = cz$, where c is the speed of light. Even though we know you can't really speak of the relative velocities between two objects at different points of a curved spacetime, the fiction works well over sufficiently short distances. Within this approximation, the "distance" d from us to a galaxy can be taken to be the **instantaneous physical distance** d_P (the distance, in physical units such as centimeters, between us and the location of the galaxy along our current spatial hypersurface). Let's write the RW metric in the form

$$ds^2 = -dt^2 + a^2(t)R_0^2\left[d\chi^2 + S_k{}^2(\chi)d\Omega^2\right], \tag{8.106}$$

where $S_k(\chi)$ is defined by (8.33), and $k \in \{+1, 0, -1\}$. In this form, the instantaneous physical distance as measured at time t between us ($\chi = 0$) and a galaxy at comoving radial coordinate χ is

$$d_P(t) = a(t)R_0\chi, \tag{8.107}$$

where χ remains constant because we assume both we and the observed galaxy are perfectly comoving. (They might not be, in which case it is trivial to include the corrections due to so-called "peculiar velocities.") Of course "distance" is in quotes because there are several inequivalent useful notions of distance once we leave this approximation, but they all agree when d_P is small. Then the observed velocity (as inferred from the redshift) is simply

$$v = \dot{d}_P = \dot{a}R_0\chi = \frac{\dot{a}}{a}d_P. \tag{8.108}$$

Evaluated today, this becomes

$$\boxed{v = H_0 d_P,} \tag{8.109}$$

the famous **Hubble law**: the observed recession velocity is directly proportional to the distance, for galaxies that are not too far away.

If the redshift is not very small, we have to think more carefully about what we mean by "distance" in cosmology. The instantaneous physical distance is a convenient construct, but not itself observable, since observations always refer to events on our past light cone, not our current spatial hypersurface. In Euclidean space there are a number of different ways to infer the distance of an object; we could for example compare its apparent brightness to its intrinsic luminosity, or its apparent angular velocity to its intrinsic transverse speed, or its apparent angular size to its physical extent. For each of these cases, we can define a kind of distance that is what we *would* infer if space were Euclidean and the universe were not expanding.

Let's start with the **luminosity distance** d_L, defined to satisfy

$$d_L^2 = \frac{L}{4\pi F}, \tag{8.110}$$

where L is the absolute luminosity of the source and F is the flux measured by the observer (the energy per unit time per unit area of some detector). This definition comes from the fact that in flat space, for a source at distance d the flux over the luminosity is just one over the area of a sphere centered around the source, $F/L = 1/A(d) = 1/4\pi d^2$. In an FRW universe, however, the flux will be diluted. Conservation of photons tells us that all of the photons emitted by the source will eventually pass through a sphere at comoving distance χ from the emitter. But the flux is diluted by two additional effects: the individual photons redshift by a factor $(1+z)$, and the photons hit the sphere less frequently, since two photons emitted a time δt apart will be measured at a time $(1+z)\delta t$ apart. Therefore we will have

$$\frac{F}{L} = \frac{1}{(1+z)^2 A}. \tag{8.111}$$

The area A of a sphere centered at comoving distance χ can be derived from the coefficient of $d\Omega^2$ in (8.106), yielding

$$A = 4\pi R_0^2 {S_k}^2(\chi), \tag{8.112}$$

where we have set $a(t) = 1$ because we are observing the photons today. Putting it all together yields

$$d_L = (1 + z) R_0 S_k(\chi). \tag{8.113}$$

The luminosity distance d_L is something we might hope to measure, since there are some astrophysical sources whose absolute luminosities are known. But χ is not observable, so we have to remove that from our equation. On a null geodesic (chosen to be radial for convenience) we have

$$0 = ds^2 = -\mathrm{d}t^2 + a^2 R_0^2 \mathrm{d}\chi^2, \tag{8.114}$$

or

$$\chi = R_0^{-1} \int \frac{dt}{a} = R_0^{-1} \int \frac{da}{a^2 H(a)}, \tag{8.115}$$

where we have used $H = \dot{a}/a$. It is conventional to convert the scale factor to redshift using $a = 1/(1 + z)$, so we have

$$\chi(z) = R_0^{-1} \int_0^z \frac{dz'}{H(z')}. \tag{8.116}$$

In order to evaluate the Hubble parameter in this integral we use the Friedmann equation (8.67), which we write as in the previous section as

$$H^2 = \frac{8\pi G}{3} \sum_{i(\mathrm{c})} \rho_i. \tag{8.117}$$

To simplify things, we may again assume that each density component evolves as a power law,

$$\rho_i(z) = \rho_{i0} a^{-n_i} = \rho_{i0}(1 + z)^{n_i}, \tag{8.118}$$

Then we can write

$$H(z) = H_0 E(z), \tag{8.119}$$

where

$$E(z) = \left[\sum_{i(\mathrm{c})} \Omega_{i0} (1 + z)^{n_i} \right]^{1/2}, \tag{8.120}$$

where the density parameters Ω_i are defined by (8.74). The equations below involving $E(z)$ will be true whether or not the energy sources evolve as power laws; if they do not, simply use $E(z) = H(z)/H_0$ [where $H(z)$ is determined by the Friedmann equation] rather than (8.120).

So the luminosity distance is

$$d_L(z) = (1+z)R_0 S_k\left[R_0^{-1}H_0^{-1}\int \frac{dz'}{E(z')}\right]. \tag{8.121}$$

Note that R_0 drops out when $k = 0$, which is good, because in that case it is a completely arbitrary parameter. Even when it is not arbitrary, it is still more common to speak in terms of $\Omega_{c0} = -k/R_0^2H_0^2$, which can be measured either directly through determinations of the spatial curvature, or by measuring the density parameter and using $\Omega_{c0} = 1 - \Omega_0$. In terms of this parameter we have

$$R_0 = H_0^{-1}\sqrt{-k\Omega_{c0}} = \frac{H_0^{-1}}{\sqrt{|\Omega_{c0}|}}. \tag{8.122}$$

We therefore write the luminosity distance in terms of measurable cosmological parameters as

$$\boxed{d_L(z) = (1+z)\frac{H_0^{-1}}{\sqrt{|\Omega_{c0}|}}\,S_k\left[\sqrt{|\Omega_{c0}|}\int \frac{dz'}{E(z')}\right].} \tag{8.123}$$

Although it appears unwieldy, this equation is of central importance in cosmology. Given the observables H_0 and Ω_{i0}, we can straightforwardly calculate the luminosity distance to an object at any redshift z; equally well, we can measure $d_L(z)$ for objects at a range of redshifts, and from that information extract H_0 and/or the Ω_{i0}'s.

Along with the luminosity distance are two other related distance measures. Just as the luminosity distance is the distance we infer from the intrinsic and observed luminosity of the source if we were in flat space, the **proper motion distance** d_M is the distance we infer from the intrinsic and observed motion of the source. It is defined to be

$$d_M = \frac{u}{\dot{\theta}}, \tag{8.124}$$

where u is the proper transverse velocity (something you would measure, for example, in k/s) and $\dot{\theta}$ is the observed angular velocity. The **angular diameter distance**, meanwhile, is the distance we infer from the intrinsic and observed size of the source; it is defined to be

$$d_A = \frac{R}{\theta}, \tag{8.125}$$

where R is the proper size of the object and θ is its observed angular diameter. In both cases we can derive formulas analogous to (8.123); fortunately, the unwieldy dependence on the cosmological parameters is common to all the distance measures, and we are left with a simple dependence on redshift:

$$d_L = (1+z)d_M = (1+z)^2 d_A, \tag{8.126}$$

as you are encouraged to check. So if we measure one such distance, we can easily convert to any other; or we can measure different distances independently and use (8.126) to test the consistency of the RW framework.

While we're contemplating distances, let's also consider the elapsed time between now and when the light from an object at redshift z was emitted. If the age of the universe today is t_0 and the age when the photon was emitted is t_*, the **lookback time** is

$$\begin{aligned} t_0 - t_* &= \int_{t_*}^{t_0} dt \\ &= \int_{a_*}^{1} \frac{da}{aH(a)} \\ &= H_0^{-1} \int_0^{z_*} \frac{dz'}{(1+z')E(z')}. \end{aligned} \tag{8.127}$$

For example, consider a flat ($k = 0$) matter-dominated ($\rho = \rho_{\mathrm{M}} = \rho_{\mathrm{M}0}a^{-3}$) universe. Then

$$E(z) = (1+z)^{3/2}, \tag{8.128}$$

so

$$\begin{aligned} t_0 - t_* &= H_0^{-1} \int_0^{z_*} \frac{dz'}{(1+z')^{5/2}} \\ &= \frac{2}{3} H_0^{-1} \left[1 - (1+z_*)^{-3/2}\right]. \end{aligned} \tag{8.129}$$

The total age of a matter-dominated universe is obtained by letting $t_* \to 0$ ($z_* \to \infty$),

$$t_0(\mathrm{MD}) = \tfrac{2}{3} H_0^{-1}. \tag{8.130}$$

For universes that are not completely matter-dominated, the factor of $\frac{2}{3}$ will be not quite right, but for reasonable values of the cosmological parameters we usually get $t_0 \sim H_0^{-1}$.

8.6 ■ GRAVITATIONAL LENSING

In Chapter 7 we introduced the concept of gravitational lensing: the deflection and time delay of light by a Newtonian gravitational field. In addition to providing a

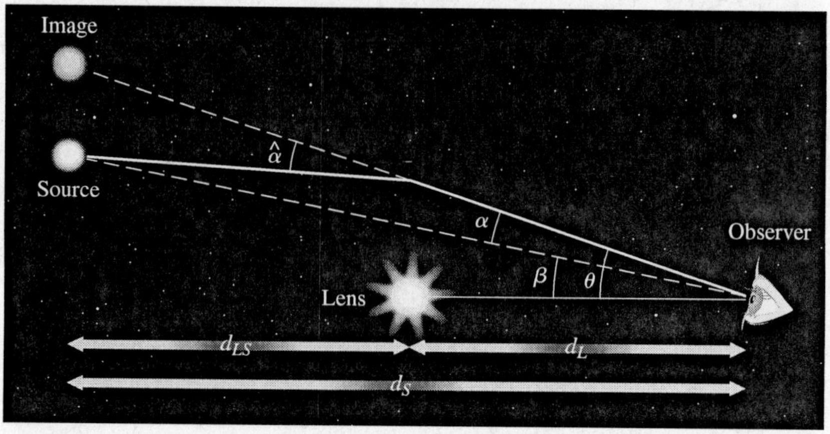

FIGURE 8.5 The geometry of gravitational lensing, encapsulated in the lens equation (8.132). The effect of the lens is to distort the angles β that would be observed in a flat Minkowski background into the angles θ.

test of GR in the Solar System, lensing occurs in numerous astrophysical contexts, and has become an indispensable part of modern cosmology.[2]

Two important features distinguish cosmological lensing from the case we discussed earlier: a Robertson–Walker metric replaces the Minkowski background, and the lenses themselves are often more complex than simple point masses. A typical lensing geometry is portrayed in Figure 8.5. Throughout this discussion we will assume that the lens is "thin"—much smaller in spatial extent than the distances between the source, lens, and observer. In this case we can sensibly speak of a unique distance to the lens, d_L, and between the lens and the source, d_{LS}.

We describe a (possibly complicated) image on the sky by a set of angles between different components of the image. These angles can be thought of as two-dimensional vectors on the sky. The effect of the lens is to distort the angles that would be observed in the absence of any deflection, such as the angle $\vec{\beta}$ between the source and the lens, into a new image characterized by a set of angles $\vec{\theta}$. We assume that the angles are small throughout. This map is described by the **reduced lensing angle** $\vec{\alpha} = \vec{\theta} - \vec{\beta}$. According to the geometry shown in Figure 8.5, it is related to the actual deflection angle $\widehat{\alpha}$ by

$$\vec{\alpha} = \frac{d_{LS}}{d_S}\widehat{\alpha}. \tag{8.131}$$

[2]An excellent overview of gravitational lensing, from which our discussion borrows, can be found in R. Narayan and M. Bartelmann, "Lectures on Gravitational Lensing," 13th Jerusalem Winter School in Theoretical Physics, `http://arxiv.org/astro-ph/9606001`.

We therefore get the **lens equation**

$$\boxed{\vec{\beta} = \vec{\theta} - \frac{d_{LS}}{d_S}\widehat{\vec{\alpha}}.} \tag{8.132}$$

The lens equation simply describes ray-tracing in a perturbed spacetime.

Of course, we should think carefully about the "distances" d_i portrayed in the figure. Lensing occurs in an expanding universe, which might also have spatial curvature. The lens equation will nevertheless hold if we *define* the distances d_i to be such that the geometrical relations described by the lens equation hold. In other words, these are the distances that we would infer, given the angles and transverse physical sizes, in a static Euclidean spatial background. But this is precisely the definition of the angular diameter distance (8.125). We therefore take all distances in this section to be angular-diameter distances. Note that angular-diameter distances do not necessarily add, so that $d_S \neq d_L + d_{LS}$.

As a simple example, consider a point mass lens. In our investigation of the Newtonian limit in Chapter 7, we found that the deflection angle for a photon traveling through a gravitational potential Φ is given by

$$\widehat{\vec{\alpha}} = 2\int \vec{\nabla}_{\perp}\Phi\, ds, \tag{8.133}$$

which for a point mass M at an impact parameter b becomes

$$\widehat{\alpha} = \frac{4GM}{b}. \tag{8.134}$$

The impact parameter can be expressed as $b = d_L\theta$. The lens equation (8.132) becomes

$$\beta = \theta - \frac{d_{LS}}{d_S d_L}\frac{4GM}{\theta}. \tag{8.135}$$

It is illuminating to consider the simplest situation, in which the source and lens are collinear ($\beta = 0$). In that case, the source will be lensed into an **Einstein ring** surrounding the lens, at an angular separation given by the **Einstein angle:**

$$\theta_{\rm E} = \sqrt{\frac{4GMd_{LS}}{d_L d_S}}. \tag{8.136}$$

The Einstein angle sets a characteristic scale for lensing, even in more complicated configurations. We can also define an associated distance scale, the **Einstein radius:**

$$R_{\rm E} = \sqrt{\frac{4GMd_L d_{LS}}{d_S}}. \tag{8.137}$$

When converting to centimeters or other physical units, don't forget that $c = 1$ in all of our equations. To get a feeling for the amount of lensing in typical astrophysical situations, we can consider two common occurrences: "microlensing" by approximately solar-mass objects within our galaxy, and cosmological lensing by galaxies or clusters. In the former case the Einstein angle will be of order milliarcseconds, while in the latter case it will be of order arcseconds:

$$\begin{aligned}\theta_{\rm E} &= 0.9\sqrt{\left(\frac{M}{M_\odot}\right)\left(\frac{10\ \text{kpc}}{D}\right)}\ \text{milliarcsecs}\\ &= 0.9\sqrt{\left(\frac{M}{10^{11}M_\odot}\right)\left(\frac{\text{Gpc}}{D}\right)}\ \text{arcsecs}.\end{aligned} \tag{8.138}$$

Sticking for the moment with the point-mass lens, most often we will not be lucky enough to have source and lens perfectly aligned, although a number of spectacular examples of Einstein rings have been observed. Then we can solve (8.135) to obtain two values of the image angle,

$$\theta_\pm = \tfrac{1}{2}\left(\beta \pm \sqrt{\beta^2 + 4\theta_{\rm E}^2}\right). \tag{8.139}$$

The image at θ_+ will always be outside the Einstein angle, while θ_- will be inside. In fact this formula is somewhat misleading, as there will always be an odd number of images; for a point mass lens, the third image would be located at the same position as the lens itself.

Now let's consider more general lenses than point masses. We know that the deflection angle will be given in terms of the Newtonian gravitational potential by (8.133). We can define the **lensing potential** by integrating over past-directed geodesic paths emanating from the observer, as

$$\psi(\vec{\theta}) = 2\frac{d_{LS}}{d_L d_S}\int \Phi(d_L\vec{\theta}, s)\, ds. \tag{8.140}$$

In terms of the lensing potential, we can straightforwardly derive the reduced lensing angle by taking the gradient,

$$\begin{aligned}\vec{\alpha} &= \vec{\nabla}_\theta \psi\\ &= 2\frac{d_{LS}}{d_S}\int \vec{\nabla}_\perp \Phi\, ds.\end{aligned} \tag{8.141}$$

Notice that the angular gradient $\vec{\nabla}_\theta$ is related to $\vec{\nabla}_\perp$, the gradient with respect to transverse distance at the location of the lens, by a factor of d_L. The thin-lens approximation allows us to collapse the integral to quantities evaluated at the location of the lens. We can also take the (two-dimensional) Laplacian of the lensing potential to obtain the **convergence** κ, via

$$\kappa(\vec{\theta}) \equiv \frac{1}{2}\nabla_\theta^2 \psi$$
$$= \frac{d_L d_{LS}}{d_S} \int \nabla^2 \Phi \, ds. \tag{8.142}$$

The convergence can be thought of as a measure of the integrated mass density. We can invert the above expressions to write both the lensing potential and the reduced deflection angle in terms of the convergence, as

$$\psi(\vec{\theta}) = \frac{1}{\pi} \int \kappa(\vec{\theta}') \ln |\vec{\theta} - \vec{\theta}'| \, d^2\theta' \tag{8.143}$$

and

$$\vec{\alpha}(\vec{\theta}) = \frac{1}{\pi} \int \kappa(\vec{\theta}') \frac{\vec{\theta} - \vec{\theta}'}{|\vec{\theta} - \vec{\theta}'|} \, d^2\theta'. \tag{8.144}$$

To check these equations, remember that the vectors are defined only in the two transverse dimensions.

The convergence describes the focusing of light rays by the gravitational lens. This focusing causes the source to appear larger (just as in a magnifying glass). According to Liouville's theorem of conservation of phase-space density for the photons emitted by the source, the surface brightness of the source will be conserved under lensing; the increase in size therefore leads to magnification of the brightness. At the same time, we can have distortion caused by twisting of the light rays through the lens, which leads to shear of the shape of the image. To describe both phenomena, we consider the 2×2 matrix of derivatives of the lens map,

$$A_{ij} \equiv \frac{\partial \beta^i}{\partial \theta^j}. \tag{8.145}$$

Note that there is no real distinction between upper and lower indices, as they are defined in a two-dimensional Euclidean plane. Since $\vec{\beta} = \vec{\theta} - \vec{\alpha}$, we have

$$A_{ij} = \delta_{ij} - \frac{\partial \alpha^i}{\partial \theta^j}$$
$$= \delta_{ij} - \psi_{ij}, \tag{8.146}$$

where we have introduced the notation

$$\psi_{ij} \equiv \frac{\partial^2 \psi}{\partial \theta^i \partial \theta^j}. \tag{8.147}$$

This matrix A encodes the local properties of the lensing map. Its inverse matrix is known as the *magnification tensor*,

$$M = \frac{\partial \vec{\theta}}{\partial \vec{\beta}} = A^{-1}. \tag{8.148}$$

Why does it get this name? The lens distorts an area element described by $\vec{\beta}$ into one described by $\vec{\theta}$, and the change in area is described by the Jacobian of this map, which is simply the determinant of M. This determinant is defined as the **magnification** μ,

$$\mu = |M| = \frac{1}{|A|}. \tag{8.149}$$

The absolute magnitude of μ tells us the actual change in brightness of the source; μ may be negative, which means that the parity of the image has been flipped. We speak of magnification because lensing is only noticeable if the lens and source are near to each other on the sky, in which case the focusing effect leads only to increases in the apparent brightness; a lens far away from the source (in position on the sky) would lead to a miniscule decrease in the luminosity that will never be noticed. (If there are multiple images, the sum of the brightnesses of all the images will exceed that of the undistorted source.)

The components of A can be decomposed into the effects of convergence and shear. For the convergence, from $\kappa = \frac{1}{2}\nabla^2_\theta \psi$ we have

$$\kappa = \tfrac{1}{2}(\psi_{11} + \psi_{22}). \tag{8.150}$$

The **shear**, meanwhile, distorts the shape of the source; if an initially circular source is distorted into an ellipse of ellipticity γ and position angle ϕ, we define the two components of the shear to be

$$\begin{aligned} \gamma_1 &= \gamma \cos(2\phi) \\ \gamma_2 &= \gamma \sin(2\phi), \end{aligned} \tag{8.151}$$

so that the total shear is $\gamma = \sqrt{\gamma_1^2 + \gamma_2^2}$. In terms of the lensing potential the components are given by

$$\begin{aligned} \gamma_1 &= \tfrac{1}{2}(\psi_{11} - \psi_{22}) \\ \gamma_2 &= \psi_{12} = \psi_{21}. \end{aligned} \tag{8.152}$$

Inverting these relationships to find the components of A yields

$$A = \begin{pmatrix} 1 - \kappa - \gamma_1 & -\gamma_2 \\ -\gamma_2 & 1 - \kappa + \gamma_1 \end{pmatrix}. \tag{8.153}$$

We can therefore express the magnification in terms of the convergence and shear, as

$$\mu = \frac{1}{(1-\kappa)^2 - \gamma^2}. \tag{8.154}$$

These features of lensing are becoming increasingly important in observational cosmology. The obvious case of interest is so-called "strong lensing," when the source is within the Einstein radius of the lens, and multiple images are possible. By observing several images of a single source, we can infer properties of the lens mass distribution (for example, to search for dark matter); we can also use the time delay along different paths to measure the Hubble constant, and the statistical frequency of lensing to constrain other cosmological parameters. However, lensing need not be strong to have an important effect. "Weak lensing," when the source and lens are separated by more than an Einstein radius, will generally lead to small amounts of magnification and shear which are impossible to detect without a priori knowledge of the properties of the source. However, the shearing effect can be detected statistically, by looking at the shapes of thousands of galaxies that are assumed to be intrinsically random in their orientations. Shearing by weak lensing leads to correlated distortions in the shapes, which can reveal a great deal about the distribution of matter between the observer and the distant sources.

8.7 ■ OUR UNIVERSE

Throughout our discussion of the behavior of FRW cosmologies, we have alluded to the actual values of the cosmological parameters corresponding to the universe in which we live. Let us now be more systematic, and discuss both the universe we see today and a plausible extrapolation back to early times. Our discussion will necessarily be brief, both for reasons of space and because cosmology is an active area of research; look for recent review articles to get up-to-date descriptions of current views.

Many of our direct determinations of the expansion rate rely on the luminosity-distance formula (8.123) applied to some type of object whose intrinsic luminosity is assumed to be known, which we call **standard candles**. (Occasionally we measure the angular diameters of objects whose intrinsic size is assumed to be known: standard rulers.) The Hubble constant, for example, is measured with a variety of standard candles, and a consensus of different methods has converged on the value $H_0 = 70 \pm 10$ km/sec/Mpc, mentioned above. Deviations at high redshift from the linear Hubble law (8.109) can yield information about the density parameters Ω_{i0}, but only if we have very bright objects whose intrinsic luminosity is accurately known. These are provided by Type Ia supernovae, which are thought to be explosions of white dwarf stars that have accreted enough mass to surpass the Chandrasekhar limit. Since the Chandrasekhar limit is close to universal, the associated explosions are essentially of equal brightness (and some of the intrinsic variability can actually be accounted for by following the evolution of the brightness through time). It was measurements of SNe Ia at redshifts $z > 0.3$ that provided the first direct evidence for a nonzero cosmological constant; these observations imply that Ω_Λ is actually larger than Ω_M. Recall that matter is pressureless, $p_M = 0$, whereas vacuum energy is associated with a negative pressure, $p_\Lambda = -\rho_\Lambda$. Plugging into the second Friedmann equation (8.68) we find that a universe with both matter and Λ obeys

$$\frac{\ddot{a}}{a} = -\frac{4\pi G}{3}(\rho_M - 2\rho_\Lambda). \tag{8.155}$$

Thus, if ρ_Λ is sufficiently large compared to ρ_M (as the supernova observations indicate), we can have $\ddot{a} > 0$, an accelerating universe (in the sense described in Section 8.4).

The matter density itself is measured by a variety of methods, often involving measuring the density ρ_M by looking for the gravitational effects of clustered matter and then extrapolating to large scales. Because $\rho_M = (3H^2/8\pi G)\Omega_M$, limits obtained in this way are often quoted in terms of $\Omega_M h^2$, where h is defined in (8.70). These days the uncertainty on H_0 appears to be small enough that it is fairly safe to take $h^2 \approx 0.5$, which we do henceforth. Most contemporary methods are consistent with the result

$$\Omega_{M0} = 0.3 \pm 0.1. \tag{8.156}$$

Before there was good evidence for a cosmological constant, this low matter density was sometimes taken as an indication that space was negatively curved, $\kappa < 0$.

In addition to matter and cosmological constant, we also have radiation in the universe. Ordinary photons are the most obvious component of the radiation density, but any relativistic particle would contribute. For photons, most of the energy density resides in the cosmic microwave background, the leftover radiation from the Big Bang. Besides photons, the only obvious candidates for a radiation component are neutrinos. We expect that the number density of relic background neutrinos is comparable to that of photons; the photon density is likely to be somewhat larger, as photons can still be created after the number of neutrinos has become fixed. However, if the mass of the neutrinos is sufficiently large (greater than about 10^{-4} eV), they will have become nonrelativistic today, and contribute to matter rather than to radiation. Current ideas about neutrino masses suggest that this probably is the case, but it is not perfectly clear. Furthermore, it is conceivable that there are as-yet-undetected massless particles in addition to the ones we know about (although they can't be too abundant, or they would suppress the formation of large-scale structure.) Altogether, it seems likely that the total radiation density is of the same order of magnitude as the photon density; in this case we would have

$$\Omega_{R0} \sim 10^{-4}. \tag{8.157}$$

As mentioned before, it is not surprising that the radiation density is lower than the matter density, as the former decays more rapidly as the universe expands. The radiation density goes as a^{-4}, while that in matter goes as a^{-3}; so the epoch of matter-radiation equality occurred at a redshift

$$z_{eq} \approx \frac{\Omega_{M0}}{\Omega_{R0}} \sim 3 \times 10^3. \tag{8.158}$$

A further crucial constraint on the cosmological parameters comes from anisotropies in the temperature of the microwave background. The average temperature is $T_{\rm CMB} = 2.74{\rm K}$, but in 1992 the COBE satellite discovered fluctuations from place to place at a level of $\Delta T/T \sim 10^{-5}$. These anisotropies arise from a number of sources, including gravitational redshift/blueshift from photons moving out of potential wells at recombination (the Sachs–Wolfe effect, dominant on large angular scales), intrinsic temperature fluctuations at the surface of last scattering (dominant on small angular scales), and the Doppler effect from motions of the plasma. The physics describing the evolution of CMB anisotropies is outside the scope of this book. A map of the CMB temperature over the entire sky clearly contains a great deal of information, but no theory predicts what the temperature at any given point is supposed to be. Instead, modern theories generally predict the expectation value of the amount of anisotropy on any given angular scale. We therefore decompose the anisotropy field into spherical harmonics,

$$\frac{\Delta T}{T}(\theta, \phi) = \sum_{lm} a_{lm} Y_{lm}(\theta, \phi). \tag{8.159}$$

The expectation value of $|a_{lm}|^2$ is likely to be independent of m; otherwise the statistical characteristics of the anisotropy will change from place to place on the sky (although we should keep an open mind). The relevant parameters to be measured are therefore

$$C_l = \langle |a_{lm}|^2 \rangle. \tag{8.160}$$

Since for any fixed l, there are $2l + 1$ possible values of m (from $-l$ to l), at all but the lowest l's there are enough independent measurements of the a_{lm}'s to accurately determine their expectation values. The irreducible uncertainty at very small l is known as cosmic variance.

Numerous experiments have measured the C_l's (the so-called CMB power spectrum), and improving these measurements is likely to be an important task for a number of years. (In addition to the temperature anisotropy, a great deal of information is contained in the polarization of the CMB, which is another target of considerable experimental effort.) To turn these observations into useful information, we need a specific theory to predict the CMB power spectrum as a function of the cosmological parameters. There are two leading possibilities (although one is much more leading than the other): either density perturbations are imprinted on all scales at extremely early times even modes for which the physical wavelength λ was much larger than the Hubble radius H^{-1}, or local dynamical mechanisms act as sources for anisotropies at all epochs. The latter possibility has essentially been ruled out by the CMB data; if anisotropies are produced continuously, we expect a relatively smooth, featureless spectrum of C_l's, whereas the observations indicate a significant amount of structure. It is therefore much more popular to imagine a primordial source of perturbations, such as inflation (discussed in the next section). Inflationary perturbations are adiabatic—perturbations in the matter density are correlated with those in the radiation density—and of nearly equal magnitude at all scales. With this input, we can make definite predictions for the C_l's as a function of all the cosmological parameters. Perhaps the most significant constraint obtained from experiments thus far is that universe is spatially flat, or nearly so; $|\Omega_{c0}| < 0.1$. Combined with the measurements of the matter density $\Omega_M \approx 0.3$, we conclude that the vacuum energy density parameter should be

$$\Omega_{\Lambda 0} = 0.7 \pm 0.1. \tag{8.161}$$

This is nicely consistent with the Type Ia supernova results described above; the concordance picture described here is that indicated in Figure 8.4. Converting from density parameter to physical energy density using $H_0 = 70$ km/sec/Mpc yields

$$\rho_{\rm vac} \approx 10^{-8}\ {\rm erg/cm^3}, \tag{8.162}$$

as mentioned in our discussion of vacuum energy in Section 4.5.

One more remarkable feature completes our schematic picture of the present-day universe. We have mentioned that about 30% of the energy density in our universe consists of matter. But to a cosmologist, "matter" is any collection of nonrelativistic particles; the matter we infer from its gravitational influence need not be the same kind of ordinary matter we are familiar with from our experience on Earth. By **ordinary matter** we mean anything made from atoms and their constituents (protons, neutrons, and electrons); this would include all of the stars, planets, gas, and dust in the universe, immediately visible or otherwise. Occasionally such matter is referred to as *baryonic* matter, where baryons include protons, neutrons, and related particles (strongly interacting particles carrying a conserved quantum number known as baryon number). Of course electrons are conceptually an important part of ordinary matter, but by mass they are negligible compared to protons and neutrons:

$$\begin{aligned} m_p &= 0.938\ {\rm GeV} \\ m_n &= 0.940\ {\rm GeV} \\ m_e &= 0.511 \times 10^{-3}\ {\rm GeV}. \end{aligned} \tag{8.163}$$

In other words, the mass of ordinary matter comes overwhelmingly from baryons.

Ordinary baryonic matter, it turns out, is not nearly enough to account for the observed density $\Omega_M \approx 0.3$. Our current best estimates for the baryon density yield

$$\Omega_b = 0.04 \pm 0.02, \tag{8.164}$$

where these error bars are conservative by most standards. This determination comes from a variety of methods: direct counting of baryons (the least precise method), consistency with the CMB power spectrum (discussed above), and agreement with the predictions of the abundances of light elements for Big-Bang nucleosynthesis (discussed below). Most of the matter density must therefore be in the form of **nonbaryonic dark matter**, which we will abbreviate to simply "dark matter." (Baryons can be dark, but it is increasingly common to reserve the terminology for the nonbaryonic component.) Essentially every known particle in the Standard Model of particle physics has been ruled out as a candidate for this dark matter. Fortunately, there are a number of plausible candidates beyond the Standard Model, including neutralinos (the lightest of the additional stable particles predicted by supersymmetry, with masses ≥ 100 GeV) and axions (light pseudoscalar particles arising from spontaneous breakdown of a hypothetical Peccei-Quinn symmetry invoked to explain conservation of CP in the strong interactions, with masses $\sim 10^{-4}$ eV). One of the few things we know about the dark matter is that it must be cold—not only is it nonrelativistic today, but it must have been that way for a very long time. If the dark matter were hot, it would have free-streamed out of overdense regions, suppressing the formation of galaxies. The other thing we know about cold dark matter (CDM) is that it should interact very weakly with ordinary matter, so as to have escaped detection thus far. Nevertheless, ambient dark matter particles may occasionally scatter off carefully shielded detectors in terrestrial laboratories; the attempt to directly detect dark matter by searching for the effects of such scatterings will be another significant experimental effort in the years to come.

The picture in which $\Omega_M = 0.3$ and $\Omega_\Lambda = 0.7$ seems to fit an impressive variety of observational data. The most surprising part of the picture is the cosmological constant. In Chapter 4 we mentioned that a naïve estimate of the vacuum energy yields a result many orders of magnitude larger than what has been measured. In fact there are three related puzzles: Why is the cosmological constant so much smaller than we expect? What is the origin of the small nonzero energy that comprises 70% of the current universe? And, why is the current value of the vacuum energy of the same order of magnitude as the matter density? The last problem is especially severe, as the vacuum energy and matter density evolve rapidly with respect to each other:

$$\frac{\Omega_\Lambda}{\Omega_M} \propto a^3. \tag{8.165}$$

If Ω_M and Ω_Λ are comparable today, in the past the vacuum energy would have been undetectably small, while in the future the matter density will be negligible. This "coincidence problem" has thus far proven to be a complete mystery. One suggested solution involves the "anthropic principle." If there are many distinct parts of the universe (in space, or even in branches of the wavefunction) in which the cosmological constant takes on very different values, intelligent life is most likely to arise in those places where the absolute magnitude is not too large—a large positive Λ would tear particles apart before galaxies could form, while a large negative Λ would cause the universe to recollapse before life could evolve. The anthropic explanation of the observed vacuum energy provides a good fit to the data, although the need to invoke such an elaborate scheme to explain this one quantity strikes some as slightly extravagant.

Another possibility that may (or may not) bear on the coincidence problem is the idea that we have not detected a nonzero cosmological constant, but rather a dynamical component that closely mimics the properties of vacuum energy. Consideration of this possibility has led cosmologists to coin the term **dark energy** to describe whatever it is that has been detected, whether it is dynamical or turns out to be a cosmological constant after all. What we know about the dark energy is that it is relatively smoothly distributed through space (or it would have been detected through its local gravitational field, just like dark matter) and is evolving slowly with time (or it would not make the universe accelerate, as indicated by the supernova data). A simple candidate for a dynamical source of dark energy is provided by a slowly-rolling scalar field. Consider a field ϕ with the usual action

$$S = \int d^4x\sqrt{-g}\left[-\frac{1}{2}g^{\mu\nu}\nabla_\mu\phi\nabla_\nu\phi - V(\phi)\right], \tag{8.166}$$

for which the energy-momentum tensor is

$$T_{\mu\nu} = \nabla_\mu\phi\nabla_\nu\phi + \left[\tfrac{1}{2}g^{\rho\sigma}\nabla_\rho\phi\nabla_\sigma\phi - V(\phi)\right]g_{\mu\nu} \tag{8.167}$$

and the equation of motion is

$$\Box\phi - \frac{dV}{d\phi} = 0. \tag{8.168}$$

Assume that the field is completely homogeneous through space ($\partial_i\phi = 0$). Then using the Christoffel symbols (8.44), we may express the d'Alembertian in terms of time derivatives and the Hubble constant to write (8.168) as

$$\ddot{\phi} + 3H\dot{\phi} + \frac{dV}{d\phi} = 0. \tag{8.169}$$

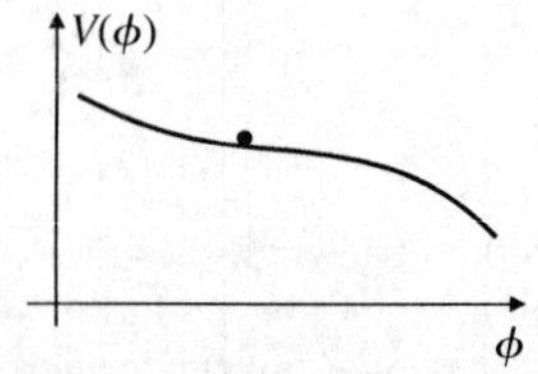

FIGURE 8.6 Potential energy for a slowly-rolling scalar field.

We see that the Hubble parameter acts as a friction term; the field will tend to roll down the potential, but when H is too large the motion will be damped. Therefore, a scalar field with a sufficiently shallow potential (as portrayed in Figure 8.6) will roll very slowly, leading to a kinetic energy much smaller than the potential energy $V(\phi)$. The energy-momentum tensor is then

$$T_{\mu\nu} \approx -V(\phi)g_{\mu\nu}, \tag{8.170}$$

where $\phi \approx$ constant. Comparing to (4.96), we see that the scalar field potential is mimicking a vacuum energy. As a simple example consider a quadratic potential, $V(\phi) = \frac{1}{2}m^2\phi^2$. Then (8.169) describes a damped harmonic oscillator, and overdamping will occur if $H > m$. But in particle-physics units, the Hubble constant today is $H_0 \approx 10^{-33}$ eV, so the mass of this scalar field would have to be incredibly tiny compared to the masses of the familiar elementary particles in equation (8.163). This seems to be an unnatural fine-tuning. Nevertheless, models of dynamical dark energy are being actively explored, partially in the hope that they will lead somehow to a solution of the coincidence problem.

With this view of the contemporary situation, we can imagine what the early universe must have been like to have produced what we see today. For purposes of physical intuition it is often more helpful to keep track of the era under consideration by indicating the temperature rather than the redshift or time since the Big Bang. The temperature today is

$$T_0 = 2.74 \text{ K} = 2.4 \times 10^{-4} \text{ eV}. \tag{8.171}$$

Of course, by "temperature" we mean the apparent blackbody temperature of the cosmic microwave background; in fact the CMB has not been in thermal equilibrium since recombination, so one should be careful in taking this concept too literally. Under adiabatic expansion, the temperature decreases as each relativistic particle redshifts, and we have $T \propto a^{-1}$. But there will be nonadiabatic phase transitions at specific moments in the early universe; in such circumstances the temperature doesn't actually increase, but decreases more gradually. To help relate the temperature, density, and scale factor, we introduce two different measures of the **effective number of relativistic degrees of freedom**: g_* and g_{*S} (where S stands for entropy). Consider a set of bosonic and fermionic species, each with their own effective temperature T_i, and number of spin states g_i. For example, a massless photon has two spin states, so $g_\gamma = 2$; a massive spin-$\frac{1}{2}$ fermion also has two spin states, so $g_{e^-} = g_{e^+} = 2$. The two different versions of the effective number of relativistic degrees of freedom obey

$$g_* = \sum_{\text{bosons}} g_i \left(\frac{T_i}{T}\right)^4 + \frac{7}{8} \sum_{\text{fermions}} g_i \left(\frac{T_i}{T}\right)^4 \tag{8.172}$$

and

$$g_{*S} = \sum_{\text{bosons}} g_i \left(\frac{T_i}{T}\right)^3 + \frac{7}{8} \sum_{\text{fermions}} g_i \left(\frac{T_i}{T}\right)^3. \tag{8.173}$$

The mysterious factors of $\frac{7}{8}$ arise from the difference between Bose and Fermi statistics when calculating the equilibrium distribution function. For any species in thermal equilibrium, the temperature T_i will be equal to the background temperature T; but we might have decoupled species at a lower temperature, which contribute less to the effective number of relativistic degrees of freedom. The reason why we need to define two different measures is that they play different roles; the first relates the temperature to the energy density (in relativistic species) via

$$\rho_R = \frac{\pi^2}{30} g_* T^4, \tag{8.174}$$

while the second relates the temperature to the scale factor,

$$T \propto g_{*S}^{-1/3} a^{-1}. \tag{8.175}$$

In fact, g_* and g_{*S} are expected to be approximately equal so long as the relativistic degrees of freedom are those of the Standard Model of particle physics. A very rough guide is given by

$$g_* \approx g_{*S} \sim \begin{cases} 100 & T > 300 \text{ MeV} \\ 10 & 300 \text{ MeV} > T > 1 \text{ MeV} \\ 3 & T < 1 \text{ MeV}. \end{cases} \tag{8.176}$$

As we will discuss shortly, the events that change the effective number of relativistic degrees of freedom are the QCD phase transition at 300 MeV, and the annihilation of electron/positron pairs at 1 MeV.

With this background, let us consider the evolution of the universe from early times to today. To begin we imagine a Robertson–Walker metric with matter fields in thermal equilibrium at a temperature of 1 TeV = 1000 GeV. The high-temperature plasma is a complicated mixture of elementary particles (quarks, leptons, gauge and Higgs bosons). The dominant form of energy density will be relativistic particles, so the early universe is radiation-dominated. It is also very close to flat, since the curvature term in the Friedmann equation evolves more slowly than the matter and radiation densities. The Friedmann equation is therefore

$$\begin{aligned} H^2 &= \frac{8\pi G}{3}\rho_R \\ &\approx 0.1 g_* \frac{T^4}{\bar{m}_{\rm P}^2}, \end{aligned} \tag{8.177}$$

where the reduced Planck scale is $\bar{m}_{\rm P} = (8\pi G)^{-1/2} \approx 10^{18}$ GeV. If the radiation-dominated phase extends back to very early times, the age of the universe will be approximately $t \sim H^{-1}$, or

$$t \sim \frac{\bar{m}_{\rm P}}{T^2}. \tag{8.178}$$

In conventional units this becomes

$$t \sim 10^{-6} \left(\frac{\text{GeV}}{T}\right)^2 \text{ sec.} \tag{8.179}$$

Current experiments at particle accelerators have provided an accurate picture of what physics is like up to perhaps 100 GeV, so an additional order of magnitude is within the realm of reasonable extrapolation. At higher temperatures we are less sure what happens; there might be nothing very interesting between 1 TeV and the Planck scale, or this regime could be filled with all manner of surprises. Of course it is also conceivable that cosmology provides surprises at even lower temperatures, even though the Standard Model physics is well understood; in this section we are describing a conservative scenario, but as always it pays to keep an open mind.

A crucial feature of the Standard Model is the spontaneously broken symmetry of the electroweak sector. In cosmology, this symmetry breaking occurs at the electroweak phase transition, at $T \sim 200$ GeV. Above this temperature the symmetry is unbroken, so that elementary fermions (quarks and leptons) and the weak interaction gauge bosons are all massless, while below this temperature we have the pattern of masses familiar from low-energy experiments. The electroweak phase transition is not expected to leave any discernible impact on the late universe; one possible exception is baryogenesis, discussed below.

At these temperatures the strong interactions described by quantum chromodynamics (QCD) are not so strong. At low energies/temperatures, QCD exhibits "confinement"—quarks and gluons are bound into composite particles such as baryons and mesons. But above the QCD scale $\Lambda_{\rm QCD} \sim 300$ MeV, quarks and gluons are free particles. As the universe expands and cools, the confinement of strongly-interacting particles into bound states is responsible for the first drop in the effective number of relativistic degrees of freedom noted in (8.176). The QCD phase transition is not expected to leave a significant imprint on the observable universe.

Just as the strong interactions are not very strong at high temperatures, the weak interactions are not as weak as you might think; they are still weak in the sense of being accurately described by perturbation theory, but they occur rapidly enough to keep weakly-interacting particles, such as neutrinos, in thermal equilibrium. This ceases to be the case when $T \sim 1$ MeV. This is also approximately the temperature at which electrons and positrons become nonrelativistic and annihilate, decreasing the effective number of relativistic degrees of freedom, but the two events are unrelated. For temperatures below 1 MeV, we say the weak interactions are "frozen out"—the interaction rate drops below the expansion rate of the universe, so interactions happen too infrequently to keep particles in equilibrium. It may be the case that cold dark matter particles decouple from the plasma at this temperature. More confidently, we can infer that neutrons and protons cease to interconvert. The equilibrium abundance of neutrons at this temperature is about $\frac{1}{6}$ the abundance of protons (due to the slightly larger neutron mass). The neutrons have a finite lifetime ($\tau_n = 890$ sec) that is somewhat larger than the age of the universe at this epoch, $t(1 \text{ MeV}) \approx 1$ sec, but they begin to gradually decay into protons and leptons. Soon thereafter, however, we reach a temperature somewhat below 100 keV, and **Big-Bang Nucleosynthesis** (BBN) begins.

The nuclear binding energy per nucleon is typically of order 1 MeV, so you might expect that nucleosynthesis would occur earlier; however, the large number of photons per nucleon prevents nucleosynthesis from taking place until the temperature drops below 100 keV. At that point the neutron/proton ratio is approximately $\frac{1}{7}$. Of all the light nuclei, it is energetically favorable for the nucleons to reside in ^{4}He, and indeed that is what most of the free neutrons are converted into; for every two neutrons and fourteen protons, we end up with one helium nucleus and twelve protons. Thus, about 25% of the baryons by mass are converted to helium. In addition, there are trace amounts of deuterium (approximately 10^{-5} deuterons per proton), ^{3}He (also $\sim 10^{-5}$), and ^{7}Li ($\sim 10^{-10}$).

Of course these numbers are predictions, which are borne out by observations of the primordial abundances of light elements. (Heavier elements are not synthesized in the Big Bang, but require stellar processes in the later universe.) We have glossed over numerous crucial details, especially those that explain how the different abundances depend on the cosmological parameters. For example, imagine that we deviate from the Standard Model by introducing more than three light neutrino species. This would increase the radiation energy density at a fixed temperature through (8.174), which in turn decreases the timescales associated with a given temperature (since $t \sim H^{-1} \propto \rho_R^{-1/2}$). Nucleosynthesis would therefore happen somewhat earlier, resulting in a higher abundance of neutrons, and hence in a larger abundance of ^{4}He. Observations of the primordial helium abundance, which are consistent with the Standard Model prediction, provided the first evidence that the number of light neutrinos is close to three. Similarly, all of the temperatures and timescales associated with nucleosynthesis depend on the baryon-to-photon ratio; agreement with the observed abundances requires that there be approximately 5×10^{-10} baryons per photon, which is the origin of the estimate (8.164) of the baryonic density parameter, and the associated need for nonbaryonic dark matter.

For our present purposes, perhaps the most profound feature of primordial nucleosynthesis is its sensitive dependence on the Friedmann relation between temperature and expansion rate, and hence on Einstein's equation. The success of BBN provides a stringent test of GR in a regime very far from our everyday experience. The fact that Einstein's theory, derived primarily from a need to reconcile gravitation with invariance under the Lorentz symmetries of electromagnetism, successfully describes the expansion of the universe when it was only one second old is a truly impressive accomplishment. To this day, BBN provides one of the most powerful constraints on alternative theories of gravity; in particular, it is the earliest epoch about which we have any direct observational signature.

Subsequent to nucleosynthesis, we have a plasma dominated by protons, electrons, and photons, with some helium and other nuclei. There is also dark matter, but it is assumed not to interact with the ordinary matter by this epoch. The next important event isn't until **recombination**, when electrons combine with protons (they combine with helium slightly earlier). Recombination happens at a temperature $T \approx 0.3$ eV; at this point the universe is matter-dominated. Again, since the binding energy of hydrogen is 13.6 eV, you might expect recombination to occur earlier, but the large photon/baryon ratio delays it. The crucial importance of recombination is that it marks the epoch at which the universe becomes transparent. The ambient photons interact strongly with free electrons, so that the photon mean free path is very short prior to recombination, but it becomes essentially infinite once the electrons and protons combine into neutral hydrogen. These ambient photons are visible today as the cosmic microwave background, which provides a snapshot of the universe at $T \approx 0.3$ eV, or a redshift $z \approx 1200$. Recombination is a somewhat gradual process, so any specification of when it happens is necessarily approximate.

Subsequent to recombination, the universe passes through a long period known as the "dark ages," as galaxies are gradually assembled through gravitational instability, but there are as yet no visible stars to light up the universe. The dark ages are a mysterious time; the processes by which stars and galaxies form are highly complicated and nonlinear, and new kinds of observations will undoubtedly be necessary before this era is well understood.

Our story has now brought us to the present day, but there are a couple of missing points we should go back and fill in. One is the asymmetry between matter and antimatter in the universe. Essentially all of the visible matter in the universe seems to be composed of protons, neutrons, and electrons, rather than their antiparticles; if distant galaxies were primarily antimatter, we would expect to observe high-energy photons from the occasional annihilation of protons with antiprotons at the boundaries of the matter/antimatter domains. While it is possible to build in an asymmetry as an initial condition, this seems somehow unsatisfying, and most physicists would prefer to find a dynamical mechanism of baryogenesis by which an initially matter/antimatter symmetric state could evolve into our present universe. Such broken symmetries are common in particle physics, and indeed numerous mechanisms for baryogenesis have been proposed (generally at temperatures at or above the electroweak scale). None of these specific schemes, however, has proven sufficiently compelling to be adopted as a standard scenario. It seems probable that we will need a better understanding of physics beyond the Standard Model to understand the origin of the baryon asymmetry.

The other missing feature we need to mention is that the universe is not, of course, perfectly homogeneous and isotropic; the current large-scale structure in the universe seems to have evolved from adiabatic and nearly scale-free perturbations present at very early times at the level of $\delta\rho/\rho \sim 10^{-5}$. Evidence for the adiabatic and scale-free nature of these perturbations comes from a combination of observations of the CMB and large-scale structure. Both the high degree of isotropy and homogeneity, and the small deviations therefrom, are simply imposed as mysterious initial conditions in the conventional cosmology. A possible dynamical origin for both is provided by the inflationary-universe scenario, to which we now turn.

8.8 ■ INFLATION

In the conventional understanding of the Big-Bang model, the universe is taken to be radiation-dominated at early times and matter-dominated at late times, with, as we now suspect, a very late transition to vacuum-domination. This picture has met with great success in describing a wide variety of observational data; nevertheless, we may still ask whether the initial conditions giving rise such a universe seem natural. This is the kind of question one might ask in cosmology but not in other sciences. Typically, as physicists we look for laws of nature, and imagine that we are free to specify initial conditions and ask how they evolve under such laws. But the universe seems to have only one set of initial conditions, so it seems sensible to wonder if they are relatively generic or finely-tuned. Within the conventional picture, the early universe is indeed finely tuned to incredible precision. In particular, two features of our universe seem highly nongeneric: its spatial flatness, and its high degree of isotropy and homogeneity. It might be that this is just the universe we are stuck with, and it makes no sense to ask about the likelihood of different initial conditions. Alternatively, it might be that these conditions are more likely than they appear at first, if there is some dynamical mechanism that can take a wide spectrum of initial conditions and evolve them toward flatness and homogeneity/isotropy. The inflationary universe scenario provides such a mechanism (and more, besides), and has become a central organizing principle of modern cosmology, even if we are still far from demonstrating its truth.

Before describing inflation, let's describe the two problems of unnaturalness it claims to solve: the flatness problem and the horizon problem associated with homogeneity/isotropy. The **flatness problem** comes from considering the Friedmann equation in a universe with matter and radiation but no vacuum energy, which for later convenience we write in terms of the reduced Planck mass $\bar{m}_P = (8\pi G)^{-1/2}$ as

$$H^2 = \frac{1}{3\bar{m}_P^2}(\rho_M + \rho_R) - \frac{\kappa}{a^2}. \tag{8.180}$$

The curvature term $-\kappa/a^2$ is proportional to a^{-2} (obviously), while the energy density terms fall off faster with increasing scale factor, $\rho_M \propto a^{-3}$ and $\rho_R \propto a^{-4}$. This raises the question of why the ratio $(\kappa a^{-2})/(\rho/3\bar{m}_P^2)$ isn't much larger than unity, given that a has increased by a factor of perhaps 10^{30} since the Planck epoch. Said another way, the point $\Omega = 1$ is a repulsive fixed point in a matter/radiation dominated universe—any deviation from this value will grow with time, so why do we observe $\Omega \sim 1$ today?

The **horizon problem** stems from the existence of particle horizons in FRW cosmologies, as illustrated in Figure 8.7. Horizons exist because there is only a finite amount of time since the Big Bang singularity, and thus only a finite distance that photons can travel within the age of the universe, as we briefly discussed in Chapter 2. Consider a photon moving along a radial trajectory in a flat universe (the generalization to nonflat universes is straightforward). A radial null path obeys

$$0 = ds^2 = -dt^2 + a^2\, dr^2, \tag{8.181}$$

so the comoving (coordinate) distance traveled by such a photon between times t_1 and t_2 is

$$\Delta r = \int_{t_1}^{t_2} \frac{dt}{a(t)}. \tag{8.182}$$

To get the physical distance as it would be measured by an observer at any time t, simply multiply by $a(t)$. For simplicity let's imagine we are in a matter-dominated

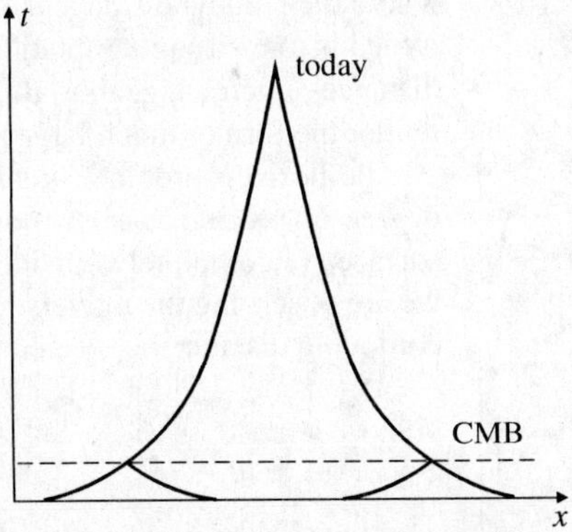

FIGURE 8.7 Past light cones in a universe expanding from a Big Bang singularity, illustrating particle horizons in cosmology. Points at recombination, observed today as parts of the cosmic microwave background on opposite sides of the sky, have nonoverlapping past light cones (in conventional cosmology); no causal signal could have influenced them to have the same temperature.

universe, for which

$$a = \left(\frac{t}{t_0}\right)^{2/3}. \tag{8.183}$$

Remember $a_0 = 1$. The Hubble parameter is therefore given by

$$\begin{aligned} H &= \tfrac{2}{3}t^{-1} \\ &= a^{-3/2}H_0. \end{aligned} \tag{8.184}$$

Then the photon travels a comoving distance

$$\Delta r = 2H_0^{-1}\left(\sqrt{a_2} - \sqrt{a_1}\right). \tag{8.185}$$

The comoving horizon size at any fixed value of the scale factor $a = a_*$ is the distance a photon travels since the Big Bang,

$$r_{\rm hor}(a_*) = 2H_0^{-1}\sqrt{a_*}. \tag{8.186}$$

The physical horizon size, as measured on the spatial hypersurface at a_*, is therefore simply

$$d_{\rm hor}(a_*) = a_* r_{\rm hor}(a_*) = 2H_*^{-1}. \tag{8.187}$$

Indeed, for any nearly-flat universe containing a mixture of matter and radiation, at any one epoch we will have

$$d_{\rm hor}(a_*) \sim H_*^{-1} = d_H(a_*), \tag{8.188}$$

where the Hubble distance d_H was introduced in (8.71). This approximate equality leads to a strong temptation to use the terms "horizon distance" and "Hubble distance" interchangeably; this temptation should be resisted, since inflation can render the former much larger than the latter, as we will soon demonstrate.

The horizon problem is simply the fact that the CMB is isotropic to a high degree of precision, even though widely separated points on the last scattering surface are completely outside each others' horizons. When we look at the CMB we are observing the universe at a scale factor $a_{\rm CMB} \approx 1/1200$; from (8.185), the comoving distance between a point on the CMB and an observer on Earth is

$$\begin{aligned}\Delta r &= 2H_0^{-1}\left(1 - \sqrt{a_{\rm CMB}}\right)\\ &\approx 2H_0^{-1}.\end{aligned} \tag{8.189}$$

However, the comoving horizon distance for such a point is

$$\begin{aligned}r_{\rm hor}(a_{\rm CMB}) &= 2H_0^{-1}\sqrt{a_{\rm CMB}}\\ &\approx 6\times 10^{-2}H_0^{-1}.\end{aligned} \tag{8.190}$$

Hence, if we observe two widely-separated parts of the CMB, they will have nonoverlapping horizons; distinct patches of the CMB sky were causally disconnected at recombination. Nevertheless, they are observed to be at the same temperature to high precision. The question then is, how did they know ahead of time to coordinate their evolution in the right way, even though they were never in causal contact? We must somehow modify the causal structure of the conventional FRW cosmology.

Let's consider modifying the conventional picture by positing a period of **inflation**: an era of acceleration ($\ddot{a} > 0$) in the very early universe, driven by some component other than matter or radiation that redshifts away slowly as the universe expands. Then the flatness and horizon problems can be simultaneously solved. For simplicity consider the case where inflation is driven by a constant vacuum energy, leading to exponential expansion. Then, during the vacuum-dominated era, $\rho/3\bar{m}_{\rm P}^2 \propto a^0$ grows rapidly with respect to $-\kappa/a^2$, so the universe becomes flatter with time (Ω is driven to unity). If this process proceeds for a sufficiently long period, after which the vacuum energy is converted into matter and radiation, the density parameter will be sufficiently close to unity that it will not have had a chance to noticeably change into the present era. The horizon problem, meanwhile, can be traced to the fact that the physical distance between any two comoving objects grows as the scale factor, while the physical horizon size in a matter- or radiation-dominated universe grows more rapidly, as $d_{\rm hor} \sim a^{n/2}H_0^{-1}$. This can again be solved by an early period of exponential expansion, in which the true horizon size grows to a fantastic amount, so that our horizon today is actually much larger than the naïve estimate that it is equal to the Hubble radius H_0^{-1}.

In fact, a truly exponential expansion is not necessary; for any accelerated expansion, the spatial curvature will diminish with respect to the energy density, and the horizon distance will grow rapidly. Typically we require that this accelerated period be sustained for 60 or more e-folds (where the number of e-folds is $N = \Delta \ln a$) which is what is needed to solve the horizon problem. It is easy to overshoot, and inflation generally makes the present-day universe spatially flat to incredible precision.

Now let's consider how we can get an inflationary phase in the early universe. The most straightforward way is to use the vacuum energy provided by the potential of a scalar field, the **inflaton**. Imagine a universe dominated by the energy of a spatially homogeneous scalar. The relevant equations of motion are precisely those of our discussion of dynamical dark energy in Section 8.7; the only difference is that the energy scale of inflation is much higher. We have the equation of motion for a scalar field in an RW metric,

$$\ddot{\phi} + 3H\dot{\phi} + V'(\phi) = 0, \tag{8.191}$$

as well as the Friedmann equation,

$$H^2 = \frac{1}{3\bar{m}_{\rm P}^2}\left(\frac{1}{2}\dot{\phi}^2 + V(\phi)\right). \tag{8.192}$$

We have ignored the curvature term, since inflation will flatten the universe anyway. Inflation can occur if the evolution of the field is sufficiently gradual that the potential energy dominates the kinetic energy, and the second derivative of ϕ is small enough to allow this state of affairs to be maintained for a sufficient period. Thus, we want

$$\begin{aligned} \dot{\phi}^2 &\ll V(\phi), \\ |\ddot{\phi}| &\ll |3H\dot{\phi}|,\ |V'|. \end{aligned} \tag{8.193}$$

Satisfying these conditions requires the smallness of two dimensionless quantities known as **slow-roll parameters**:

$$\begin{aligned} \epsilon &= \frac{1}{2}\bar{m}_{\rm P}^2\left(\frac{V'}{V}\right)^2, \\ \eta &= \bar{m}_{\rm P}^2\left(\frac{V''}{V}\right). \end{aligned} \tag{8.194}$$

Note that $\epsilon \geq 0$, while η can have either sign. Note also that these definitions are not universal; some people like to define them in terms of the Hubble parameter rather than the potential. Our choice describes whether a field has a chance to roll slowly for a while; the description in terms of the Hubble parameter describes whether the field actually is rolling slowly. When both of these quantities are small we can have a prolonged inflationary phase. They are not sufficient, however; no matter what the potential looks like, we can always choose initial conditions with $|\dot{\phi}|$ so large that slow-roll is never applicable. However, most initial conditions are attracted to an inflationary phase if the slow-roll parameters are small.

It isn't hard to invent potentials that satisfy the slow-roll conditions. Consider perhaps the simplest possible example,[3] $V(\phi) = \frac{1}{2}m^2\phi^2$. In this case

$$\epsilon = \eta = \frac{2\bar{m}_{\rm P}^2}{\phi^2}. \tag{8.195}$$

Clearly, for large enough ϕ, we can get the slow-roll parameters to be as small as we like. However, we have the constraint that the energy density should not be as high as the Planck scale, so that our classical analysis makes sense; this implies $\phi \ll \bar{m}_{\rm P}^2/m$. If we start the field at a value ϕ_i, the number of e-folds before inflation ends (that is, before the slow-roll parameters become of order unity) will be

$$\begin{aligned} N &= \int_{t_i}^{t_e} H\,dt \\ &\approx -\bar{m}_{\rm P}^{-2}\int_{\phi_i}^{\phi_e} \frac{V}{V'}\,d\phi \\ &\approx \frac{\phi_i^2}{4\bar{m}_{\rm P}^2} - \frac{1}{2}. \end{aligned} \tag{8.196}$$

The first equality is always true, the second uses the slow-roll approximation, and the third is the result for this particular model. To get 60 e-folds we therefore need $\phi_i > 16\bar{m}_{\rm P}$. Together with the upper limit on the energy density, we find that there is an upper limit on the mass parameter, $m \ll \bar{m}_{\rm P}/16$. In fact the size of the observed density fluctuations puts a more stringent upper limit on m, as we will discuss below. But there is no lower limit on m, so it is easy to obtain appropriate inflationary potentials only if we are willing to posit large hierarchies $m \ll \bar{m}_{\rm P}$, or equivalently a small dimensionless number $m/\bar{m}_{\rm P}$. Going through the same exercise with a $\lambda\phi^4$ potential would have yielded a similar conclusion, that λ would have had to be quite small; we often say that the inflaton must be weakly coupled. Of course, there is a sense in which we are cheating, since for field values $\phi > \bar{m}_{\rm P}$ we should expect additional terms in the effective potential, of the form $\bar{m}_{\rm P}^{4-n}\phi^n$ with $n > 4$, to become important. So in a *realistic* model it can be quite hard to get an appropriate potential.

At some point inflation ends, and the energy in the inflaton potential is converted into a thermalized gas of matter and radiation, a process known as "reheating." A proper understanding of the reheating process is of utmost importance, as it controls the production of various relics that we may or may not want in our universe. For example, one important beneficial aspect of inflation is that it can "inflate away" various relics that could be produced in the early universe, but are not observed today. A classic example occurs in the context of grand unified theories of particle physics, which generically predict the existence of super-

[3]We follow the exposition in A.R. Liddle, "An Introduction to Cosmological Inflation," `http://arxiv.org/astro-ph/9901124`.

heavy magnetic monopoles, with an abundance many orders of magnitude greater than allowed by observations. Historically, the monopole problem was the primary motivation for the invention of inflation by Guth; solutions to the flatness and horizon problems were considered a bonus. Inflation can dilute the monopole abundance appropriately, but they will be produced anew if the universe reheats to above the temperature of the grand-unification phase transition; fortunately, this is not a stringent constraint on most models. Similar considerations apply to other unwanted relics; in supersymmetric models, an especially worrisome problem is raised by the abundance of gravitinos (supersymmetric partners of the graviton). At the same time, it is necessary to reheat to a sufficiently high temperature to allow for some sort of baryogenesis scenario. For any specific implementation of inflation within a particle-physics model, it is crucial to check that unwanted relics are dispersed while wanted relics (such as baryons) are preserved.

A crucial element of inflationary scenarios is the production of density perturbations, which may be the origin of the CMB temperature anisotropies and the large-scale structure in galaxies that we observe today. The idea behind density perturbations generated by inflation is fairly straightforward. Inflation will attenuate any ambient particle density rapidly to zero, leaving behind only the vacuum. But the vacuum state in an accelerating universe has a nonzero temperature, the Gibbons–Hawking temperature, analogous to the Hawking temperature of a black hole. We won't be able to explore this subject in detail; here we simply outline the basic results.

For a universe dominated by a potential energy V the Gibbons–Hawking temperature is given by

$$T_{\mathrm{GH}} = \frac{H}{2\pi} \sim \frac{V^{1/2}}{\bar{m}_{\mathrm{P}}}. \tag{8.197}$$

Corresponding to this temperature are fluctuations in the inflaton field ϕ at each wavenumber k, with magnitude

$$|\Delta\phi|_k = T_{\mathrm{GH}}. \tag{8.198}$$

Since the potential is by hypothesis nearly flat, the fluctuations in ϕ lead to small fluctuations in the energy density,

$$\delta\rho = V'(\phi)\delta\phi. \tag{8.199}$$

Inflation therefore produces density perturbations on every scale. The amplitude of the perturbations is nearly equal at each wavenumber, but there will be slight deviations due to the gradual change in V as the inflaton rolls. Describing the perturbations is a messy subject, involving countless different notations. A sensible place to start is root-mean-square (RMS) density fluctuation,

$$\left.\frac{\delta\rho}{\rho}\right|_{\mathrm{rms}} = \sqrt{\left\langle\left(\frac{\delta\rho}{\rho}\right)^2\right\rangle}, \tag{8.200}$$

where the angle brackets represent an average over spatial locations. For statistically isotropic perturbations (the expected amplitude is independent of direction), a bit of Fourier analysis allows us to write

$$\left(\frac{\delta\rho}{\rho}\bigg|_{\rm rms}\right)^2 = \int \Delta^2(k)\, d(\ln k), \tag{8.201}$$

where we have introduced the dimensionless power spectrum,

$$\Delta^2(k) \equiv \frac{k^3|\delta_k|^2}{2\pi^2}, \tag{8.202}$$

and δ_k is the expectation value of the Fourier transform of the fractional density perturbation,

$$\delta_{\mathbf{k}} = \frac{1}{(2\pi)^{3/2}} \int e^{-i\mathbf{k}\cdot\mathbf{x}} \frac{\delta\rho}{\rho}\, d^3x, \tag{8.203}$$

which we've assumed to be isotropic. The dimensionless power spectrum is a function of time, as the amplitude for each mode evolves; it is most common to express the predictions of any specific model in terms of the amplitude of the perturbations at the moment when the physical wavelength of the mode, $\lambda = a/k$, is equal to the Hubble radius H^{-1},

$$A_{\rm S}^2(k) \equiv \Delta^2(k)\Big|_{k=aH}. \tag{8.204}$$

Thus, $A_{\rm S}(k)$ measures the amplitude for different modes at different times. For inflation driven by a slowly-rolling scalar field, $A_{\rm S}(k)$ is related to the potential via

$$A_{\rm S}^2(k) \sim \frac{V^3}{\bar{m}_{\rm P}^6 (V')^2}\bigg|_{k=aH} \sim \frac{V}{\bar{m}_{\rm P}^4 \epsilon}\bigg|_{k=aH}. \tag{8.205}$$

We have intentionally suppressed dimensionless numerical factors, which differ widely from reference to reference, in favor of highlighting the dependence on the potential.

The spectrum is given the subscript "S" because it describes scalar fluctuations in the metric. These are tied to the energy-momentum distribution, and the density fluctuations produced by inflation are adiabatic—fluctuations in the density of all species are correlated. The fluctuations are also Gaussian, in the sense that the phases of the Fourier modes describing fluctuations at different scales are uncorrelated. These aspects of inflationary perturbations—a nearly scale-free spectrum of adiabatic density fluctuations with a Gaussian distribution—are all consistent with current observations of the CMB and large-scale structure, and new data scheduled to be collected in years to come should greatly improve the precision of these tests.

It is not only the nearly-massless inflaton that is excited during inflation, but also any other nearly-massless particle. The other important example is the graviton, which corresponds to tensor perturbations in the metric (propagating excitations of the gravitational field). Tensor fluctuations have a spectrum

$$A_{\mathrm{T}}^2(k) \sim \left.\frac{V}{\bar{m}_{\mathrm{P}}^4}\right|_{k=aH} . \tag{8.206}$$

Importantly, the tensor amplitude depends only on the potential, not on its derivatives; observations of tensor perturbations would therefore give direct information about the energy scale of inflation.

For purposes of understanding observations, it is useful to parameterize the perturbation spectra in terms of observable quantities. We therefore write

$$A_{\mathrm{S}}^2(k) \propto k^{n_{\mathrm{S}}-1} \tag{8.207}$$

and

$$A_{\mathrm{T}}^2(k) \propto k^{n_{\mathrm{T}}}, \tag{8.208}$$

where n_{S} and n_{T} are the spectral indices. They are related to the slow-roll parameters of the potential by

$$n_{\mathrm{S}} = 1 - 6\epsilon + 2\eta \tag{8.209}$$

and

$$n_{\mathrm{T}} = -2\epsilon. \tag{8.210}$$

In models of the type we have considered (driven by single slowly-rolling scalar fields), there is a consistency relation relating the amplitudes and spectral indices of the scalar and tensor modes. It can be expressed in a convention-independent way as a relation between observable quantities, temperature fluctuations ΔT due to the different perturbations, as

$$\frac{(\Delta T/T)_{\mathrm{T}}^2}{(\Delta T/T)_{\mathrm{S}}^2} = -7n_{\mathrm{T}}. \tag{8.211}$$

The existence of tensor perturbations is a crucial prediction of inflation that may in principle be verifiable through observations of the polarization of the CMB. Polarization is also induced by ordinary density fluctuations, through the anisotropy of the Thompson scattering cross-section in an inhomogeneous plasma. Fortunately, we can imagine decomposing the polarization vector field on the sky into a curl-free part (E-modes) and a curl part (B-modes); the scalar perturbations lead to E-mode polarization, whereas tensor perturbations lead to B-modes (up to some inevitable processing in the post-recombination universe). CMB polarization has been detected; the challenge for the future will be to separate out the scalar and tensor contributions, to test the prediction (8.211) of simple inflationary models. Of course this requires not only detecting the tensor-induced polarization, but measuring its spectral index with some precision.

Our current knowledge of the amplitude of the perturbations already gives us important information about the energy scale of inflation. The tensor perturbations depend on V alone, not its derivatives; if the CMB anisotropies seen by COBE are due to tensor fluctuations (possible, although unlikely), we can instantly derive $V_{\text{inflation}} \sim (10^{16}\ \text{GeV})^4$. Here, the value of V being constrained is that which was responsible for creating the observed fluctuations; namely, 60 e-folds before the end of inflation. This is remarkably reminiscent of the grand unification scale, which is very encouraging. Even in the more likely case that the perturbations observed in the CMB are scalar in nature, we can still write

$$V_{\text{inflation}}^{1/4} \sim \epsilon^{1/4} 10^{16}\ \text{GeV}, \tag{8.212}$$

where ϵ is the slow-roll parameter defined in (8.194). Although we expect ϵ to be small, the $1/4$ in the exponent means that the dependence on ϵ is quite weak; unless this parameter is extraordinarily tiny, it is very likely that $V_{\text{inflation}}^{1/4} \sim 10^{15}$-$10^{16}$ GeV. The fact that we can have such information about such tremendous energy scales is a cause for great wonder.

8.9 ■ EXERCISES

1. Consider an $(N+n+1)$-dimensional spacetime with coordinates $\{t, x^I, y^i\}$, where I goes from 1 to N and i goes from 1 to n. Let the metric be

$$ds^2 = -dt^2 + a^2(t)\delta_{IJ}\, dx^I dx^J + b^2(t)\gamma_{ij}(y)\, dy^i dy^j, \tag{8.213}$$

where δ_{IJ} is the usual Kronecker delta and $\gamma_{ij}(y)$ is the metric on an n-dimensional maximally symmetric spatial manifold. Imagine that we normalize the metric γ such that the curvature parameter

$$k = \frac{R(\gamma)}{n(n-1)} \tag{8.214}$$

is either $+1$, 0, or -1, where $R(\gamma)$ is the Ricci scalar corresponding to the metric γ_{ij}.

(a) Calculate the Ricci tensor for this metric.

(b) Define an energy-momentum tensor in terms of an energy density ρ and pressure in the x^I and y^i directions, $p^{(N)}$ and $p^{(n)}$:

$$T_{00} = \rho \tag{8.215}$$

$$T_{IJ} = a^2 p^{(N)} \delta_{IJ} \tag{8.216}$$

$$T_{ij} = b^2 p^{(n)} \gamma_{ij}. \tag{8.217}$$

Plug the metric and $T_{\mu\nu}$ into Einstein's equations to derive Friedmann-like equations for a and b (three independent equations in all).

(c) Derive equations for the energy density and the two pressures at a static solution where $\dot{a} = \dot{b} = \ddot{a} = \ddot{b} = 0$, in terms of k, n, and N. Use these to derive expressions for the equation-of-state parameters $w^{(N)} = p^{(N)}/\rho$ and $w^{(n)} = p^{(n)}/\rho$, valid at the static solution.

2. Consider de Sitter space in coordinates where the metric takes the form

$$ds^2 = -dt^2 + e^{2Ht}[dx^2 + dy^2 + dz^2]. \tag{8.218}$$

Solve the geodesic equation for noncomoving observers (x^i not constant) to find the affine parameter as a function of t. Show that the geodesics reach $t = -\infty$ in a finite affine parameter, demonstrating that these coordinates fail to cover the entire manifold.

3. In Appendix F we discuss Raychaudhuri's equation. Show that, applied to a Robertson–Walker cosmology, the Raychaudhuri equation is equivalent to the second Friedmann equation, (8.68).

4. Consider the best-fit universe, with density parameters $\Omega_{R0} = 10^{-4}$, $\Omega_{M0} = 0.3$, $\Omega_{\Lambda 0} = 0.7$. Make a plot of the three Ω_i's as a function of the scale factor a, on a log scale, from $a = 10^{-35}$ to $a = 10^{35}$. Indicate the Planck time, nucleosynthesis, and today.

5. In a flat spacetime, objects of a fixed physical size subtend smaller and smaller angles as they are further and further away; in an expanding universe this is not necessarily so. Consider the angular size $\theta(z)$ of an object of physical size L at redshift z. In a matter-dominated flat universe, at what redshift is $\theta(z)/L$ a minimum? If all galaxies are at least 10 kpc across (and always have been), what is the minimum angular size of a galaxy in such a universe? Express your result both in terms of H_0, and plugging in $H_0 = 70$ km/s/Mpc.

6. In cosmology we tend to idealize nonrelativistic particles as having zero temperature T and pressure p. In reality, random motions will give them some temperature and pressure, satisfying $p \propto T\rho$.
   (a) How does the pressure of a gas of massive particles decay as a function of the scale factor?
   (b) Suppose neutrinos have a mass $m_\nu = 0.1$ eV, and a current temperature $T_{\nu 0} = 2$K. At about what redshift did the neutrinos go from being relativistic to nonrelativistic?

7. Suppose that the universe started out in a state of equipartition at the Planck time (so that the energy density in matter and radiation are of order the Planck density, and the temporal and spatial curvature radii are of order the Planck length). Neglecting any spatial inhomogeneity, calculate how long a positively curved universe will last, and how old a negatively curved universe would be when the temperature reaches 3K. How old would a flat universe be when the temperature reaches 3K? How old would a flat universe be by the time the expansion rate slows to $H_0 = 70$ km s^{-1} Mpc^{-1}?

CHAPTER

9 Quantum Field Theory in Curved Spacetime

9.1 ■ INTRODUCTION

Nobody believes that general relativity is the final word as far as gravity is concerned. The singularity theorems provide internal evidence that the theory is somehow incomplete; more convincing, however, is the fact that GR is a classical theory, while the world is fundamentally quantum-mechanical. The search for a working theory of quantum gravity drives a great deal of research in theoretical physics today, and much has been learned along the way, but convincing success remains elusive.

There are two parts to general relativity: the framework of spacetime curvature and its influence on matter, and the dynamics of the metric in response to energy-momentum (as described by Einstein's equation). Lacking a true theory of quantum gravity, we may still take the first part of GR—the idea that matter fields propagate on a curved spacetime background—and consider the case where those matter fields are quantum-mechanical. In other words, we take the metric to be fixed, rather than obeying some dynamical equations, and study quantum field theory (QFT) in that curved spacetime.

The epochal event in the study of QFT in curved spacetime was Hawking's realization in 1976 that black holes are not really black, but instead emit thermal radiation at a **Hawking temperature** proportional to the surface gravity κ,

$$\boxed{T = \frac{\kappa}{2\pi}.} \tag{9.1}$$

(Recall that our units set $\hbar = c = k = 1$; the Hawking temperature is actually proportional to $\hbar$ and inversely proportional to Boltzmann's constant k.) Since this remarkable discovery, QFT in curved spacetime has been put on a fairly rigorous theoretical footing, although its range of applicability is generally thought to be quite far away from any possible experimental probes. The Hawking temperature of a Schwarzschild black hole, for which $\kappa = 1/4GM$, can be written

$$T = \frac{1}{8\pi GM} = 1.2 \times 10^{26} K \left(\frac{1\ \text{g}}{M}\right) = 6.0 \times 10^{-8} K \left(\frac{M_\odot}{M}\right), \tag{9.2}$$

where $M_\odot \sim 10^{33}$ g is the mass of the Sun. So the radiation from a realistic astrophysical black hole is at a much lower temperature even than the $3K$ cosmic microwave background, and thus would be hopelessly unobservable.

Recent observations in cosmology, however, have changed this situation somewhat. One example is the apparent discovery that the universe is accelerating, which is most readily interpreted as evidence for a nonzero vacuum energy (as discussed in Chapter 8). Although the magnitude of the vacuum energy remains a profound mystery, it seems clear that an understanding of how quantum-mechanical matter behaves in curved spacetime will play an important role in any eventual resolution to the puzzle. The other example comes from cosmological perturbations. Observations of the microwave background and large-scale structure provide strong evidence in favor of a nearly scale-free spectrum of primordial perturbations, including at wavelengths that would be much larger than the horizon size in a conventional cosmology. The leading theory for the origin of these perturbations comes from inflation. In the inflationary scenario, cosmological perturbations originate in the vacuum fluctuations of quantum fields in an inflating universe. If this picture is correct, what we are seeing in maps of the CMB is the imprint of primordial quantum fluctuations, greatly stretched by the expansion of the universe, and it is these fluctuations which eventually grew via gravitational instability into the galaxies and clusters we see today. At the very least, then, cosmological observations provide strong incentive for the study of QFT in curved spacetime.

Even without this empirical motivation, thought experiments based on QFT in curved spacetime have proven very fruitful in our tentative explorations of quantum gravity. In particular, the evaporation of black holes as predicted by Hawking radiation has led to the information-loss paradox, which we will discuss below. Since it is so difficult to do real experiments that bear directly on questions of quantum gravity, we must rely on thought experiments that focus on the tension between GR and quantum mechanics, much as Einstein used thought experiments in his attempts to reconcile classical dynamics with the Lorentz invariance of electromagnetism.

With these considerations in mind, the goal of the present chapter is to provide a brief introduction to some of the ideas and results of QFT in curved spacetime. Many introductory GR books do not cover this subject, usually because familiarity with ordinary QFT in flat spacetime should not be a prerequisite for studying GR. The happy fact is, however, that a familiarity with QFT in flat spacetime is by no means necessary for studying QFT in curved spacetime. This is because the features of QFT that are most interesting and useful in flat spacetime are almost completely distinct from those that are interesting and useful in curved spacetime. Deep down, a quantum field theory is simply an example of a quantum-mechanical system, just like a square well or a helium atom. Once a field theory is defined, applications in flat spacetime (to particle physics or condensed matter) will naturally focus on the issue of interactions between the various fields, often treated as perturbations around some natural vacuum state. In curved spacetime, however, we are generally interested in the effects of spacetime itself on the fields, for which the interactions are beside the point. We therefore can consider free (noninteracting) fields, but we will have to take great care in defining what an appropriate vacuum state should be. (Indeed, as we will see, almost all of the states we deal with will be vacuum states!) Consequently, knowledge of QFT in flat spacetime is not only unnecessary for the present discussion, it probably won't even be of much help; the only prerequisite is a familiarity with the basics of ordinary quantum mechanics.

We will gradually work our way up to quantum field theory in curved spacetime, beginning with a review of the quantum mechanics of the system to which every physicist turns when the going gets rough: the simple harmonic oscillator. This is, of course, a paradigmatic example of the principles of the workings of quantum mechanics, but there is a bonus: When we next turn to field theory, we will find that the quantum mechanics of a free field in flat spacetime is precisely that of an infinite number of harmonic oscillators. (It is not that there is one oscillator at every point in space, but that each mode in the Fourier transform of the field acts like an harmonic oscillator.) The transition to field theory is then fairly straightforward. Once we grasp the basics of field theory, given our previous study of GR, it is not very difficult to generalize to curved spacetime, although a number of subtleties are encountered along the way. Our discussion will necessarily be somewhat superficial, focused on the goal of understanding the physical basis of Hawking radiation through an understanding of the Unruh effect in flat spacetime. In particular, we won't be discussing the important applications of QFT in curved spacetime to cosmology, nor will we be entering into detailed examination of renormalization and related issues. We will largely follow the discussion in Birrell and Davies (1982); look there or in Wald (1994) or in the review by Ford[1] for further discussion.

9.2 ■ QUANTUM MECHANICS

A quantum field theory is just a particular example of a quantum-mechanical system, so we can begin by reminding ourselves what that means. Of course, although the world is fundamentally quantum-mechanical, our intuition tends to align more readily with classical physics, so let's set the stage by thinking about classical mechanics. Any physical theory describing a certain system, classical or quantum, consists of the answers to three questions:

1. What are the possible states of the system? In classical mechanics, the space of states is typically given by a set of coordinates and momenta (what we might think of as "initial conditions" for the system). They can be specified exactly, and that is all there is to know about the state of the system.
2. What can we observe about the system? This question is often addressed only implicitly in classical mechanics, since the answer is trivial: any function of the coordinates and momenta qualifies as an observable.
3. How does the system evolve? This is usually expressed by a set of equations of motion. Given the state and the equations of motion, the subsequent

[1]L. H. Ford, "Quantum field theory in curved spacetime," (1997), `http://arxiv.org/gr-qc/9707062`.

evolution is uniquely defined; as a result, the space of initial conditions is equivalent to the space of classical solutions to the theory.

To make these ideas more concrete, and also because it will be directly relevant to our study of field theory, let's consider the simple harmonic oscillator. A simple harmonic oscillator may be thought of as a particle in one dimension subject to a quadratic potential. The state is specified by a single coordinate x, and a single momentum p. To get the equations of motion, we could start with the Lagrangian, which is written in terms of x and its time derivative $\dot{x}$ as

$$L = \tfrac{1}{2}\dot{x}^2 - \tfrac{1}{2}\omega^2 x^2, \tag{9.3}$$

where we have set the mass of the oscillator to unity for convenience. We can immediately derive the equation of motion

$$\ddot{x} + \omega^2 x = 0. \tag{9.4}$$

For the transition to quantum mechanics, however, it is more convenient to work in terms of the Hamiltonian, which is a function of x and p rather than x and $\dot{x}$. The Hamiltonian is related to the Lagrangian by a Legendre transformation,

$$H = p\dot{x} - L, \tag{9.5}$$

where the momentum satisfies

$$p = \frac{\partial L}{\partial \dot{x}} = \dot{x}. \tag{9.6}$$

We therefore have the Hamiltonian for the oscillator,

$$H = \tfrac{1}{2}p^2 + \tfrac{1}{2}\omega^2 x^2, \tag{9.7}$$

and Hamilton's equations

$$\frac{dx}{dt} = \partial_p H = p, \qquad \frac{dp}{dt} = -\partial_x H = -\omega^2 x, \tag{9.8}$$

serve as equations of motion. The solutions are, of course, straightforward; it is useful to express them as complex numbers

$$x(t) = x_0 e^{i(\omega t + \alpha_0)}, \tag{9.9}$$

where x_0 is the amplitude and α_0 is a phase. We can take the real part at the end of the day to get the physical answer.

Now we turn to quantum mechanics. Although quantum mechanics is profoundly different from classical mechanics, a given theory still consists of the answers to the same three questions listed above, with the answers taking somewhat different forms.

1. The state of the system is represented as an element of a **Hilbert space**. Mathematically, a Hilbert space is just a complex vector space equipped with a complex-valued inner product with the property that taking the inner product of two states in the opposite order is equivalent to complex conjugation. We denote elements of the Hilbert space as $|\psi\rangle$ and elements of the dual space as $\langle\psi|$, so that the inner product of $|\psi_1\rangle$ and $|\psi_2\rangle$ is $\langle\psi_2|\psi_1\rangle$, and obeys
$$\langle\psi_2|\psi_1\rangle^* = \langle\psi_1|\psi_2\rangle. \tag{9.10}$$
(We are glossing over technical requirements concerning completeness of the space.) In quantum mechanics the Hilbert spaces of interest are very often infinite-dimensional. For example, if a classical system is represented by coordinate x and momentum p, the Hilbert space could be taken to consist of all square-integrable complex-valued functions of x, or equivalently all square-integrable complex-valued functions of p (but not both at once).
2. Observables are represented by **self-adjoint operators** on the Hilbert space. The definition of "self-adjoint" is actually very subtle, but in simple circumstances amounts to our usual understanding of an Hermitian operator,
$$A^\dagger = A, \tag{9.11}$$
where $A^\dagger$ obeys
$$\langle\psi_2|A\psi_1\rangle = \langle A^\dagger\psi_2|\psi_1\rangle \tag{9.12}$$
for all states $|\psi_1\rangle$, $|\psi_2\rangle$. Of course many operators will not be Hermitian, but observables should have this property. In general such operators do not commute, so we cannot simultaneously specify the precise values of everything we might want to measure about the system; there will be a complete set of commuting observables that represents all we can say about a system at once.
3. Evolution of the system may be represented in one of two ways: as unitary evolution of the state vector in Hilbert space (the **Schrödinger picture**), or by keeping the state fixed and allowing the observables to evolve according to equations of motion (the **Heisenberg picture**).

Strictly speaking, quantum mechanics is just different from classical mechanics; it is by no means necessary to start with a classical model and "quantize" it. Nevertheless, we usually do exactly that. Even for simple classical models, there is more than one way to construct a quantized version; these include canonical quantization and path-integral quantization, as well as more exotic procedures. What is worse, there is no simple map between classical and quantum theories; there are classical theories with no well-defined quantum counterpart, classical theories with multiple quantum versions, and quantum theories without any classical analogue. For our present purposes, we may blithely ignore all of these subtleties, and proceed directly with canonical quantization.

Once again, the simple harmonic oscillator provides a useful example. Consider first the familiar Schrödinger picture, in which states are represented by complex-valued wave functions that evolve with time, such as $\psi(x,t)$. The wave function is really the set of components of the state vector $|\psi\rangle$, expressed in the "delta-function position basis" $|x\rangle$, so that $|\psi(t)\rangle = \int dx\,\psi(x,t)|x\rangle$. Canonical quantization consists of imposing the canonical commutation relation,

$$[\hat{x}, \hat{p}] = i, \tag{9.13}$$

on the coordinate operator $\hat{x}$ and its conjugate momentum $\hat{p}$. For states represented as wave functions depending on x and t, $\hat{x}$ is simply multiplication by x, so (9.13) can be implemented by setting

$$\hat{p} = -i\partial_x. \tag{9.14}$$

The Hamiltonian operator is

$$H = -\tfrac{1}{2}\partial_x^2 + \tfrac{1}{2}\omega^2 x^2, \tag{9.15}$$

and the equation of motion is the Schrödinger equation,

$$H\psi = i\partial_t\psi. \tag{9.16}$$

Since the Hamiltonian is time-independent, solutions to this equation separate into functions of space and functions of time, $\psi(x,t) = f(t)g(x)$. The solutions then come in a discrete set labeled by an integer $n \geq 0$, and we find (up to normalization)

$$\psi_n(x,t) = e^{-(1/2)\omega x^2} H_n(\sqrt{\omega}x)e^{-iE_n t}, \tag{9.17}$$

where H_n is a Hermite polynomial of degree n, and

$$E_n = \left(n + \tfrac{1}{2}\right)\omega. \tag{9.18}$$

These states are all eigenfunctions of H, and E_n is the energy eigenvalue. An arbitrary state of the oscillator will simply be a superposition of the energy eigenstates,

$$\psi(x,t) = \sum_n c_n\psi_n(x,t), \tag{9.19}$$

for some set of appropriately normalized coefficients c_n.

A number of important features of the quantum-mechanical oscillator are contained in this brief overview. There is a discrete spectrum of energy eigenstates; this is why it's called "quantum" mechanics (even though it is not hard to find systems with continuous spectra). There is a ground state of lowest energy, plus a set of excited states uniquely labeled by their energy eigenvalue. The ground state has a nonvanishing energy,

$$E_0 = \tfrac{1}{2}\omega, \tag{9.20}$$

sometimes called the "zero-point" energy. It is interesting to note that the minimum energy of the classical system would have been zero, representing a particle with $x = 0$ and $p = 0$. The quantum zero-point energy can be traced to the Heisenberg uncertainty principle, which forbids us from localizing a state simultaneously in both position and momentum; there is consequently a minimum amount of "jiggle" in the oscillator, leading to a nonzero ground-state energy. On the other hand, we could certainly have chosen to examine an oscillator with a potential given by $V(x) = \frac{1}{2}\omega^2 x^2 - \frac{1}{2}\omega$; our analysis would have been identical, except that the factor of $\frac{1}{2}$ in (9.18) would have been missing, and the ground-state energy would have been zero. Quantum mechanics does not insist on a nonvanishing zero-point energy, it simply displaces the energy from the classical value.

An alternative way to solve the simple harmonic oscillator is to introduce creation and annihilation operators $\hat{a}^\dagger$ and $\hat{a}$ (often called raising and lowering operators), defined by

$$\hat{a} = \frac{1}{\sqrt{2\omega}}\left(\omega\hat{x} + i\hat{p}\right), \qquad \hat{a}^\dagger = \frac{1}{\sqrt{2\omega}}\left(\omega\hat{x} - i\hat{p}\right), \tag{9.21}$$

so that

$$\hat{x} = \frac{1}{\sqrt{2\omega}}(\hat{a} + \hat{a}^\dagger), \qquad \hat{p} = -i\sqrt{\frac{\omega}{2}}(\hat{a} - \hat{a}^\dagger). \tag{9.22}$$

Given our previous expressions for the commutation relations (9.13) and Hamiltonian (9.7), we can easily calculate the commutation relation for the creation and annihilation operators,

$$[\hat{a}, \hat{a}^\dagger] = 1, \tag{9.23}$$

and the new expression for the Hamiltonian,

$$H = \left(\hat{a}^\dagger\hat{a} + \tfrac{1}{2}\right)\omega. \tag{9.24}$$

The creation/annihilation operators commute with the Hamiltonian via

$$\begin{aligned} [H, \hat{a}] &= -\omega\hat{a} \\ [H, \hat{a}^\dagger] &= \omega\hat{a}^\dagger. \end{aligned} \tag{9.25}$$

Comparing this version of the Hamiltonian to the energy eigenvalues (9.18), we are inspired to define a number operator

$$\hat{n} = \hat{a}^\dagger\hat{a}. \tag{9.26}$$

Let's think about why the creation/annihilation operators and the number operator deserve their names. Consider an eigenstate $|n\rangle$ of the number operator,

$$\hat{n}|n\rangle = n|n\rangle, \tag{9.27}$$

where the $\hat{n}$ on the left stands for the number operator, while the first n on the right stands for the actual number n. (This formula is the most charming in all of quantum mechanics.) By playing with the commutation relations, it is easy to show that

$$\begin{aligned} \hat{n}\hat{a}^\dagger|n\rangle &= (n+1)\hat{a}^\dagger|n\rangle \\ \hat{n}\hat{a}|n\rangle &= (n-1)\hat{a}|n\rangle. \end{aligned} \tag{9.28}$$

Thus, when $\hat{a}^\dagger$ acts on $|n\rangle$, it gives another eigenstate of $\hat{n}$ with eigenvalue raised by 1, while $\hat{a}$ gives an eigenstate with eigenvalue lowered by 1. As before we can show that n takes integral values from 0 to ∞, so there must be a vacuum state $|0\rangle$ satisfying

$$\hat{a}|0\rangle = 0. \tag{9.29}$$

From this state we can construct all of the eigenstates by successive operation by creation operators,

$$|n\rangle = \frac{1}{\sqrt{n!}}\left(\hat{a}^\dagger\right)^n|0\rangle. \tag{9.30}$$

The number operator counts the number of excitations above the ground state. The set of eigenstates $|n\rangle$ acts as a basis; any state is an appropriate linear combination of these states. The creation and annihilation operators act on them according to

$$\begin{aligned} \hat{a}|n\rangle &= \sqrt{n}|n-1\rangle \\ \hat{a}^\dagger|n\rangle &= \sqrt{n+1}|n+1\rangle, \end{aligned} \tag{9.31}$$

and the energy of each state is of course given by (9.18). The basis states are taken to be time-independent, so a physical system obeying Schrödinger's equation will be described by a state

$$|\psi(t)\rangle = \sum_n c_n e^{-iE_n t}|n\rangle, \tag{9.32}$$

where again the c_n's are constant coefficients.

For purposes of smoothing the transition to field theory, it is useful to translate this Schrödinger-picture description into the Heisenberg picture, in which the states are fixed and the operators evolve with time. Given Schrödinger's equation (9.16), any state can be written formally as some fixed initial state acted on by a unitary time-evolution operator

$$|\psi(t)\rangle = U(t)|\psi(0)\rangle, \tag{9.33}$$

where

$$U(t) = \mathcal{P}e^{-i\int H_t\,dt}, \tag{9.34}$$

(by unitary we mean $U^\dagger U = 1$.) The symbol $\mathcal{P}$ stands for path-ordering, as discussed in Appendix I. If the Hamiltonian is time-independent, of course, we simply have $U(t) = e^{-iHt}$. The Schrödinger-picture expression for the matrix element of a time-independent operator A between time-dependent states $|\psi_1(t)\rangle$ and $|\psi_2(t)\rangle$ can then be written as a Heisenberg-picture expression in terms of a time-dependent operator $A(t)$ and time-independent states as

$$\begin{aligned}\langle\psi_2(t)|A|\psi_1(t)\rangle &= \langle\psi_2(0)|U^\dagger(t)AU(t)|\psi_1(0)\rangle \\ &= \langle\psi_2|A(t)|\psi_1\rangle,\end{aligned} \tag{9.35}$$

where clearly the Heisenberg-picture operator is given by

$$A(t) = U^\dagger(t)AU(t). \tag{9.36}$$

Such an operator satisfies the **Heisenberg equation of motion,**

$$\frac{dA(t)}{dt} = i[H, A(t)], \tag{9.37}$$

which takes the place of Schrödinger's equation in this picture. For the harmonic oscillator, we would find

$$\frac{d\hat{a}}{dt} = -i\omega\hat{a}, \qquad \frac{d\hat{a}^\dagger}{dt} = i\omega\hat{a}^\dagger, \tag{9.38}$$

with solutions

$$\hat{a}(t) = e^{-i\omega t}\hat{a}(0), \qquad \hat{a}(t)^\dagger = e^{i\omega t}\hat{a}(0)^\dagger. \tag{9.39}$$

From this we immediately find

$$\hat{n}(t) = \hat{a}(t)^\dagger\hat{a}(t) = \hat{a}(0)^\dagger\hat{a}(0), \tag{9.40}$$

which reflects the fact that the number operator is conserved.

It is common to say that in the Heisenberg picture the states are time-independent; this is somewhat confusing, if nevertheless true. It might be better to say that the states extend throughout time, rather than only being defined at a fixed time. To make this more clear, consider a simple harmonic oscillator subject to an external influence, for example by simply adding a forcing term to the Hamiltonian,

$$H = \tfrac{1}{2}p^2 + \tfrac{1}{2}\omega^2x^2 + F(t), \tag{9.41}$$

where the function $F(t)$ vanishes outside an interval,

$$F(t) = \begin{cases} 0 & t < t_1 \\ F(t) & t_1 \le t \le t_2 \\ 0 & t_2 < t. \end{cases} \tag{9.42}$$

We can think of someone coming along and shaking our oscillator for a short while, and then leaving it alone after that. In the Schrödinger picture, we would say that an oscillator that started in its ground state would be excited by the external force, and the final state would not be the ground state. In the Heisenberg picture, however, we take the state to be a solution to the equation of motion for all times, and say that the number operator went from being zero to some other value.

For the oscillator subject to a transient external force, there are clearly a set of states that look like energy eigenstates at early times, although they don't look that way in the future; we might call such states the "in states" $|n_{\rm in}\rangle$, with the property that

$$\hat{n}(t < t_1)|n_{\rm in}\rangle = n|n_{\rm in}\rangle. \tag{9.43}$$

There is also a separate set of states that look like energy eigenstates at late times, correspondingly called "out states" $|n_{\rm out}\rangle$, and obeying

$$\hat{n}(t > t_2)|n_{\rm out}\rangle = n|n_{\rm out}\rangle. \tag{9.44}$$

Both sets of states exist at all times, but they look like energy eigenstates only in the appropriate asymptotic regime. Either set forms a basis for the entire Hilbert space, so in particular we could decompose one set in terms of the other. For example, by multiplying by a complete set of in states, we can write

$$|n_{\rm out}\rangle = \sum_m \langle m_{\rm in}|n_{\rm out}\rangle|m_{\rm in}\rangle. \tag{9.45}$$

The complex numbers $\langle m_{\rm in}|n_{\rm out}\rangle$ are matrix elements, which could, in principle, be calculated from the Hamiltonian (9.41); together they comprise the ***S*-matrix**. An observer equipped with a way to detect excitations of the oscillator would find that the number of excitations was changed by the applied force, and the S-matrix encodes the information necessary to characterize these changes between the asymptotic past and future. All of this discussion, needless to say, carries over essentially without modification to field theory. For particle physics, the role of the external force is played by the interactions between different particles, whereas for our purposes it will be played by the curvature of spacetime.

9.3 ■ QUANTUM FIELD THEORY IN FLAT SPACETIME

As we have already mentioned, quantum field theory is just a particular example of a quantum-mechanical system, in which we are quantizing a field (a function, or more generally some tensor field, defined on spacetime) rather than a single oscillator. We begin with the simplest possible example, of a free scalar field in flat spacetime; only a couple of generalizations are necessary to make the transition from a single oscillator to this field theory. Extending the theory to curved spacetime is straightforward as usual, involving writing the theory in a covariant form and declaring it to be true. Once we lose the symmetries of Minkowski space, however, some of the ideas we think of as central in a quantum field theory will no longer seem so crucial; in particular, the notions of "vacuum" and "particles" will lose their privileged positions. (Expositions of quantum mechanics will occasionally make the point that waves and particles are complementary notions with different domains of validity, but don't be misled; in quantum field theory it is the fields that are truly fundamental, while the particles are approximate notions useful in certain restricted circumstances.) In this section we study QFT in flat spacetime, before generalizing to curved spacetime in the next section.

We start with the classical theory, in this case a real scalar field $\phi(x^\mu)$ in flat spacetime, just as we considered in Chapter 1, this time generalized to n dimensions. The action is the spacetime integral of the Lagrange density, $S = \int d^n x\, \mathcal{L}$; we will consider the Klein–Gordon Lagrangian

$$\mathcal{L} = -\tfrac{1}{2}\eta^{\mu\nu}\partial_\mu\phi\partial_\nu\phi - \tfrac{1}{2}m^2\phi^2. \tag{9.46}$$

It is not necessary to include the volume-element factor $\sqrt{|g|}$, since we are using inertial coordinates in Minkowski space, with metric

$$ds^2 = -dt^2 + (d\mathbf{x})^2. \tag{9.47}$$

The equation of motion is the Klein–Gordon equation,

$$\Box\phi - m^2\phi = 0. \tag{9.48}$$

Translation into a Hamiltonian description for the field theory is straightforward. The conjugate momentum for a field is simply the derivative of the Lagrange density with respect to the time derivative of that field,

$$\pi = \frac{\partial\mathcal{L}}{\partial(\partial_0\phi)}. \tag{9.49}$$

For the Klein–Gordon Lagrangian (9.46), this is

$$\pi = \dot{\phi}. \tag{9.50}$$

Of course, referring to the time derivative assumes that we have chosen a particular inertial frame; consequently, the Hamiltonian procedure necessarily violates manifest Lorentz invariance. If we are careful, however, observable quantities in the resulting theory will still be Lorentz-invariant. The Hamiltonian itself can be expressed as an integral over space of a Hamiltonian density,

$$H = \int d^{n-1}x\, \mathcal{H}, \tag{9.51}$$

which is related to the Lagrangian by a Legendre transformation,

$$\begin{aligned}\mathcal{H}(\phi,\pi) &= \pi\dot{\phi} - \mathcal{L}(\phi,\partial_\mu\phi)\\ &= \tfrac{1}{2}\pi^2 + \tfrac{1}{2}(\nabla\phi)^2 + \tfrac{1}{2}m^2\phi^2,\end{aligned} \tag{9.52}$$

where $(\nabla\phi)^2 = \delta^{ij}(\partial_i\phi)(\partial_j\phi)$. The correspondence between this field theory and the harmonic oscillator should be clear: the field value $\phi(x)$ plays the role of the coordinate x, with momentum field $\pi(x)$ instead of a single momentum p. Instead of the state being specified by two numbers (x and p) at some fixed time, we would have to give field values [$\phi(x^i)$ and $\pi(x^i)$] all over space at some fixed time as initial data, and there is an additional gradient term that was missing in the oscillator case; but otherwise the formalism is very similar.

We should emphasize that $\phi(x^\mu)$ is *not* a wave function; it is a dynamical variable, generalizing the single degree of freedom x in the case of the harmonic oscillator. In a Schrödinger-picture quantization of the field theory, we would define a complex wave functional $\Psi[\phi(x^\mu)]$, which would represent the probability amplitude for finding the field in each configuration. Instead, however, we will use the Heisenberg picture, so that our primary concern will be to promote ϕ to a quantum operator.

First, we should complete the classical analysis by actually solving this theory. It is not hard to write down solutions to the Klein–Gordon equation. One good example is a plane wave,

$$\phi(x^\mu) = \phi_0 e^{ik_\mu x^\mu} = \phi_0 e^{-i\omega t + i\mathbf{k}\cdot\mathbf{x}}, \tag{9.53}$$

where the wave vector has components

$$k^\mu = (\omega, \mathbf{k}), \tag{9.54}$$

and the frequency must satisfy the dispersion relation

$$\omega^2 = \mathbf{k}^2 + m^2. \tag{9.55}$$

There is a clear similarity between such a solution and that for the simple harmonic oscillator, given by (9.9). But there is also an important difference: For the oscillator, there is only one independent solution. Because the oscillator has a unique frequency, when we add two solutions with specified amplitude x_0 and phase α_0, they combine to give a third solution with the same frequency but different amplitude and phase. This is no longer true in field theory. Given (9.55), the frequency is determined by the spatial wave vector $\mathbf{k}$, at least up to sign. Therefore, instead of a single kind of solution, we have a set parameterized by $\mathbf{k}$ and the sign of ω.

However, we can still write down the most general solution by constructing a complete, orthonormal set of modes in terms of which any solution may be expressed. To make sense of "orthonormal," we need to define an inner product on the space of solutions to the Klein–Gordon equation. Although the modes themselves are functions of spacetime, the appropriate inner product can be expressed as an integral over a constant-time hypersurface Σ_t,

$$(\phi_1, \phi_2) = -i \int_{\Sigma_t} \left(\phi_1 \partial_t \phi_2^* - \phi_2^* \partial_t \phi_1\right) d^{n-1}x. \tag{9.56}$$

As we would hope, the inner product is actually *independent* of the hypersurface Σ_t over which the integral is taken, as you can easily check by using Stokes's theorem and the Klein–Gordon equation. Applying this inner product to two plane waves of different wave vectors gives

$$\begin{aligned}
&(e^{ik_1^\mu x_\mu}, e^{ik_2^\nu x_\nu}) \\
&\quad = -i \int_{\Sigma_t} \left(e^{-i\omega_1 t + i\mathbf{k}_1\cdot\mathbf{x}} \partial_t e^{i\omega_2 t - i\mathbf{k}_2\cdot\mathbf{x}} - e^{i\omega_2 t - i\mathbf{k}_2\cdot\mathbf{x}} \partial_t e^{-i\omega_1 t + i\mathbf{k}_1\cdot\mathbf{x}}\right) d^{n-1}x \\
&\quad = (\omega_2 + \omega_1) e^{-i(\omega_1 - \omega_2)t} \int_{\Sigma_t} e^{i(\mathbf{k}_1 - \mathbf{k}_2)\cdot\mathbf{x}}\, d^{n-1}x \\
&\quad = (\omega_2 + \omega_1) e^{-i(\omega_1 - \omega_2)t} (2\pi)^{n-1} \delta^{(n-1)}(\mathbf{k}_1 - \mathbf{k}_2),
\end{aligned} \tag{9.57}$$

where we have used

$$\int e^{i\mathbf{k}\cdot\mathbf{x}}\, d^{n-1}x = (2\pi)^{n-1}\delta^{(n-1)}(\mathbf{k}). \tag{9.58}$$

The inner product thus vanishes unless the spatial wave vectors $\mathbf{k}$, and hence the frequencies ω, are equal for both modes. An orthonormal set of mode solutions is thus given by

$$f_{\mathbf{k}}(x^\mu) = \frac{e^{ik_\mu x^\mu}}{[(2\pi)^{n-1} 2\omega]^{1/2}}, \tag{9.59}$$

with k^μ obeying (9.55), so that

$$(f_{\mathbf{k}_1}, f_{\mathbf{k}_2}) = \delta^{(n-1)}(\mathbf{k}_1 - \mathbf{k}_2). \tag{9.60}$$

Given the dispersion relation (9.55), $\mathbf{k}$ only determines the frequency up to an overall sign. Our strategy will be to insist that ω always be a positive number, and complete the set of modes by including the complex conjugates $f^*_{\mathbf{k}}(x^\mu)$. (Complex conjugation changes the sign of the $\mathbf{k}$ term in the exponent as well as the ω term, but the components of $\mathbf{k}$ are defined from $-\infty$ to ∞ already.) The $f_{\mathbf{k}}$ modes are said to be positive-frequency, meaning they satisfy

$$\partial_t f_{\mathbf{k}} = -i\omega f_{\mathbf{k}}, \qquad \omega > 0, \tag{9.61}$$

while the $f^*_{\mathbf{k}}$ modes are negative-frequency, satisfying

$$\partial_t f^*_{\mathbf{k}} = i\omega f^*_{\mathbf{k}}, \qquad \omega > 0. \tag{9.62}$$

(Be careful; these modes are called negative-frequency even though $\omega > 0$, because the time derivative pulls down a factor $+i\omega$ rather than $-i\omega$.) The complex conjugate modes are orthogonal to the original modes,

$$(f_{\mathbf{k}_1}, f^*_{\mathbf{k}_2}) = 0, \tag{9.63}$$

and orthonormal with each other but with a negative norm,

$$(f_{\mathbf{k}_1}^*, f_{\mathbf{k}_2}^*) = -\delta^{(n-1)}(\mathbf{k}_1 - \mathbf{k}_2). \tag{9.64}$$

Together, the modes $f_{\mathbf{k}}$ and $f_{\mathbf{k}}^*$ form a complete set, in terms of which we can expand any solution to the Klein–Gordon equation.

To canonically quantize this theory, we promote our classical variables (the fields and their conjugate momenta) to operators acting on a Hilbert space, and impose the canonical commutation relations on equal-time hypersurfaces:

$$\begin{aligned}
[\phi(t,\mathbf{x}), \phi(t,\mathbf{x}')] &= 0 \\
[\pi(t,\mathbf{x}), \pi(t,\mathbf{x}')] &= 0 \\
[\phi(t,\mathbf{x}), \pi(t,\mathbf{x}')] &= i\delta^{(n-1)}(\mathbf{x} - \mathbf{x}').
\end{aligned} \tag{9.65}$$

In field theory we need to state explicitly that the field and its momentum commute with themselves throughout space; for a single oscillator this is implicit, since there is only a single coordinate and momentum, each of which will necessarily commute with itself. The delta function implies that operators at equal times commute everywhere except at coincident spatial points; this feature arises from the demands of causality (operators at spacelike separation cannot influence each other).

Just as classical solutions to the Klein–Gordon equation can be expanded in terms of the modes (9.59), so can the quantum operator field $\phi(t, \mathbf{x})$. Denoting the coefficients of the mode expansion of the field operator by $\hat{a}_{\mathbf{k}}^\dagger$ and $\hat{a}_{\mathbf{k}}$, we have

$$\phi(t,\mathbf{x}) = \int d^{n-1}k \left[\hat{a}_{\mathbf{k}} f_{\mathbf{k}}(t,\mathbf{x}) + \hat{a}_{\mathbf{k}}^\dagger f_{\mathbf{k}}^*(t,\mathbf{x})\right]. \tag{9.66}$$

Plugging this expansion into (9.65), we find that the operators $\hat{a}_{\mathbf{k}}^\dagger$ and $\hat{a}_{\mathbf{k}}$ obey commutation relations

$$\begin{aligned}
[\hat{a}_{\mathbf{k}}, \hat{a}_{\mathbf{k}'}] &= 0 \\
[\hat{a}_{\mathbf{k}}^\dagger, \hat{a}_{\mathbf{k}'}^\dagger] &= 0 \\
[\hat{a}_{\mathbf{k}}, \hat{a}_{\mathbf{k}'}^\dagger] &= \delta^{(n-1)}(\mathbf{k} - \mathbf{k}').
\end{aligned} \tag{9.67}$$

These operators thus obey the commutation relations characteristic of creation and annihilation operators, familiar from (9.23) for the simple harmonic oscillator. The difference, of course, is that there are an infinite number of such operators, indexed by $\mathbf{k}$. We can see the relevance of dividing the modes into positive- and negative-frequency; the positive-frequency modes are coefficients of annihilation operators, while negative-frequency modes are coefficients of creation operators. The idea of positive- and negative-frequency modes will turn out to generalize to static spacetimes, although not to arbitrary spacetimes.

In the case of the harmonic oscillator, we used the creation and annihilation operators to define a basis for the Hilbert space in which the basis states were eigenstates of the number operator. The same procedure works for the free scalar field, although now we have to keep track of separate numbers of excitations for each spatial wave vector $\mathbf{k}$. There will be a single vacuum state $|0\rangle$, characterized by the fact that it is annihilated by each $\hat{a}_{\mathbf{k}}$,

$$\hat{a}_{\mathbf{k}}|0\rangle = 0 \qquad \text{for all } \mathbf{k}. \tag{9.68}$$

A state with $n_{\mathbf{k}}$ particles with identical momenta $\mathbf{k}$ is created by repeated action by $\hat{a}^\dagger_{\mathbf{k}}$,

$$|n_{\mathbf{k}}\rangle = \frac{1}{\sqrt{n_{\mathbf{k}}!}}\left(\hat{a}^\dagger_{\mathbf{k}}\right)^{n_{\mathbf{k}}}|0\rangle, \tag{9.69}$$

while a state with n_i excitations of various momenta $\mathbf{k}_i$ would be

$$|n_1, n_2, \ldots, n_j\rangle = \frac{1}{\sqrt{n_1! n_2! \cdots n_j!}}\left(\hat{a}^\dagger_{\mathbf{k}_1}\right)^{n_1}\left(\hat{a}^\dagger_{\mathbf{k}_2}\right)^{n_2}\cdots\left(\hat{a}^\dagger_{\mathbf{k}_j}\right)^{n_j}|0\rangle. \tag{9.70}$$

Acting on such a state, the creation and annihilation operators change the number of excitations, as expected:

$$\begin{aligned}\hat{a}_{\mathbf{k}_i}|n_1, n_2, \ldots, n_i, \ldots, n_j\rangle &= \sqrt{n_i}\,|n_1, n_2, \ldots, n_i - 1, \ldots, n_j\rangle \\ \hat{a}^\dagger_{\mathbf{k}_i}|n_1, n_2, \ldots, n_i, \ldots, n_j\rangle &= \sqrt{n_i + 1}\,|n_1, n_2, \ldots, n_i + 1, \ldots, n_j\rangle.\end{aligned} \tag{9.71}$$

We can define a number operator for each wave vector,

$$\hat{n}_{\mathbf{k}} = \hat{a}^\dagger_{\mathbf{k}}\hat{a}_{\mathbf{k}}, \tag{9.72}$$

which obeys

$$\hat{n}_{\mathbf{k}_i}|n_1, n_2, \ldots, n_i, \ldots, n_j\rangle = n_i|n_1, n_2, \ldots, n_i, \ldots, n_j\rangle. \tag{9.73}$$

The states that are eigenstates of the number operators form a basis for the entire Hilbert space, known as the **Fock basis**; the space constructed from this basis is often called "Fock space," but of course it is just the original Hilbert space.

One thing we might want to investigate is how our Fock basis behaves under Lorentz transformations. We have clearly been taking advantage of the symmetries of Minkowski space, for example in using plane waves as a basis for solutions to the Klein–Gordon equation. The crucial aspect of these modes is our ability to distinguish between positive and negative frequencies, allowing for an interpretation of their coefficients in the mode expansion of ϕ as annihilation and creation operators. Now consider a boost by velocity $\mathbf{v} = d\mathbf{x}/dt$, leading to new coordinates $x^{\mu'}$ given by

$$t' = \gamma t - \gamma \mathbf{v}\cdot\mathbf{x}, \qquad \mathbf{x}' = \gamma\mathbf{x} - \gamma\mathbf{v}t, \tag{9.74}$$

where $\gamma = 1/\sqrt{1 - v^2}$, and the inverse transformation is given by

$$t = \gamma t' + \gamma\mathbf{v}\cdot\mathbf{x}', \qquad \mathbf{x} = \gamma\mathbf{x}' + \gamma\mathbf{v}t'. \tag{9.75}$$

The time derivative of our mode functions in the boosted frame is

$$\begin{aligned}\partial_{t'} f_{\mathbf{k}} &= \frac{\partial x^\mu}{\partial t'}\partial_\mu f_{\mathbf{k}} \\ &= \gamma(-i\omega) f_{\mathbf{k}} + \gamma \mathbf{v}\cdot(i\mathbf{k}) f_{\mathbf{k}} \\ &= -i\omega' f_{\mathbf{k}}\end{aligned} \tag{9.76}$$

where

$$\omega' = \gamma\omega - \gamma\mathbf{v}\cdot\mathbf{k} \tag{9.77}$$

is simply the frequency in the boosted frame. Clearly, then, a state describing a collection of particles with certain momenta is boosted into a state describing the same particles, but with boosted momenta. Thus, the total number operator in the two frames will coincide, and in particular the vacuum state will coincide. In this sense, our original choice of inertial frame was irrelevant. In the next section we will see that our ability to find positive- and negative-frequency solutions can be traced to the existence of a timelike Killing vector ∂_t in Minkowski spacetime, while the invariance of the Fock space under changes of basis can be traced to the fact that all such timelike Killing vectors are related by Lorentz transformations. Therefore, even if the frequency of a mode depends on the choice of inertial frame, the decomposition into positive and negative frequencies is invariant.

We would like to express the Hamiltonian

$$H = \int d^{n-1}x \left[\frac{1}{2}\dot{\phi}^2 + \frac{1}{2}(\nabla\phi)^2 + \frac{1}{2}m^2\phi^2\right] \tag{9.78}$$

in terms of the creation and annihilation operators, just as we did for the harmonic oscillator. We can analyze this expression term-by-term, starting with the ϕ^2 term for simplicity:

$$\begin{aligned}&\frac{1}{2}m^2\int d^{n-1}x\,\phi^2 \\ &= \frac{1}{2}m^2\int d^{n-1}x\, d^{n-1}k\, d^{n-1}k' \left(\hat{a}_{\mathbf{k}} f_{\mathbf{k}} + \hat{a}^\dagger_{\mathbf{k}} f^*_{\mathbf{k}}\right)\left(\hat{a}_{\mathbf{k}'} f_{\mathbf{k}'} + \hat{a}^\dagger_{\mathbf{k}'} f^*_{\mathbf{k}'}\right) \\ &= \frac{1}{2}m^2\int d^{n-1}x\, d^{n-1}k\, d^{n-1}k' \Big(\hat{a}_{\mathbf{k}}\hat{a}_{\mathbf{k}'} f_{\mathbf{k}} f_{\mathbf{k}'} + \hat{a}^\dagger_{\mathbf{k}}\hat{a}_{\mathbf{k}'} f^*_{\mathbf{k}} f_{\mathbf{k}'} \\ &\quad + \hat{a}_{\mathbf{k}}\hat{a}^\dagger_{\mathbf{k}'} f_{\mathbf{k}} f^*_{\mathbf{k}'} + \hat{a}^\dagger_{\mathbf{k}}\hat{a}^\dagger_{\mathbf{k}'} f^*_{\mathbf{k}} f^*_{\mathbf{k}'}\Big).\end{aligned} \tag{9.79}$$

Zooming in on the first term in parentheses, and ignoring for the moment the integral over **k**, we can plug in the explicit form of the mode functions (9.59) to obtain

$$\int d^{n-1}x\, d^{n-1}k'\, \hat{a}_{\mathbf{k}}\hat{a}_{\mathbf{k}'} f_{\mathbf{k}} f_{\mathbf{k}'} = \int d^{n-1}x\, d^{n-1}k'\, \hat{a}_{\mathbf{k}}\hat{a}_{\mathbf{k}'} \frac{e^{-i(\omega+\omega')t} e^{i(\mathbf{k}+\mathbf{k}')\cdot\mathbf{x}}}{2(2\pi)^{n-1}\sqrt{\omega\omega'}}$$

$$
\begin{aligned}
&= \int d^{n-1}k'\, \hat{a}_{\mathbf{k}}\hat{a}_{\mathbf{k}'} \frac{e^{-i(\omega+\omega')t}}{2\sqrt{\omega\omega'}} \delta^{(n-1)}(\mathbf{k}+\mathbf{k}') \\
&= \hat{a}_{\mathbf{k}}\hat{a}_{-\mathbf{k}} \frac{e^{-2i\omega t}}{2\omega},
\end{aligned} \tag{9.80}
$$

where we have used (9.58) again. Evaluating the other terms in (9.79) similarly, we find that the potential-energy contribution to the Hamiltonian therefore becomes

$$
\begin{aligned}
\frac{1}{2}m^2 \int d^{n-1}x\, \phi^2 = \frac{1}{2}m^2 \int d^{n-1}k \left(\frac{1}{2\omega}\right) \Big[& \hat{a}_{\mathbf{k}}\hat{a}_{-\mathbf{k}} e^{-2i\omega t} \\
& + \hat{a}^\dagger_{\mathbf{k}}\hat{a}_{\mathbf{k}} + \hat{a}_{\mathbf{k}}\hat{a}^\dagger_{\mathbf{k}} + \hat{a}^\dagger_{\mathbf{k}}\hat{a}^\dagger_{-\mathbf{k}} e^{2i\omega t}\Big].
\end{aligned} \tag{9.81}
$$

For the kinetic-energy and gradient-energy pieces, the derivatives pull down factors of ω and $\mathbf{k}$ respectively; we obtain

$$
\frac{1}{2}\int d^{n-1}x\, \dot{\phi}^2 = \frac{1}{2}\int d^{n-1}k \left(\frac{\omega}{2}\right) \left[-\hat{a}_{\mathbf{k}}\hat{a}_{-\mathbf{k}} e^{-2i\omega t} + \hat{a}^\dagger_{\mathbf{k}}\hat{a}_{\mathbf{k}} + \hat{a}_{\mathbf{k}}\hat{a}^\dagger_{\mathbf{k}} - \hat{a}^\dagger_{\mathbf{k}}\hat{a}^\dagger_{-\mathbf{k}} e^{2i\omega t}\right] \tag{9.82}
$$

and

$$
\begin{aligned}
\frac{1}{2}\int d^{n-1}x\, (\nabla\phi)^2 = \frac{1}{2}\int d^{n-1}k \left(\frac{\mathbf{k}^2}{2\omega}\right) \Big[& \hat{a}_{\mathbf{k}}\hat{a}_{-\mathbf{k}} e^{-2i\omega t} \\
& + \hat{a}^\dagger_{\mathbf{k}}\hat{a}_{\mathbf{k}} + \hat{a}_{\mathbf{k}}\hat{a}^\dagger_{\mathbf{k}} + \hat{a}^\dagger_{\mathbf{k}}\hat{a}^\dagger_{-\mathbf{k}} e^{2i\omega t}\Big].
\end{aligned} \tag{9.83}
$$

Using $\omega^2 = \mathbf{k}^2 + m^2$, we can put it all together to write the Hamiltonian for the scalar field theory as

$$
\begin{aligned}
H &= \frac{1}{2}\int d^{n-1}k \left[\hat{a}^\dagger_{\mathbf{k}}\hat{a}_{\mathbf{k}} + \hat{a}_{\mathbf{k}}\hat{a}^\dagger_{\mathbf{k}}\right]\omega \\
&= \int d^{n-1}k \left[\hat{n}_{\mathbf{k}} + \frac{1}{2}\delta^{(n-1)}(0)\right]\omega,
\end{aligned} \tag{9.84}
$$

where the last step invokes the commutation relation (9.67) and the number operator $\hat{n}_{\mathbf{k}} = \hat{a}^\dagger_{\mathbf{k}}\hat{a}_{\mathbf{k}}$. By similar logic, we can construct an operator corresponding to the spatial components of the total momentum, which works out to be

$$
P^i = \int d^{n-1}k\, \hat{n}_{\mathbf{k}} k^i. \tag{9.85}
$$

As we might expect, energy eigenstates will be those with fixed numbers of excitations, each of which carries an energy ω. The excitations in the Fock basis are interpreted as particles. This is how particles arise in a quantum field theory: energy eigenstates are collections of particles with definite momenta. Of course, our modes are plane waves that extend throughout space, not the localized tracks in bubble chambers that come to mind when we think of particles. What is worse, in a curved spacetime the wave equation will not have plane-wave solutions of definite frequency that we can interpret as particles. The solution to both issues is to think operationally, in terms of what would be observed by an experimental apparatus. The best strategy is to define a sensible notion of a particle detector that reduces to our intuitive picture in flat spacetime, and then define "particles" as "what a particle detector detects." For a properly defined particle detector, our plane wave modes can be shown to "leave tracks" in the way we would hope; in an array of such detectors, if a plane wave sets off one detector, there is a high probability that it will set off other detectors along a path from the first one in a direction given by the wave vector. (We should point out that, if you visit an actual particle accelerator at a place like Fermilab or CERN, the detectors bear little resemblance to those invented by theorists studying quantum field theory in curved spacetime; deep down, however, there is a fundamental similarity.) For a discussion of particle detectors see Birrell and Davies (1982).

You might worry about the factor $\delta^{(n-1)}(0)$ in the Hamiltonian (9.84), and well you should. It means that the Hamiltonian is infinite even when measured in the vacuum state $|0\rangle$. This term is the field-theory analogue of the harmonic-oscillator zero-point energy (9.20). In our discussion of the cosmological constant in Chapter 4, we mentioned that quantum fluctuations induced a formally infinite displacement of the classical vacuum energy; this infinite contribution to the scalar-field Hamiltonian will, when gravity is included, show up as a divergent cosmological constant. The fact that it is an integral over an infinite range of $\mathbf{k}$ of the infinite quantity $\delta^{(n-1)}(0)$ can be translated into the statement that the total energy is an integral over an infinitely big space of an infinite energy density. But the energy density contributed by high-frequency modes is the real problem, not the infinite volume; if we regularized the calculation by performing it in a box of volume L^{n-1}, we would find

$$\frac{1}{2}\int d^{n-1}k\, \delta^{(n-1)}(0)\omega \rightarrow \frac{1}{2}\left(\frac{L}{2\pi}\right)^{n-1}\sum_{\mathbf{k}}\omega, \tag{9.86}$$

which diverges even for finite L, since $\mathbf{k}$ (and thus ω) can be arbitrarily large. Putting a cutoff at some high momentum k_{max} would recover (4.104).

In the case of the simple harmonic oscillator, we pointed out that the zero-point energy could have been avoided had we chosen a classical potential with a negative minimum; the quantum-mechanical contribution does not necessarily represent the true answer, only the displacement of the energy from its classical value. The same holds in field theory; we are free to define our original classical scalar field theory so that the quantum-mechanical vacuum energy vanishes. However, we cannot simply subtract off a finite energy mode by mode, since our freedom is only to add a single constant to the potential, and thus to the Hamiltonian density (9.52). To obtain a finite Hamiltonian for the vacuum state, this constant would have to be infinite. There is nothing wrong with subtracting off an infinite constant; it is a venerable technique in quantum field theory, known as "renormalization." At times renormalization can seem scary or somehow illegitimate, but in truth it is perfectly sensible; infinities only arise in the relationship between quantum theories and their classical counterparts, not in any observable quantities. Since Nature presumably doesn't know or care about our fondness for classical mechanics, there should be nothing deeply disturbing about renormalization.

Of course, once we renormalize to obtain a finite vacuum energy, this energy could be anything we like; it is completely arbitrary. This continues to hold for quantum field theory in curved spacetime; we might not be able to decompose the field into modes of definite frequency, and it is therefore impossible to assign a vacuum energy contribution to each mode, but a careful analysis allows one to renormalize the vacuum energy to whatever number you like. Again, nothing profound has happened; the vacuum energy was completely arbitrary in our classical model in the first place, we simply chose it to be zero for convenience. The cosmological constant problem does not arise because quantum mechanics contributes a huge amount of vacuum energy, since this contribution can be straightforwardly renormalized away; the problem arises because there is no reason for the resulting arbitrary number to be close to zero. As discussed before, from the point of view of effective field theory the problem is somewhat sharper, since there is a logical expectation for the scale of the vacuum energy, namely the Planck scale at which unknown quantum-gravity effects should be contributing. Throughout this chapter, however, we will only be concerned with the propagation of quantum fields in fixed spacetime backgrounds, not in using the quantum energy-momentum tensor as a source for Einstein's equation; we can therefore choose to ignore the cosmological constant problem.

9.4 ■ QUANTUM FIELD THEORY IN CURVED SPACETIME

In Chapter 4 we discussed how easy it is to generalize physical theories from flat to curved spacetime—we simply express the theories in a coordinate-invariant form, and assert that they remain true when spacetime is curved. This procedure remains valid for quantum field theory, although we will need to give up on some of the concepts that seemed indispensable in flat spacetime.

We start with the Lagrange density of a scalar field in curved spacetime,

$$\mathcal{L} = \sqrt{-g}\left(-\tfrac{1}{2}g^{\mu\nu}\nabla_\mu\phi\nabla_\nu\phi - \tfrac{1}{2}m^2\phi^2 - \xi R\phi^2\right). \tag{9.87}$$

Aside from the predictable appearance of the metric $g_{\mu\nu}$ and its determinant, we have also included a direct coupling to the curvature scalar R, parameterized by a constant ξ. Since ξ is dimensionless, there is no reason to expect that it is small;

indeed, it should naturally be of order unity. In the literature there are two favorite choices for the value of ξ: **minimal coupling** simply turns off the direct interaction with R,

$$\xi = 0, \tag{9.88}$$

while **conformal coupling** sets

$$\xi = \frac{(n-2)}{4(n-1)}, \tag{9.89}$$

which is $\xi = \frac{1}{6}$ in four dimensions. Using the formulas in Appendix G, it is easy to check that when ξ takes on this value and $m = 0$, the scalar field theory is invariant under conformal transformations $g_{\mu\nu} \to \omega^2(x) g_{\mu\nu}$. In fact, there is no good reason to choose either minimal or conformal coupling in the real world; no symmetry is enhanced by minimal coupling, and conformal invariance is certainly not a symmetry of most physical theories. (Since conformal transformations are local changes of scale, theories characterized by dimensionful parameters such as masses will generally not be conformally invariant.) Even if a classical theory is conformally invariant, quantization can break this symmetry, which happens for example in the theory of quantum chromodynamics (QCD) coupled to massless quarks. Generally, in four dimensions it is difficult to find exactly conformally invariant interacting theories, although some models with high degrees of supersymmetry are known to be conformally invariant.

We may proceed to quantize the theory as before. The conjugate momentum is

$$\pi = \frac{\partial \mathcal{L}}{\partial(\nabla_0 \phi)}, \tag{9.90}$$

which for the Lagrangian (9.87) is

$$\pi = \sqrt{-g}\nabla_0 \phi. \tag{9.91}$$

We can impose canonical commutation relations

$$\begin{aligned} [\phi(t, \mathbf{x}), \phi(t, \mathbf{x}')] &= 0 \\ [\pi(t, \mathbf{x}), \pi(t, \mathbf{x}')] &= 0 \\ [\phi(t, \mathbf{x}), \pi(t, \mathbf{x}')] &= \frac{i}{\sqrt{-g}} \delta^{(n-1)}(\mathbf{x} - \mathbf{x}'). \end{aligned} \tag{9.92}$$

The equation of motion for the scalar field is

$$\Box \phi - m^2 \phi - \xi R \phi = 0. \tag{9.93}$$

For a spacelike hypersurface Σ with induced metric γ_{ij} and unit normal vector n^μ, the inner product on solutions to this equation is

$$(\phi_1, \phi_2) = -i \int_\Sigma \left(\phi_1 \nabla_\mu \phi_2^* - \phi_2^* \nabla_\mu \phi_1\right) n^\mu \sqrt{\gamma}\, d^{n-1}x, \tag{9.94}$$

which is independent of the choice of Σ.

So far, so good. To continue the steps we took in flat space, we would now introduce a set of positive- and negative-frequency modes forming a complete basis for solutions to (9.93), expand the field operator ϕ in terms of these modes, and interpret the operator coefficients as creation and annihilation operators. It is at this point where our procedure breaks down. Since there will generically not be any timelike Killing vector, we will not in general be able to find solutions to the wave equation that separate into time-dependent and space-dependent factors, and correspondingly cannot classify modes as positive- or negative-frequency. We can find a set of basis modes, but the problem is that there will generally be many such sets, with no way to prefer one over any others, and the notion of a vacuum or number operator will depend sensitively on which set we choose.

Let's see what we can do. We will always be able to find a set of solutions $f_i(x^\mu)$ to (9.93) that are orthonormal,

$$(f_i, f_j) = \delta_{ij}, \tag{9.95}$$

and corresponding conjugate modes with negative norm,

$$(f_i^*, f_j^*) = -\delta_{ij}. \tag{9.96}$$

The index i may be continuous or discrete; for the moment we will adopt notation appropriate to the discrete case. These modes can be chosen to be a complete set, so that we may expand our field as

$$\phi = \sum_i \left(\hat{a}_i f_i + \hat{a}_i^\dagger f_i^*\right). \tag{9.97}$$

The coefficients $\hat{a}_i$ and $\hat{a}_i^\dagger$ have commutation relations

$$\begin{aligned} [\hat{a}_i, \hat{a}_j] &= 0 \\ [\hat{a}_i^\dagger, \hat{a}_j^\dagger] &= 0 \\ [\hat{a}_i, \hat{a}_j^\dagger] &= \delta_{ij}. \end{aligned} \tag{9.98}$$

There will be a vacuum state $|0_f\rangle$ that is annihilated by all the annihilation operators,

$$\hat{a}_i |0_f\rangle = 0 \qquad \text{for all } i. \tag{9.99}$$

From this vacuum state we can define an entire Fock basis for the Hilbert space. As before, a state with n_i excitations is created by repeated action by $\hat{a}_i^\dagger$,

$$|n_i\rangle = \frac{1}{\sqrt{n_i!}}\left(\hat{a}_i^\dagger\right)^{n_i}|0_f\rangle, \tag{9.100}$$

and likewise for states with different kinds of excitations. We can even define a number operator for each mode,

$$\hat{n}_{fi} = \hat{a}_i^\dagger \hat{a}_i. \tag{9.101}$$

The subscript f on the vacuum state and the number operator reminds us that they are defined with respect to the set of modes f_i.

This apparatus seems quite similar to what we had in flat space; why can't we declare the excitations created by $\hat{a}_i^\dagger$ to be particles and be done with it? We could, but we must face the fact that there are other choices we could have made; the basis modes $f_i(x^\mu)$ are highly nonunique. Consider an alternative set of modes $g_i(x^\mu)$ with all of the properties that our original modes possessed, including forming (along with conjugate modes g_i^*) a complete basis with respect to which we can expand our field operator,

$$\phi = \sum_i \left(\hat{b}_i g_i + \hat{b}_i^\dagger g_i^*\right). \tag{9.102}$$

The annihilation and creation operators $\hat{b}_i$ and $\hat{b}_i^\dagger$ have commutation relations

$$\begin{aligned}
[\hat{b}_i, \hat{b}_j] &= 0 \\
[\hat{b}_i^\dagger, \hat{b}_j^\dagger] &= 0 \\
[\hat{b}_i, \hat{b}_j^\dagger] &= \delta_{ij},
\end{aligned} \tag{9.103}$$

and there will be a vacuum state $|0_g\rangle$ that is annihilated by all the annihilation operators,

$$\hat{b}_i|0_g\rangle = 0 \qquad \text{for all } i. \tag{9.104}$$

We can construct a Fock basis by repeated application of creation operators on this vacuum, and define a number operator

$$\hat{n}_{gi} = \hat{b}_i^\dagger \hat{b}_i. \tag{9.105}$$

What we have lost in the transition from flat to curved spacetime is any reason to prefer one set of modes over any other. In flat spacetime, we were able to pick out a natural set of modes by demanding that they be positive-frequency with respect to the time coordinate, as defined by (9.61). The time coordinate is not unique, since we are free to perform Lorentz transformations; but we saw that the vacuum state and total number operators are invariant under such transformations. Thus, every inertial observer will agree on what is the vacuum state, and how many particles are around.

In the more general context we are considering now, if one observer defines particles with respect to a set of modes f_i and another observer uses a set of modes g_i, they will typically disagree on how many particles are observed (or even if particles are observed at all). To see this, it is convenient to expand each set of modes in terms of the other,

$$g_i = \sum_j \left(\alpha_{ij} f_j + \beta_{ij} f_j^*\right)$$

$$f_i = \sum_j \left(\alpha_{ji}^* g_j - \beta_{ji} g_j^*\right). \qquad (9.106)$$

The transformation from one set of basis modes into another is known as a **Bogolubov transformation**, and the matrices α_{ij} and β_{ij} implementing the transformation are Bogolubov coefficients. Using the orthonormality of the mode functions, they can be expressed as

$$\alpha_{ij} = (g_i, f_j)$$

$$\beta_{ij} = -(g_i, f_j^*). \qquad (9.107)$$

They satisfy their own normalization conditions,

$$\sum_j \left(\alpha_{ik}\alpha_{jk}^* - \beta_{ik}\beta_{jk}^*\right) = \delta_{ij}$$

$$\sum_j \left(\alpha_{ik}\beta_{jk} - \beta_{ik}\alpha_{jk}\right) = 0. \qquad (9.108)$$

As well as describing a transformation between modes, the Bogolubov coefficients can be used to transform between the operators,

$$\hat{a}_i = \sum_j \left(\alpha_{ji}\hat{b}_j + \beta_{ji}^*\hat{b}_j^\dagger\right)$$

$$\hat{b}_i = \sum_j \left(\alpha_{ij}^*\hat{a}_j - \beta_{ij}^*\hat{a}_j^\dagger\right). \qquad (9.109)$$

Now imagine that the system is in the f-vacuum $|0_f\rangle$, in which no f-particles would be observed; we would like to know how many particles are observed by an observer using the g-modes. We therefore calculate the expectation value of the g number operator in the f-vacuum:

$$\begin{aligned}\langle 0_f|\hat{n}_{gi}|0_f\rangle &= \langle 0_f|b_i^\dagger b_i|0_f\rangle \\ &= \left\langle 0_f\middle|\sum_{jk}\left(\alpha_{ij}\hat{a}_j^\dagger - \beta_{ij}\hat{a}_j\right)\left(\alpha_{ik}^*\hat{a}_k - \beta_{ik}^*\hat{a}_k^\dagger\right)\middle|0_f\right\rangle \\ &= \sum_{jk}(-\beta_{ij})\left(-\beta_{ik}^*\right)\langle 0_f|\hat{a}_j\hat{a}_k^\dagger|0_f\rangle\end{aligned}$$

$$
\begin{aligned}
&= \sum_{jk} \beta_{ij}\beta^*_{ik}\langle 0_f|\left(\hat{a}^\dagger_k\hat{a}_j + \delta_{jk}\right)|0_f\rangle \\
&= \sum_{jk} \beta_{ij}\beta^*_{ik}\delta_{jk}\langle 0_f|0_f\rangle \\
&= \sum_{j} \beta_{ij}\beta^*_{ij}.
\end{aligned}
\tag{9.110}
$$

The number of g-particles in the f-vacuum can thus be expressed in terms of the Bogolubov coefficients as

$$
\langle 0_f|\hat{n}_{gi}|0_f\rangle = \sum_j |\beta_{ij}|^2. \tag{9.111}
$$

There is no reason for this to vanish; what looks like an empty vacuum from one perspective will be bubbling with particles according to another. If any of the β_{ij} are nonvanishing, the vacuum states will not coincide. We can understand why this is by looking at (9.109), where we see that β_{ij} describes the admixture of creation operators from one basis into the annihilation operators in the other basis.

This talk about modes and number operators may seem unnecessarily abstract; certainly, if an actual particle detector is traveling along some trajectory in a possibly-curved spacetime, it will either detect particles or not, without knowing what set of basis modes we are using for field theory. How do we know what definition of "particles" is actually being used by such a detector? The answer is that a detector measures the proper time τ along its trajectory, and will define positive- and negative-frequency with respect to that proper time. Thus, if a set of modes f_i can be found that obey

$$
\frac{D}{d\tau} f_i = -i\omega f_i, \tag{9.112}
$$

we can use these modes to calculate how many particles the detector will see. Of course, it will generally not be possible to find such modes all over the spacetime. The one time that it might be possible is in a *static* spacetime, when we have a hypersurface-orthogonal timelike Killing vector K^μ. In that case we can choose coordinates in which the metric components are independent of the time coordinate t, and there are no time-space cross terms:

$$
\partial_0 g_{\mu\nu} = 0, \qquad g_{0i} = 0. \tag{9.113}
$$

(Indices i, j are now spatial components, not mode labels.) For such a metric, the d'Alembertian acting on some mode function $f(t, \mathbf{x})$ works out to be

$$
\Box f = \left[g^{00}\partial_0^2 + \tfrac{1}{2}g^{00}g^{ij}(\partial_i g_{00})\partial_j + g^{ij}\partial_i\partial_j - g^{ij}\Gamma^k_{ij}\partial_k\right] f. \tag{9.114}
$$

The equation of motion (9.93) can thus be written in the form

$$\partial_0^2 f = -\left(g^{00}\right)^{-1}\left[g^{ij}\partial_i\partial_j + \tfrac{1}{2}g^{00}g^{ij}(\partial_i g_{00})\partial_j - g^{ij}\Gamma^k_{ij}\partial_k - (m^2 + \xi R)\right] f. \tag{9.115}$$

The operator on the left is a pure time derivative, while the operator on the right involves only spatial derivatives and functions of space alone. We can therefore find separable solutions

$$f_\omega(t, \mathbf{x}) = e^{-i\omega t}\,\bar{f}_\omega(\mathbf{x}), \tag{9.116}$$

which can be described as positive-frequency,

$$\partial_t f_\omega(t, \mathbf{x}) = -i\omega f_\omega(t, \mathbf{x}), \qquad \omega > 0. \tag{9.117}$$

This relation can be recast in a coordinate-invariant form as

$$\mathcal{L}_K f_\omega = K^\mu \partial_\mu f_\omega = -i\omega f_\omega, \qquad \omega > 0, \tag{9.118}$$

where $\mathcal{L}_K f_\omega$ denotes the Lie derivative of f_ω along K. There will also be negative-frequency conjugate modes,

$$\mathcal{L}_K f^*_\omega = K^\mu \partial_\mu f^*_\omega = i\omega f^*_\omega, \qquad \omega > 0. \tag{9.119}$$

Together, the modes (f_ω, f^*_ω) will form a basis for solutions to the wave equation in a static background. The existence of such modes won't help us unless they are relevant for our detector; if the detector's trajectory follows along orbits of the Killing field (the four-velocity $U^\mu = dx^\mu/d\tau$ is proportional to K^μ), the proper time will be proportional to the Killing time t, and modes that are positive-frequency with respect to this Killing vector will serve as a natural basis for describing Fock space. We will see this phenomenon at work in our discussion of the Unruh effect in the next section.

In the last section we mentioned the need to renormalize the vacuum energy in quantum field theory. This requirement still exists in curved spacetime, but an appropriate renormalization procedure is harder to construct, since there is no preferred mode basis. Nevertheless, algebraic methods have been developed to define a renormalized energy-momentum tensor rigorously, at least in certain cases; we won't delve into this subject in detail, but should at least present some of the underlying philosophy. The basic idea is that, even in the presence of curvature, spacetime should look Minkowskian on small enough scales. Because the vacuum-energy divergence we found in flat spacetime was due to short-wavelength modes, we should be able to match the behavior of fields in curved spacetime on very small scales to those in flat spacetime, and subtract off any divergences that appear. In particular, we consider the two-point function of a quantum field ϕ in some state $|\psi\rangle$,

$$G(x_1, x_2) = \langle \psi | \phi(x_1) \phi(x_2) | \psi \rangle, \tag{9.120}$$

where x_1 and x_2 are two spacetime points. The two-point function in the Minkowski vacuum becomes singular as x_1 and x_2 are brought close to each other. We would like to characterize this singularity, and insist that it hold for any regular state in curved spacetime. By "brought close to each other" we mean that $\sigma(x_1, x_2)$, the squared distance along the shortest geodesic connecting the two points, goes to zero. In the limit as x_1 and x_2 are very close, the squared geodesic distance is simply

$$\sigma(x_1, x_2) = g_{\mu\nu}(x_1^\mu - x_2^\mu)(x_1^\nu - x_2^\nu), \qquad x_1 \to x_2. \tag{9.121}$$

Of course, in a Lorentzian manifold, the geodesic distance will vanish when points are null separated, not only when they are coincident. We therefore include a small imaginary part and take the limit as it goes to zero, by defining

$$\sigma_\epsilon(x_1, x_2) = \sigma(x_1, x_2) + 2i\epsilon(t_1 - t_2) + \epsilon^2. \tag{9.122}$$

Here, t is the timelike coordinate, and the limit as $\epsilon \to 0^+$ is assumed. (The manifest coordinate-dependence of this formula will be irrelevant in this limit.) Then it turns out that there is a unique singularity structure for the natural vacuum in Minkowski spacetime, such that the two-point function (in four dimensions) contains a leading singularity of the form $1/(4\pi^2\sigma_\epsilon)$ and a subleading one proportional to $\ln \sigma_\epsilon$, with all other terms being regular. We therefore require that any physically reasonable quantum state in curved spacetime obey

$$G(x_1, x_2) = \frac{U(x_1, x_2)}{4\pi^2\sigma_\epsilon} + V(x_1, x_2) \ln \sigma_\epsilon + W(x_1, x_2), \tag{9.123}$$

where the functions $U(x_1, x_2)$, $V(x_1, x_2)$, and $W(x_1, x_2)$ are all regular at $x_1 = x_2$, and $U(x, x) = 1$. A state with this property is said to be a **Hadamard state.** It can be shown that the renormalized energy-momentum tensor is well-defined and nonsingular in all Hadamard states, and furthermore that it will be singular in any non-Hadamard state. If the Hadamard condition is obeyed on some partial Cauchy surface, it will also be obeyed everywhere in the domain of dependence; in other words, the energy-momentum tensor may become singular on a horizon, but not within the Cauchy development of some well-posed initial data. States of this form, therefore, seem appropriate for consideration in QFT on curved spacetime. For details see Wald (1994).

We see that QFT in curved spacetime shares most of the basic features of QFT in flat spacetime; the crucial difference involves what we cannot do, namely decide on a natural set of basis modes that all inertial observers would identify as particles. At the end of Section 9.2 we briefly discussed an oscillator subject to a transient force, and how to define an S-matrix relating number eigenstates at early times to number eigenstates at late times. The same set of ideas translates directly to quantum field theory. If we have a situation in which spacetime is static in the asymptotic past and future, but with some disturbance in between, we can define in- and out-states that are energy eigenstates at early and late times, and a set of Bogolubov coefficients describing how the in-vacuum (for example) will be described as a multiparticle configuration in terms of the out-states. This phenomenon goes by the name of particle production by gravitational fields; relevant physical examples include the early universe and black holes.[2]

9.5 ■ THE UNRUH EFFECT

We must admit that, having put so much effort into understanding the basics of quantum field theory in curved spacetime, we won't actually do any detailed calculations in a curved background. Instead, we will investigate a phenomenon that relies on the ideas we have introduced, but is manifested even in flat spacetime: the Unruh effect, which states that an accelerating observer in the traditional Minkowski vacuum state will observe a thermal spectrum of particles. Historically, the Unruh effect was discovered in an attempt to understand the physics underlying the Hawking effect (thermal radiation in the presence of a black hole event horizon). Our strategy will be to carefully derive the Unruh effect, and in the next section argue under reasonable assumptions that this implies the Hawking effect, which is more difficult to derive directly just because it's harder to solve wave equations in curved spacetime than in flat spacetime.

The basic idea of the Unruh effect is simple: it is a manifestation of the idea that observers with different notions of positive- and negative-frequency modes will disagree on the particle content of a given state. For a uniformly accelerated observer in Minkowski space, the trajectory will move along orbits of a timelike Killing vector, but not that of the usual time-translation symmetry. We can therefore expand the field in modes appropriate to the accelerated observer, and calculate the number operator in the ordinary Minkowski vacuum, where we will find a thermal spectrum of particles. Different sets of explanatory words can be attached to this result; the basic lesson to learn is that what we think of as an inert vacuum actually has the character of a thermal state.

In the interest of discarding all possible complications to get at the underlying phenomenon, we consider a quantum field theory that is as simple as it can be without becoming completely trivial: a massless ($m = 0$) scalar field in two spacetime dimensions ($n = 2$). In two dimensions, conformal coupling and minimal coupling coincide, so we do not include any direct interaction with the curvature scalar. (We're in flat spacetime, so such a coupling wouldn't have any effect anyway.) The relevant wave equation is thus

$$\Box\phi = 0. \tag{9.124}$$

[2]Interestingly, the first discussion of particle production in curved spacetime was given by Schrödinger himself; see E. Schrödinger (1939), *Physica* (Utrecht) **6**, 899.

Before diving into the quantization of this field theory, let's think about two-dimensional Minkowski space as seen by a uniformly accelerating observer. We know that the metric can be written in inertial coordinates as

$$ds^2 = -dt^2 + dx^2. \tag{9.125}$$

Consider an observer moving at a uniform acceleration of magnitude α in the x-direction. We claim that the resulting trajectory $x^\mu(\tau)$ will be given by

$$\begin{aligned} t(\tau) &= \frac{1}{\alpha}\sinh(\alpha\tau) \\ x(\tau) &= \frac{1}{\alpha}\cosh(\alpha\tau). \end{aligned} \tag{9.126}$$

Let's verify that this path corresponds to constant acceleration. The acceleration two-vector is given in the globally inertial coordinate system by

$$a^\mu = \frac{D^2x^\mu}{d\tau^2} = \frac{d^2x^\mu}{d\tau^2}, \tag{9.127}$$

where the covariant derivative along the path is equal to the ordinary derivative because the Christoffel symbols vanish in these coordinates. The components of a^μ are thus

$$\begin{aligned} a^t &= \alpha\sinh(\alpha\tau) \\ a^x &= \alpha\cosh(\alpha\tau), \end{aligned} \tag{9.128}$$

and the magnitude is

$$\sqrt{a_\mu a^\mu} = \sqrt{-\alpha^2\sinh^2(\alpha\tau) + \alpha^2\cosh^2(\alpha\tau)} = \alpha. \tag{9.129}$$

The path therefore corresponds to a constant acceleration of magnitude α, as desired. The trajectory of our accelerated observer obeys the relation

$$x^2(\tau) = t^2(\tau) + 1/\alpha^2, \tag{9.130}$$

and thus describes an hyperboloid asymptoting to null paths $x = -t$ in the past and $x = t$ in the future. The accelerated observer travels from past null infinity to future null infinity, rather than timelike infinity as would be reached by geodesic observers.

We can choose new coordinates (η, ξ) on two-dimensional Minkowski space that are adapted to uniformly accelerated motion. Let

$$t = \frac{1}{a}e^{a\xi}\sinh(a\eta), \qquad x = \frac{1}{a}e^{a\xi}\cosh(a\eta) \qquad (x > |t|). \tag{9.131}$$

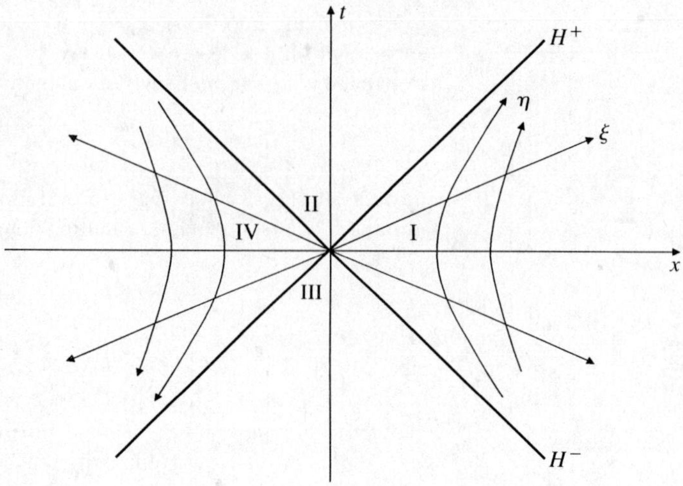

FIGURE 9.1 Minkowski spacetime in Rindler coordinates. Region I is the region accessible to an observer undergoing constant acceleration in the $+x$-direction. The coordinates (η, ξ) can be used in region I, or separately in region IV, where they point in the opposite sense. The vector field ∂_η corresponds to the generator of Lorentz boost symmetry. The horizons $H^\pm$ are Killing horizons for this vector field, and also represent boundaries of the past and future as witnessed by the Rindler observer.

The new coordinates have ranges

$$-\infty < \eta, \xi < +\infty, \tag{9.132}$$

and cover the wedge $x > |t|$, labeled as region I in Figure 9.1. In these coordinates, the constant-acceleration path (9.126) is given by

$$\begin{aligned} \eta(\tau) &= \frac{\alpha}{a}\tau \\ \xi(\tau) &= \frac{1}{a}\ln\left(\frac{a}{\alpha}\right), \end{aligned} \tag{9.133}$$

so that the proper time is proportional to η and the spatial coordinate ξ is constant. In particular, an observer with $\alpha = a$ moves along the path

$$\eta = \tau, \qquad \xi = 0. \tag{9.134}$$

The metric in these coordinates takes the form

$$ds^2 = e^{2a\xi}(-d\eta^2 + d\xi^2). \tag{9.135}$$

Region I, with this metric, is known as **Rindler space**, even though it is obviously just a part of Minkowski space. A **Rindler observer** is one moving along a constant-acceleration path, as in (9.133). The causal structure of Rindler space resembles the region $r > 2GM$ of the maximally extended Schwarzschild solution of Figure 5.12. In particular, the null line $x = t$, labeled H^+ in Figure 9.1, is a future Cauchy horizon for any $\eta =$ constant spacelike hypersurface in region I; similarly, H^- is a past Cauchy horizon. These horizons are reminiscent of the event horizons in the Kruskal diagram, with static observers ($r =$ constant) in Schwarzschild being related to constant-acceleration paths in Rindler space.

The metric components in (9.135) are independent of η, so we immediately know that ∂_η is a Killing vector. But of course this is just Minkowski spacetime, so we think we know what all of the Killing vectors are. Indeed, if we express ∂_η in the (t, x) coordinates, we find

$$\begin{aligned}\partial_\eta &= \frac{\partial t}{\partial \eta}\partial_t + \frac{\partial x}{\partial \eta}\partial_x \\ &= e^{a\xi}\left[\cosh(a\eta)\partial_t + \sinh(a\eta)\partial_x\right] \\ &= a(x\partial_t + t\partial_x).\end{aligned} \tag{9.136}$$

This is nothing more or less than the Killing field associated with a boost in the x-direction. It is clear from this expression that this Killing field naturally extends throughout the spacetime; in regions II and III it is spacelike, while in region IV it is timelike but past-directed. The horizons we have identified are actually Killing horizons for ∂_η. The redshift factor, defined in (6.12) as the magnitude of the norm of the Killing vector, is

$$V = e^{a\xi}. \tag{9.137}$$

The surface gravity $\kappa = \sqrt{\nabla_\mu V \nabla^\mu V}$ of this Killing horizon is thus

$$\kappa = a. \tag{9.138}$$

There is no real gravitational force, since we're in flat space; but this surface gravity characterizes the acceleration of Rindler observers.

We can also define coordinates (η, ξ) in region IV by flipping the signs in (9.131),

$$t = -\frac{1}{a}e^{a\xi}\sinh(a\eta), \qquad x = -\frac{1}{a}e^{a\xi}\cosh(a\eta) \qquad (x < |t|). \tag{9.139}$$

The sign guarantees that ∂_η and ∂_t point in opposite directions in region IV. Strictly speaking, we cannot use (η, ξ) simultaneously in regions I and IV, since the ranges of these coordinates are the same in each region, but we will be okay so long as we explicitly indicate to which region we are referring. The reason why it's better to use the same set of coordinate labels twice, rather than simply introducing new coordinates, is that the metric (9.135) will apply to both region I and region IV.

Along the surface $t = 0$, ∂_η is a hypersurface-orthogonal timelike Killing vector, except for the single point $x = 0$ where it vanishes. This vector can therefore be used to define a set of positive- and negative-frequency modes, on which we can build a Fock basis for the scalar-field Hilbert space. The massless Klein–Gordon equation in Rindler coordinates takes the form

$$\Box\phi = e^{-2a\xi}(-\partial_\eta^2 + \partial_\xi^2)\phi = 0. \tag{9.140}$$

A normalized plane wave $g_k = (4\pi\omega)^{-1/2}e^{-i\omega\eta+ik\xi}$, with $\omega = |k|$, solves this equation and apparently has positive frequency, in the sense that $\partial_\eta g_k = -i\omega g_k$. But we need our modes to be positive-frequency with respect to a future-directed Killing vector, and in region IV that role is played by $\partial_{(-\eta)} = -\partial_\eta$ rather than ∂_η. To deal with this annoyance, we introduce two sets of modes, one with support in region I and the other in region IV:

$$g_k^{(1)} = \begin{cases} \dfrac{1}{\sqrt{4\pi\omega}}e^{-i\omega\eta+ik\xi} & \text{I} \\ 0 & \text{IV} \end{cases}$$
$$g_k^{(2)} = \begin{cases} 0 & \text{I} \\ \dfrac{1}{\sqrt{4\pi\omega}}e^{+i\omega\eta+ik\xi} & \text{IV} \end{cases} \tag{9.141}$$

We take $\omega = |k|$ in each case; in two dimensions, the spatial wave vector is just the single number k. Each set of modes is positive-frequency with respect to the appropriate future-directed timelike Killing vector,

$$\begin{aligned} \partial_\eta g_k^{(1)} &= -i\omega g_k^{(1)} \\ \partial_{(-\eta)} g_k^{(2)} &= -i\omega g_k^{(2)}, \qquad \omega > 0. \end{aligned} \tag{9.142}$$

These two sets, along with their conjugates, form a complete set of basis modes for any solutions to the wave equation throughout the spacetime. (The single point $x = t = 0$ is a set of measure zero, so we shouldn't have to worry about it.) Both sets are nonvanishing in regions II and III of the Rindler diagram; this is obscured by writing them in terms of the coordinates η and ξ, but these functions can be analytically extended into the future and past regions. Denoting the associated annihilation operators as $\hat{b}_k^{(1,2)}$, we can write

$$\phi = \int dk\,\left(\hat{b}_k^{(1)}g_k^{(1)} + \hat{b}_k^{(1)\dagger}g_k^{(1)*} + \hat{b}_k^{(2)}g_k^{(2)} + \hat{b}_k^{(2)\dagger}g_k^{(2)*}\right). \tag{9.143}$$

This expansion is an alternative to our expression (9.66) in terms of the original Minkowski modes, which in two dimensions takes the form

$$\phi = \int dk\,\left(\hat{a}_k f_k + \hat{a}_k^\dagger f_k^*\right). \tag{9.144}$$

It is straightforward to check that the modes (9.141) are properly normalized with respect to the inner product (9.94). In the metric (9.135), the future-directed unit normal to the surface $\eta = 0$ is normalized to

$$-1 = g_{\mu\nu} n^\mu n^\nu = -e^{2a\xi}(n^0)^2, \tag{9.145}$$

or

$$n^0 = e^{-a\xi}. \tag{9.146}$$

Meanwhile, the spatial metric determinant satisfies

$$\sqrt{\gamma} = e^{a\xi}. \tag{9.147}$$

We therefore have $n^0\sqrt{\gamma} = 1$, and the calculation of the inner product of the Rindler modes follows precisely that of ordinary Minkowski modes. We end up with

$$\begin{aligned}
(g^{(1)}_{k_1}, g^{(1)}_{k_2}) &= \delta(k_1 - k_2) \\
(g^{(2)}_{k_1}, g^{(2)}_{k_2}) &= \delta(k_1 - k_2) \\
(g^{(1)}_{k_1}, g^{(2)}_{k_2}) &= 0,
\end{aligned} \tag{9.148}$$

and similarly for the conjugate modes.

There are thus two sets of modes, Minkowski and Rindler, with which we can expand solutions to the Klein–Gordon equation in a flat two-dimensional spacetime. Although the Hilbert space for the theory is the same in either representation, its interpretation as a Fock space will be different; in particular, the vacuum states will be different. The Minkowski vacuum $|0_{\rm M}\rangle$, satisfying

$$\hat{a}_k|0_{\rm M}\rangle = 0, \tag{9.149}$$

will be described as a multi-particle state in the Rindler representation; likewise, the Rindler vacuum $|0_{\rm R}\rangle$, satisfying

$$\hat{b}^{(1)}_k|0_{\rm R}\rangle = \hat{b}^{(2)}_k|0_{\rm R}\rangle = 0, \tag{9.150}$$

will be described as a multi-particle state in the Minkowski representation. At a practical level, the difference arises because an individual Rindler mode can never be written as a sum of positive-frequency Minkowski modes; at $t = 0$ the Rindler modes only have support on the half-line, and such a function cannot be expanded in purely positive-frequency plane waves. Thus, the Rindler annihilation operators used to define $|0_{\rm R}\rangle$ are necessarily superpositions of Minkowski creation and annihilation operators, so the two vacua cannot coincide.

A Rindler observer will be static with respect to orbits of the boost Killing vector ∂_η. Such an observer in region I will therefore describe particles in terms of the Rindler modes $g_k^{(1)}$, and in particular will observe a state in the Rindler vacuum to be devoid of particles, a state $\hat{b}_k^{(1)\dagger}|0_R\rangle$ to contain a single particle of frequency $\omega = |k|$, and so on. Conversely, a Rindler observer traveling through the Minkowski vacuum state will detect a background of particles, even though an inertial observer would describe the state as being completely empty. What kind of particles would the Rindler observer detect? We know how to answer this question: Calculate the Bogolubov coefficients relating the Minkowski and Rindler modes, and use them to determine the expectation value of the Rindler number operator in the Minkowski vacuum. This is straightforward but tedious, so we will take a shortcut due to Unruh. We will find a set of modes that share the same vacuum state as the Minkowski modes (although the description of excited states may be different), but for which the overlap with the Rindler modes is more direct. The way to do this is to start with the Rindler modes, analytically extend them to the entire spacetime, and express this extension in terms of the original Rindler modes.

To see how this works, notice from (9.131) and (9.139) that we have the following relationships between the Minkowski coordinates (t, x) and Rindler coordinates (η, ξ) in regions I and IV:

$$
\begin{aligned}
e^{-a(\eta-\xi)} &= \begin{cases} a(-t+x) & \text{I} \\ a(t-x) & \text{IV} \end{cases} \\
e^{a(\eta+\xi)} &= \begin{cases} a(t+x) & \text{I} \\ a(-t-x) & \text{IV} \end{cases}
\end{aligned}
\tag{9.151}
$$

We can therefore express the spacetime dependence of a mode $g_k^{(1)}$ with $k > 0$ (so $\omega = k$) in terms of Minkowski coordinates in region I as

$$
\begin{aligned}
\sqrt{4\pi\omega}\, g_k^{(1)} &= e^{-i\omega\eta+ik\xi} \\
&= e^{-i\omega(\eta-\xi)} \\
&= a^{i\omega/a}(-t+x)^{i\omega/a}.
\end{aligned}
\tag{9.152}
$$

The analytic extension of this function throughout spacetime is straightforward; we simply use this final expression for any values of (t, x). But we would like to express the result in terms of the original Rindler modes everywhere; since the $g_k^{(1)}$ modes vanish in region IV, we need to bring the modes $g_k^{(2)}$ into play. When we express them in terms of the Minkowski coordinates in region IV, for $k > 0$ we obtain

$$
\begin{aligned}
\sqrt{4\pi\omega}\, g_k^{(2)} &= e^{+i\omega\eta+ik\xi} \\
&= e^{+i\omega(\eta+\xi)} \\
&= a^{-i\omega/a}(-t-x)^{-i\omega/a}.
\end{aligned}
\tag{9.153}
$$

This doesn't match the behavior of (9.152) that we want. But if we take the complex conjugate and reverse the wave number, we obtain

$$\begin{aligned}\sqrt{4\pi\omega}\, g_{-k}^{(2)*} &= e^{-i\omega\eta+ik\xi} \\ &= e^{-i\omega(\eta-\xi)} \\ &= a^{i\omega/a}(t-x)^{i\omega/a} \\ &= a^{i\omega/a}[e^{-i\pi}(-t+x)]^{i\omega/a} \\ &= a^{i\omega/a}e^{\pi\omega/a}(-t+x)^{i\omega/a}. \end{aligned} \tag{9.154}$$

The combination

$$\sqrt{4\pi\omega}\left(g_k^{(1)} + e^{-\pi\omega/a}g_{-k}^{(2)*}\right) = a^{i\omega/a}(-t+x)^{i\omega/a} \tag{9.155}$$

is therefore well-defined along the whole surface $t = 0$. We have explicitly examined the case $k > 0$, but an identical result obtains for $k < 0$.

A properly normalized version of this mode is given by

$$h_k^{(1)} = \frac{1}{\sqrt{2\sinh\left(\frac{\pi\omega}{a}\right)}}\left(e^{\pi\omega/2a}g_k^{(1)} + e^{-\pi\omega/2a}g_{-k}^{(2)*}\right). \tag{9.156}$$

This is an appropriate analytic extension of the $g_k^{(1)}$ modes; to get a complete set, we need to include the extensions of the $g_k^{(2)}$ modes, which by an analogous argument are given by

$$h_k^{(2)} = \frac{1}{\sqrt{2\sinh\left(\frac{\pi\omega}{a}\right)}}\left(e^{\pi\omega/2a}g_k^{(2)} + e^{-\pi\omega/2a}g_{-k}^{(1)*}\right). \tag{9.157}$$

To verify the normalization, for example for $h_k^{(1)}$, we use (9.148):

$$\begin{aligned}\left(h_{k_1}^{(1)}, h_{k_2}^{(1)}\right) &= \frac{1}{2\sqrt{\sinh\left(\frac{\pi\omega_1}{a}\right)\sinh\left(\frac{\pi\omega_2}{a}\right)}}\Big[e^{\pi(\omega_1+\omega_2)/2a}\left(g_{k_1}^{(1)}, g_{k_2}^{(1)}\right) \\ &\quad + e^{-\pi(\omega_1+\omega_2)/2a}\left(g_{-k_1}^{(2)*}, g_{-k_2}^{(2)*}\right)\Big] \\ &= \frac{1}{2\sqrt{\sinh\left(\frac{\pi\omega_1}{a}\right)\sinh\left(\frac{\pi\omega_2}{a}\right)}}\Big[e^{\pi(\omega_1+\omega_2)/2a}\delta(k_1-k_2) \\ &\quad + e^{-\pi(\omega_1+\omega_2)/2a}\delta(-k_1+k_2)\Big] \\ &= \frac{e^{\pi\omega_1/a} - e^{-\pi\omega_1/a}}{2\sinh\left(\frac{\pi\omega_1}{a}\right)}\delta(k_1-k_2)\end{aligned}$$

$$= \delta(k_1 - k_2), \tag{9.158}$$

just as we would like.

We can now expand our field in these modes,

$$\phi = \int dk \left(\hat{c}_k^{(1)} h_k^{(1)} + \hat{c}_k^{(1)\dagger} h_k^{(1)*} + \hat{c}_k^{(2)} h_k^{(2)} + \hat{c}_k^{(2)\dagger} h_k^{(2)*} \right). \tag{9.159}$$

From our discussion of Bogolubov transformations in Section 9.4, we know that the expressions (9.156) and (9.157) for the $h_k^{(1,2)}$ modes in terms of the $g_k^{(1,2)}$ modes implies corresponding expressions for the Rindler operators $\hat{b}_k^{(1,2)}$ in terms of the operators $\hat{c}_k^{(1,2)}$, as

$$\begin{aligned} \hat{b}_k^{(1)} &= \frac{1}{\sqrt{2\sinh\left(\frac{\pi\omega}{a}\right)}} \left(e^{\pi\omega/2a} \hat{c}_k^{(1)} + e^{-\pi\omega/2a} \hat{c}_{-k}^{(2)\dagger} \right) \\ \hat{b}_k^{(2)} &= \frac{1}{\sqrt{2\sinh\left(\frac{\pi\omega}{a}\right)}} \left(e^{\pi\omega/2a} \hat{c}_k^{(2)} + e^{-\pi\omega/2a} \hat{c}_{-k}^{(1)\dagger} \right). \end{aligned} \tag{9.160}$$

We can therefore express the Rindler number operator in region I,

$$\hat{n}_{\mathrm{R}}^{(1)}(k) = \hat{b}_k^{(1)\dagger} \hat{b}_k^{(1)}, \tag{9.161}$$

in terms of the new operators $\hat{c}_k^{(1,2)}$.

The original positive-frequency Minkowski plane-wave modes with $k > 0$, $f_k \propto e^{-i\omega(t-x)}$, are analytic and bounded for complex (t, x) so long as $\mathrm{Im}(t - x) \leq 0$. (Such modes are called "right-moving," as they describe waves propagating to the right.) The same holds for our new modes $h_k^{(1)}$ so long as we take the branch cut for the imaginary power to lie in the upper-half complex $(t - x)$ plane, as we can see from examination of (9.152) and (9.154); this is consistent with our setting $-1 = e^{-i\pi}$ in (9.154). Similar considerations apply to the $h_k^{(2)}$ modes, which are analytic and bounded in the lower-half complex $(t + x)$ plane, as are the positive-frequency Minkowski plane-wave modes with $k < 0$ (left-moving). Consequently, unlike the original Rindler modes $g_k^{(1,2)}$, we know that the modes $h_k^{(1,2)}$ can be expressed purely in terms of positive-frequency Minkowski modes f_k. They therefore share the same vacuum state $|0_{\mathrm{M}}\rangle$, so that

$$\hat{c}_k^{(1)}|0_{\mathrm{M}}\rangle = \hat{c}_k^{(2)}|0_{\mathrm{M}}\rangle = 0. \tag{9.162}$$

The excited states will not coincide, but that won't bother us, since we are interested in what a Rindler observer sees when the state is precisely in the Minkowski vacuum. An observer in region I, for example, will observe particles defined by the operators $\hat{b}_k^{(1)}$; the expected number of such particles of frequency ω will be given by

$$\langle 0_{\mathrm{M}}|\hat{n}_{\mathrm{R}}^{(1)}(k)|0_{\mathrm{M}}\rangle = \langle 0_{\mathrm{M}}|\hat{b}_k^{(1)\dagger}\hat{b}_k^{(1)}|0_{\mathrm{M}}\rangle$$

$$
\begin{aligned}
&= \frac{1}{2\sinh\left(\frac{\pi\omega}{a}\right)} \langle 0_{\rm M} | e^{-\pi\omega/a} \hat{c}^{(1)}_{-k} \hat{c}^{(1)\dagger}_{-k} | 0_{\rm M} \rangle \\
&= \frac{e^{-\pi\omega/a}}{2\sinh\left(\frac{\pi\omega}{a}\right)} \delta(0) \\
&= \frac{1}{e^{2\pi\omega/a} - 1} \delta(0), \qquad (9.163)
\end{aligned}
$$

where we have used the fact that a $\hat{c}^{(1)\dagger}_k |0_{\rm M}\rangle$ is a normalized one-particle state,

$$
\langle 0_{\rm M} | \hat{c}^{(1)}_k \hat{c}^{(1)\dagger}_k | 0_{\rm M} \rangle = \delta(0). \qquad (9.164)
$$

The delta function in (9.163) is merely an artifact of our use of (nonsquare-integrable) plane wave basis modes; had we constructed normalized wave packets, we would have obtained a finite result with an identical spectrum.

The result (9.163) is a Planck spectrum with temperature

$$
T = \frac{a}{2\pi}. \qquad (9.165)
$$

Thus, *an observer moving with uniform acceleration through the Minkowski vacuum observes a thermal spectrum of particles.* This is the **Unruh effect**. Of course, there is more to thermal radiation than just the spectrum (9.163); to be truly thermal, we should check that there are no hidden correlations in the observed particles. This has been verified; the radiation detected by a Rindler observer is truly thermal. At the most basic level, the Unruh effect shows how two different sets of observers (inertial and Rindler) will describe the same state in very different terms; at a slightly deeper level, it reveals the essentially thermal nature of the vacuum in quantum field theory.

The temperature $T = a/2\pi$ is what would be measured by an observer moving along the path $\xi = 0$, which feels an acceleration $\alpha = a$. Using (9.133), we know that any other path with $\xi =$ constant feels an acceleration

$$
\alpha = ae^{-a\xi} \qquad (9.166)
$$

and thus should measure thermal radiation with a temperature $\alpha/2\pi$. This is consistent with our discussion in Chapter 6 of the redshift witnessed by static observers moving along orbits of some Killing vector K^μ; we found that radiation emitted with frequency ω_1 at a point x_1 would be observed at a point x_2 with a frequency

$$
\omega_2 = \frac{V_1}{V_2}\omega_1, \qquad (9.167)
$$

where the redshift factor V is the norm of the Killing vector. In (9.137) we found that the redshift factor associated with ∂_η is $V = e^{a\xi}$, so that

$$
\omega_2 = e^{a(\xi_1 - \xi_2)}\omega_1. \qquad (9.168)
$$

Thus, if an observer at $\xi_1 = 0$ detects a temperature $T = a/2\pi$, the observer at $\xi_2 = \xi$ will see it to be redshifted to a temperature $T = ae^{-a\xi}/2\pi$, just as in (9.166). In particular, the temperature redshifts all the way to zero as $\xi \to +\infty$. This makes sense, since a Rindler observer at infinity will be nearly inertial, and will define the same notion of vacuum and particles as an ordinary Minkowski observer.

The Unruh effect tells us that an accelerated observer will detect particles in the Minkowski vacuum state. An inertial observer, of course, would describe the same state as being completely empty; indeed, the expectation value of the energy-momentum tensor would be $\langle T_{\mu\nu} \rangle = 0$. But if there is no energy-momentum, how can the Rindler observers detect particles? This is a subtle issue, but by no means a contradiction. If the Rindler observer is to detect background particles, she must carry a detector—some sort of apparatus coupled to the particles being detected. But if a detector is being maintained at constant acceleration, energy is not conserved; we need to do work constantly on the detector to keep it accelerating. From the point of view of the Minkowski observer, the Rindler detector *emits* as well as absorbs particles; once the coupling is introduced, the possibility of emission is unavoidable. When the detector registers a particle, the inertial observer would say that it had emitted a particle and felt a radiation-reaction force in response. Ultimately, then, the energy needed to excite the Rindler detector does not come from the background energy-momentum tensor, but from the energy we put into the detector to keep it accelerating.

9.6 ■ THE HAWKING EFFECT AND BLACK HOLE EVAPORATION

Even though it occurs in flat spacetime, the Unruh effect teaches us the most important lesson of QFT in curved spacetime, the idea that "vacuum" and "particles" are observer-dependent notions rather than fundamental concepts. In fact, given our understanding of the Unruh effect, we can see almost immediately how the Hawking effect arises. This should not be too surprising, as we have already noted the similarity between the causal structure of Rindler space and that of the maximally-extended Schwarzschild spacetime describing an eternal black hole. We will therefore be able to argue in favor of Hawking radiation without ever doing an explicit calculation in curved spacetime; of course, there are many features that you might like to investigate in more detail, for which the full power of the curved metric is necessary. In addition to Birrell and Davies (1982) and Wald (1994), there are good review articles where you can find a more full discussion of the issues discussed here.[3] Our derivation of Hawking radiation follows that of Jacobson.

[3]T.A. Jacobson, "Introductory Lectures on Black Hole Thermodynamics," Lectures at University of Utrecht (1996), `http://www.fys.ruu.nl/~wwthe/lectures/itfuu-0196.ps`; R.M. Wald, "The thermodynamics of black holes," *Living Rev. Rel.* **4**, 6 (2001), `http://arxiv.org/gr-qc/9912119`; J. Traschen, "An introduction to black hole evaporation" (2000), `http://arxiv.org/gr-qc/0010055`.

Consider a static observer at radius $r_1 > 2GM$ outside a Schwarzschild black hole. Such an observer moves along orbits of the timelike Killing vector $K = \partial_t$. In Chapter 6 we showed that the redshift factor $V = \sqrt{-K_\mu K^\mu}$ for static observers in Schwarzschild is given by

$$V = \sqrt{1 - \frac{2GM}{r}}, \tag{9.169}$$

with a corresponding magnitude of the acceleration given by

$$a = \frac{GM}{r\sqrt{r - 2GM}}. \tag{9.170}$$

For observers very close to the event horizon, $r_1 - 2GM \ll 2GM$, this acceleration becomes very large compared to the scale set by the Schwarzschild radius,

$$a_1 \gg \frac{1}{2GM}. \tag{9.171}$$

The Schwarzschild radius in turn sets the radius of curvature of spacetime near the horizon. Therefore, as observed over length- and timescales set by $a_1^{-1} \ll 2GM$, spacetime looks essentially flat. Let us make the crucial assumption that the quantum state of some scalar field ϕ looks like the Minkowski vacuum (free of any particles) as seen by *freely-falling* observers near the black hole. This assumption is reasonable, since the event horizon is not a local barrier; a freely-falling observer sees nothing special happen when crossing the horizon. Then the static observer looks just like a constant-acceleration observer in flat spacetime, and will detect Unruh radiation at a temperature $T_1 = a_1/2\pi$.

Now consider a static observer at infinity, or at least a distance r_2 large compared to $2GM$. In that case there is no sense in which the spacetime curvature can be neglected over timescales $a_2^{-1} \gg 2GM$, so there is no reason to expect that they will see radiation with a temperature $a_2/2\pi$, where a_2 is evaluated at r_2. But the radiation observed near the horizon will propagate to infinity with an appropriate redshift. We can apply the argument used at the end of the last section to determine what such an observer should see; they should detect thermal radiation redshifted to a temperature

$$T_2 = \frac{V_1}{V_2} T_1 = \frac{V_1}{V_2} \frac{a}{2\pi}. \tag{9.172}$$

At infinity we have $V_2 \to 1$, so the observed temperature is

$$T = \lim_{r_1 \to 2GM} \frac{V_1 a_1}{2\pi} = \frac{\kappa}{2\pi}, \tag{9.173}$$

where $\kappa = \lim(Va)$ is the surface gravity; for Schwarzschild, $\kappa = 1/4GM$. Unlike for accelerating observers in flat spacetime, in Schwarzschild the static Killing vector has finite norm at infinity, and the radiation near the horizon redshifts to a finite value rather than all the way to zero. Observers far from the black hole thus see a flux of thermal radiation emitted from the black hole at a temperature proportional to its surface gravity. This is the celebrated **Hawking effect**, and the radiation itself is known as Hawking radiation.

Despite its slickness, there is nothing dishonest about this derivation of the Hawking effect. In particular, the relation to acceleration makes it clear why the temperature is proportional to the black hole surface gravity (which continues to hold for more general black holes, not only Schwarzschild). However, we need to be clear about the assumption we made that the vacuum state near the horizon looks nonsingular to freely-falling observers. In technical terms, the renormalized energy-momentum tensor is taken to be finite at the horizon, or equivalently, the two-point function obeys the Hadamard condition (9.123).

The meaning of this assumption becomes more clear by considering possible vacuum states in the maximally extended Schwarzschild geometry. Such states are not necessarily physically relevant to a realistic black hole formed by gravitational collapse, but the possibilities that arise in the idealized case carry instructive lessons for the real world. We will only describe the states, not specify them quantitatively or derive any of their properties; for more details see the references above.

In searching for a vacuum state, we might begin by looking for a state that is regular [in the Hadamard sense, (9.123)] throughout spacetime. For maximally extended Schwarzschild, such a state was found by Hartle and Hawking, so we call it the **Hartle–Hawking vacuum**; indeed, this is the unique vacuum state that is regular everywhere and invariant under the Schwarzschild Killing vector ∂_t representing time translations at infinity. In particular, recalling the conformal diagram of Schwarzschild shown in Figure 5.16, the Hartle–Hawking vacuum is regular on the past and future event horizons $H^{\pm}$ at $r = 2GM$, and also on past and future null infinity $\mathscr{I}^{\pm}$. From the consideration of static observers as outlined above, we should then expect that the Hartle–Hawking vacuum features thermal radiation being emitted from the black hole, and indeed this turns out to be true. However, a close examination of this state reveals that there is an equal flux of thermal radiation coming in from past null infinity ($\mathscr{I}^-$) toward the black hole; in other words, it represents a black hole in thermal equilibrium with its environment. This is not what we would use to model a realistic black hole in our universe. Another vacuum, more closely analogous to that of a black hole formed via gravitational collapse, is the **Unruh vacuum**, which is nonsingular on H^+ (and therefore predicts outgoing Hawking radiation), but exhibits no incoming radiation from $\mathscr{I}^-$. The Unruh vacuum turns out to be singular on the past horizon H^- of Schwarzschild; this doesn't bother us if we are only using it as a model for realistic black holes, since a spacetime featuring gravitational collapse as in Figure 5.17 would not have a white hole or any past horizons. Finally, we might look for a vacuum state in which no particles come into the black hole, nor escape to infinity; in other words, vanishing flux at $\mathscr{I}^{\pm}$. There is such a state, called the **Boulware vacuum**. The existence of such a state seems to be in conflict with our argument for the Hawking effect from the Unruh effect, except that a careful analysis reveals that the Boulware vacuum is singular both on H^- and H^+. Thus, the assumption that the vacuum is regular as seen by freely-falling observers near the horizon is violated in this state.

So a careful examination of vacuum states in an eternal Schwarzschild metric is consistent with our reasoning from the Unruh effect; states that are regular on H^+ predict Hawking radiation of the expected form. Note that the existence of an event horizon is crucial to the argument; without such an horizon, the requirement that the state be regular on the horizon has no force. Consider for example a neutron star, whose radius may be close to the Schwarzschild radius but for which the spacetime is free of any horizons. Neutron stars do not emit any Hawking radiation. One way to understand this is to recognize that a static neutron-star metric features a Killing vector that is timelike everywhere, and can be used to define positive-frequency modes that extend throughout the spacetime and match the Minkowski modes at infinity. The resulting vacuum state would actually resemble the Boulware vacuum, free of flux at $\mathscr{I}^{\pm}$; the fact that the full Boulware vacuum is singular on the horizon doesn't bother us in the neutron-star case, since there aren't any horizons.

To be absolutely sure that we have correctly chosen a vacuum state appropriate to realistic black holes, we should consider gravitational collapse in a spacetime that is nearly Minkowskian in the past and Schwarzschild in the future, as in Figure 5.17. If the vacuum takes the standard Minkowski form on $\mathscr{I}^-$, we can ask how the modes propagate through the collapse geometry to $\mathscr{I}^+$, defining an S-matrix as in (9.45) to determine what would be seen by asymptotic observers. This is in fact what Hawking did when he first discovered black hole radiation; the calculations involve some messy algebra but are basically straightforward, with the same answer for the temperature as we derived above.

Of course, from a complete calculation we can learn more than just the blackbody temperature; we might ask, for example, what happens when the wavelength of the emitted radiation is comparable to the Schwarzschild radius, in which case our approximations clearly break down. If we were to carefully investigate the emission of arbitrary species of particles from any kind of black hole (that is, allowing for both charge and spin), we would find that the spectrum of emitted radiation takes the form

$$\langle \hat{n}_\omega \rangle = \frac{\Gamma(\omega)}{e^{2\pi(\omega-\mu)/\kappa} \pm 1}. \tag{9.174}$$

Here, κ is of course the surface gravity. The parameter μ is a chemical potential, characterizing the tendency of the black hole to shed its conserved quantum numbers; a charged black hole preferentially emits particles with the same-sign charge as the hole, while a rotating black hole preferentially emits particles with the same-sign angular momentum as the hole. Hawking radiation therefore tends to bring black holes to a Schwarzschild state. $\Gamma(\omega)$ is a greybody factor, which can be thought of as arising from backscattering of wavepackets off of the gravitational field and into the black hole. In the high-frequency limit the wavelength is very small and backscattering can be neglected; at very low frequencies the wavelength becomes greater than the Schwarzschild radius and backscattering becomes important. Although an analytic expression for the greybody factor is hard to derive, in the limiting cases of large and small frequencies the greybody factor for a scalar field obeys

$$\Gamma(\omega) \to 1, \qquad \omega \gg \frac{1}{GM}$$
$$\Gamma(\omega) \to \frac{A}{4\pi}\omega^2, \qquad \omega \ll \frac{1}{GM}, \tag{9.175}$$

where A is the area of the black hole.

The discovery that black holes emit thermal radiation is certainly surprising from the point of view of classical general relativity, where we emphasized the impossibility of escape to infinity from points inside the event horizon. One picturesque way to understand what is going on is to think of vacuum fluctuations in terms of Feynman diagrams, with the fluctuations being represented by virtual particle/antiparticle pairs popping in and out of existence. This picture is also helpful, for example, in understanding observed phenomena such as the Lamb shift, in which atomic spectra are affected by the interaction of photons with virtual electron/positron pairs. Normally, the pairs will always annihilate, and their effect is only indirect, through a renormalization of processes coupled to the virtual particles. In the presence of an event horizon, however, occasionally one member of a virtual pair will fall into the black hole while its partner escapes to infinity, as depicted in Figure 9.2. In this picture, it is these escaping virtual particles that we observe as Hawking radiation. The total energy of the virtual pair must add to zero, but the infalling particle can have a negative energy as viewed from infinity, because the asymptotically-timelike Killing vector is spacelike inside the horizon. The picture is somewhat informal, but provides a useful heuristic for what is going on.

Once we know the formula for the temperature of a black hole we can fix the proportionality constants in the relationships between black hole parameters and

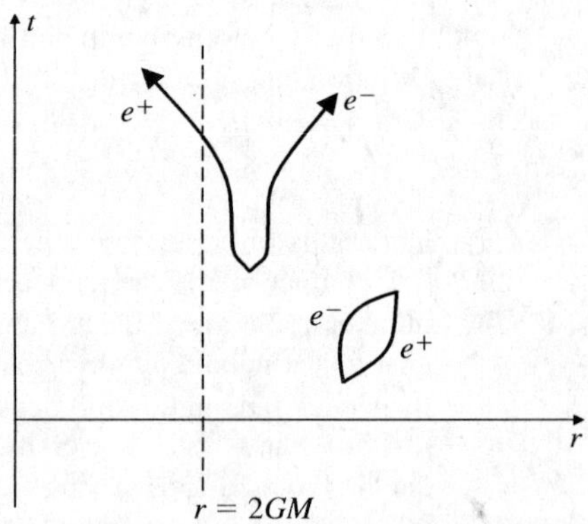

FIGURE 9.2 Vacuum fluctuations occasionally result in one of a particle/antiparticle pair falling into the event horizon, and the other escaping to infinity as Hawking radiation.

thermodynamic variables, as listed in (6.118). Hawking radiation essentially consummates the marriage of black hole mechanics and thermodynamics; stationary black holes act just like bodies of energy $E = M$ in thermal equilibrium with temperature $T = \kappa/2\pi$ and entropy $S = A/4G$. This is a very large entropy indeed. For matter fields in the universe, the entropy is approximately equal to the number of relativistic particles; within one Hubble radius, this number works out to be

$$S_M \sim 10^{88}. \tag{9.176}$$

Meanwhile, the entropy of a black hole is the area of its horizon measured in Planck units (remember that we have been setting $\hbar = 1$ all along). We can convert to astrophysical units to obtain

$$S_{BH} \sim 10^{90} \left(\frac{M}{10^6 M_\odot}\right)^2. \tag{9.177}$$

Thus, a single million-solar-mass black hole (such as can be found at the center of our galaxy, and many other galaxies) has more entropy than all of the matter in the visible universe. The total entropy of the universe is much smaller than we could make it, just by putting more mass into black holes. (When cosmologists say that the entropy S_M is large, they mean it is surprising that so much entropy is found within one curvature radius.) Presumably the reason why we are in such a low-entropy state has to do with initial conditions, and perhaps with inflation.

Coming back to black hole mechanics, we see a puzzle: The entropy of a macroscopic black hole will be huge, but from a statistical-mechanical point of view the entropy is supposed to measure the logarithm of the number of accessible states. A classical black hole is specified by a small number of parameters (mass, charge, and spin), so it is hard to know what those states could be. Nevertheless, we could take the attitude that this discrepancy doesn't really matter, since any information about the state of a black hole would presumably be hidden behind the event horizon.

The inclusion of quantum mechanics makes the puzzle worse rather than better, because black holes will not only radiate but also evaporate. When we started our investigation of QFT in curved spacetime, one of the rules we set was that we would assume a fixed background metric, and not worry about the effect of the energy-momentum tensor of the quantum fields themselves. Nevertheless, even in quantum mechanics we have conservation of energy (in the sense, for example, of a conserved ADM mass in an asymptotically flat spacetime). Hence, when Hawking radiation escapes to infinity, we may safely conclude that it will carry energy away from the black hole, which must therefore shrink in mass. (This phenomenon does not violate the area theorem, since the quantum field energy-momentum tensor will not obey the weak energy condition near the horizon.) As the mass shrinks, the surface gravity increases, and with it the temperature; there is a runaway process in which the entire mass evaporates away in a finite time.

Plugging in the numbers gives a lifetime of order

$$\tau_{\rm BH} \sim \left(\frac{M}{m_{\rm P}}\right)^3 t_{\rm P} \sim \left(\frac{M}{M_\odot}\right)^3 \times 10^{71}\ \text{sec}, \tag{9.178}$$

where $m_{\rm P} \sim 10^{-5}$ g is the Planck mass and $t_{\rm P} \sim 10^{-43}$ sec is the Planck time. Since the Hubble time is $H_0^{-1} \sim 10^{18}$ sec, a solar-mass black hole has a lifetime of order 10^{53} times the age of the universe. This seems like a long time, but we are speaking of questions of principle here.

You can see why the question of the black hole entropy has become so severe: Once the black hole has evaporated, we can no longer appeal to the event horizon as a way to hide purported states of the black hole. There is no black hole any more, just the Hawking radiation it produced. The fact that this radiation is supposed to be precisely thermal (no hidden correlations in the outgoing particles) means that it has no way of conveying the vast amount of information needed to specify the states implied by our entropy calculation. Thus, if we assemble two very different original states and collapse them into two black holes of the same mass, charge, and spin, they will radiate away into two indistinguishable clouds of Hawking particles. The information that went into the specification of the system before it became a black hole seems to have been erased; this is the **information loss paradox**. Both quantum field theory and general relativity feature unitary evolution—the information required to specify a state at early times is precisely equal to that needed to specify a state at later times, since they are connected by the equations of motion. But in the process of combining QFT with GR this unitarity has apparently been violated. It seems likely that we have made an inappropriate assumption somewhere in our argument, but it is hard to see where.

One way of conveying the essence of the information loss paradox is to consider a hypothetical conformal diagram for an evaporating black hole, shown in Figure 9.3. We don't really know what the full spacetime should look like, but here we have made the plausible assumptions that a singularity forms, along with an associated event horizon, both of which disappear when the black hole has fully evaporated, leaving behind a spacetime with a Minkowskian causal structure. The problem is then obvious if we think in terms of Cauchy surfaces. The future domain of dependence of an achronal surface stretching from spacelike infinity i^0 to a point with $r = 0$ to the past of the singularity would be the entire spacetime, so such a surface would be a Cauchy surface. But a similar surface stretching to a point with $r = 0$ to the future of the singularity would not be a Cauchy surface, since the region behind the event horizon would not be in its domain of dependence. Thus, the past cannot be retrodicted from the future, due to the disappearance of information into the singularity. In other words, this process seems to be time-irreversible (in a microscopic sense, not merely a statistical sense), even though the dynamical laws that were used to predict it were fully invariant under time reversal.

In addressing the information loss paradox, keep in mind that our analysis of black-hole evaporation has only been in the context of a hybrid theory of quantum

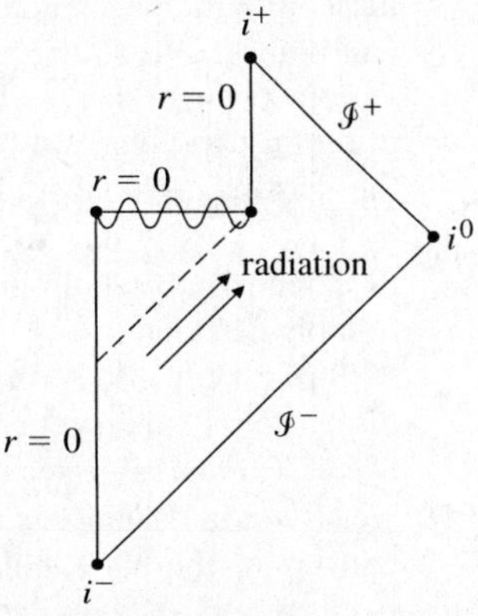

FIGURE 9.3 Hypothetical conformal diagram for an evaporating black hole. Energy is carried away by the Hawking radiation, so that the black hole eventually evaporates away entirely, leaving a future with the causal structure of Minkowski space. Information that falls past the event horizon into the singularity appears to be lost.

field theory coupled to general relativity, not in a realistic theory of quantum gravity. What might be going on in the real world? One possibility is that information really is lost, unitarity is violated, and we just have to learn to live with it. Many physicists find the introduction of such a fundamental breakdown of predictability to be unpalatable, and arguments have also been made that unitarity violations would necessarily lead to violations of energy conservation. Another possibility is that unitarity appears to be violated in our world, but only because the information that entered the black hole has somehow escaped to a disconnected region of space (a baby universe). General relativity predics a singularity at the center of the black hole, not creation of a disconnected region, but clearly we are in a regime where quantum effects will dramatically alter our classical expectations, so we should keep an open mind.

Some evidence against information loss comes from string theory. String theory is naturally defined in 10 or 11 spacetime dimensions, and features not only one-dimensional extended objects (strings), but also various types of higher-dimensional extended objects known collectively as "branes." A crucial aspect of string theory is a high degree of supersymmetry relating bosons to fermions. In the real world supersymmetry must be spontaneously broken if it exists at all, since we don't observe a bosonic version of the electron with the same mass and charge. But as a tool for thought experiments, supersymmetry is invaluable. Supersymmetric configurations of strings and branes can be assembled that describe black hole geometries in various dimensions. In string theory there is a free parameter (really a scalar field), the string coupling, that controls the strength of gravity as well as the strength of other forces. If we consider a configuration describing a black hole at a certain value of the string coupling, as we decrease the coupling the Schwarzschild radius will eventually shrink below the size of the configuration, which thus turns into a collection of weakly-coupled strings and branes. Due to the high degree of supersymmetry, we can be confident that various characteristics of the state remain unchanged as we vary the string coupling; in particular, we expect that the number of degrees of freedom (and thus the entropy) is unaltered. But in the weakly-coupled regime there is no black hole, we simply have a "gas" of conventional degrees of freedom (admittedly, of extended objects in higher dimensions), whose entropy we should be able to reliably calculate.

Strominger and Vafa considered this process for a particular type of five-dimensional supersymmetric black hole with different kinds of charges.[4] They found a remarkable result: the number of degrees of freedom of the system at weak coupling matches precisely that which would be predicted based on the entropy of the black hole at strong coupling. Since the black hole entropy depends nontrivially on the charges of the configuration, it seems unlikely that this agreement is simply an accident. Subsequent investigations have extended this analysis to other kinds of black holes, for which agreement continues to be found. Furthermore, we can even calculate the greybody factors expected for the black hole by considering scattering off of the weakly-coupled system; again, the result matches the strong-coupling expectation. Thus, in string theory at least, there is excellent reason to believe that the degrees of freedom implied by black hole radiation are really there.

Unfortunately, the string theory counting of states provides little direct understanding of how information about the black hole state could somehow be conveyed to the outgoing Hawking radiation. Nevertheless, we should certainly take seriously the possibility that this is what happens, even if there are severe difficulties in imagining how such a process might actually work. The difficulties arise when considering some information, perhaps in the form of a volume of an encyclopedia, being tossed into a large black hole, long before it has evaporated away. At this stage the black hole temperature is low, there is very little surface gravity, and the spacetime curvature near the event horizon is quite small. From the point of view of the encyclopedia, nothing special happens at the horizon, and we should expect it to fall through essentially unmolested. In particular, it is hard to imagine how the information in the encyclopedia can be transferred to the Hawking radiation being emitted at early times. In unitary evolution, the information cannot be duplicated; either it falls past the horizon with the encyclopedia, or it needs to be effectively extracted just before the horizon is crossed, which seems implausible. We might hope that the information accompanies the encyclopedia into a region near the singularity, and is somehow preserved there until late times when the hole is very small. But by then most of the radiating particles have already been emitted, and the number of states accessible to the final burst of radiation will generally be smaller than required to describe the different states that could have fallen into the hole.

[4]A. Strominger and C. Vafa, "Microscopic origin of the Bekenstein-Hawking entropy," *Phys. Lett.* B **379**, 99 (1996), `http://arxiv.org/hep-th/9601029`. For reviews see Johnson (2003) or A.W. Peet, "TASI lectures on black holes in string theory," (2001), `http://arxiv.org/hep-th/0008241`.

To imagine that the information is somehow encoded in the outgoing radiation, it therefore seems necessary to encode correlations in the Hawking particles even at early times. We just argued that this is hard to do, given that the horizon is an unremarkable place when the black hole is large. One conceivable way out of this dilemma is to take the dramatic step of giving up on local quantum field theory. In other words, we have been making the implicit assumption that information can be sensibly described as being located in some region of space; this is an indisputable feature of ordinary quantum field theories. But perhaps quantum gravity is different, and the information contained in the black hole is somehow spread out nonlocally across the horizon. By itself this suggestion doesn't lead directly to a mechanism for getting the information into the outgoing Hawking radiation, but it does call into question some of the arguments we have given for why it would be difficult to do so.

A particular realization of nonlocality goes under the name of the **holographic principle**. This is the idea, suggested originally by 't Hooft and Susskind, that the number of degrees of freedom in a region of space is not proportional to the volume of the region (as would be expected in a local field theory), but rather to the area of the boundary of the region.[5] The inspiration comes of course from black hole entropy, which scales as the area of the event horizon; if the entropy counts the number of accessible states, holography would account for why it is the area rather than the enclosed volume that matters. You might worry about how to deal with closed universes, in which a region might consist of almost all of space but have a very small boundary, but a more covariant version of the holographic principle may be formulated by replacing the region of space by a set of "light-sheets" extending inward from the boundary. The great triumph of holography has been in the AdS/CFT correspondence, mentioned in Chapter 8. There, the physics of quantum gravity in an anti-de Sitter background is equivalent to a conformal field theory without gravity defined on the boundary of AdS, which has one lower dimension. One can imagine that all of the physical phenomena we observe in the universe could be described by the nonlocal holographic projection of some ordinary nongravitational theory defined in lower dimensions; it is by no means clear how we should go about constructing such a correspondence or connecting it with observations, but considerations of cosmology and the large-scale structure of the universe might be a promising place to start.

These remarks about black hole entropy, string theory, and holography are obviously not intended as a careful introduction to what is a very active area of research. Rather, they are meant to indicate some of the possibilities being explored at the forefront of gravitational physics. Classical general relativity is the most beautiful physical theory invented to date, but we have every right to expect that a synthesis of GR with other areas of physics will reveal layers of beauty we can only now imagine.

[5]For a review see R. Bousso, "The Holographic Principle" (2002), `http://arxiv.org/hep-th/0203101`.

APPENDIX

A Maps between Manifolds

When we discussed manifolds in Chapter 2, we introduced maps between two different manifolds and how maps could be composed. Here we will investigate such maps in much greater detail, focusing on the use of such maps in carrying along tensor fields from one manifold to another. The manifolds in question might end up being a submanifold and the bigger space in which it is embedded, or we might just have two different copies of the same abstract manifold being mapped to each other.

Consider two manifolds M and N, possibly of different dimension, with coordinate systems x^μ and y^α, respectively. We imagine that we have a map $\phi : M \rightarrow N$ and a function $f : N \rightarrow \mathbf{R}$. Obviously we can compose ϕ with f to construct a map $(f \circ \phi) : M \rightarrow \mathbf{R}$, which is simply a function on M. Such a construction is sufficiently useful that it gets its own name; we define the **pullback** of f by ϕ, denoted $\phi^* f$, by

$$\phi^* f = (f \circ \phi). \tag{A.1}$$

The name makes sense, since we think of ϕ^* as "pulling back" the function f from N to M (see Figure A.1).

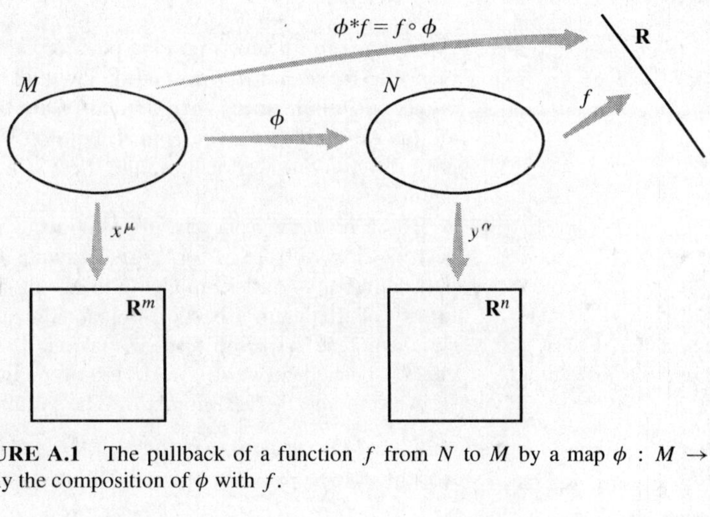

FIGURE A.1 The pullback of a function f from N to M by a map $\phi : M \rightarrow N$ is simply the composition of ϕ with f.

We can pull functions back, but we cannot push them forward. If we have a function $g : M \to \mathbf{R}$, there is no way we can compose g with ϕ to create a function on N; the arrows don't fit together correctly. But recall that a vector can be thought of as a derivative operator that maps smooth functions to real numbers. This allows us to define the **pushforward** of a vector; if $V(p)$ is a vector at a point p on M, we define the pushforward vector $\phi_* V$ at the point $\phi(p)$ on N by giving its action on functions on N:

$$(\phi_* V)(f) = V(\phi^* f). \tag{A.2}$$

So to push forward a vector field we say "the action of $\phi_* V$ on any function is simply the action of V on the pullback of that function."[1]

This discussion is a little abstract, and it would be nice to have a more concrete description. We know that a basis for vectors on M is given by the set of partial derivatives $\partial_\mu = \partial/\partial x^\mu$, and a basis on N is given by the set of partial derivatives $\partial_\alpha = \partial/\partial y^\alpha$. Therefore we would like to relate the components of $V = V^\mu \partial_\mu$ to those of $(\phi_* V) = (\phi_* V)^\alpha \partial_\alpha$. We can find the sought-after relation by applying the pushed-forward vector to a test function and using the chain rule (2.12):

$$\begin{aligned}(\phi_* V)^\alpha \partial_\alpha f &= V^\mu \partial_\mu (\phi^* f) \\ &= V^\mu \partial_\mu (f \circ \phi) \\ &= V^\mu \frac{\partial y^\alpha}{\partial x^\mu} \partial_\alpha f.\end{aligned} \tag{A.3}$$

This simple formula makes it irresistible to think of the pushforward operation ϕ_* as a matrix operator, $(\phi_* V)^\alpha = (\phi_*)^\alpha{}_\mu V^\mu$, with the matrix being given by

$$(\phi_*)^\alpha{}_\mu = \frac{\partial y^\alpha}{\partial x^\mu}. \tag{A.4}$$

The behavior of a vector under a pushforward thus bears an unmistakable resemblance to the vector transformation law under change of coordinates. In fact it is a generalization, since when M and N are the same manifold the constructions are (as we shall discuss) identical; but don't be fooled, since in general μ and α have different allowed values, and there is no reason for the matrix $\partial y^\alpha / \partial x^\mu$ to be invertible.

It is a rewarding exercise to convince yourself that, although you can push vectors forward from M to N (given a map $\phi : M \to N$), you cannot in general pull them back—just keep trying to invent an appropriate construction until the futility of the attempt becomes clear. Since one-forms are dual to vectors, you should not be surprised to hear that one-forms can be pulled back (but not in general pushed forward). To do this, remember that one-forms are linear maps from vectors to the real numbers. The pullback $\phi^* \omega$ of a one-form ω on N can

[1]Unfortunately the location of the asterisks is not completely standard; some references use a superscript * for pushforward and a subscript $_*$ for pullback, so be careful.

therefore be defined by its action on a vector V on M, by equating it with the action of ω itself on the pushforward of V:

$$(\phi^*\omega)(V) = \omega(\phi_* V). \tag{A.5}$$

Once again, there is a simple matrix description of the pullback operator on forms, $(\phi^*\omega)_\mu = (\phi^*)_\mu{}^\alpha \omega_\alpha$, which we can derive using the chain rule. It is given by

$$(\phi^*)_\mu{}^\alpha = \frac{\partial y^\alpha}{\partial x^\mu}. \tag{A.6}$$

That is, it is the same matrix as the pushforward (A.4), but of course a different index is contracted when the matrix acts to pull back one-forms.

There is a way of thinking about why pullbacks and pushforwards work on some objects but not others, which may be helpful. If we denote the set of smooth functions on M by $\mathcal{F}(M)$, then a vector $V(p)$ at a point p on M (that is, an element of the tangent space T_pM) can be thought of as an operator from $\mathcal{F}(M)$ to $\mathbf{R}$. But we already know that the pullback operator on functions maps $\mathcal{F}(N)$ to $\mathcal{F}(M)$, just as ϕ itself maps M to N, but in the opposite direction. Therefore we can define the pushforward ϕ_* acting on vectors simply by composing maps, as we first defined the pullback of functions; this is shown in Figure A.2. Similarly, if T_qN is the tangent space at a point q on N, then a one-form ω at q (that is, an element of the cotangent space T_q^*N) can be thought of as an operator from T_qN to $\mathbf{R}$. Since the pushforward ϕ_* maps T_pM to $T_{\phi(p)}N$, the pullback ϕ^* of a one-form can also be thought of as mere composition of maps, as indicated in Figure A.3. If this is not helpful, don't worry about it. But do keep straight what exists and what doesn't; the actual concepts are simple, it's just forgetting which map goes what way that leads to confusion.

You will recall further that a $(0, l)$ tensor—one with l lower indices and no upper ones—is a linear map from the direct product of l vectors to $\mathbf{R}$. We can therefore pull back not only one-forms, but tensors with an arbitrary number of lower indices. The definition is simply the action of the original tensor on the pushed-forward vectors:

$$(\phi^*T)(V^{(1)}, V^{(2)}, \ldots, V^{(l)}) = T(\phi_* V^{(1)}, \phi_* V^{(2)}, \ldots, \phi_* V^{(l)}), \tag{A.7}$$

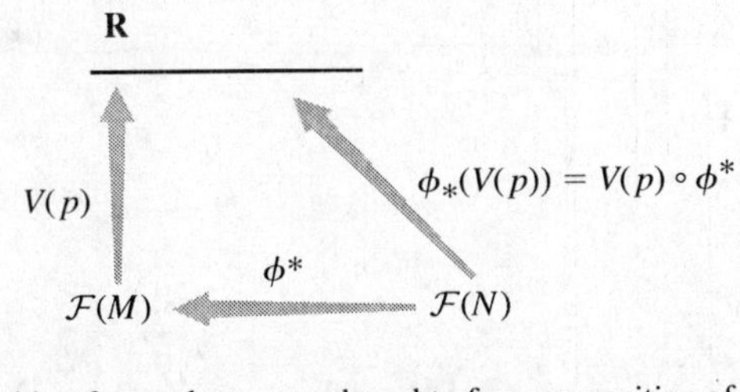

FIGURE A.2 Pushing forward a vector, thought of as composition of a map between the spaces of functions on N and M, and a map from functions on M to $\mathbf{R}$.

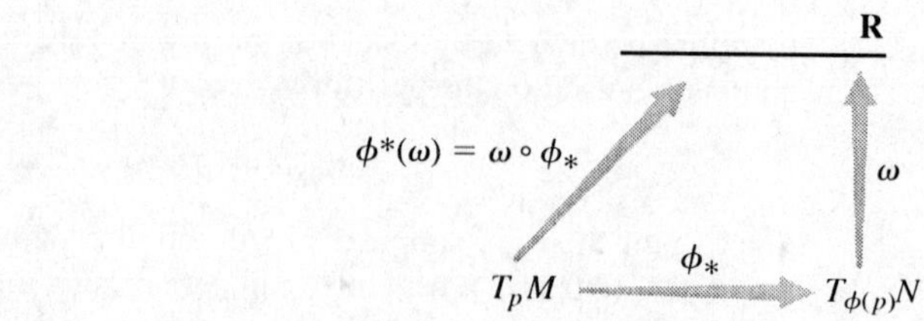

FIGURE A.3 Pulling back a one-form, thought of as composition of a map between tangent spaces T_pM and $T_{\phi(p)}N$ and a map from $T_{\phi(p)}N$ to $\mathbf{R}$.

where $T_{\alpha_1\cdots\alpha_l}$ is a $(0, l)$ tensor on N. We can similarly push forward any $(k, 0)$ tensor $S^{\mu_1\cdots\mu_k}$ by acting it on pulled-back one-forms:

$$(\phi_* S)(\omega^{(1)}, \omega^{(2)}, \ldots, \omega^{(k)}) = S(\phi^*\omega^{(1)}, \phi^*\omega^{(2)}, \ldots, \phi^*\omega^{(k)}). \tag{A.8}$$

Fortunately, the matrix representations of the pushforward (A.4) and pullback (A.6) extend to the higher-rank tensors simply by assigning one matrix to each index; thus, for the pullback of a $(0, l)$ tensor, we have

$$\boxed{(\phi^* T)_{\mu_1\cdots\mu_l} = \frac{\partial y^{\alpha_1}}{\partial x^{\mu_1}} \cdots \frac{\partial y^{\alpha_l}}{\partial x^{\mu_l}} T_{\alpha_1\cdots\alpha_l},} \tag{A.9}$$

while for the pushforward of a $(k, 0)$ tensor we have

$$\boxed{(\phi_* S)^{\alpha_1\cdots\alpha_k} = \frac{\partial y^{\alpha_1}}{\partial x^{\mu_1}} \cdots \frac{\partial y^{\alpha_k}}{\partial x^{\mu_k}} S^{\mu_1\cdots\mu_k}.} \tag{A.10}$$

Our complete picture is therefore as portrayed in Figure A.4. Note that tensors with both upper and lower indices can generally be neither pushed forward nor pulled back.

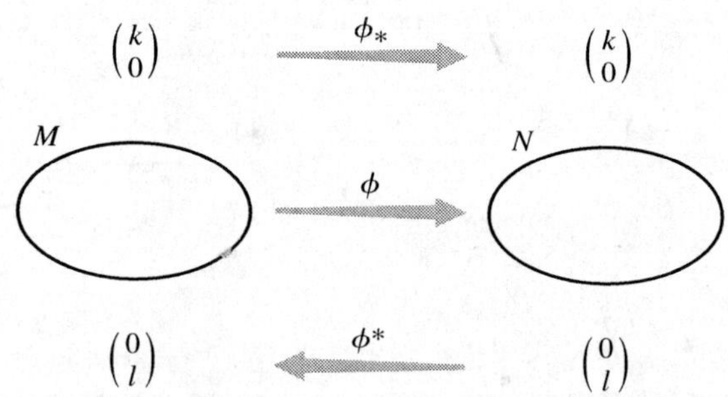

FIGURE A.4 A map $\phi : M \to N$ allows us to pull back $(0, l)$ tensors and push forward $(k, 0)$ tensors.

This machinery becomes somewhat less imposing once we see it at work in a simple example. One common occurrence of a map between two manifolds is when M is actually a submanifold of N, which we will discuss more carefully in Appendix C. The basic idea is that there is a map from M to N that just takes an element of M to the "same" element of N. Consider the two-sphere embedded in $\mathbf{R}^3$, thought of as the locus of points a unit distance from the origin. If we put coordinates $x^\mu = (\theta, \phi)$ on $M = S^2$ and $y^\alpha = (x, y, z)$ on $N = \mathbf{R}^3$, the map $\phi : M \to N$ is given by

$$\phi(\theta, \phi) = (\sin\theta \cos\phi, \sin\theta \sin\phi, \cos\theta). \tag{A.11}$$

Sticking the sphere into $\mathbf{R}^3$ in this way induces a metric on S^2, which is just the pullback of the flat-space metric. The simple-minded way to find this is to start with the metric $ds^2 = \mathrm{d}x^2 + \mathrm{d}y^2 + \mathrm{d}z^2$ on $\mathbf{R}^3$ and substitute (A.11) into this expression, yielding a metric $\mathrm{d}\theta^2 + \sin^2\theta\, \mathrm{d}\phi^2$ on S^2. Let's see how this answer comes about using the more respectable formalism. (Of course it would be easier if we worked in spherical coordinates on $\mathbf{R}^3$, but doing it the hard way is more illustrative.) The matrix of partial derivatives is given by

$$\frac{\partial y^\alpha}{\partial x^\mu} = \begin{pmatrix} \cos\theta\cos\phi & \cos\theta\sin\phi & -\sin\theta \\ -\sin\theta\sin\phi & \sin\theta\cos\phi & 0 \end{pmatrix}. \tag{A.12}$$

The metric on S^2 is obtained by simply pulling back the metric from $\mathbf{R}^3$,

$$\begin{aligned} (\phi^* g)_{\mu\nu} &= \frac{\partial y^\alpha}{\partial x^\mu}\frac{\partial y^\beta}{\partial x^\nu} g_{\alpha\beta} \\ &= \begin{pmatrix} 1 & 0 \\ 0 & \sin^2\theta \end{pmatrix}, \end{aligned} \tag{A.13}$$

as you can easily check. So the answer really is the same as you would get by naive substitution, but now we know why.

APPENDIX B

Diffeomorphisms and Lie Derivatives

In this Appendix we continue the explorations of the previous one, now focusing on the special case when the two manifolds are actually the same. Thus far, we have been careful to emphasize that a map $\phi : M \to N$ can be used to pull certain things back (A.9) and push other things forward (A.10). The reason why it generally doesn't work both ways can be traced to the fact that ϕ might not be invertible. If ϕ is invertible (and both ϕ and ϕ^{-1} are smooth, which we always implicitly assume), then it defines a diffeomorphism between M and N. This can only be the case if M and N are actually the same abstract manifold; indeed, the existence of a diffeomorphism is the definition of two manifolds being the same. The beauty of diffeomorphisms is that we can use both ϕ and ϕ^{-1} to move tensors from M to N; this will allow us to define the pushforward and pullback of arbitrary tensors. Specifically, for a (k, l) tensor field $T^{\mu_1\cdots\mu_k}{}_{\nu_1\cdots\nu_l}$ on M, we define the pushforward by

$$\begin{aligned}(\phi_* T)(\omega^{(1)}, \ldots, \omega^{(k)}, V^{(1)}, \ldots, V^{(l)})\\ = T(\phi^*\omega^{(1)}, \ldots, \phi^*\omega^{(k)}, [\phi^{-1}]_* V^{(1)}, \ldots, [\phi^{-1}]_* V^{(l)}),\end{aligned} \qquad \text{(B.1)}$$

where the $\omega^{(i)}$'s are one-forms on N and the $V^{(i)}$'s are vectors on N. In components this becomes

$$(\phi_* T)^{\alpha_1\cdots\alpha_k}{}_{\beta_1\cdots\beta_l} = \frac{\partial y^{\alpha_1}}{\partial x^{\mu_1}} \cdots \frac{\partial y^{\alpha_k}}{\partial x^{\mu_k}} \frac{\partial x^{\nu_1}}{\partial y^{\beta_1}} \cdots \frac{\partial x^{\nu_l}}{\partial y^{\beta_l}} T^{\mu_1\cdots\mu_k}{}_{\nu_1\cdots\nu_l}. \qquad \text{(B.2)}$$

The appearance of the inverse matrix $\partial x^\nu / \partial y^\beta$ is legitimate because ϕ is invertible. Note that we could also define the pullback in the obvious way, but there is no need to write separate equations because the pullback ϕ^* is the same as the pushforward via the inverse map, $[\phi^{-1}]_*$.

We are now in a position to explain the relationship between diffeomorphisms and coordinate transformations: they are two different ways of doing precisely the same thing. If you like, diffeomorphisms are "active coordinate transformations," while traditional coordinate transformations are "passive." Consider an n-dimensional manifold M with coordinate functions $x^\mu : M \to \mathbf{R}^n$. To change coordinates we can either simply introduce new functions $y^\mu : M \to \mathbf{R}^n$ ("keep the manifold fixed, change the coordinate maps"), or we could just as well introduce a diffeomorphism $\phi : M \to M$, after which the coordinates would just be the pullbacks $(\phi^* x)^\mu : M \to \mathbf{R}^n$ ("move the points on the manifold, and then evaluate the coordinates of the new points"), as shown in Figure B.1. In this sense, (B.2) really is the tensor transformation law, just thought of from a different point of view.

Since a diffeomorphism allows us to pull back and push forward arbitrary tensors, it provides another way of comparing tensors at different points on a manifold. Given a diffeomorphism $\phi : M \to M$ and a tensor field $T^{\mu_1 \cdots \mu_k}{}_{\nu_1 \cdots \nu_l}(x)$, we can form the difference between the value of the tensor at some point p and $\phi^*[T^{\mu_1 \cdots \mu_k}{}_{\nu_1 \cdots \nu_l}(\phi(p))]$, its value at $\phi(p)$ pulled back to p. This suggests that we could define another kind of derivative operator on tensor fields, one that categorizes the rate of change of the tensor under the flow of the diffeomorphism. For that, however, a single discrete diffeomorphism is insufficient; we require a one-parameter family of diffeomorphisms, ϕ_t. This family can be thought of as a smooth map $\mathbf{R} \times M \to M$, such that for each $t \in \mathbf{R}$ we have a diffeomorphism ϕ_t, satisfying $\phi_s \circ \phi_t = \phi_{s+t}$. This last condition implies that ϕ_0 is the identity map.

One-parameter families of diffeomorphisms can be thought of as arising from vector fields (and vice-versa). If we consider what happens to the point p under the entire family ϕ_t, it is clear that it describes a curve in M; since the same thing will be true of every point on M, these curves fill the manifold (although there can be degeneracies where the diffeomorphisms have fixed points). We can define a vector field $V^\mu(x)$ to be the set of tangent vectors to each of these curves at every point, evaluated at $t = 0$. An example on S^2 is provided by the diffeomorphism $\phi_t(\theta, \phi) = (\theta, \phi + t)$, shown in Figure B.2. We can reverse the construction to define a one-parameter family of diffeomorphisms from any vector field. Given a vector field $V^\mu(x)$, we define the **integral curves** of the vector field to be those curves $x^\mu(t)$ that solve

$$\frac{dx^\mu}{dt} = V^\mu. \tag{B.3}$$

Note that this familiar-looking equation is now to be interpreted in the opposite sense from our usual way; we are given the vectors, from which we define the curves. Solutions to (B.3) are guaranteed to exist as long as we don't do anything silly like run into the edge of our manifold; the proof amounts to finding a coordinate system in which the problem reduces to the fundamental theorem

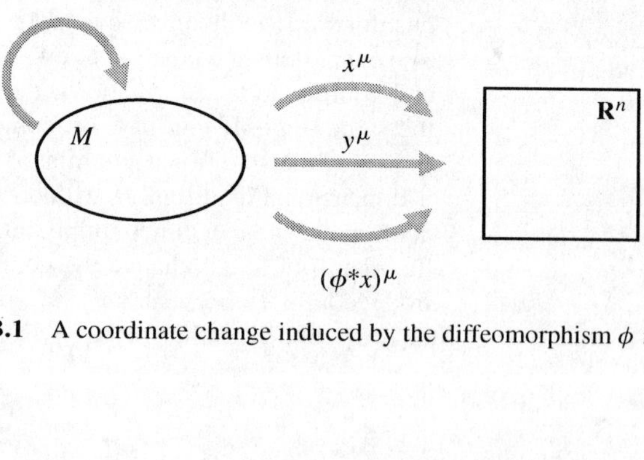

FIGURE B.1 A coordinate change induced by the diffeomorphism $\phi : M \to M$.

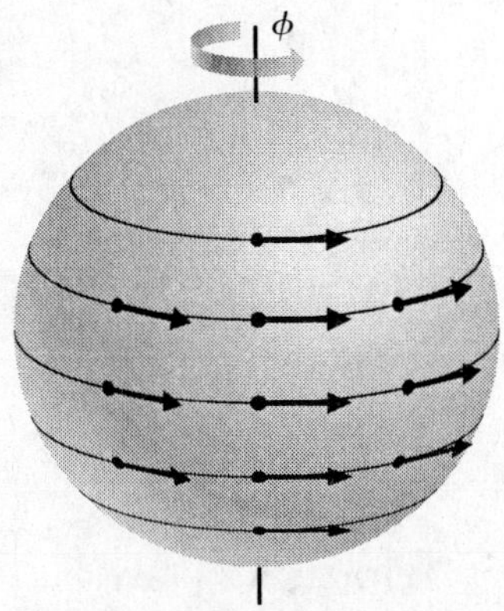

FIGURE B.2 A diffeomorphism on the two-sphere, given by a rotation about its axis.

of ordinary differential equations. Our diffeomorphisms ϕ_t represent "flow down the integral curves," and the associated vector field is referred to as the **generator** of the diffeomorphism. (Confusingly, vector fields and their integral curves also appear in the context of null hypersurfaces, where it is the curves rather than the vector fields that are called "generators.") Integral curves are used all the time in elementary physics, just not given the name. The "lines of magnetic flux" traced out by iron filings in the presence of a magnet are simply the integral curves of the magnetic field vector **B**.

Given a vector field $V^\mu(x)$, then, we have a family of diffeomorphisms parameterized by t, and we can ask how fast a tensor changes as we travel down the integral curves. For each t we can define this change as the difference between the pullback of the tensor to p and its original value at p,

$$\Delta_t T^{\mu_1\cdots\mu_k}{}_{\nu_1\cdots\nu_l}(p) = \phi_t^*[T^{\mu_1\cdots\mu_k}{}_{\nu_1\cdots\nu_l}(\phi_t(p))] - T^{\mu_1\cdots\mu_k}{}_{\nu_1\cdots\nu_l}(p). \tag{B.4}$$

Note that both terms on the right-hand side are tensors at p, as shown in Figure B.3. We then define the **Lie derivative** of the tensor along the vector field as

$$\mathcal{L}_V T^{\mu_1\cdots\mu_k}{}_{\nu_1\cdots\nu_l} = \lim_{t\to 0}\left(\frac{\Delta_t T^{\mu_1\cdots\mu_k}{}_{\nu_1\cdots\nu_l}}{t}\right). \tag{B.5}$$

The Lie derivative is a map from (k, l) tensor fields to (k, l) tensor fields, which is manifestly independent of coordinates. Since the definition essentially amounts to the conventional definition of an ordinary derivative applied to the component functions of the tensor, it should be clear that it is linear,

$$\mathcal{L}_V(aT + bS) = a\mathcal{L}_V T + b\mathcal{L}_V S, \tag{B.6}$$

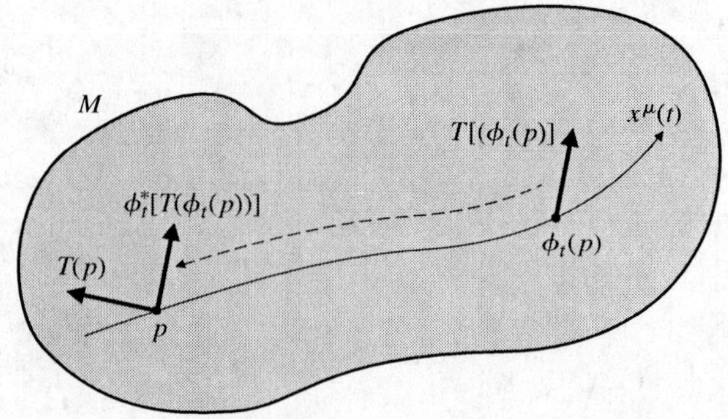

FIGURE B.3 The rate of change of a tensor along the integral curves of a vector field is computed by comparing the original tensor $T(p)$ at a point p to the value of T at a point $\phi_t(p)$ by pulling $T(\phi_t(p))$ back to p.

and obeys the Leibniz rule,

$$\mathcal{L}_V(T \otimes S) = (\mathcal{L}_V T) \otimes S + T \otimes (\mathcal{L}_V S), \tag{B.7}$$

where S and T are tensors and a and b are constants. The Lie derivative is in fact a more primitive notion than the covariant derivative, since it does not require specification of a connection (although it does require a vector field, of course). A moment's reflection will convince you that it reduces to the ordinary directional derivative on functions,

$$\mathcal{L}_V f = V(f) = V^\mu \partial_\mu f. \tag{B.8}$$

To discuss the action of the Lie derivative on tensors in terms of other operations we know, it is convenient to choose a coordinate system adapted to our problem. Specifically, we will work in coordinates $x^\mu = (x^1, \ldots x^n)$, such that x^1 is the parameter along the integral curves and the other coordinates are chosen any way we like. Then the vector field takes the form $V = \partial/\partial x^1$; that is, it has components $V^\mu = (1, 0, 0, \ldots, 0)$. The magic of this coordinate system is that a diffeomorphism by t amounts to a coordinate transformation from x^μ to $y^\mu = (x^1 + t, x^2, \ldots, x^n)$. Thus, from (A.6) the pullback matrix is simply

$$(\phi_t^*)_\mu{}^\nu = \delta_\mu^\nu, \tag{B.9}$$

and the components of the tensor pulled back from $\phi_t(p)$ to p are simply

$$\phi_t^*[T^{\mu_1\cdots\mu_k}{}_{\nu_1\cdots\nu_l}(\phi_t(p))] = T^{\mu_1\cdots\mu_k}{}_{\nu_1\cdots\nu_l}(x^1 + t, x^2, \ldots, x^n). \tag{B.10}$$

In this coordinate system, then, the Lie derivative becomes

$$\mathcal{L}_V T^{\mu_1\cdots\mu_k}{}_{\nu_1\cdots\nu_l} = \frac{\partial}{\partial x^1} T^{\mu_1\cdots\mu_k}{}_{\nu_1\cdots\nu_l}, \tag{B.11}$$

and in particular the derivative of a vector field $U^\mu(x)$ is

$$\mathcal{L}_V U^\mu = \frac{\partial U^\mu}{\partial x^1}. \tag{B.12}$$

Although this expression is clearly not covariant, we know that the commutator $[V, U]$ is a well-defined tensor, and in this coordinate system

$$\begin{aligned} [V, U]^\mu &= V^\nu \partial_\nu U^\mu - U^\nu \partial_\nu V^\mu \\ &= \frac{\partial U^\mu}{\partial x^1}. \end{aligned} \tag{B.13}$$

Therefore the Lie derivative of U with respect to V has the same components in this coordinate system as the commutator of V and U; but since both are vectors, they must be equal in any coordinate system:

$$\mathcal{L}_V U^\mu = [V, U]^\mu. \tag{B.14}$$

As an immediate consequence, we have $\mathcal{L}_V U = -\mathcal{L}_U V$. It is because of (B.14) that the commutator is sometimes called the **Lie bracket.**

To derive the action of $\mathcal{L}_V$ on a one-form ω_μ, begin by considering the action on the scalar $\omega_\mu U^\mu$ for an arbitrary vector field U^μ. First use the fact that the Lie derivative with respect to a vector field reduces to the action of the vector itself when applied to a scalar:

$$\begin{aligned} \mathcal{L}_V(\omega_\mu U^\mu) &= V(\omega_\mu U^\mu) \\ &= V^\nu \partial_\nu(\omega_\mu U^\mu) \\ &= V^\nu(\partial_\nu \omega_\mu) U^\mu + V^\nu \omega_\mu (\partial_\nu U^\mu). \end{aligned} \tag{B.15}$$

Then use the Leibniz rule on the original scalar:

$$\begin{aligned} \mathcal{L}_V(\omega_\mu U^\mu) &= (\mathcal{L}_V \omega)_\mu U^\mu + \omega_\mu (\mathcal{L}_V U)^\mu \\ &= (\mathcal{L}_V \omega)_\mu U^\mu + \omega_\mu V^\nu \partial_\nu U^\mu - \omega_\mu U^\nu \partial_\nu V^\mu. \end{aligned} \tag{B.16}$$

Setting these expressions equal to each other and requiring that equality hold for arbitrary U^μ, we see that

$$\mathcal{L}_V \omega_\mu = V^\nu \partial_\nu \omega_\mu + (\partial_\mu V^\nu)\omega_\nu, \tag{B.17}$$

which (like the definition of the commutator) is completely covariant, although not manifestly so.

By a similar procedure we can define the Lie derivative of an arbitrary tensor field. The answer can be written

$$\boxed{\begin{aligned}\mathcal{L}_V T^{\mu_1\mu_2\cdots\mu_k}{}_{\nu_1\nu_2\cdots\nu_l} = {}& V^\sigma \partial_\sigma T^{\mu_1\mu_2\cdots\mu_k}{}_{\nu_1\nu_2\cdots\nu_l} \\ & - (\partial_\lambda V^{\mu_1}) T^{\lambda\mu_2\cdots\mu_k}{}_{\nu_1\nu_2\cdots\nu_l} \\ & - (\partial_\lambda V^{\mu_2}) T^{\mu_1\lambda\cdots\mu_k}{}_{\nu_1\nu_2\cdots\nu_l} - \cdots \\ & + (\partial_{\nu_1} V^{\lambda}) T^{\mu_1\mu_2\cdots\mu_k}{}_{\lambda\nu_2\cdots\nu_l} \\ & + (\partial_{\nu_2} V^{\lambda}) T^{\mu_1\mu_2\cdots\mu_k}{}_{\nu_1\lambda\cdots\nu_l} + \cdots .\end{aligned}} \tag{B.18}$$

Once again, this expression is covariant, despite appearances. It would undoubtedly be comforting, however, to have an equivalent expression that looked manifestly tensorial. In fact it turns out that we can write

$$\begin{aligned}\mathcal{L}_V T^{\mu_1\mu_2\cdots\mu_k}{}_{\nu_1\nu_2\cdots\nu_l} = {}& V^\sigma \nabla_\sigma T^{\mu_1\mu_2\cdots\mu_k}{}_{\nu_1\nu_2\cdots\nu_l} \\ & - (\nabla_\lambda V^{\mu_1}) T^{\lambda\mu_2\cdots\mu_k}{}_{\nu_1\nu_2\cdots\nu_l} \\ & - (\nabla_\lambda V^{\mu_2}) T^{\mu_1\lambda\cdots\mu_k}{}_{\nu_1\nu_2\cdots\nu_l} - \cdots \\ & + (\nabla_{\nu_1} V^{\lambda}) T^{\mu_1\mu_2\cdots\mu_k}{}_{\lambda\nu_2\cdots\nu_l} \\ & + (\nabla_{\nu_2} V^{\lambda}) T^{\mu_1\mu_2\cdots\mu_k}{}_{\nu_1\lambda\cdots\nu_l} + \cdots ,\end{aligned} \tag{B.19}$$

where ∇_μ represents *any* symmetric (torsion-free) covariant derivative (including, of course, one derived from a metric). You can check that all of the terms that would involve connection coefficients if we were to expand (B.19) would cancel, leaving only (B.18). Both versions of the formula for a Lie derivative are useful at different times. A particularly useful formula is for the Lie derivative of the metric:

$$\begin{aligned}\mathcal{L}_V g_{\mu\nu} &= V^\sigma \nabla_\sigma g_{\mu\nu} + (\nabla_\mu V^\lambda) g_{\lambda\nu} + (\nabla_\nu V^\lambda) g_{\mu\lambda} \\ &= \nabla_\mu V_\nu + \nabla_\nu V_\mu ,\end{aligned} \tag{B.20}$$

or

$$\boxed{\mathcal{L}_V g_{\mu\nu} = 2\nabla_{(\mu} V_{\nu)} ,} \tag{B.21}$$

where ∇_μ is the covariant derivative derived from $g_{\mu\nu}$.

Let's put some of these ideas into the context of general relativity. You will often hear it proclaimed that GR is a "diffeomorphism invariant" theory. What this means is that, if the universe is represented by a manifold M with metric $g_{\mu\nu}$ and matter fields ψ, and $\phi : M \to M$ is a diffeomorphism, then the sets $(M, g_{\mu\nu}, \psi)$ and $(M, \phi^* g_{\mu\nu}, \phi^* \psi)$ represent the same physical situation. Since diffeomorphisms are just active coordinate transformations, this is a highbrow way of saying that the theory is coordinate invariant. Although such a statement is true, it is a source of great misunderstanding, for the simple fact that it conveys very little information. Any semi-respectable theory of physics is coordinate invariant, including those based on special relativity or Newtonian mechanics; GR is not unique in this regard. When people say that GR is diffeomorphism invariant, more likely than not they have one of two (closely related) concepts in mind: the theory is free of "prior geometry," and there is no *preferred* coordinate system for spacetime. The first of these stems from the fact that the metric is a dynamical variable, and along with it the connection and volume element and so forth. Nothing is given to us ahead of time, unlike in classical mechanics or SR. As a consequence, there is no way to simplify life by sticking to a specific coordinate system adapted to some absolute elements of the geometry. This state of affairs forces us to be very careful; it is possible that two purportedly distinct configurations (of matter and metric) in GR are actually "the same," related by a diffeomorphism. In a path integral approach to quantum gravity, where we would like to sum over all possible configurations, special care must be taken not to overcount by allowing physically indistinguishable configurations to contribute more than once. In SR or Newtonian mechanics, meanwhile, the existence of a preferred set of coordinates saves us from such ambiguities. The fact that GR has no preferred coordinate system is often garbled into the statement that it is coordinate invariant (or "generally covariant," or "diffeomorphism invariant"); both things are true, but one has more content than the other.

On the other hand, the fact of diffeomorphism invariance can be put to good use. Recall that the complete action for gravity coupled to a set of matter fields ψ^i is given by a sum of the Hilbert action for GR plus the matter action,

$$S = \frac{1}{16\pi G} S_H[g_{\mu\nu}] + S_M[g_{\mu\nu}, \psi^i]. \tag{B.22}$$

The Hilbert action S_H is diffeomorphism invariant when considered in isolation, so the matter action S_M must also be if the action as a whole is to be invariant. We can write the variation in S_M under a diffeomorphism as

$$\delta S_M = \int d^n x \frac{\delta S_M}{\delta g_{\mu\nu}} \delta g_{\mu\nu} + \int d^n x \frac{\delta S_M}{\delta \psi^i} \delta \psi^i. \tag{B.23}$$

We are not considering arbitrary variations of the fields, only those that result from a diffeomorphism. Nevertheless, the matter equations of motion tell us that the variation of S_M with respect to ψ^i will vanish for any variation, since the gravitational part of the action doesn't involve the matter fields. Hence, for a diffeomorphism invariant theory the first term on the right-hand side of (B.23) must also vanish. If the diffeomorphism is generated by an infinitesimal vector field $V^\mu(x)$, the infinitesimal change in the metric is simply given by its Lie derivative along V^μ; by (B.20) we have

$$\begin{aligned}\delta g_{\mu\nu} &= \mathcal{L}_V g_{\mu\nu}\\ &= 2\nabla_{(\mu} V_{\nu)}.\end{aligned} \tag{B.24}$$

Setting $\delta S_M = 0$ then implies

$$\begin{aligned}0 &= \int d^n x \frac{\delta S_M}{\delta g_{\mu\nu}} \nabla_\mu V_\nu\\ &= -\int d^n x \sqrt{-g}\, V_\nu \nabla_\mu \left(\frac{1}{\sqrt{-g}} \frac{\delta S_M}{\delta g_{\mu\nu}}\right),\end{aligned} \tag{B.25}$$

where we are able to drop the symmetrization of $\nabla_{(\mu} V_{\nu)}$ since $\delta S_M / \delta g_{\mu\nu}$ is already symmetric. Demanding that (B.25) hold for diffeomorphisms generated by arbitrary vector fields V^μ, and using the definition (4.75) of the energy-momentum tensor, we obtain precisely the law of energy-momentum conservation,

$$\nabla_\mu T^{\mu\nu} = 0. \tag{B.26}$$

Conservation of $T_{\mu\nu}$ is a powerful statement, and it might seem surprising that we derived it from as weak a requirement as diffeomorphism invariance. Actually we sneaked in a much stronger assumption, namely that there is a clean separation between the "matter" and "gravitational" actions (in the sense that no matter fields appeared in the gravitational action). If there were, for example, a scalar field multiplying the curvature scalar and also appearing in the matter action (as in the scalar-tensor theories discussed in Chapter 4), this assumption would have been violated, and $T_{\mu\nu}$ would not be conserved by itself.

Recall that in Chapter 3 we spoke of symmetries and Killing vectors, with repeated appeals to look in the Appendices. Now that we understand more about diffeomorphisms, it is perfectly straightforward to understand symmetries. We say that a diffeomorphism ϕ is a **symmetry** of some tensor T if the tensor is invariant after being pulled back under ϕ:

$$\phi^* T = T. \tag{B.27}$$

Although symmetries may be discrete, it is also common to have a one-parameter family of symmetries ϕ_t. If the family is generated by a vector field $V^\mu(x)$, then (B.27) amounts to

$$\mathcal{L}_V T = 0. \tag{B.28}$$

By (B.12), one implication of a symmetry is that, if T is symmetric under some one-parameter family of diffeomorphisms, we can always find a coordinate system in which the components of T are all independent of one of the coordinates (the integral curve coordinate of the vector field). The converse is also true; if all of the components are independent of one of the coordinates, then the partial derivative vector field associated with that coordinate generates a symmetry of the tensor.

The most important symmetries are those of the metric, for which $\phi^* g_{\mu\nu} = g_{\mu\nu}$. A diffeomorphism of this type is called an isometry. If a one-parameter family of isometries is generated by a vector field $K^\mu(x)$, then K^μ turns out to be a Killing vector field. The condition that K^μ be a Killing vector is thus

$$\mathcal{L}_K g_{\mu\nu} = 0, \tag{B.29}$$

or from (B.20),

$$\nabla_{(\mu} K_{\nu)} = 0. \tag{B.30}$$

We recognize this last version as Killing's equation, (3.174). From our discussion in Chapter 3 we know that, if a spacetime has a Killing vector, we can find a coordinate system in which the metric is independent of one of the coordinates, and the quantity $p_\mu K^\mu$ will be constant along geodesics with tangent vector p^μ. Once we have set up the machinery of diffeomorphisms and Lie derivatives, the derivation of Killing vectors proceeds much more elegantly.

B.1 ■ EXERCISES

1. In Euclidean three-space, find and draw the integral curves of the vector fields

$$A = \frac{y-x}{r}\frac{\partial}{\partial x} - \frac{x+y}{r}\frac{\partial}{\partial y}$$

and

$$B = xy\frac{\partial}{\partial x} - y^2\frac{\partial}{\partial y}.$$

Calculate $C = \mathcal{L}_A B$ and draw the integral curves of C.

APPENDIX

C
Submanifolds

The notion of a submanifold, some subset of another manifold which might be (and usually is) of lower dimension, is intuitively straightforward; it should come as no surprise, however, to learn that a certain amount of formalism comes along for the ride. Submanifolds arise all the time in general relativity—as boundaries of spacetimes, hypersurfaces at fixed time, spaces into which larger spaces are foliated by the action of symmetries—so it is worth our effort to understand how they work.

Consider an n-dimensional manifold M and an m-dimensional manifold S, with $m \leq n$, and a map $\phi : S \to M$. If the map ϕ is both C^∞ and one-to-one, and the inverse $\phi^{-1} : \phi[S] \to S$ is also C^∞, then we say that the image $\phi[S]$ is an **embedded submanifold** of M. If ϕ is one-to-one locally but not necessarily globally (that is, there may be self-intersections of $\phi[S]$ in M), then we say that $\phi[S]$ is an **immersed submanifold** of M. When we speak of "submanifolds" without any particular modifier, we are imagining that they are embedded. An m-dimensional submanifold of an n-dimensional manifold is said to be of **codimension** $n - m$.

As discussed in Appendix A, the map $\phi : S \to M$ can be used to push forward $(k, 0)$ tensors from S to M, and to pull back $(0, l)$ tensors from M to S. In particular, given a point $q \in S$ and its image $\phi(q) \in M$, the tangent space $T_{\phi(q)}\phi[S]$ is naturally identified as an m-dimensional subspace of the n-dimensional vector space $T_{\phi(q)}M$. If you think about the definition of a vector as the directional derivative along a curve, this makes perfect sense; any curve $\gamma : \mathbf{R} \to S$ clearly defines a curve in M via composition ($\phi \circ \gamma : \mathbf{R} \to M$), which in turn defines a directional derivative. Similarly, differential forms in M can be pulled back to S by restricting their action to vectors in the subspace $T_{\phi(q)}\phi[S]$.

Another way to define submanifolds is as places where a collection of functions takes on some specified fixed set of values. An m-dimensional submanifold of M can be specified in terms of $n - m$ functions $f^a(x)$, where a runs from 1 to $n - m$, as the set of points x, where the f^a's are equal to some constants f^a_*:

$$\begin{aligned} f^1(x) &= f^1_* \\ f^2(x) &= f^2_* \\ &\vdots \\ f^{n-m}(x) &= f^{n-m}_* . \end{aligned} \tag{C.1}$$

The functions should be nondegenerate, so that the submanifold really is of dimension m. Notice that the submanifold defined in this way is an actual subset of M; it is equivalent to what we called $\phi[S]$ in our previous definition. For convenience, we will henceforth tend to blur the distinction between the original space and its embedding as a submanifold, and simply refer to "the submanifold S."

To see the relationship between the two definitions of a submanifold, imagine constructing a set of coordinates $x^\mu = \{f^a, y^\alpha\}$ in a neighborhood of $\phi[S] \subset M$, consisting of the $n - m$ functions f^a and an additional m functions y^α. Then we can pull back the functions y^α to serve as coordinates on S, and the map $\phi : S \to M$ is simply given by

$$\phi : (y^\alpha) \to (f^a_*, y^\alpha). \tag{C.2}$$

A simple example is the two-sphere S^2, which in fact we defined as the set of points a unit distance from the origin in $\mathbf{R}^3$. In polar coordinates (r, θ, ϕ), this is equivalent to the requirement $r = 1$, so the coordinate r plays the role of the function $f(x)$, while θ and ϕ are induced coordinates on S^2.

We have already mentioned in (B.3) that specifying a single vector field leads to a family of integral curves, which are simply one-dimensional submanifolds. We might imagine generalizing this construction by using a set of several vector fields to define higher-dimensional submanifolds. Imagine we have an n-dimensional manifold M, an m-dimensional submanifold S, and a set of p linearly independent vector fields $V^\mu_{(a)}$, with $p \geq m$. Then the notion that these vector fields "fit together to define S" means that each vector is tangent to S everywhere, so that the $V^\mu_{(a)}$'s span each tangent space T_pS; we say that S is an **integral submanifold** of the vector fields. However, any given set of vector fields may or may not actually fit together to define such submanifolds. Whether they do or not is revealed by **Frobenius's theorem**: a set of vector fields $V^\mu_{(a)}$ fit together to define integral submanifolds if and only if all of their commutators are in the space spanned by the $V^\mu_{(a)}$'s; that is, if

$$[V_{(a)}, V_{(b)}]^\mu = \alpha^c V^\mu_{(c)} \tag{C.3}$$

for some set of coefficients $\alpha^c(x)$. (In the language of group theory, this means that the vector fields form a Lie algebra.) We won't provide a proof, but hopefully the result makes some mathematical sense. If the vector fields are going to fit together to form a submanifold S, they must remain tangent to S everywhere. But the commutator $[V, W]$ is equivalent to the Lie derivative $\mathcal{L}_V W$, which measures how W changes as we travel along V. If this Lie derivative doesn't remain in the space defined by the vectors, it means that W starts sticking out of the submanifold S. Examples of vector fields fitting together to form submanifolds are easy to come by; in Section 5.2 we discussed how the three Killing vectors associated with spherical symmetry define a foliation of a three-dimensional space into two-spheres. (Notice that the dimensionality of the integral submanifold can be less than the number of vector fields.) For a discussion of Frobenius's theorem, see Schutz (1980).

An interesting alternative formulation of Frobenius's theorem uses differential forms. First notice that any set of p linearly independent one-forms $\omega_\mu^{(a)}$ defines an $(n-p)$-dimensional vector subspace of T_pM, called the **annihilator** of the set of forms, consisting of those vectors $V^\mu \in T_pM$ satisfying

$$\omega_\mu^{(a)} V^\mu = 0 \tag{C.4}$$

for all $\omega_\mu^{(a)}$. So instead of asking whether a collection of vector fields fit together to define a set of submanifolds, we could ask whether a collection of one-forms $\omega_\mu^{(a)}$ define a set of vector subspaces that fit together as tangent spaces to a set of submanifolds. To understand when this happens, recall the definition (C.1) of an m-dimensional submanifold as a place where a set of $p = n - m$ functions $f^a(x)$ are set equal to constants. A constant function is one for which the exterior derivative $(\mathrm{d}f^a)_\mu = \nabla_\mu f^a$ vanishes; but if a function is constant only along some submanifold, that means that

$$\mathrm{d}f^a(V) = V^\mu \nabla_\mu f^a = 0 \tag{C.5}$$

for all vectors V^μ tangent to the submanifold, $V^\mu \in T_pS$. It also goes the other way; if a vector V^μ is annihilated by all of the gradients $\nabla_\mu f^a$, it is necessarily tangent to the corresponding submanifold S. Therefore, if a set of one-forms are each exact, $\omega_\mu^{(a)} = \nabla_\mu f^a$, the vector spaces they annihilate will certainly define submanifolds, namely those along which the f^a's are constant. But if a set of p one-forms annihilates a certain subspace, so will any other set of p one-forms that are linear combinations of the originals. We therefore say that a set of one-forms $\omega_\mu^{(a)}$ is **surface-forming** if every member can be expressed as a linear combination of a set of exact forms; that is, if there exist functions $g^a{}_b(x)$ and $f^a(x)$ such that

$$\omega_\mu^{(a)} = \sum_b g^a{}_b \nabla_\mu f^b. \tag{C.6}$$

Of course, when handed a set of forms, it might be hard to tell whether there exist functions such that this condition is satisfied; this is where the dual formulation of Frobenius's theorem comes in. This version of the theorem states that a set of one-forms $\omega_\mu^{(a)}$ is surface-forming if and only if every pair of vectors in the annihilator of the set is also annihilated by the exterior derivatives $\mathrm{d}\omega^{(a)}$. In other words, the set $\omega_\mu^{(a)}$ will satisfy (C.6) if and only if, for every pair of vectors V^μ and W^μ satisfying $\omega_\mu^{(a)} V^\mu = 0$ and $\omega_\mu^{(a)} W^\mu = 0$ for all a, we also have

$$\nabla_{[\mu}\omega_{\nu]}^{(a)} V^\mu W^\nu = 0. \tag{C.7}$$

A set of forms $\omega_\mu^{(a)}$ satisfying this condition is sometimes called "closed," which is obviously a generalization of the notion of a single form being closed (namely, that its exterior derivative vanishes). We won't prove the equivalence of the dual formulation of Frobenius' theorem with the vector-field version, but it clearly involves acting our set of forms on the vector-field commutator (C.3).

APPENDIX

D Hypersurfaces

A **hypersurface** is an $(n-1)$-dimensional (codimension one) submanifold Σ of an n-dimensional manifold M. (Of course if $n = 3$, Σ might as well just be called a "surface," but we'll continue to use "hyper-" for consistency.) Hypersurfaces are of great utility in general relativity, and a lot of formalism goes along with them. In this Appendix we collect a set of results in the study of hypersurfaces: normal vectors, generators of null hypersurfaces, Frobenius's theorem for hypersurfaces, Gaussian normal coordinates, induced metrics, projection tensors, extrinsic curvature, and manifolds with boundary. It's something of a smorgasbord, with all the messiness that implies, but hopefully appetizing and nutritious as well.

One way to specify a hypersurface Σ is by setting single function to a constant,

$$f(x) = f_*. \tag{D.1}$$

The vector field

$$\zeta^\mu = g^{\mu\nu}\nabla_\nu f \tag{D.2}$$

will be normal to the surface, in the sense that it is orthogonal to all vectors in $T_p\Sigma \subset T_pM$. If ζ^μ is timelike, the hypersurface is said to be spacelike; if ζ^μ is spacelike the hypersurface is timelike, and if ζ^μ is null the hypersurface is also null. Any vector field proportional to a normal vector field,

$$\xi^\mu = h(x)\nabla^\mu f \tag{D.3}$$

for some function $h(x)$, will itself be a normal vector field; since the normal vector is unique up to scaling, any normal vector can be written in this form. For timelike and spacelike hypersurfaces, we can therefore define a normalized version of the normal vector,

$$n^\mu = \pm\frac{\zeta^\mu}{|\zeta_\mu\zeta^\mu|^{1/2}}. \tag{D.4}$$

Then $n^\mu n_\mu = -1$ for spacelike surfaces and $n^\mu n_\mu = +1$ for timelike surfaces; up to an overall orientation, such a normal vector field is unique. For spacelike surfaces the sign is typically chosen so as to make n^μ be future-directed.

Null hypersurfaces have a special feature: they can be divided into a set of null geodesics, called **generators** of the hypersurface. Let's see how this works.

Notice that the normal vector ζ^μ is tangent to Σ as well as normal to it, since null vectors are orthogonal to themselves. Therefore the integral curves $x^\mu(\alpha)$, satisfying

$$\zeta^\mu = \frac{dx^\mu}{d\alpha}, \tag{D.5}$$

will be null curves contained in the hypersurface. These curves $x^\mu(\alpha)$ necessarily turn out to be geodesics, although α might not be an affine parameter. To verify this claim, recall that the general form of the geodesic equation can be expressed as

$$\zeta^\mu \nabla_\mu \zeta_\nu = \eta(\alpha)\zeta_\nu, \tag{D.6}$$

where $\eta(\alpha)$ is a function that will vanish if α is an affine parameter. We simply plug in (D.2) and calculate:

$$\begin{aligned} \zeta^\mu \nabla_\mu \zeta_\nu &= \zeta^\mu \nabla_\mu \nabla_\nu f \\ &= \zeta^\mu \nabla_\nu \nabla_\mu f \\ &= \zeta^\mu \nabla_\nu \zeta_\mu \\ &= \tfrac{1}{2}\nabla_\nu(\zeta^\mu \zeta_\mu). \end{aligned} \tag{D.7}$$

In the second line we used the torsion-free condition, that covariant derivatives acting on scalars commute. Note that, even though $\zeta^\mu\zeta_\mu = 0$ on Σ itself, we can't be sure that $\nabla_\nu(\zeta^\mu\zeta_\mu)$ vanishes, since $\zeta^\mu\zeta_\mu$ might be nonzero off the hypersurface. If the gradient vanishes, (D.7) is the geodesic equation, and we're done. But if it doesn't vanish, we can use $\zeta^\mu\zeta_\mu = 0$ as an alternative way to define the submanifold Σ, and its derivative defines a normal vector. Therefore, we must have

$$\nabla_\mu(\zeta^\nu\zeta_\nu) = g\nabla_\mu f = g\zeta_\mu, \tag{D.8}$$

where $g(x)$ is some scalar function. We then plug into (D.7) to get

$$\zeta^\mu \nabla_\mu \zeta_\nu = \tfrac{1}{2} g\zeta_\nu, \tag{D.9}$$

which is equivalent to the geodesic equation (D.6). Of course, once we know that a path $x^\mu(\alpha)$ is a geodesic, we are free to re-parameterize it with an affine parameter $\lambda(\alpha)$. Equivalently, we scale the normal vector field by a scalar function $h(x)$,

$$\xi^\mu = h\zeta^\mu, \tag{D.10}$$

such that $\xi^\mu \nabla_\mu \xi^\nu = 0$. It is conventional to do exactly this, and use the corresponding tangent vectors

$$\xi^\mu = \frac{dx^\mu}{d\lambda} \tag{D.11}$$

as normal vectors to Σ. The null geodesics $x^\mu(\lambda)$, whose union is the null hypersurface Σ, are the generators of Σ.

From (D.3) we know that a vector field normal to a hypersurface can be written in the form $\xi^\mu = h\nabla^\mu f$. In the exercises for Chapter 4 you were asked to show that this implies

$$\xi_{[\mu}\nabla_\nu\xi_{\sigma]} = 0, \tag{D.12}$$

or in differential forms notation,

$$\xi \wedge \mathrm{d}\xi = 0. \tag{D.13}$$

The converse, that any vector field satisfying this equation is orthogonal to a hypersurface, is harder to show from first principles, but is a direct consequence of the dual formulation of Frobenius's theorem. Imagine we have two vectors V^μ and W^μ, both of which are annihilated by a one-form ξ_μ obeying (D.12). From Frobenius's theorem (C.7), ξ_μ will define a hypersurface if and only if

$$\nabla_{[\mu}\xi_{\nu]}V^\mu W^\nu = 0. \tag{D.14}$$

Applying the expression in (D.12) to $V^\mu W^\nu$ and expanding the antisymmetrization brackets, we get

$$\begin{aligned}\xi_{[\mu}\nabla_\nu\xi_{\sigma]}V^\mu W^\nu &= \tfrac{1}{3}(\xi_\mu\nabla_{[\nu}\xi_{\sigma]} + \xi_\nu\nabla_{[\sigma}\xi_{\mu]})V^\mu W^\nu + \tfrac{1}{3}\xi_\sigma\nabla_{[\mu}\xi_{\nu]}V^\mu W^\nu \\ &= \tfrac{1}{3}\xi_\sigma\nabla_{[\mu}\xi_{\nu]}V^\mu W^\nu,\end{aligned} \tag{D.15}$$

where in the last line we used the fact that V^μ and W^μ are annihilated by ξ_μ. But since $\nabla_{[\mu}\xi_{\nu]}V^\mu W^\nu$ is a scalar and ξ_σ is a nonvanishing one-form, the only way (D.15) can vanish is if (D.14) holds. Therefore, (D.12) will be true if and only if ξ_μ is hypersurface-orthogonal.

It is often convenient to put a coordinate system on a manifold (or part of it) that is naturally adapted to some hypersurface Σ; Gaussian normal coordinates provide a convenient way to do just that. First choose coordinates $y^i = \{y^1, \ldots, y^{n-1}\}$ on Σ. At each point $p \in \Sigma$, construct the (unique) geodesic for which n^μ is the tangent vector at p. Let z be the affine parameter on each geodesic. [This parameter is unique if n^μ is normalized and $z(p) = 0$.] Any point q in a neighborhood of Σ lives on one such geodesic. To each such point we assign coordinates $\{z, y^1, \ldots, y^{n-1}\}$, where the y^i's are the coordinates of the point p connected to q by the geodesic we have constructed. These coordinates $\{z, y^1, \ldots, y^{n-1}\}$ are **Gaussian normal coordinates** (not to be confused with "Riemann normal coordinates," constructed by following geodesics in all directions from a single point p). These coordinates will eventually fail to be well-defined if we reach a point where geodesics focus and intersect, but they will always exist in some region including Σ. All of our statements about Gaussian normal coordinates should be taken as applying in the region where they are well-defined.

Associated with the coordinate functions $\{z, y^1, \ldots, y^{n-1}\}$ are coordinate-basis vector fields $\{\partial_z, \partial_1, \ldots, \partial_{n-1}\}$. For notational convenience let's label these vector fields

$$(\partial_z)^\mu = n^\mu,$$
$$(\partial_i)^\mu = Y^\mu_{(i)}, \tag{D.16}$$

where the first line makes sense because ∂_z is simply the extension along the geodesics of the original normal vector n^μ. With respect to these basis vectors, the metric takes on a simple form. To start, we know that

$$g_{zz} = ds^2(\partial_z, \partial_z) = n_\mu n^\mu = \pm 1, \tag{D.17}$$

since n^μ is just the normalized tangent vector to the geodesics emanating from Σ. To encapsulate the sign ambiguity, let's label this σ:

$$\sigma = n_\mu n^\mu = \pm 1. \tag{D.18}$$

But it is also the case that $g_{zi} = n_\mu Y^\mu_{(i)} = 0$, as we can straightforwardly check. Start at the original surface Σ, where $n_\mu Y^\mu_{(i)} = 0$ by hypothesis (since n^μ is normal to Σ). Then we calculate

$$\begin{aligned}\frac{D}{dz}(n_\mu Y^\mu_{(i)}) &= n^\nu \nabla_\nu (n_\mu Y^\mu_{(i)})\\ &= n^\nu n_\mu \nabla_\nu Y^\mu_{(i)}\\ &= Y^\nu_{(i)} n_\mu \nabla_\nu n^\mu\\ &= \frac{1}{2} Y^\nu_{(i)} \nabla_\nu (n_\mu n^\mu)\\ &= 0.\end{aligned} \tag{D.19}$$

Let's explain this derivation line-by-line. The first line is simply the definition of the directional covariant derivative D/dz. The second uses the Leibniz rule, plus the fact that n_μ is parallel-transported along the geodesic ($n^\nu \nabla_\nu n_\mu = 0$). The third line uses the fact that n^μ and $Y^\nu_{(i)}$ are both coordinate basis vectors, so their Lie bracket vanishes: $[n, Y_{(i)}]^\mu = n^\nu \nabla_\nu Y^\mu_{(i)} - Y^\nu_{(i)} \nabla_\nu n^\mu = 0$. The fourth line again uses Leibniz and the fact that n_μ is parallel-transported, while the fifth simply reflects the fact that $n_\mu n^\mu = \sigma$ is a constant.

We can therefore write the metric in Gaussian normal coordinates as

$$\boxed{ds^2 = \sigma \mathrm{d}z^2 + \gamma_{ij} \mathrm{d}y^i \mathrm{d}y^j,} \tag{D.20}$$

where $\gamma_{ij} = g(\partial_i, \partial_j)$ will in general be a function of all the coordinates $\{z, y^1, \ldots, y^{n-1}\}$. We haven't made any assumptions whatsoever about the geometry; we have simply chosen a coordinate system in which the metric takes a certain form. Notice that setting z = constant defines a family of hypersurfaces diffeomorphic to the original surface Σ; the lack of off-diagonal terms g_{zi} in (D.20) reflects the fact that the vector field n^μ is orthogonal to all of these surfaces, not just the original one. Gaussian normal coordinates are by no means exotic; we use them all the time. Simple examples include inertial coordinates on Minkowski space,

$$ds^2 = -\mathrm{d}t^2 + \mathrm{d}x^2 + \mathrm{d}y^2 + \mathrm{d}z^2, \tag{D.21}$$

or polar coordinates in Euclidean 3-space,

$$ds^2 = \mathrm{d}r^2 + r^2\mathrm{d}\theta^2 + r^2\sin^2\theta\,\mathrm{d}\phi^2. \tag{D.22}$$

Ordinary Robertson–Walker coordinates in cosmology provide a slightly less trivial example,

$$ds^2 = -\mathrm{d}t^2 + a^2(t)\left[\frac{\mathrm{d}r^2}{1-\kappa r^2} + r^2 d\Omega^2\right]. \tag{D.23}$$

Of course, the RW geometries are highly symmetric (homogeneous and isotropic). But, since we have just seen that Gaussian normal coordinates can always be defined, we know that we can describe a perfectly general geometry by altering the spatial components of the metric. This provides one popular way of describing cosmological perturbations; we define "synchronous gauge" for flat spatial sections as

$$ds^2 = -\mathrm{d}t^2 + a^2(t)(\delta_{ij} + h_{ij})\mathrm{d}x^i\mathrm{d}x^j, \tag{D.24}$$

where $h_{ij}(t, \mathbf{x})$ is the metric perturbation. (The generalization to curved spatial sections is immediate.) Again, we have not made any assumptions about the geometry, only chosen a potentially convenient coordinate system.

Recall that the map $\phi : \Sigma \to M$ that embeds any submanifold allows us to pull back the metric from M to Σ. Given coordinates y^i on Σ and x^μ on M, we define the **induced metric** on the submanifold as

$$(\phi^* g)_{ij} = \frac{\partial x^\mu}{\partial y^i}\frac{\partial x^\nu}{\partial y^j} g_{\mu\nu}. \tag{D.25}$$

In the case where the submanifold is a hypersurface, this induced metric is precisely the same as the γ_{ij} appearing in (D.20). To see this, notice that Gaussian normal coordinates are a special case of the natural embedding coordinates described by (C.2). We have a hypersurface Σ defined by $z = z_*$ on M, with coordinates y^i defined on Σ, and a map $\phi : \Sigma \to M$ given by

$$\phi : y^i \to x^\mu = (z_*, y^i). \tag{D.26}$$

Given the form of the metric (D.20) on M, it is immediate that under this map the pullback (D.25) is simply

$$(\phi^* g)_{ij} = \gamma_{ij}. \tag{D.27}$$

Keep in mind that this equation should only be evaluated in Gaussian normal coordinates; otherwise the right-hand side doesn't even make sense.

Along with an induced metric, submanifolds inherit an induced volume element from the manifold in which they are embedded. Recall that a volume element on an n-dimensional manifold with metric $g_{\mu\nu}$ is given by the Levi–Civita tensor, which can be expressed as

$$\epsilon = \sqrt{|g|}\, dx^1 \wedge \cdots \wedge dx^n. \tag{D.28}$$

To get a volume element on a submanifold Σ, it is convenient to introduce Gaussian normal coordinates $(z, y^1, \ldots y^{n-1})$, in which the metric takes the form (D.20). The volume element $\widehat{\epsilon}$ on Σ will then take the form

$$\widehat{\epsilon} = \sqrt{|\gamma|}\, dy^1 \wedge \cdots \wedge dy^{n-1}. \tag{D.29}$$

(By choosing the first coordinate to be the one normal to the hypersurface, we have implicitly chosen a convention for how the orientation on M defines an orientation on Σ.) In these coordinates we have

$$\sqrt{|g|} = \sqrt{|\gamma|}, \tag{D.30}$$

and the volume element on M therefore becomes

$$\epsilon = \sqrt{|\gamma|}\, dz \wedge dy^1 \wedge \cdots \wedge dy^{n-1}. \tag{D.31}$$

We can relate the two volume elements by using the normal vector to Σ, which has components

$$n^\mu = (1, 0, \ldots, 0). \tag{D.32}$$

The contraction of ϵ with n^μ can be denoted

$$[\epsilon(n)]_{\mu_1 \cdots \mu_{n-1}} = n^\lambda \epsilon_{\lambda\mu_1 \cdots \mu_{n-1}}. \tag{D.33}$$

It is then clear that, in these coordinates, we have

$$\begin{aligned} \epsilon(n) &= \sqrt{|\gamma|}\, dy^1 \wedge \cdots \wedge dy^{n-1} \\ &= \widehat{\epsilon}. \end{aligned} \tag{D.34}$$

Thus, the induced volume element has components

$$\widehat{\epsilon}_{\mu_1 \cdots \mu_{n-1}} = n^\lambda \epsilon_{\lambda\mu_1 \cdots \mu_{n-1}}. \tag{D.35}$$

But this is a relation between tensors, so will be true in any coordinate system. We can also reconstruct ϵ from $\widehat{\epsilon}$ and n^μ, via

$$\frac{1}{n}\epsilon_{\nu\mu_1 \cdots \mu_{n-1}} = n_{[\nu}\widehat{\epsilon}_{\mu_1 \cdots \mu_{n-1}]}, \tag{D.36}$$

as can easily be checked by contracting with n^ν. The notion of a submanifold volume element will be crucial in our discussion of Stokes's theorem below.

Another concept closely related to the induced metric on a hypersurface is that of the **projection tensor** for a hypersurface Σ with unit normal vector n^μ, given by

$$P_{\mu\nu} = g_{\mu\nu} - \sigma n_\mu n_\nu, \tag{D.37}$$

where $\sigma = n_\mu n^\mu$. Let's collect some useful properties of this object. Given any vector V^μ in T_pM, $P_{\mu\nu}$ will project it tangent to the hypersurface (that is, orthogonal to n^μ):

$$\begin{aligned}(P_{\mu\nu}V^\mu)n^\nu &= g_{\mu\nu}V^\mu n^\nu - \sigma n_\mu n_\nu V^\mu n^\nu \\ &= V^\mu n_\mu - \sigma^2 V^\mu n_\mu \\ &= 0.\end{aligned} \tag{D.38}$$

Acting on any two vectors V^μ and W^ν that are already tangent to Σ, the projection tensor acts like the metric:

$$\begin{aligned}P_{\mu\nu}V^\mu W^\nu &= g_{\mu\nu}V^\mu W^\nu - \sigma n_\mu n_\nu V^\mu W^\nu \\ &= g_{\mu\nu}V^\mu W^\nu.\end{aligned} \tag{D.39}$$

Finally, the projection tensor is idempotent; acting two (or more) times produces the same result as only acting once:

$$\begin{aligned}P^\mu{}_\lambda P^\lambda{}_\nu &= (\delta^\mu_\lambda - \sigma n^\mu n_\lambda)(\delta^\lambda_\nu - \sigma n^\lambda n_\nu) \\ &= \delta^\mu_\nu - \sigma n^\mu n_\nu - \sigma n^\mu n_\nu + \sigma^3 n^\mu n_\nu \\ &= P^\mu{}_\nu.\end{aligned} \tag{D.40}$$

$P_{\mu\nu}$ is sometimes called the **first fundamental form** of the hypersurface. Because it really does act like the metric for vectors tangent to Σ, and hypersurfaces are often spacelike, you will sometimes see it referred to as the "spatial metric."

Long ago when we first spoke of manifolds and curvature, we were careful to distinguish between the "intrinsic" curvature of a space, as measured by the Riemann tensor, and the "extrinsic" curvature, which depends on how the space is embedded in some larger space. For example, a two-torus can have a flat metric, but any embedding in $\mathbf{R}^3$ makes it look curved. We are now in a position to give a formal definition of this notion, which makes sense for hypersurfaces. Let's assume we have a family of hypersurfaces Σ with unit vector field n^μ, and we extend n^μ through a region (any way we like). Then the **extrinsic curvature** of Σ is simply given by the Lie derivative of the projection tensor along the normal vector field,

$$K_{\mu\nu} = \tfrac{1}{2}\mathcal{L}_n P_{\mu\nu}. \tag{D.41}$$

The extrinsic curvature, sometimes called the **second fundamental form** of the submanifold, is thus interpreted as the rate of change of the projection tensor (the spatial metric, if Σ is spacelike) as we travel along the normal vector field; it is independent of the extension of n^μ away from Σ. It is the work of a few lines to show that this definition is equivalent to the projected Lie derivative of the metric itself,

$$K_{\mu\nu} = \frac{1}{2} P^\alpha{}_\mu P^\beta{}_\nu \mathcal{L}_n g_{\alpha\beta}. \tag{D.42}$$

We know from (B.20) that the Lie derivative of $g_{\mu\nu}$ is given by the symmetrized covariant derivative of the normal vector, so we have

$$K_{\mu\nu} = P^\alpha{}_\mu P^\beta{}_\nu \nabla_{(\alpha} n_{\beta)}. \tag{D.43}$$

Since we are not assuming that the integral curves of n^μ are geodesics, we can define the acceleration as

$$a^\mu = n^\nu \nabla_\nu n^\mu. \tag{D.44}$$

Then it is the work of a few more lines to show that (D.43) is equivalent to

$$\boxed{K_{\mu\nu} = \nabla_\mu n_\nu - \sigma n_\mu a_\nu.} \tag{D.45}$$

The extrinsic curvature has a number of nice properties. It is symmetric,

$$K_{\mu\nu} = K_{\nu\mu}, \tag{D.46}$$

which looks obvious from (D.41), although not from (D.45). You can check that (D.45) really is symmetric, taking advantage of the fact that n^μ is hypersurface-orthogonal. The extrinsic curvature is also orthogonal to the normal direction ("purely spatial"),

$$\begin{aligned} n^\mu K_{\mu\nu} &= n^\mu \nabla_\mu n_\nu - \sigma n^\mu n_\mu a_\nu \\ &= a_\nu - \sigma^2 a_\nu \\ &= 0. \end{aligned} \tag{D.47}$$

We can define a covariant derivative acting along the hypersurface, $\widehat{\nabla}_\mu$, by taking an ordinary covariant derivative and projecting it. For example, on a (1, 1) tensor $X^\mu{}_\nu$ we would have

$$\widehat{\nabla}_\sigma X^\mu{}_\nu = P^\alpha{}_\sigma P^\mu{}_\beta P^\gamma{}_\nu \nabla_\alpha X^\beta{}_\gamma. \tag{D.48}$$

From this we can construct the curvature tensor on the hypersurface $\widehat{R}^\rho{}_{\sigma\mu\nu}$, for example by considering the commutator of covariant derivatives acting on a vector field V^μ, which is tangent to the hypersurface ($P^\mu{}_\nu V^\nu = V^\mu$),

$$[\widehat{\nabla}_\mu, \widehat{\nabla}_\nu] V^\rho = \widehat{R}^\rho{}_{\sigma\mu\nu} V^\sigma. \tag{D.49}$$

Two important equations relate the n-dimensional Riemann curvature to the hypersurface Riemann curvature and the extrinsic curvature. **Gauss's equation** is

$$\widehat{R}^{\rho}{}_{\sigma\mu\nu} = P^{\rho}{}_{\alpha}P^{\beta}{}_{\rho}P^{\gamma}{}_{\mu}P^{\delta}{}_{\nu}R^{\alpha}{}_{\beta\gamma\delta} + \sigma(K^{\rho}{}_{\mu}K_{\sigma\nu} - K^{\rho}{}_{\nu}K_{\sigma\mu}). \tag{D.50}$$

We can take the appropriate traces to get the hypersurface curvature scalar,

$$\widehat{R} = P^{\sigma\nu}\widehat{R}^{\lambda}{}_{\sigma\lambda\nu} = R - \sigma(2R_{\mu\nu}n^{\mu}n^{\nu} + K^2 - K^{\mu\nu}K_{\mu\nu}), \tag{D.51}$$

where $K = g^{\mu\nu}K_{\mu\nu}$. We also have **Codazzi's equation,**

$$\widehat{\nabla}_{[\mu}K_{\nu]}{}^{\mu} = \tfrac{1}{2}P^{\sigma}{}_{\nu}R_{\rho\sigma}n^{\rho}. \tag{D.52}$$

Together, (D.50) and (D.52) are, imaginatively enough, called the Gauss–Codazzi equations.

To stave off confusion, we should note that the definition of extrinsic curvature tends to vary from reference to reference. In some sources the normal vector field is taken to be geodesic everywhere ($a^{\mu} = 0$); things then simplify considerably, and it's straightforward to show that in this case we have

$$\begin{aligned} K_{\mu\nu} &= \tfrac{1}{2}\mathcal{L}_n P_{\mu\nu} \\ &= \tfrac{1}{2}\mathcal{L}_n g_{\mu\nu} \\ &= \nabla_{\mu}n_{\nu}. \end{aligned} \tag{D.53}$$

(If we are given an entire set of hypersurfaces ahead of time, we cannot simply assume that integral curves of the unit normal vector field are geodesics. However, we are often given just a single surface, in which case we are allowed to extend the normal vector field off the surface by solving the geodesic equation.) Other references prefer to think of the extrinsic curvature as a tensor $\widehat{K}_{ij}$ living on Σ rather than in M. If we have an embedding $\phi : y^i \rightarrow x^{\mu}$, this version of the extrinsic curvature is given by the pullback,

$$\begin{aligned} \widehat{K}_{ij} &= (\phi^* K)_{ij} \\ &= \frac{\partial x^{\mu}}{\partial y^i}\frac{\partial x^{\nu}}{\partial y^j}K_{\mu\nu}. \end{aligned} \tag{D.54}$$

Finally, some sources like to define the extrinsic curvature to be minus our definition. It should be straightforward to convert back and forth between the different conventions.

To conclude our discussion, we mention that a very common appearance of hypersurfaces is as the **boundary** of a closed region N of a manifold M, conventionally denoted ∂N. If for example N consists of all the elements of $\mathbf{R}^n$ that lie at a distance from the origin $r \leq 1$, the boundary ∂N is clearly the $(n-1)$-sphere defined by $r = 1$. We may extend this notion to cases where we are not considering a closed region, but an entire manifold with a boundary attached. A **manifold with**

boundary is a set equipped with an atlas of coordinate charts, exactly as in our definition of a manifold in Chapter 2, except that the charts are taken to be maps to the upper half of $\mathbf{R}^n$: the set of n-tuples $\{x^1, \ldots, x^n\}$ with $x^1 \geq 0$. The boundary ∂M is the set of points that are mapped to $x^1 = 0$ by the charts. Then ∂M is naturally an $(n-1)$-dimensional submanifold (without boundary). An example of a boundary of a manifold will appear in our later discussion of conformal diagrams, in which conformal infinity can be thought of as a boundary to spacetime. By continuity, we can treat the boundary as a hypersurface, including inducing metrics and so on; occasionally we need to be careful in taking derivatives on the boundary, but for the most part we can trust our intuition.

APPENDIX

E Stokes's Theorem

In Section 2.10 we introduced the idea that integration on a manifold maps n-form fields to the real numbers. This point of view leads to an elegant statement of one of the most powerful theorems of differential geometry: Stokes's theorem. This theorem is the generalization of the fundamental theorem of calculus, $\int_b^a dx = a - b$. Imagine that we have an n-dimensional region M (which might be an entire manifold) with boundary ∂M, and an $(n-1)$-form ω on M. We will soon explain what is meant by the boundary of a manifold. Then $\mathrm{d}\omega$ is an n-form, which can be integrated over M, while ω itself can be integrated over ∂M. Stokes's theorem is simply

$$\boxed{\int_M \mathrm{d}\omega = \int_{\partial M} \omega.} \tag{E.1}$$

Different special cases of this theorem include not only the fundamental theorem of calculus, but also the theorems of Green, Gauss, and Stokes, familiar from vector calculus in three dimensions.

The presentation (E.1) of Stokes's theorem is extremely elegant, almost too elegant to be useful. We can, fortunately, recast it in pedestrian coordinate-and-index notation. It is convenient to first write the $(n-1)$-form ω as the Hodge dual of a one-form V,

$$\omega = *V, \tag{E.2}$$

with components

$$\begin{aligned} \omega_{\mu_1\cdots\mu_{n-1}} &= (*V)_{\mu_1\cdots\mu_{n-1}} \\ &= \epsilon^{\nu}{}_{\mu_1\cdots\mu_{n-1}} V_\nu \\ &= \epsilon_{\nu\mu_1\cdots\mu_{n-1}} V^\nu, \end{aligned} \tag{E.3}$$

where ϵ is the Levi–Civita n-form on M and we have raised the index on V in the last line. If we wanted to construct V from ω, we apply the Hodge operator again to obtain

$$V = (-1)^{s+n-1} **V = (-1)^{s+n-1} * \omega, \tag{E.4}$$

where s equals -1 for Lorentzian signatures and $+1$ for Euclidean signatures. The exterior derivative of $\omega = *V$ is an n-form, given by

$$\begin{aligned}(\mathrm{d}\omega)_{\lambda\mu_1\cdots\mu_{n-1}} &= (\mathrm{d}*V)_{\lambda\mu_1\cdots\mu_{n-1}} \\ &= n\nabla_{[\lambda}(\epsilon_{|\nu|\mu_1\cdots\mu_{n-1}]}V^\nu) \\ &= n\epsilon_{\nu[\mu_1\cdots\mu_{n-1}}\nabla_{\lambda]}V^\nu,\end{aligned} \tag{E.5}$$

where n is the dimensionality of the region, not to be confused with the normal vector n^μ to the boundary. But any n-form can be written as a function $f(x)$ times ϵ, or equivalently as the Hodge dual of $f(x)$,

$$\mathrm{d}\omega = f\epsilon = *f. \tag{E.6}$$

Taking the dual of both sides gives

$$f = (-1)^s **f = (-1)^s*\mathrm{d}\omega. \tag{E.7}$$

In our case,

$$\begin{aligned}*\mathrm{d}\omega &= *\mathrm{d}*V \\ &= \frac{1}{n!}\epsilon^{\lambda\mu_1\cdots\mu_{n-1}}\left(n\epsilon_{\nu[\mu_1\cdots\mu_{n-1}}\nabla_{\lambda]}V^\nu\right) \\ &= \frac{1}{(n-1)!}(-1)^s(n-1)!\delta^\lambda_\nu\nabla_\lambda V^\nu \\ &= (-1)^s\nabla_\nu V^\nu.\end{aligned} \tag{E.8}$$

Finally we recall that the Levi–Civita tensor is simply the volume element,

$$\begin{aligned}\epsilon &= \sqrt{|g|}\,\mathrm{d}x^1\wedge\cdots\wedge\mathrm{d}x^n \\ &= \sqrt{|g|}\,d^nx.\end{aligned} \tag{E.9}$$

Putting it all together, we find

$$\mathrm{d}\omega = \nabla_\nu V^\nu\sqrt{|g|}\,d^nx. \tag{E.10}$$

So the exterior derivative of an $(n-1)$-form on an n manifold is just a slick way of representing the divergence of a vector (times the metric volume element).

To make sense of the right-hand side of (E.1), we recall from the previous Appendix that the induced volume element on a hypersurface (such as the boundary) is given by

$$\widehat{\epsilon} = \sqrt{|\gamma|}\,d^{n-1}y, \tag{E.11}$$

where γ_{ij} is the induced metric on the boundary in coordinates y^i. The components of $\widehat{\epsilon}$ in the x^μ coordinates on M are

$$\widehat{\epsilon}_{\mu_1\cdots\mu_{n-1}} = n^\lambda\epsilon_{\lambda\mu_1\cdots\mu_{n-1}}, \tag{E.12}$$

where n^μ is the unit normal to the boundary. For a general hypersurface, the sign of n^μ is arbitrary; when the hypersurface is the boundary of a region, however, we have a notion of inward-pointing and outward-pointing. A crucial point is that, to correctly recover Stokes's theorem, n^μ should be chosen to be inward-pointing if the boundary is timelike, and outward-pointing if it's spacelike. Since ω is an $(n-1)$-form, it must be proportional to $\widehat{\epsilon}$ when restricted to the $(n-1)$-dimensional boundary. Following in the path of the previous paragraph, we derive

$$\omega = n_\mu V^\mu \sqrt{|\gamma|}\, d^{n-1}y. \tag{E.13}$$

Stokes's theorem therefore relates the divergence of the vector field to its value on the boundary:

$$\boxed{\int_M d^n x \sqrt{|g|}\, \nabla_\mu V^\mu = \int_{\partial M} d^{n-1}y \sqrt{|\gamma|}\, n_\mu V^\mu.} \tag{E.14}$$

This is the most common version of Stokes's theorem in general relativity.

You shouldn't get the impression that we need to descend to index notation to put Stokes's theorem to use. As a simple counterexample, let's show that the charge associated with a conserved current is "conserved" in a very general sense: Not only is it independent of time in some specific coordinate system, but also the charge passing through a spacelike hypersurface Σ is (under reasonable assumptions) completely independent of the choice of hypersurface. Start by imagining that we have a current J^μ that is conserved, by which we mean

$$\nabla_\mu J^\mu = 0. \tag{E.15}$$

In terms of the one-form $J_\mu = g_{\mu\nu} J^\nu$, we can translate the conservation condition into

$$\mathrm{d}(*J) = 0. \tag{E.16}$$

We then define the charge passing through a hypersurface Σ via

$$Q_\Sigma = -\int_\Sigma *J. \tag{E.17}$$

Typically we will choose Σ to be a hypersurface of constant time, so that Q_Σ is the total charge throughout space at that moment in time; but the formula is applicable more generally. The minus sign is a convention, which can be understood by converting (temporarily) to components. Comparing to (E.2) and (E.13), we can turn (E.17) into

$$Q_\Sigma = -\int_\Sigma d^{n-1}y \sqrt{|\gamma|}\, n_\mu J^\mu. \tag{E.18}$$

We see that the minus sign serves to compensate for the minus sign that the time component of n^μ picks up when we lower the index, so that a positive charge density $\rho = J^0$ yields a positive integrated total charge. Next imagine a four-dimensional spacetime region R, defined as the region between two spatial hypersurfaces Σ_1 and Σ_2, as shown in Figure E.1; the part of the boundary connecting these two hypersurfaces is assumed to be off at infinity where all of the fields vanish, and can be ignored. The conservation law (E.16) and Stokes's theorem (E.1) then give us

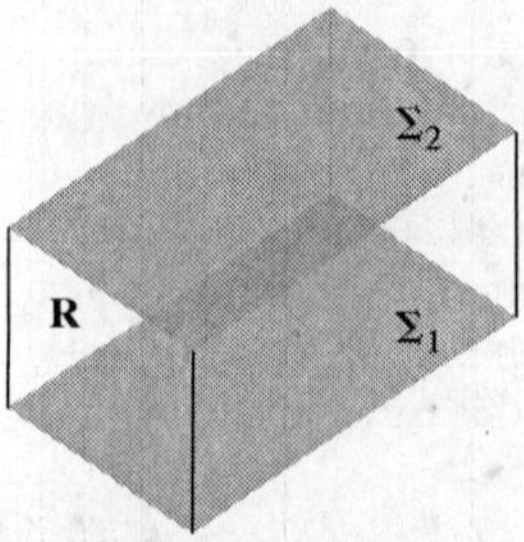

FIGURE E.1 A region R of spacetime with spatial boundaries at infinity; the future and past boundaries include the two spatial hypersurfaces Σ_1 and Σ_2.

$$\begin{aligned} 0 &= \int_R \mathrm{d}(*J) \\ &= \int_{\partial R} *J \\ &= \int_{\Sigma_1} *J - \int_{\Sigma_2} *J \\ &= Q_1 - Q_2. \end{aligned} \tag{E.19}$$

The minus sign in the third line is due to the orientation on Σ_2 inherited from R; the normal vector is pointing inward, which is opposite from what would be the conventional choice in an integral over Σ_2. We see that Q_Σ will be the same over any spacelike hypersurface Σ chosen such that the current vanishes at infinity. Thus, Stokes's theorem shows how the existence of a divergenceless current implies the existence of a conserved charge.

Another use of Stokes's theorem (corresponding to the conventional use of Gauss's theorem in three-dimensional Euclidean space) is to actually calculate this charge Q by integrating over the hypersurface. Thinking momentarily about the real world, let's consider Maxwell's equations in a four-dimensional spacetime. These equations describe how the electromagnetic field strength tensor $F_{\mu\nu}$ responds to the conserved current four-vector,

$$\nabla_\mu F^{\nu\mu} = J^\nu. \tag{E.20}$$

We can therefore plug $\nabla_\mu F^{\nu\mu}$ into (E.18) to calculate the charge:

$$Q = -\int_\Sigma d^3y\sqrt{|\gamma|}\, n_\mu \nabla_\nu F^{\mu\nu}. \tag{E.21}$$

Whenever we are faced with the divergence of an antisymmetric tensor field $F^{\mu\nu} = -F^{\nu\mu}$ integrated over a hypersurface Σ, we can follow similar steps to those used to arrive at (E.14), to relate the divergence to the value of $F^{\mu\nu}$ on the boundary, this time at spatial infinity (if the hypersurface is timelike):

$$\boxed{\int_\Sigma d^{n-1}y\,\sqrt{|\gamma|}n_\mu \nabla_\nu F^{\mu\nu} = \int_{\partial\Sigma} d^{n-2}z\,\sqrt{|\gamma^{(\partial\Sigma)}|}n_\mu \sigma_\nu F^{\mu\nu},} \tag{E.22}$$

where the z^a's are coordinates on $\partial\Sigma$, $\gamma_{ab}^{(\partial\Sigma)}$ is the induced metric on $\partial\Sigma$, and σ^μ is the unit normal to $\partial\Sigma$. You might worry about the integral over $\partial\Sigma$, since the boundary of a boundary is zero; but Σ is not the entire boundary of any region, just a piece of one, so it can certainly have a boundary of its own.

Just to make sure we know what we're doing, let's verify that we can actually recover the charge of a point particle in Minkowski space. We write the metric in polar coordinates,

$$ds^2 = -dt^2 + dr^2 + r^2 d\theta^2 + r^2 \sin^2\theta \, d\phi^2. \tag{E.23}$$

The electric field of a charge q in our units (Lorentz–Heaviside conventions, where there are no 4π's in Maxwell's equations) is

$$E^r = \frac{q}{4\pi r^2}, \tag{E.24}$$

with other components vanishing; this is related to the field strength tensor by

$$F^{tr} = -F^{rt} = E^r. \tag{E.25}$$

The unit normal vectors are

$$n^\mu = (1, 0, 0, 0), \quad \sigma^\mu = (0, 1, 0, 0), \tag{E.26}$$

so that

$$n_\mu \sigma_\nu F^{\mu\nu} = -E^r = -\frac{q}{4\pi r^2}. \tag{E.27}$$

The metric on the two-sphere at spatial infinity is

$$\gamma^{(S^2)}_{ab} dz^a dz^b = r^2 d\theta^2 + r^2 \sin^2\theta \, d\phi^2, \tag{E.28}$$

so the volume element is

$$d^2z \sqrt{\gamma^{(S^2)}} = r^2 \sin\theta \, d\theta \, d\phi. \tag{E.29}$$

Plugging (E.27), (E.29), and (E.21) into (E.22) gives

$$\begin{aligned} Q &= -\lim_{r\to\infty} \int_{S^2} d\theta \, d\phi \, r^2 \sin\theta \left(-\frac{q}{4\pi r^2}\right) \\ &= q, \end{aligned} \tag{E.30}$$

which is just the answer we're looking for.

APPENDIX

F Geodesic Congruences

In Section 3.10 we derived the geodesic deviation equation, governing the evolution of a separation vector connecting a one-parameter family of neighboring geodesics. A more comprehensive picture of the behavior of neighboring geodesics comes from considering not just a one-parameter family, but an entire **congruence** of geodesics. A congruence is a set of curves in an open region of spacetime such that every point in the region lies on precisely one curve. We can think of a geodesic congruence as tracing the paths of a set of noninteracting particles moving through spacetime with nonintersecting paths. If the geodesics cross, the congruence necessarily comes to an end at that point. Clearly, in a multi-dimensional congruence there is a lot of information to keep track of; we will be interested in the local behavior in the neighborhood of a single geodesic, for which things become quite tractable.

Let $U^\mu = dx^\mu/d\tau$ be the tangent vector field to a four-dimensional timelike geodesic congruence; equivalently, the four-velocity field of some pressureless fluid. (If the fluid were not pressureless, integral curves of U^μ would not in general describe geodesics.) Null geodesics present special problems, which we will return to later; for now stick with the timelike case. For reference we recall that the tangent field is normalized and obeys the geodesic equation:

$$U_\mu U^\mu = -1, \quad U^\lambda \nabla_\lambda U^\mu = 0. \tag{F.1}$$

When we discussed the geodesic deviation equation in Section 3.10, we considered a separation vector V^μ pointing from one geodesic to a neighboring one, and found that it obeyed

$$\frac{DV^\mu}{d\tau} \equiv U^\nu \nabla_\nu V^\mu = B^\mu{}_\nu V^\nu, \tag{F.2}$$

where

$$B^\mu{}_\nu = \nabla_\nu U^\mu. \tag{F.3}$$

(In Chapter 3 we used T instead of U, and S instead of V.) The tensor $B_{\mu\nu}$ therefore can be thought of as measuring the failure of V^μ to be parallel-transported along the congruence; in other words, it describes the extent to which neighboring geodesics deviate from remaining perfectly parallel.

To deal with an entire congruence, rather than just a one-parameter family of curves, we can imagine setting up a set of three normal vectors orthogonal to our timelike geodesics, and following their evolution. The failure of this set of vectors to be parallel-transported will tell us how nearby geodesics in the congruence are evolving. Equivalently, we can imagine a small sphere of test particles centered at some point, and we want to describe quantitatively the evolution of these particles with respect to their central geodesic. Fortunately, all we have to keep track of is the behavior of $B_{\mu\nu}$.

Given the vector field U^μ, at each point p we consider the subspace of T_pM corresponding to vectors normal to U^μ. Any vector in T_pM can be projected into this subspace via the projection tensor

$$P^\mu{}_\nu = \delta^\mu_\nu + U^\mu U_\nu, \tag{F.4}$$

familiar from our discussion of submanifolds in Appendix D. In this case we are not projecting onto a submanifold, only onto a vector subspace of the tangent space, but the idea is the same. We notice that $B_{\mu\nu}$ is already in the normal subspace, since

$$\begin{aligned} U^\mu B_{\mu\nu} &= U^\mu \nabla_\nu U_\mu = 0 \\ U^\nu B_{\mu\nu} &= U^\nu \nabla_\nu U_\mu = 0. \end{aligned} \tag{F.5}$$

The first of these follows from $\nabla_\nu(U^\mu U_\mu) = \nabla_\nu(-1) = 0$, while the second follows from the geodesic equation. We should not confuse $B_{\mu\nu}$ with the extrinsic curvature $K_{\mu\nu}$ from (D.53); the difference is that our tangent vector field U^μ will generally not be orthogonal to any hypersurface.

As a (0, 2) tensor, $B_{\mu\nu}$ can be decomposed into symmetric and antisymmetric parts, and the symmetric part can further be decomposed into a trace and a trace-free part. Since $B_{\mu\nu}$ is in the normal subspace, we can use $P_{\mu\nu}$ to take the trace in this decomposition. The result can be written

$$B_{\mu\nu} = \tfrac{1}{3}\theta P_{\mu\nu} + \sigma_{\mu\nu} + \omega_{\mu\nu}. \tag{F.6}$$

Here we have introduced three quantities describing the decomposition, starting with the **expansion** θ of the congruence,

$$\theta = P^{\mu\nu} B_{\mu\nu} = \nabla_\mu U^\mu, \tag{F.7}$$

which is simply the trace of $B_{\mu\nu}$. The expansion describes the change in volume of the sphere of test particles centered on our geodesic. It is clearly a scalar, which makes sense, since the overall expansion/contraction of the volume is described by a single number. The **shear** $\sigma_{\mu\nu}$ is given by

$$\sigma_{\mu\nu} = B_{(\mu\nu)} - \tfrac{1}{3}\theta P_{\mu\nu}. \tag{F.8}$$

It is symmetric and traceless. The shear represents a distortion in the shape of our collection of test particles, from an initial sphere into an ellipsoid; symmetry represents the fact that elongation along (say) the x-direction is the same as elongation along the $-x$ direction. Finally we have the **rotation** $\omega_{\mu\nu}$, given by

$$\omega_{\mu\nu} = B_{[\mu\nu]}. \tag{F.9}$$

It is an antisymmetric tensor, which also makes sense; the xy component (for example) describes a rotation about the z axis, while the yx component describes a rotation in the opposite sense around the same axis.

The evolution of our congruence is described by the covariant derivative of these quantities along the path, $D/d\tau = U^\sigma \nabla_\sigma$. We can straightforwardly calculate this for the entire tensor $B_{\mu\nu}$, and then take the appropriate decomposition. We have

$$\begin{aligned}
\frac{DB_{\mu\nu}}{d\tau} &\equiv U^\sigma \nabla_\sigma B_{\mu\nu} = U^\sigma \nabla_\sigma \nabla_\nu U_\mu \\
&= U^\sigma \nabla_\nu \nabla_\sigma U_\mu + U^\sigma R^\lambda{}_{\mu\nu\sigma} U_\lambda \\
&= \nabla_\nu (U^\sigma \nabla_\sigma U_\mu) - (\nabla_\nu U^\sigma)(\nabla_\sigma U_\mu) - R_{\lambda\mu\nu\sigma} U^\sigma U^\lambda \\
&= -B^\sigma{}_\nu B_{\mu\sigma} - R_{\lambda\mu\nu\sigma} U^\sigma U^\lambda.
\end{aligned} \tag{F.10}$$

Taking the trace of this equation yields an evolution equation for the expansion,

$$\boxed{\frac{d\theta}{d\tau} = -\frac{1}{3}\theta^2 - \sigma_{\mu\nu}\sigma^{\mu\nu} + \omega_{\mu\nu}\omega^{\mu\nu} - R_{\mu\nu}U^\mu U^\nu.} \tag{F.11}$$

This is **Raychaudhuri's equation**, and plays a crucial role in the proofs of the singularity theorems. [Sometimes the demand that the congruence obey the geodesic equation is dropped; this simply adds a term $\nabla_\mu(U^\nu \nabla_\nu U^\mu)$ to the right-hand side.] Similarly, the symmetric trace-free part of (F.10) is

$$\begin{aligned}
\frac{D\sigma_{\mu\nu}}{d\tau} &= -\frac{2}{3}\theta\sigma_{\mu\nu} - \sigma_{\mu\alpha}\sigma^\alpha{}_\nu - \omega_{\mu\rho}\omega^\rho{}_\nu + \frac{1}{3}P_{\mu\nu}(\sigma_{\alpha\beta}\sigma^{\alpha\beta} - \omega_{\alpha\beta}\omega^{\alpha\beta}) \\
&\quad + C_{\alpha\nu\mu\beta}U^\alpha U^\beta + \frac{1}{2}\tilde{R}_{\mu\nu},
\end{aligned} \tag{F.12}$$

where $\tilde{R}_{\mu\nu}$ is the spatially-projected trace-free part of $R_{\mu\nu}$,

$$\tilde{R}_{\mu\nu} = P^\alpha{}_\mu P^\beta{}_\nu R_{\alpha\beta} - \tfrac{1}{3}P_{\mu\nu}P^{\alpha\beta}R_{\alpha\beta}, \tag{F.13}$$

and the antisymmetric part of (F.10) is

$$\frac{D\omega_{\mu\nu}}{d\tau} = -\frac{2}{3}\theta\omega_{\mu\nu} + \sigma_\mu{}^\alpha\omega_{\nu\alpha} - \sigma_\nu{}^\alpha\omega_{\mu\alpha}. \tag{F.14}$$

These equations do not get used as frequently as Raychaudhuri's equation, but they're nice to have around.

Let's give a brief example of the way in which Raychaudhuri's equation gets used. First notice that, since the shear and rotation are both "spatial" tensors, we have

$$\sigma_{\mu\nu}\sigma^{\mu\nu} \geq 0, \quad \omega_{\mu\nu}\omega^{\mu\nu} \geq 0. \tag{F.15}$$

Next, notice that the last term in (F.11) is just what appears if we combine Einstein's equation with the Strong Energy Condition; from Einstein's equation we know

$$R_{\mu\nu}U^\mu U^\nu = 8\pi G\left(T_{\mu\nu} - \tfrac{1}{2}Tg_{\mu\nu}\right)U^\mu U^\nu, \tag{F.16}$$

and the SEC demands that the right-hand side of this expression be nonnegative for any timelike U^μ. We therefore have

$$R_{\mu\nu}U^\mu U^\nu \geq 0 \tag{F.17}$$

if the SEC holds. Finally, note that $\omega_{\mu\nu} = 0$ if and only if the vector field U^μ is orthogonal to a family of hypersurfaces. This follows straightforwardly from the facts that the rotation is a spatial tensor ($U^\mu\omega_{\mu\nu} = 0$), and by Frobenius's theorem a necessary and sufficient condition for a vector field U^μ to be hypersurface-orthogonal is $U_{[\mu}\nabla_\nu U_{\rho]}$; the details are left for you to check. Therefore, if we have a congruence whose tangent field is hypersurface-orthogonal, in a spacetime obeying Einstein's equations and the SEC, Raychaudhuri's equation implies

$$\frac{d\theta}{d\tau} \leq -\frac{1}{3}\theta^2. \tag{F.18}$$

This equation is easily integrated to obtain

$$\theta^{-1}(\tau) \geq \theta_0^{-1} + \tfrac{1}{3}\tau. \tag{F.19}$$

Consider a hypersurface-orthogonal congruence, which is initially converging ($\theta_0 < 0$) rather than expanding. Then (F.19) tells us convergence will continue, and we must hit a caustic (a place where geodesics cross) in a finite proper time $\tau \leq -3\theta_0^{-1}$. In other words, matter obeying the SEC can never begin to push geodesics apart, it can only increase the rate at which they are converging. Of course, this result only applies to some arbitrarily-chosen congruence, and the appearance of caustics certainly doesn't indicate any singularity in the spacetime (geodesics cross all the time, even in flat spacetime). But many of the proofs of the singularity theorems take advantage of this property of the Raychaudhuri equation to show that spacetime must be geodesically incomplete in some way.

We turn next to the behavior of congruences of null geodesics. These are trickier, essentially because our starting point (studying the evolution of vectors in a three-dimensional subspace normal to the tangent field) doesn't make as much sense, since the tangent vector of a null curve is normal to itself. Instead, in the null case what we care about is the evolution of vectors in a *two-dimensional* subspace of "spatial" vectors normal to the null tangent vector field $k^\mu = dx^\mu/d\lambda$.

Unfortunately, there is no unique way to define this subspace, as observers in different Lorentz frames will have different notions of what constitutes a spatial vector. Faced with this dilemma, we have two sensible approaches. A slick approach would be to define an abstract two-dimensional vector space by starting with the three-dimensional space of vectors orthogonal to k^μ, and then taking equivalence classes where two vectors are equivalent if they differ by a multiple of k^μ. The grungier approach, which we will follow, is simply to choose a second "auxiliary" null vector l^μ, which (in some frame) points in the opposite spatial direction to k^μ, normalized such that

$$l^\mu l_\mu = 0, \quad l^\mu k_\mu = -1. \tag{F.20}$$

We furthermore demand that the auxiliary vector be parallel-transported,

$$k^\mu \nabla_\mu l^\nu = 0, \tag{F.21}$$

which is compatible with (F.20) because parallel transport preserves inner products. The auxiliary null vector l^μ is by no means unique, since as we've just noted the idea of pointing in opposite spatial directions is frame-dependent. Nevertheless, we can make a choice and hope that important quantities are independent of the arbitrary choice. Having done so, the two-dimensional space of normal vectors we are interested in, called $T_\perp$, consists simply of those vectors V^μ that are orthogonal to both k^μ and l^μ,

$$T_\perp = \{V^\mu | V^\mu k_\mu = 0, V^\mu l_\mu = 0\}. \tag{F.22}$$

Our task now is to follow the evolution of deviation vectors living in this subspace, which represent a family of neighboring null geodesics.

Projecting into the normal subspace $T_\perp$ requires a slightly modified definition of the projection tensor; it turns out that

$$Q_{\mu\nu} = g_{\mu\nu} + k_\mu l_\nu + k_\nu l_\mu \tag{F.23}$$

does the trick. Namely, $Q_{\mu\nu}$ will act like the metric when acting on vectors V^μ, W^μ in $T_\perp$, while annihilating anything proportional to k^μ or l^μ. Some useful properties of this projection tensor include

$$\begin{aligned}
Q_{\mu\nu} V^\nu W^\nu &= g_{\mu\nu} V^\mu W^\nu \\
Q^\mu{}_\nu V^\nu &= V^\mu \\
Q^\mu{}_\nu k^\nu &= 0 \\
Q^\mu{}_\nu l^\nu &= 0 \\
Q^\mu{}_\nu Q^\nu{}_\sigma &= Q^\mu{}_\sigma \\
k^\sigma \nabla_\sigma Q^\mu{}_\nu &= 0.
\end{aligned} \tag{F.24}$$

Just as for timelike geodesics, the failure of a normal deviation vector V^μ to be parallel-propagated is governed by the tensor $B^\mu{}_\nu = \nabla_\nu k^\mu$, in the sense that

$$\frac{DV^\mu}{d\lambda} \equiv k^\nu \nabla_\nu V^\mu = B^\mu{}_\nu V^\nu. \tag{F.25}$$

However, in the null case the tensor $B_{\mu\nu}$ is actually more than we need; the relevant information is completely contained in the projected version,

$$\widehat{B}^\mu{}_\nu = Q^\mu{}_\alpha Q^\beta{}_\nu B^\alpha{}_\beta. \tag{F.26}$$

To see this, we simply play around with (F.25), using the various properties in (F.24):

$$\begin{aligned}
\frac{DV^\mu}{d\lambda} &= k^\nu \nabla_\nu V^\mu \\
&= k^\nu \nabla_\nu (Q^\mu{}_\rho V^\rho) \\
&= Q^\mu{}_\rho k^\nu \nabla_\nu V^\rho \\
&= Q^\mu{}_\rho B^\rho{}_\nu V^\nu \\
&= Q^\mu{}_\rho B^\rho{}_\nu Q^\nu{}_\sigma V^\sigma \\
&= \widehat{B}^\mu{}_\sigma V^\sigma.
\end{aligned} \tag{F.27}$$

So we only have to keep track of the evolution of this projected tensor, not the full $B_{\mu\nu}$.

To understand that evolution, we again decompose into the expansion, shear, and rotation:

$$\widehat{B}_{\mu\nu} = \tfrac{1}{2}\theta Q_{\mu\nu} + \widehat{\sigma}_{\mu\nu} + \widehat{\omega}_{\mu\nu}, \tag{F.28}$$

where

$$\begin{aligned}
\theta &= Q^{\mu\nu}\widehat{B}_{\mu\nu} = \widehat{B}^\mu{}_\mu \\
\widehat{\sigma}_{\mu\nu} &= \widehat{B}_{(\mu\nu)} - \tfrac{1}{2}\theta Q_{\mu\nu} \\
\widehat{\omega}_{\mu\nu} &= \widehat{B}_{[\mu\nu]}.
\end{aligned} \tag{F.29}$$

We find factors of $\frac{1}{2}$ rather than $\frac{1}{3}$ because our normal space $T_\perp$ is two-dimensional, reflected in the fact that $Q^{\mu\nu}Q_{\mu\nu} = 2$. As in the timelike case, $\widehat{\omega}_{\mu\nu} = 0$ is a necessary and sufficient condition for the congruence to be hypersurface-orthogonal. The evolution of $\widehat{B}_{\mu\nu}$ along the path is given by

$$\begin{aligned}
\frac{D\widehat{B}_{\mu\nu}}{d\lambda} &\equiv k^\sigma \nabla_\sigma \widehat{B}_{\mu\nu} = k^\sigma \nabla_\sigma (Q^\alpha{}_\mu Q^\beta{}_\nu \nabla_\alpha k_\beta) \\
&= Q^\alpha{}_\mu Q^\beta{}_\nu k^\sigma \nabla_\sigma \nabla_\alpha k_\beta
\end{aligned}$$

$$= -Q^{\alpha}{}_{\mu} Q^{\beta}{}_{\nu} (B_{\alpha}{}^{\sigma} B_{\beta\sigma} + R_{\alpha\lambda\beta\sigma} k^{\lambda} k^{\sigma})$$
$$= -\widehat{B}_{\mu}{}^{\sigma} \widehat{B}_{\nu\sigma} - Q^{\alpha}{}_{\mu} Q^{\beta}{}_{\nu} R_{\mu\lambda\nu\sigma} k^{\lambda} k^{\sigma}. \qquad \text{(F.30)}$$

Continuing to follow our previous logic, we can take the trace of this equation to find an evolution equation for the expansion of null geodesics,

$$\frac{d\theta}{d\lambda} = -\frac{1}{2}\theta^2 - \widehat{\sigma}_{\mu\nu}\widehat{\sigma}^{\mu\nu} + \widehat{\omega}_{\mu\nu}\widehat{\omega}^{\mu\nu} - R_{\mu\nu} k^{\mu} k^{\nu}. \qquad \text{(F.31)}$$

Happily, this equation turns out to be completely independent of our arbitrarily chosen auxiliary vector l^{μ}. First, the expansion itself is independent of l^{μ}, as we easily verify:

$$\theta = Q^{\mu\nu} \widehat{B}_{\mu\nu}$$
$$= Q^{\mu\nu} B_{\mu\nu}$$
$$= g^{\mu\nu} B_{\mu\nu}, \qquad \text{(F.32)}$$

where the second line follows from $Q^{\mu\nu} Q^{\alpha}{}_{\nu} = Q^{\mu\alpha}$, and the third from $k^{\mu} B_{\mu\nu} = k^{\nu} B_{\mu\nu} = 0$. (This is why we never put a hat on θ to begin with.) Second, both $\widehat{\sigma}_{\mu\nu}\widehat{\sigma}^{\mu\nu}$ and $\widehat{\omega}_{\mu\nu}\widehat{\omega}^{\mu\nu}$ are likewise independent of l^{μ} (as you are welcome to verify), even though $\widehat{\sigma}_{\mu\nu}$ and $\widehat{\omega}_{\mu\nu}$ themselves are not. Finally, the projection tensors dropped out of the curvature-tensor piece when we took the trace. We therefore have a well-defined notion of the evolution of the expansion, independent of any arbitrary choices we made. Notice that, because k^{μ} is null, Einstein's equation implies

$$R_{\mu\nu} k^{\mu} k^{\nu} = 8\pi G \left(T_{\mu\nu} - \tfrac{1}{2} T g_{\mu\nu} \right) k^{\mu} k^{\nu}$$
$$= 8\pi G T_{\mu\nu} k^{\mu} k^{\nu}. \qquad \text{(F.33)}$$

For this to be nonnegative, we need only invoke the Null Energy Condition, which is the least restrictive of all the energy conditions we discussed in Chapter 3. Thus, null geodesics tend to converge to caustics under more general circumstances than timelike ones.

We can continue on to get evolution equations for the shear,

$$\frac{D\widehat{\sigma}_{\mu\nu}}{d\lambda} = -\theta\widehat{\sigma}_{\mu\nu} - Q^{\alpha}{}_{\mu} Q^{\beta}{}_{\nu} C_{\mu\lambda\nu\sigma} k^{\lambda} k^{\sigma}, \qquad \text{(F.34)}$$

and for the rotation,

$$\frac{D\widehat{\omega}_{\mu\nu}}{d\lambda} = -\theta\widehat{\omega}_{\mu\nu}. \qquad \text{(F.35)}$$

These equations are less natural than the one for the expansion, since the shear and rotation do depend on our choice of l^{μ}; nevertheless, they can be useful in specific circumstances.

APPENDIX

G Conformal Transformations

A **conformal transformation** is essentially a local change of scale. Since distances are measured by the metric, such transformations are implemented by multiplying the metric by a spacetime-dependent (nonvanishing) function:

$$\tilde{g}_{\mu\nu} = \omega^2(x) g_{\mu\nu}, \tag{G.1}$$

or equivalently

$$\widetilde{ds}^2 = \omega^2(x) ds^2, \tag{G.2}$$

for some nonvanishing function $\omega(x)$. (Here x is used to denote the collection of spacetime coordinates x^μ.) Note that the inverse conformal transformation is trivial: $g_{\mu\nu} = \omega^{-2}\tilde{g}_{\mu\nu}$. Transformations of this sort have a number of uses in GR; our favorite purposes will be to change dynamical variables in scalar-tensor theories (as in Section 4.8), and to remap spacetimes into convenient conformal diagrams (as in the following Appendix).

We first mention one critical fact: *null curves are left invariant by conformal transformations*. By this we mean simply that, if $x^\mu(\lambda)$ is a curve that is null with respect to $g_{\mu\nu}$, it will also be null with respect to $\tilde{g}_{\mu\nu}$. This follows immediately once we understand that a curve $x^\mu(\lambda)$ is null if and only if its tangent vector $dx^\mu/d\lambda$ is null,

$$g_{\mu\nu}\frac{dx^\mu}{d\lambda}\frac{dx^\nu}{d\lambda} = 0. \tag{G.3}$$

Then in the conformally-related metric we have

$$\tilde{g}_{\mu\nu}\frac{dx^\mu}{d\lambda}\frac{dx^\nu}{d\lambda} = \omega^2(x) g_{\mu\nu}\frac{dx^\mu}{d\lambda}\frac{dx^\nu}{d\lambda} = 0. \tag{G.4}$$

Thus, curves that are null as defined by one metric will also be null as defined by any conformally-related metric. We may say that "conformal transformations leave light cones invariant." (Indeed, you can check that they leave angles between any two four-vectors invariant, a feature that our conformal transformations share with the familiar conformal transformations of complex analysis.)

Let us next consider how geometrical quantities change under conformal transformations. A conformal transformation is not a change of coordinates, but an actual change of the geometry—timelike geodesics of $\tilde{g}_{\mu\nu}$, for example, will generally differ from timelike geodesics of $g_{\mu\nu}$. However, we can use conformal transformations to change our dynamical variables: anything that is a function of $g_{\mu\nu}$

can be equally well thought of as a function of $\tilde{g}_{\mu\nu}$ and $\omega(x)$. We then say that the quantities are expressed in the **conformal frame**. In this Appendix we collect some expressions for how quantities in the original metric $g_{\mu\nu}$ are related to those in the conformal metric $\tilde{g}_{\mu\nu}$.

We begin by considering the Christoffel symbols. Because the connection coefficients are linear in derivatives of the metric and also linear in the inverse metric, the conformally-transformed connection takes the form

$$\widetilde{\Gamma}^{\rho}_{\mu\nu} = \Gamma^{\rho}_{\mu\nu} + C^{\rho}{}_{\mu\nu}. \tag{G.5}$$

$C^{\rho}{}_{\mu\nu}$ is clearly a tensor, as it is the difference of two connections. An explicit calculation reveals it to be given by

$$C^{\rho}{}_{\mu\nu} = \omega^{-1}\left(\delta^{\rho}_{\mu}\nabla_{\nu}\omega + \delta^{\rho}_{\nu}\nabla_{\mu}\omega - g_{\mu\nu}g^{\rho\lambda}\nabla_{\lambda}\omega\right). \tag{G.6}$$

This formula immediately becomes useful when we consider how the Riemann tensor behaves under conformal transformations. In fact under any change of connection of the form (G.5), we have

$$\widetilde{R}^{\rho}{}_{\sigma\mu\nu} = R^{\rho}{}_{\sigma\mu\nu} + \nabla_{\mu}C^{\rho}{}_{\nu\sigma} - \nabla_{\nu}C^{\rho}{}_{\mu\sigma} + C^{\rho}{}_{\mu\lambda}C^{\lambda}{}_{\nu\sigma} - C^{\rho}{}_{\nu\lambda}C^{\lambda}{}_{\mu\sigma}. \tag{G.7}$$

Thus it is a matter of simply plugging in and grinding away to get

$$\begin{aligned}\widetilde{R}^{\rho}{}_{\sigma\mu\nu} = R^{\rho}{}_{\sigma\mu\nu} &- 2\left(\delta^{\rho}_{[\mu}\delta^{\alpha}_{\nu]}\delta^{\beta}_{\sigma} - g_{\sigma[\mu}\delta^{\alpha}_{\nu]}g^{\rho\beta}\right)\omega^{-1}(\nabla_{\alpha}\nabla_{\beta}\omega)\\ &+ 2\left(2\delta^{\rho}_{[\mu}\delta^{\alpha}_{\nu]}\delta^{\beta}_{\sigma} - 2g_{\sigma[\mu}\delta^{\alpha}_{\nu]}g^{\rho\beta} + g_{\sigma[\mu}\delta^{\rho}_{\nu]}g^{\alpha\beta}\right)\omega^{-2}(\nabla_{\alpha}\omega)(\nabla_{\beta}\omega).\end{aligned} \tag{G.8}$$

Contracting the first and third indices yields the Ricci tensor,

$$\begin{aligned}\widetilde{R}_{\sigma\nu} = R_{\sigma\nu} &- \left[(n-2)\delta^{\alpha}_{\sigma}\delta^{\beta}_{\nu} + g_{\sigma\nu}g^{\alpha\beta}\right]\omega^{-1}(\nabla_{\alpha}\nabla_{\beta}\omega)\\ &+ \left[2(n-2)\delta^{\alpha}_{\sigma}\delta^{\beta}_{\nu} - (n-3)g_{\sigma\nu}g^{\alpha\beta}\right]\omega^{-2}(\nabla_{\alpha}\omega)(\nabla_{\beta}\omega),\end{aligned} \tag{G.9}$$

where n is the number of dimensions. Raising an index (with $\tilde{g}^{\mu\nu} = \omega^{-2}g^{\mu\nu}$) and contracting again gets us the curvature scalar,

$$\widetilde{R} = \omega^{-2}R - 2(n-1)g^{\alpha\beta}\omega^{-3}(\nabla_{\alpha}\nabla_{\beta}\omega) - (n-1)(n-4)g^{\alpha\beta}\omega^{-4}(\nabla_{\alpha}\omega)(\nabla_{\beta}\omega). \tag{G.10}$$

Another useful quantity is the covariant derivative of a scalar field ϕ. The first covariant derivative is equal in the original or conformal frame, since they are both equal to the partial derivative:

$$\widetilde{\nabla}_{\mu}\phi = \nabla_{\mu}\phi = \partial_{\mu}\phi. \tag{G.11}$$

The second derivative, however, involves the Christoffel symbol, and therefore has a nontrivial transformation:

$$\widetilde{\nabla}_\mu \widetilde{\nabla}_\nu \phi = \nabla_\mu \nabla_\nu \phi - \left(\delta^\alpha_\mu \delta^\beta_\nu + \delta^\beta_\mu \delta^\alpha_\nu - g_{\mu\nu} g^{\alpha\beta}\right) \omega^{-1} (\nabla_\alpha \omega)(\nabla_\beta \phi). \tag{G.12}$$

We can contract this with $\tilde{g}^{\mu\nu}$ to obtain the D'Alembertian,

$$\widetilde{\Box}\phi = \omega^{-2}\Box\phi + (n-2) g^{\alpha\beta} \omega^{-3} (\nabla_\alpha \omega)(\nabla_\beta \phi). \tag{G.13}$$

Finally, we may want to go backward, and express quantities in the original metric in terms of the conformal metric. This is simply a matter of tedious computation, the answers to which are reproduced here for convenience. The curvature tensor and its contractions are

$$\begin{aligned} R^\rho{}_{\sigma\mu\nu} = {} & \widetilde{R}^\rho{}_{\sigma\mu\nu} + 2\left(\delta^\rho_{[\mu} \delta^\alpha_{\nu]} \delta^\beta_\sigma - \tilde{g}_{\sigma[\mu} \delta^\alpha_{\nu]} \tilde{g}^{\rho\beta}\right) \omega^{-1} (\widetilde{\nabla}_\alpha \widetilde{\nabla}_\beta \omega) \\ & + 2\tilde{g}_{\sigma[\mu} \delta^\rho_{\nu]} \tilde{g}^{\alpha\beta} \omega^{-2} (\widetilde{\nabla}_\alpha \omega)(\widetilde{\nabla}_\beta \omega), \end{aligned} \tag{G.14}$$

$$\begin{aligned} R_{\sigma\nu} = {} & \widetilde{R}_{\sigma\nu} + \left[(n-2)\delta^\alpha_\sigma \delta^\beta_\nu + \tilde{g}_{\sigma\nu} \tilde{g}^{\alpha\beta}\right] \omega^{-1} (\widetilde{\nabla}_\alpha \widetilde{\nabla}_\beta \omega) \\ & - (n-1) \tilde{g}_{\sigma\nu} \tilde{g}^{\alpha\beta} \omega^{-2} (\widetilde{\nabla}_\alpha \omega)(\widetilde{\nabla}_\beta \omega), \end{aligned} \tag{G.15}$$

and

$$R = \omega^2 \widetilde{R} + 2(n-1) \tilde{g}^{\alpha\beta} \omega (\widetilde{\nabla}_\alpha \widetilde{\nabla}_\beta \omega) - n(n-1) \tilde{g}^{\alpha\beta} (\widetilde{\nabla}_\alpha \omega)(\widetilde{\nabla}_\beta \omega), \tag{G.16}$$

while the covariant derivatives of a scalar field are given by

$$\nabla_\mu \nabla_\nu \phi = \widetilde{\nabla}_\mu \widetilde{\nabla}_\nu \phi + \left(\delta^\alpha_\mu \delta^\beta_\nu + \delta^\beta_\mu \delta^\alpha_\nu - \tilde{g}_{\mu\nu} \tilde{g}^{\alpha\beta}\right) \omega^{-1} (\widetilde{\nabla}_\alpha \omega)(\widetilde{\nabla}_\beta \phi) \tag{G.17}$$

and

$$\Box\phi = \omega^2 \widetilde{\Box}\phi - (n-2) \tilde{g}^{\alpha\beta} \omega (\widetilde{\nabla}_\alpha \omega)(\widetilde{\nabla}_\beta \phi). \tag{G.18}$$

G.1 ■ EXERCISES

1. Show that conformal transformations leave null geodesics invariant, that is, that the null geodesics of $g_{\mu\nu}$ are the same as those of $\omega^2 g_{\mu\nu}$. (We already know that they leave null *curves* invariant; you have to show that the transformed curves still are geodesics.) What is the relationship between the affine parameters in the original and conformal metrics?

2. Show that in two dimensions, a conformal transformation can always be found (provided that the operator $\nabla^\mu \nabla_\mu$ is invertible) such that the curvature of the transformed metric vanishes, at least in some coordinate chart. (It can't in general be done simultaneously over the entire manifold.) This means that any two-dimensional metric can be written locally as a flat metric multiplied by a conformal factor.

3. Suppose that two metrics are related by an overall conformal transformation of the form

$$\tilde{g}_{\mu\nu} = e^{\alpha(x)} g_{\mu\nu}. \tag{G.19}$$

(a) Show that if ξ^μ is a Killing vector for the metric $g_{\mu\nu}$, then it is a conformal Killing vector for the metric $\widetilde{g}_{\mu\nu}$. A **conformal Killing vector** obeys the equation

$$\nabla_\mu \xi_\nu + \nabla_\nu \xi_\mu = (\nabla_\lambda \alpha)\xi^\lambda g_{\mu\nu}. \tag{G.20}$$

(b) Show that $\xi_\mu k^\mu$ is constant along photon geodesics in $\widetilde{g}_{\mu\nu}$. Here k^μ is the photon's 4-momentum.

(c) Show that the conformal time $\eta = \int dt/R(t)$ is associated with a conformal Killing vector $\xi = \partial_\eta$.

(d) Use part (c) to rederive the relationship between the scale factor and redshift.

APPENDIX H

Conformal Diagrams

Curved spacetime manifolds can in principle be impossibly complex; fortunately, we may often approximate physically realistic situations by manifolds with high degrees of symmetry (especially spherical symmetry). Even symmetric spacetimes, however, can pose formidable challenges to our powers of visualization, if we try to imagine the global structure of such manifolds. It is therefore useful to be able to draw standardized representations of spacetime diagrams that capture the global properties and causal structure of sufficiently symmetric spacetimes. (By "causal structure" we mean the relationship between the past and future of different events, as defined by their light cones.) An elegant fulfillment of this wish is provided by **conformal diagrams** (or Carter–Penrose, or just Penrose diagrams).

A conformal diagram is simply an ordinary spacetime diagram for a metric on which we have performed a particularly clever coordinate transformation. Since our goal is to portray the causal structure of the spacetime, which is defined by its light cones, "clever" means that the new coordinates $x^{\mu'}$ have a "timelike" coordinate and a "radial" one, with the feature that radial light cones can be consistently portrayed at 45° on a spacetime diagram. In addition, we aim for coordinates in which "infinity" is only a finite coordinate value away, so that the structure of the entire spacetime is immediately apparent.

As explained in the previous Appendix, conformal transformations leave light cones invariant. Since we would like to find coordinates in which light cones are at 45°, we need only find coordinates in which the metric of interest is conformally related to a different metric for which we know that the light cones are at 45°. (Of course the angle at which our light cones are drawn depends on our units, or equivalently how we draw our axes; what we really mean is a set of coordinates T, R in which radial null rays satisfy $dT/dR = \pm 1$.)

Let's begin with Minkowski space to see how the technique works. The Minkowski metric in polar coordinates is

$$ds^2 = -\mathrm{d}t^2 + \mathrm{d}r^2 + r^2 d\Omega^2, \tag{H.1}$$

where $d\Omega^2 = \mathrm{d}\theta^2 + \sin^2\theta \mathrm{d}\phi^2$ is the metric on a unit two-sphere. Here it is already true that we can draw light cones at 45° everywhere (the trajectories $t = \pm r$ are null), but we would like to make the causal structure of the entire spacetime more transparent by switching to coordinates with finite ranges. Nothing unusual will happen to the θ, ϕ coordinates, but we will want to keep careful track of the ranges of the other two coordinates. To start with of course we have

$$-\infty < t < \infty, \quad 0 \leq r < \infty. \tag{H.2}$$

Technically, the worldline $r = 0$ represents a coordinate singularity and should be covered by a different patch, but we all know what is going on so we'll just act like $r = 0$ is well-behaved.

A first guess (which turns out not to work) might be simply to rescale the timelike and radial coordinates so that they cover a finite range. A good candidate is to use the arctangent, portrayed in Figure H.1, and define $\bar{t} = \arctan t$, $\bar{r} = \arctan r$. The metric then would take the form [using $d\tan x = (1/\cos^2 x)dx$]

$$ds^2 = -\frac{1}{\cos^4 \bar{t}}\, d\bar{t}^2 + \frac{1}{\cos^4 \bar{r}}\, d\bar{r}^2 + \tan^2 \bar{r}\, d\Omega^2, \tag{H.3}$$

with

$$-\frac{\pi}{2} < \bar{t} < \frac{\pi}{2}$$
$$0 \leq \bar{r} < \frac{\pi}{2}. \tag{H.4}$$

The good news is that the new coordinates have finite ranges; the bad news is that the slope of the light cones (given by $d\bar{t}/d\bar{r} = \pm\cos^2\bar{t}/\cos^2\bar{r}$) is not equal to ± 1, as we wished. If we were to draw the appropriate spacetime diagram (which you might want to do, just for fun), it would not be clear where null rays traveled, especially at the edges of the spacetime.

The way out of this cul-de-sac is, instead of straightforwardly manipulating the original coordinates t and r, to be even more clever and switch to null coordinates:

$$u = t - r$$
$$v = t + r, \tag{H.5}$$

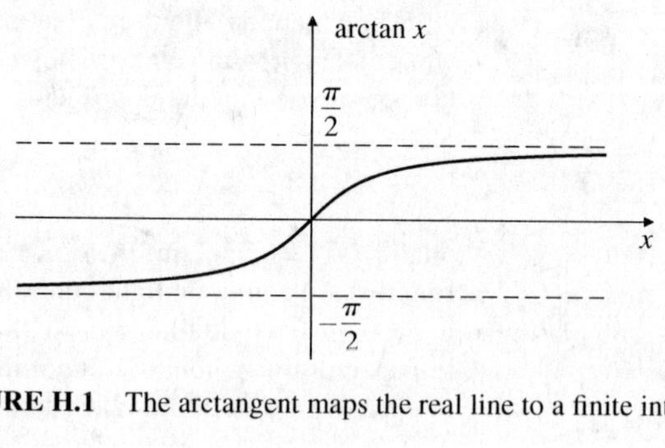

FIGURE H.1 The arctangent maps the real line to a finite interval.

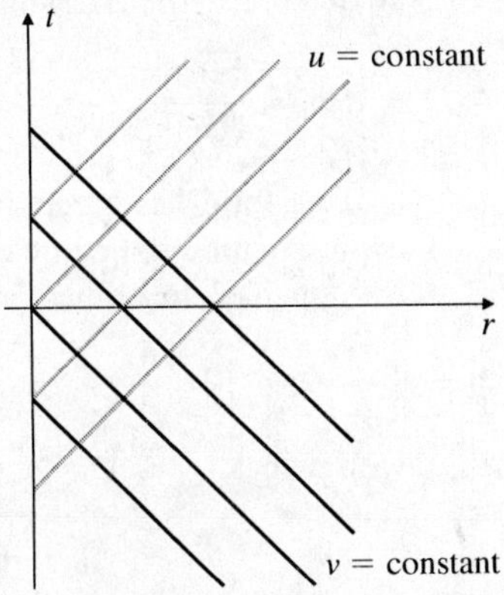

FIGURE H.2 Null radial coordinates on Minkowski space.

with corresponding ranges given by

$$-\infty < u < \infty, \quad -\infty < v < \infty, \quad u \leq v. \tag{H.6}$$

These coordinates are as portrayed in Figure H.2, on which each point represents a 2-sphere of radius $r = \frac{1}{2}(v - u)$. The Minkowski metric in null coordinates is given by

$$ds^2 = -\tfrac{1}{2}(dudv + dvdu) + \tfrac{1}{4}(v - u)^2 d\Omega^2. \tag{H.7}$$

Now we use the arctangent to bring infinity into a finite coordinate value, letting

$$\begin{aligned} U &= \arctan u \\ V &= \arctan v, \end{aligned} \tag{H.8}$$

with ranges

$$-\pi/2 < U < \pi/2, \quad -\pi/2 < V < \pi/2, \quad U \leq V. \tag{H.9}$$

We then have

$$du\, dv + dv\, du = \frac{1}{\cos^2 U \cos^2 V}(dU\, dV + dV\, dU), \tag{H.10}$$

and

$$\begin{aligned} (v - u)^2 = (\tan V - \tan U)^2 &= \frac{1}{\cos^2 U \cos^2 V}(\sin V \cos U - \cos V \sin U)^2 \\ &= \frac{1}{\cos^2 U \cos^2 V} \sin^2(V - U), \end{aligned} \tag{H.11}$$

so that the metric (H.7) in these coordinates is

$$ds^2 = \frac{1}{4\cos^2 U \cos^2 V}\left[-2(dUdV + dVdU) + \sin^2(V - U)\, d\Omega^2\right]. \quad \text{(H.12)}$$

This form has a certain appeal, since the metric appears as a fairly simple expression multiplied by an overall factor. We can make it even better by transforming back to a timelike coordinate T and a radial coordinate R, via

$$T = V + U, \quad R = V - U, \quad \text{(H.13)}$$

with ranges

$$0 \leq R < \pi, \quad |T| + R < \pi. \quad \text{(H.14)}$$

Now the metric is

$$ds^2 = \omega^{-2}(T, R)\left(-dT^2 + dR^2 + \sin^2 R\, d\Omega^2\right), \quad \text{(H.15)}$$

where

$$\begin{aligned}\omega &= 2\cos U \cos V \\ &= 2\cos\left[\tfrac{1}{2}(T - R)\right]\cos\left[\tfrac{1}{2}(T + R)\right] \\ &= \cos T + \cos R. \end{aligned} \quad \text{(H.16)}$$

The original Minkowski metric, which we denoted ds^2, may therefore be thought of as related by a conformal transformation to the "unphysical" metric

$$\begin{aligned}\widetilde{ds}^2 &= \omega^2(T, R) ds^2 \\ &= -dT^2 + dR^2 + \sin^2 R\, d\Omega^2. \end{aligned} \quad \text{(H.17)}$$

This describes the manifold $\mathbf{R} \times S^3$, where the 3-sphere is purely spacelike, perfectly round, and unchanging in time. There is curvature in this metric, unlike in Minkowski spacetime. This shouldn't bother us, since it is unphysical; the true physical metric, obtained by a conformal transformation, is simply flat spacetime, no matter what coordinates we choose. In fact the metric (H.17) is that of the "Einstein static universe," a static solution to Einstein's equation with a perfect fluid and a cosmological constant (Figure H.3). Of course, the full range of coordinates on $\mathbf{R} \times S^3$ would usually be $-\infty < T < \infty, 0 \leq R \leq \pi$, while Minkowski space is mapped into the subspace defined by (H.14). The entire $\mathbf{R} \times S^3$ can be drawn as a cylinder, in which each circle of constant T represents a 3-sphere. The shaded region represents Minkowski space. We can unroll the shaded region to portray Minkowski space as a triangle, as shown in Figure H.4. This is the conformal diagram. Each point represents a two-sphere.

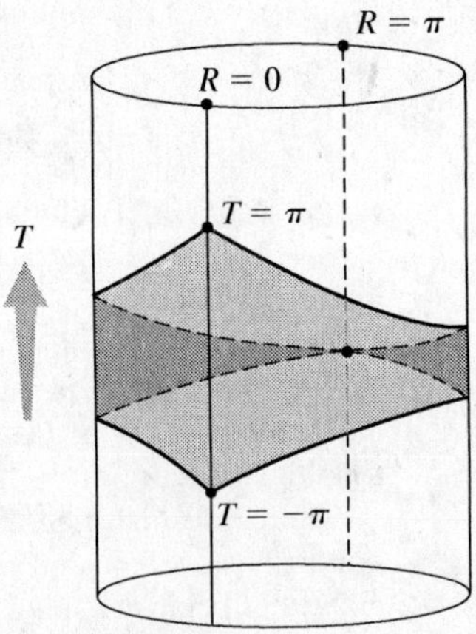

FIGURE H.3 The Einstein static universe, $\mathbf{R} \times S^3$, portrayed as a cylinder. The shaded region is conformally related to Minkowski space.

In fact Minkowski space is only the *interior* of the above diagram (including $R = 0$); the boundaries are not part of the original spacetime. The boundaries are referred to as **conformal infinity**, and the union of the original spacetime with conformal infinity is the **conformal compactification**, which is a manifold with boundary. The structure of the conformal diagram allows us to subdivide conformal infinity into a few different regions:

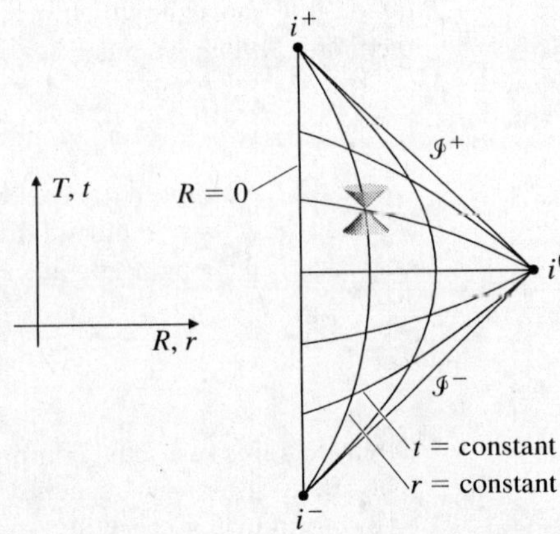

FIGURE H.4 The conformal diagram of Minkowski space. Light cones are at ±45° throughout the diagram.

$$\begin{aligned}
i^+ &= \text{future timelike infinity } (T = \pi,\ R = 0) \\
i^0 &= \text{spatial infinity } (T = 0,\ R = \pi) \\
i^- &= \text{past timelike infinity } (T = -\pi,\ R = 0) \\
\mathscr{I}^+ &= \text{future null infinity } (T = \pi - R,\ 0 < R < \pi) \\
\mathscr{I}^- &= \text{past null infinity } (T = -\pi + R,\ 0 < R < \pi)
\end{aligned}$$

($\mathscr{I}^+$ and $\mathscr{I}^-$ are pronounced as "scri-plus" and "scri-minus," respectively.) Note that i^+, i^0, and i^- are actually *points*, since $R = 0$ and $R = \pi$ are the north and south poles of S^3. Meanwhile $\mathscr{I}^+$ and $\mathscr{I}^-$ are actually null surfaces, with the topology of $\mathbf{R} \times S^2$.

The conformal diagram for Minkowski spacetime contains a number of important features. Radial null geodesics are at $\pm 45°$ in the diagram. All timelike geodesics begin at i^- and end at i^+; all null geodesics begin at $\mathscr{I}^-$ and end at $\mathscr{I}^+$; all spacelike geodesics both begin and end at i^0. On the other hand, there can be nongeodesic timelike curves that end at null infinity, if they become "asymptotically null."

It is nice to be able to fit all of Minkowski space on a small piece of paper, but we don't really learn much that we didn't already know. Conformal diagrams are more useful when we want to represent slightly more complicated spacetimes, such as those for black holes. As discussed in Chapter 6, asymptotically flat spacetimes (or regions of a spacetime) are those that share the structure of $\mathscr{I}^+$, i^0, and $\mathscr{I}^-$ with Minkowski space. Equally importantly, the conformal diagram gives us an idea of the causal structure of the spacetime, for example, whether the past or future light cones of two specified points intersect. In Minkowski space this is always true for any two points, but curved spacetimes can be more interesting, as we saw for the case of an expanding universe in Chapter 2.

Let's consider the conformal diagram for the cosmological spacetime introduced in Chapter 2, which provides a vivid illustration of the usefulness of this technique. When we put polar coordinates on space, the metric becomes

$$ds^2 = -dt^2 + t^{2q}\left(dr^2 + r^2 d\Omega^2\right), \tag{H.18}$$

where we have chosen to consider power-law behavior for the scale factor, $a(t) = t^q$, and $0 < q < 1$. A crucial difference between this metric and that of Minkowski space is the singularity at $t = 0$, which restricts the range of our coordinates:

$$0 < t < \infty \tag{H.19}$$

$$0 \le r < \infty. \tag{H.20}$$

Other than this restricted coordinate range, our analysis follows almost precisely that of the case of flat spacetime. This is because we can bring the metric (H.18) to the form of flat spacetime times a conformal factor; once done, we need only to reproduce our previous coordinate transformations to express our expanding-universe metric as a conformal factor times the Einstein static universe.

We begin by choosing a new time coordinate η, sometimes called **conformal time**, which satisfies

$$dt^2 = t^{2q} d\eta^2, \tag{H.21}$$

or

$$\eta = \frac{1}{1-q} t^{1-q}. \tag{H.22}$$

This simple choice allows us to bring out the scale factor as an overall conformal factor,

$$ds^2 = [(1-q)\eta]^{2q/(1-q)} \left(-d\eta^2 + dr^2 + r^2 d\Omega^2\right). \tag{H.23}$$

The range of η is the same as that of t,

$$0 < \eta < \infty. \tag{H.24}$$

Note that η is a timelike coordinate [in the sense that the vector ∂_η is timelike, $ds^2(\partial_\eta, \partial_\eta) < 0$], but it does not measure the proper time of a comoving clock (one with constant spatial coordinates). If we consider a trajectory $x^\mu(\lambda) = (\eta(\lambda), 0, 0, 0)$, and calculated the proper time $\tau(\eta)$, we would find that it was equal to our previous time coordinate but not our new one: $\tau \propto t \propto \eta^{1/(1-q)}$. So η is a timelike coordinate, but not the time that anyone would measure. This is perfectly okay, and simply serves as an illustration of the independence of the notions of observable quantities and spacetime coordinates.

Now that we have our expanding-universe metric in the form of a conformal factor times Minkowski, we can perform the same sequence of coordinate transformations—(H.5), (H.8), and (H.13)—where we allow η to take the place of t. These changes transform our coordinates from (η, r) to (T, R), where the ranges are now

$$0 \leq R, \quad 0 < T, \quad T + R < \pi. \tag{H.25}$$

The metric (H.23) becomes

$$ds^2 = \omega^{-2}(T, R)\left(-dT^2 + dR^2 + \sin^2 R \, d\Omega^2\right), \tag{H.26}$$

where some heroic use of trigonometric identities reveals that the conformal factor is of the form

$$\omega(T, R) = \left(\frac{\cos T + \cos R}{2 \sin T}\right)^{2q} (\cos T + \cos R). \tag{H.27}$$

The precise form of the conformal factor is actually not of primary importance; the crucial feature is that we have once again expressed our metric as a conformal factor times that of the Einstein static universe. The important distinction between

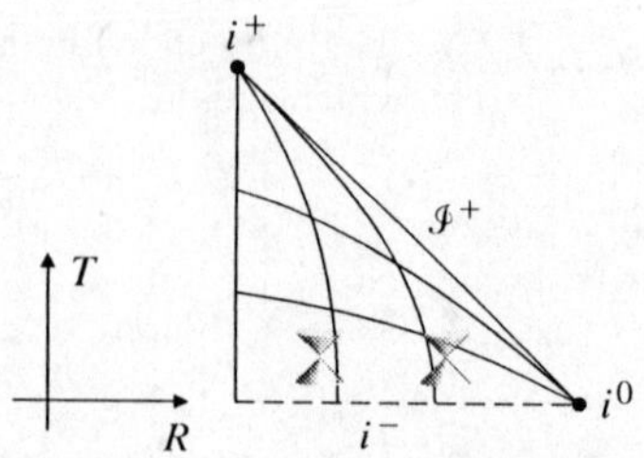

FIGURE H.5 Conformal diagram for a Robertson–Walker universe with $a(t) \propto t^q$ for $0 < q < 1$. The dashed line represents the singularity at $t = 0$ (which also corresponds to $T = 0$).

this case and that of flat spacetime is that the timelike coordinate ends at the singularity at $T = 0$; otherwise the spacetime diagram is identical. We therefore have the conformal diagram of Figure H.5, which resembles the upper half of the Minkowski diagram (Figure H.4). Once again, light cones appear at 45°. We see how the conformal diagram makes the causal structure apparent; it is straightforward to choose two events in the spacetime with the property that their past light cones will hit the singularity before they intersect (while future light cones will always overlap). For more complicated geometries, this convenient way of representing a spacetime will be even more useful.

APPENDIX I

The Parallel Propagator

The idea of parallel-transporting a tensor along a curve is obviously of central importance in GR. For a vector V^μ being transported down a path $x^\mu(\lambda)$, the equation of parallel transport is

$$\frac{dx^\mu}{d\lambda}\nabla_\mu V^\nu \equiv \frac{dx^\mu}{d\lambda}\partial_\mu V^\nu + \frac{dx^\mu}{d\lambda}\Gamma^\nu_{\mu\sigma}V^\sigma = 0. \tag{I.1}$$

It turns out to be possible to write down an explicit and general solution to this equation; it's somewhat formal, but interesting both in its own right and for its connections to techniques in quantum field theory.

We begin by noticing that for some path $\gamma : \lambda \to x^\sigma(\lambda)$, solving the parallel transport equation for a vector V^μ amounts to finding a matrix $P^\mu{}_\rho(\lambda, \lambda_0)$, which relates the vector at its initial value $V^\mu(\lambda_0)$ to its value somewhere later down the path:

$$V^\mu(\lambda) = P^\mu{}_\rho(\lambda, \lambda_0)V^\rho(\lambda_0). \tag{I.2}$$

Of course the matrix $P^\mu{}_\rho(\lambda, \lambda_0)$, known as the **parallel propagator**, depends on the path γ (although it's hard to find a notation that indicates this without making γ look like an index). If we define

$$A^\mu{}_\rho(\lambda) = -\Gamma^\mu_{\sigma\rho}\frac{dx^\sigma}{d\lambda}, \tag{I.3}$$

where the quantities on the right-hand side are evaluated at $x^\nu(\lambda)$, then the parallel transport equation becomes

$$\frac{d}{d\lambda}V^\mu = A^\mu{}_\rho V^\rho. \tag{I.4}$$

Since the parallel propagator must work for any vector, substituting (I.2) into (I.4) shows that $P^\mu{}_\rho(\lambda, \lambda_0)$ also obeys this equation:

$$\frac{d}{d\lambda}P^\mu{}_\rho(\lambda, \lambda_0) = A^\mu{}_\sigma(\lambda)P^\sigma{}_\rho(\lambda, \lambda_0). \tag{I.5}$$

To solve this equation, first integrate both sides:

$$P^\mu{}_\rho(\lambda, \lambda_0) = \delta^\mu_\rho + \int_{\lambda_0}^{\lambda} A^\mu{}_\sigma(\eta)P^\sigma{}_\rho(\eta, \lambda_0)\, d\eta. \tag{I.6}$$

The Kronecker delta, it is easy to see, provides the correct normalization for $\lambda = \lambda_0$.

We can solve (I.6) by iteration, taking the right-hand side and plugging it into itself repeatedly, giving

$$P^{\mu}{}_{\rho}(\lambda, \lambda_0) = \delta^{\mu}_{\rho} + \int_{\lambda_0}^{\lambda} A^{\mu}{}_{\rho}(\eta)\, d\eta + \int_{\lambda_0}^{\lambda}\int_{\lambda_0}^{\eta} A^{\mu}{}_{\sigma}(\eta) A^{\sigma}{}_{\rho}(\eta')\, d\eta' d\eta + \cdots . \tag{I.7}$$

The nth term in this series is an integral over an n-dimensional right triangle, or n-simplex:

$$\int_{\lambda_0}^{\lambda} A(\eta_1)\, d\eta_1$$

$$\int_{\lambda_0}^{\lambda}\int_{\lambda_0}^{\eta_2} A(\eta_2)A(\eta_1)\, d\eta_1 d\eta_2$$

$$\int_{\lambda_0}^{\lambda}\int_{\lambda_0}^{\eta_3}\int_{\lambda_0}^{\eta_2} A(\eta_3)A(\eta_2)A(\eta_1)\, d^3\eta.$$

See Figure I.1.

It would simplify things if we could consider such an integral to be over an n-cube instead of an n-simplex. Is there some way to do this? There are $n!$ such simplices in each cube, so we would have to multiply by $1/n!$ to compensate for this extra volume. But we also want to get the integrand right; using matrix notation, the integrand at nth order is $A(\eta_n)A(\eta_{n-1})\cdots A(\eta_1)$, but with the special property that $\eta_n \geq \eta_{n-1} \geq \cdots \geq \eta_1$. We therefore define the **path-ordering symbol**, $\mathcal{P}$, to ensure that this condition holds. In other words, the expression

$$\mathcal{P}[A(\eta_n)A(\eta_{n-1})\cdots A(\eta_1)] \tag{I.8}$$

stands for the product of the n matrices $A(\eta_i)$, ordered in such a way that the largest value of η_i is on the left, and each subsequent value of η_i is less than or

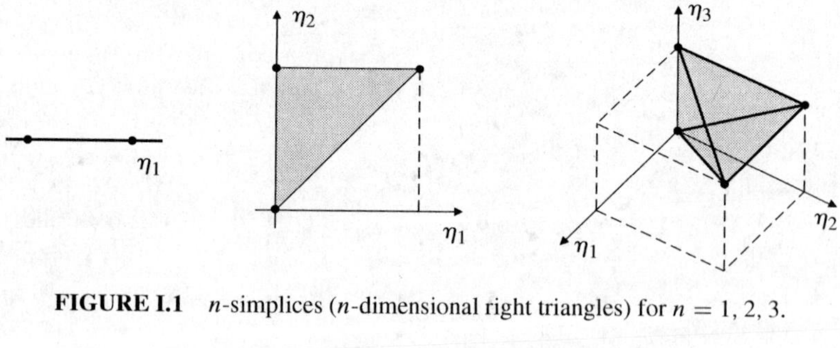

FIGURE I.1 n-simplices (n-dimensional right triangles) for $n = 1, 2, 3$.

equal to the previous one. We then can express the nth-order term in (I.7) as

$$\int_{\lambda_0}^{\lambda}\int_{\lambda_0}^{\eta_n}\cdots\int_{\lambda_0}^{\eta_2} A(\eta_n)A(\eta_{n-1})\cdots A(\eta_1)\, d^n\eta$$
$$= \frac{1}{n!}\int_{\lambda_0}^{\lambda}\int_{\lambda_0}^{\lambda}\cdots\int_{\lambda_0}^{\lambda} \mathcal{P}[A(\eta_n)A(\eta_{n-1})\cdots A(\eta_1)]\, d^n\eta. \tag{I.9}$$

This expression contains no substantive statement about the matrices $A(\eta_i)$; it is just notation. But we can now write (I.7) in matrix form as

$$P(\lambda,\lambda_0) = \mathbf{1} + \sum_{n=1}^{\infty}\frac{1}{n!}\int_{\lambda_0}^{\lambda} \mathcal{P}[A(\eta_n)A(\eta_{n-1})\cdots A(\eta_1)]\, d^n\eta. \tag{I.10}$$

This formula is just the series expression for an exponential; we therefore say that the parallel propagator is given by the path-ordered exponential

$$P(\lambda,\lambda_0) = \mathcal{P}\exp\left(\int_{\lambda_0}^{\lambda} A(\eta)\, d\eta\right), \tag{I.11}$$

where once again this is just notation; the path-ordered exponential is defined to be the right-hand side of (I.10). We can write it more explicitly as

$$\boxed{P^{\mu}{}_{\nu}(\lambda,\lambda_0) = \mathcal{P}\exp\left(-\int_{\lambda_0}^{\lambda} \Gamma^{\mu}_{\sigma\nu}\frac{dx^{\sigma}}{d\eta}\, d\eta\right).} \tag{I.12}$$

It's nice to have an explicit formula, even if it is rather abstract. The same kind of expression appears in quantum field theory as "Dyson's Formula," where it arises because the Schrödinger equation for the time-evolution operator has the same form as (I.5).

An especially interesting example of the parallel propagator occurs when the path is a loop, starting and ending at the same point. Then if the connection is metric-compatible, the resulting matrix will just be a Lorentz transformation on the tangent space at the point. This transformation is known as the "holonomy" of the loop. If you know the holonomy of every possible loop, that turns out to be equivalent to knowing the metric. One can then examine general relativity in the "loop representation," where the fundamental variables are holonomies rather than the explicit metric. A program called "loop quantum gravity" attempts to directly quantize general relativity in these variables (as opposed to something like string theory, in which GR falls out in some limit). A great deal of mathematical progress has been made in this direction, but fundamental obstacles remain.[1]

[1] For a review of this approach, see C. Rovelli, "Loop quantum gravity," *Living Rev. Rel.* **1**, 1 (1998) `http://arxiv.org/gr-qc/9710008`.

APPENDIX

J Noncoordinate Bases

Early on in our study of manifolds, we made a decision to choose bases for our tangent spaces that were adapted to coordinates. For both aesthetic and pragmatic reasons, we should consider once again the formalism of connections and curvature, but this time using sets of basis vectors in the tangent space that are *not* derived from any coordinate system. It will turn out that this slight change in emphasis reveals a different point of view on the connection and curvature, one in which the relationship to gauge theories of particle physics is much more transparent. In fact the concepts to be introduced are very straightforward, but the subject is a notational nightmare, so it looks more difficult than it really is.

Until now we have been taking advantage of the fact that a natural basis for the tangent space T_p at a point p is given by the partial derivatives with respect to the coordinates at that point, $\hat{e}_{(\mu)} = \partial_\mu$. Similarly, a basis for the cotangent space T_p^* is given by the gradients of the coordinate functions, $\hat{\theta}^{(\mu)} = dx^\mu$. Nothing stops us, however, from setting up any bases we like. Let us therefore imagine that at each point in the manifold we introduce a set of basis vectors $\hat{e}_{(a)}$ (indexed by a Latin letter rather than Greek, to remind us that they are not related to any coordinate system). We will choose these basis vectors to be "orthonormal," in a sense that is appropriate to the signature of the manifold on which we are working. That is, if the canonical form of the metric is written η_{ab}, we demand that the inner product of our basis vectors be

$$g(\hat{e}_{(a)}, \hat{e}_{(b)}) = \eta_{ab}, \tag{J.1}$$

where $g(\ ,\ )$ is the usual metric tensor. Thus, in a Lorentzian spacetime η_{ab} represents the Minkowski metric, while in a space with positive-definite metric it would represent the Euclidean metric. The set of vectors comprising an orthonormal basis is sometimes known as a **tetrad** (from Greek *tetras*, "a group of four") or **vielbein** (from the German for "many legs"). In different numbers of dimensions it occasionally becomes a *vierbein* (four), *dreibein* (three), *zweibein* (two), and so on. Just as we cannot in general find coordinate charts that cover the entire manifold, we will often not be able to find a single set of smooth basis vector fields that are defined everywhere. As usual, we can overcome this problem by working in different patches and making sure things are well-behaved on the overlaps.

The point of having a basis is that any vector can be expressed as a linear combination of basis vectors. Specifically, we can express our old basis vectors

$\hat{e}_{(\mu)} = \partial_\mu$ in terms of the new ones:

$$\hat{e}_{(\mu)} = e_\mu{}^a \hat{e}_{(a)}. \tag{J.2}$$

The components $e_\mu{}^a$ form an $n \times n$ invertible matrix. (In accord with our usual practice of blurring the distinction between objects and their components, we will refer to the $e_\mu{}^a$ as the tetrad or vielbein, and often in the plural as "vielbeins.") We denote their inverse by switching indices to obtain $e^\mu{}_a$, which satisfy

$$e^\mu{}_a e_\nu{}^a = \delta^\mu_\nu, \quad e_\mu{}^a e^\mu{}_b = \delta^a_b. \tag{J.3}$$

These serve as the components of the vectors $\hat{e}_{(a)}$ in the coordinate basis:

$$\hat{e}_{(a)} = e^\mu{}_a \hat{e}_{(\mu)}. \tag{J.4}$$

In terms of the inverse vielbeins, (J.1) becomes

$$g_{\mu\nu} e^\mu{}_a e^\nu{}_b = \eta_{ab}, \tag{J.5}$$

or equivalently

$$g_{\mu\nu} = e_\mu{}^a e_\nu{}^b \eta_{ab}. \tag{J.6}$$

This last equation sometimes leads people to say that the vielbeins are the "square root" of the metric.

We can similarly set up an orthonormal basis of one-forms in T_p, which we denote $\hat{\theta}^{(a)}$. They may be chosen to be compatible with the basis vectors, in the sense that

$$\hat{\theta}^{(a)}(\hat{e}_{(b)}) = \delta^a_b. \tag{J.7}$$

An immediate consequence is that the orthonormal one-forms are related to their coordinate-based cousins $\hat{\theta}^{(\mu)} = dx^\mu$ by

$$\hat{\theta}^{(\mu)} = e^\mu{}_a \hat{\theta}^{(a)} \tag{J.8}$$

and

$$\hat{\theta}^{(a)} = e_\mu{}^a \hat{\theta}^{(\mu)}. \tag{J.9}$$

The vielbeins $e_\mu{}^a$ thus serve double duty as the components of the coordinate basis vectors in terms of the orthonormal basis vectors, and as components of the orthonormal basis one-forms in terms of the coordinate basis one-forms; while the inverse vielbeins serve as the components of the orthonormal basis vectors in terms of the coordinate basis, and as components of the coordinate basis one-forms in terms of the orthonormal basis.

Any other vector can be expressed in terms of its components in the orthonormal basis. If a vector V is written in the coordinate basis as $V^\mu \hat{e}_{(\mu)}$ and in the orthonormal basis as $V^a \hat{e}_{(a)}$, the sets of components will be related by

$$V^a = e_\mu{}^a V^\mu. \tag{J.10}$$

So the vielbeins allow us to "switch from Latin to Greek indices and back." The nice property of tensors, that there is usually only one sensible thing to do based on index placement, is of great help here. We can go on to refer to multi-index tensors in either basis, or even in terms of mixed components:

$$V^a{}_b = e_\mu{}^a V^\mu{}_b = e^\nu{}_b V^a{}_\nu = e_\mu{}^a e^\nu{}_b V^\mu{}_\nu. \tag{J.11}$$

Looking back at (J.5), we see that the components of the metric tensor in the orthonormal basis are just those of the flat metric, η_{ab}. (For this reason the Greek indices are sometimes referred to as "curved" and the Latin ones as "flat.") In fact we can go so far as to raise and lower the Latin indices using the flat metric and its inverse η^{ab}. You can check for yourself that everything works (for example, that the lowering an index with the metric commutes with changing from orthonormal to coordinate bases). In particular, our definition of the inverse vielbeins is consistent with our usual notion of raising and lowering indices,

$$e^\mu{}_a = g^{\mu\nu}\eta_{ab} e_\nu{}^b. \tag{J.12}$$

We have introduced the vielbeins $e_\nu{}^a$ as components of a set of basis vectors, evaluated in a different basis. This is equivalent to thinking of them as the components of a $(1, 1)$ tensor,

$$e = e_\nu{}^a \mathrm{d}x^\nu \otimes \hat{e}_{(a)}. \tag{J.13}$$

But this is actually a tensor we already know and love: the identity map. If we act this tensor on a vector, we get back the same vector, just in a different basis; that's the content of (J.10). Likewise, if we use the inverse vielbein e^μ_a to convert the Latin index on $e_\nu{}^a$ to a Greek index, according to (J.3) we get the Kronecker delta δ^μ_ν, which of course is the identity map on vectors (or one-forms). This point is worth emphasizing because we could also choose to interpret $e_\nu{}^a$ as a set of vector components (and some references do so), in which case the covariant derivative would look different. By introducing a new set of basis vectors and one-forms, we necessitate a return to our favorite topic of transformation properties. We've been careful all along to emphasize that the tensor transformation law was only an indirect outcome of a coordinate transformation; the real issue was a change of basis. Now that we have noncoordinate bases, these bases can be changed independently of the coordinates. The only restriction is that the orthonormality property (J.1) be preserved. But we know what kind of transformations preserve the flat metric—in a Euclidean signature metric they are orthogonal transformations, while in a Lorentzian signature metric they are Lorentz transformations. We therefore consider changes of basis of the form

$$\hat{e}_{(a)} \to \hat{e}_{(a')} = \Lambda^a{}_{a'}(x)\hat{e}_{(a)}, \tag{J.14}$$

where the matrices $\Lambda^a{}_{a'}(x)$ represent position-dependent transformations which (at each point) leave the canonical form of the metric unaltered:

$$\Lambda^a{}_{a'}\Lambda^b{}_{b'}\eta_{ab} = \eta_{a'b'}. \tag{J.15}$$

In fact these matrices correspond to what in flat space we called the inverse Lorentz transformations (which operate on basis vectors); as before we also have ordinary Lorentz transformations $\Lambda^{a'}{}_a$, which transform the basis one-forms. As far as components are concerned, as before we transform upper indices with $\Lambda^{a'}{}_a$ and lower indices with $\Lambda^a{}_{a'}$.

So we now have the freedom to perform a Lorentz transformation (or an ordinary Euclidean rotation, depending on the signature) at every point in space. These transformations are therefore called **local Lorentz transformations**, or LLT's. We still have our usual freedom to make changes in coordinates, which are called **general coordinate transformations**, or GCT's. Both can happen at the same time, resulting in a mixed tensor transformation law:

$$T^{a'\mu'}{}_{b'\nu'} = \Lambda^{a'}{}_a \frac{\partial x^{\mu'}}{\partial x^\mu} \Lambda^b{}_{b'} \frac{\partial x^\nu}{\partial x^{\nu'}} T^{a\mu}{}_{b\nu}. \tag{J.16}$$

Translating what we know about tensors into noncoordinate bases is for the most part merely a matter of sticking vielbeins in the right places. The crucial exception comes when we begin to differentiate things. In our ordinary formalism, the covariant derivative of a tensor is given by its partial derivative plus correction terms, one for each index, involving the tensor and the connection coefficients. The same procedure will continue to be true for the noncoordinate basis, but we replace the ordinary connection coefficients $\Gamma^\lambda_{\mu\nu}$ by the **spin connection**, denoted $\omega_\mu{}^a{}_b$. Each Latin index gets a factor of the spin connection in the usual way:

$$\nabla_\mu X^a{}_b = \partial_\mu X^a{}_b + \omega_\mu{}^a{}_c X^c{}_b - \omega_\mu{}^c{}_b X^a{}_c. \tag{J.17}$$

(The name "spin connection" comes from the fact that this can be used to take covariant derivatives of spinors, which is actually impossible using the conventional connection coefficients.) In the presence of mixed Latin and Greek indices we get terms of both kinds.

The usual demand that a tensor be independent of the way it is written allows us to derive a relationship between the spin connection, the vielbeins, and the $\Gamma^\nu_{\mu\lambda}$'s. Consider the covariant derivative of a vector X, first in a purely coordinate basis:

$$\begin{aligned}\nabla X &= (\nabla_\mu X^\nu)dx^\mu \otimes \partial_\nu \\ &= (\partial_\mu X^\nu + \Gamma^\nu_{\mu\lambda}X^\lambda)dx^\mu \otimes \partial_\nu.\end{aligned} \tag{J.18}$$

Now find the same object in a mixed basis, and convert into the coordinate basis:

$$\begin{aligned}\nabla X &= (\nabla_\mu X^a)\mathrm{d}x^\mu \otimes \hat{e}_{(a)} \\ &= (\partial_\mu X^a + \omega_\mu{}^a{}_b X^b)\mathrm{d}x^\mu \otimes \hat{e}_{(a)} \\ &= (\partial_\mu(e_\nu{}^a X^\nu) + \omega_\mu{}^a{}_b e_\lambda{}^b X^\lambda)\mathrm{d}x^\mu \otimes (e^\sigma{}_a \partial_\sigma) \\ &= e^\sigma{}_a(e_\nu{}^a \partial_\mu X^\nu + X^\nu \partial_\mu e_\nu{}^a + \omega_\mu{}^a{}_b e_\lambda{}^b X^\lambda)\mathrm{d}x^\mu \otimes \partial_\sigma \\ &= (\partial_\mu X^\nu + e^\nu{}_a \partial_\mu e_\lambda{}^a X^\lambda + e^\nu{}_a e_\lambda{}^b \omega_\mu{}^a{}_b X^\lambda)\mathrm{d}x^\mu \otimes \partial_\nu. \end{aligned} \tag{J.19}$$

Comparison with (J.18) reveals

$$\Gamma^\nu_{\mu\lambda} = e^\nu{}_a \partial_\mu e_\lambda{}^a + e^\nu{}_a e_\lambda{}^b \omega_\mu{}^a{}_b, \tag{J.20}$$

or equivalently

$$\omega_\mu{}^a{}_b = e_\nu{}^a e^\lambda{}_b \Gamma^\nu_{\mu\lambda} - e^\lambda{}_b \partial_\mu e_\lambda{}^a. \tag{J.21}$$

A bit of manipulation allows us to write this relation as the vanishing of the covariant derivative of the vielbein,

$$\begin{aligned}\nabla_\mu e_\nu{}^a &= \partial_\mu e_\nu{}^a - \Gamma^\lambda_{\mu\nu} e_\lambda{}^a + \omega_\mu{}^a{}_b e_\nu{}^b \\ &= 0, \end{aligned} \tag{J.22}$$

which is sometimes known as the "tetrad postulate." Note that this is always true; we did not need to assume anything about the connection in order to derive it. Specifically, we did not need to assume that the connection was metric compatible or torsion free. We did, however, implicitly take $e_\nu{}^a$ to represent the (1, 1) tensor (J.13); since this tensor is the identity map, it is no surprise that its covariant derivative vanishes. (Not all references have this philosophy, so be careful.)

Since the connection may be thought of as something we need to introduce in order to fix up the transformation law of the covariant derivative, it should come as no surprise that the spin connection does not itself obey the tensor transformation law. Actually, under GCT's the one lower Greek index does transform in the right way, as a one-form. But under LLT's the spin connection transforms inhomogeneously, as

$$\omega_\mu{}^{a'}{}_{b'} = \Lambda^{a'}{}_a \Lambda^b{}_{b'} \omega_\mu{}^a{}_b - \Lambda^c{}_{b'} \partial_\mu \Lambda^{a'}{}_c. \tag{J.23}$$

You are encouraged to check for yourself that this results in the proper transformation of the covariant derivative.

So far we have done nothing but empty formalism, translating things we already knew into a new notation. But the work we are doing does buy us two things. The first, which we already alluded to, is the ability to describe spinor fields on spacetime and take their covariant derivatives; we won't explore this further here. The second is a change in viewpoint, in which we can think of various tensors as tensor-valued differential forms. For example, an object like $X_\mu{}^a$, which we think of as a (1, 1) tensor written with mixed indices, can also be thought of as a "vector-valued one-form." It has one lower Greek index, so we think of it as a one-form, but for each value of the lower index it is a vector. Similarly a tensor $A_{\mu\nu}{}^a{}_b$, antisymmetric in μ and ν, can be thought of as a "(1, 1)-tensor-valued two-form." Thus, any tensor with some number of antisymmetric lower Greek indices and some number of Latin indices can be thought of as a differential form, but taking values in the tensor bundle. (Ordinary differential forms are simply scalar-valued forms.) The usefulness of this viewpoint comes when we consider exterior derivatives. If we want to think of $X_\mu{}^a$ as a vector-valued one-form, we are tempted to take its exterior derivative:

$$(\mathrm{d}X)_{\mu\nu}{}^a = \partial_\mu X_\nu{}^a - \partial_\nu X_\mu{}^a. \tag{J.24}$$

It is easy to check that this object transforms like a two-form [that is, according to the transformation law for (0, 2) tensors] under GCT's, but not as a vector under LLT's (the Lorentz transformations depend on position, which introduces an inhomogeneous term into the transformation law). But we can fix this by judicious use of the spin connection, which can be thought of as a one-form, but not a tensor-valued one-form, due to the nontensorial transformation law (J.23). Thus, the object

$$(\mathrm{d}X)_{\mu\nu}{}^a + (\omega \wedge X)_{\mu\nu}{}^a = \partial_\mu X_\nu{}^a - \partial_\nu X_\mu{}^a + \omega_\mu{}^a{}_b X_\nu{}^b - \omega_\nu{}^a{}_b X_\mu{}^b, \tag{J.25}$$

as you can verify, transforms as a proper tensor.

An immediate application of this formalism is to the expressions for the torsion and curvature, the two tensors that characterize any given connection. The torsion, with two antisymmetric lower indices, can be thought of as a vector-valued two-form $T_{\mu\nu}{}^a$. The curvature, which is always antisymmetric in its last two indices, is a (1, 1)-tensor-valued two-form, $R^a{}_{b\mu\nu}$. Using our freedom to suppress indices on differential forms, we can express these in terms of the basis one-forms

$$e^a = e_\mu{}^a \mathrm{d}x^\mu \tag{J.26}$$

and the spin-connection one-forms

$$\omega^a{}_b = \omega_\mu{}^a{}_b \mathrm{d}x^\mu. \tag{J.27}$$

Notice that we have switched notations, defining $e^a \equiv \hat{\theta}^{(a)}$. This is fairly conventional, as well as cleaner. The defining relations for the torsion and curvature are then

$$T^a = \mathrm{d}e^a + \omega^a{}_b \wedge e^b \tag{J.28}$$

and

$$R^a{}_b = \mathrm{d}\omega^a{}_b + \omega^a{}_c \wedge \omega^c{}_b. \tag{J.29}$$

Keep in mind that $R^a{}_b$ represents the entire Riemann tensor, with Greek indices suppressed; don't confuse it with the Ricci tensor. These are known as the **Cartan structure equations**. They are equivalent to the usual definitions; let's go through the exercise of showing this for the torsion, and you can check the curvature for yourself. We have

$$\begin{aligned} T_{\mu\nu}{}^\lambda &= e^\lambda{}_a T_{\mu\nu}{}^a \\ &= e^\lambda{}_a(\partial_\mu e_\nu{}^a - \partial_\nu e_\mu{}^a + \omega_\mu{}^a{}_b e_\nu{}^b - \omega_\nu{}^a{}_b e_\mu{}^b) \\ &= \Gamma^\lambda_{\mu\nu} - \Gamma^\lambda_{\nu\mu}, \end{aligned} \tag{J.30}$$

which is just the original definition we gave. Here we have used (J.20), the expression for the $\Gamma^\lambda_{\mu\nu}$'s in terms of the vielbeins and spin connection. We can also express identities obeyed by these tensors as

$$\mathrm{d}T^a + \omega^a{}_b \wedge T^b = R^a{}_b \wedge e^b \tag{J.31}$$

and

$$\mathrm{d}R^a{}_b + \omega^a{}_c \wedge R^c{}_b - R^a{}_c \wedge \omega^c{}_b = 0. \tag{J.32}$$

The first of these is the generalization of $R^\rho{}_{[\sigma\mu\nu]} = 0$, while the second is the Bianchi identity $\nabla_{[\lambda|}R^\rho{}_{\sigma|\mu\nu]} = 0$. (Sometimes both equations are called Bianchi identities.)

The form of these expressions leads to an almost irresistible temptation to define a "covariant exterior derivative," which acts on a tensor-valued form by taking the ordinary exterior derivative and then adding appropriate terms with the spin connection, one for each Latin index. Although we won't do that here, it is okay to give in to this temptation, and in fact the right-hand side of (J.28) and the left-hand sides of (J.31) and (J.32) can be thought of as just such covariant-exterior derivatives. But be careful, since (J.29) cannot be; you can't take any sort of covariant derivative of the spin connection, since it's not a tensor.

So far our equations have been true for general connections; let's see what we get for the Christoffel connection. The torsion-free requirement is just that (J.28) vanish; this does not lead immediately to any simple statement about the coefficients of the spin connection. Metric compatibility is expressed as the vanishing of the covariant derivative of the metric: $\nabla g = 0$. We can see what this leads to when we express the metric in the orthonormal basis, where its components are simply η_{ab}:

$$\begin{aligned} \nabla_\mu \eta_{ab} &= \partial_\mu \eta_{ab} - \omega_\mu{}^c{}_a \eta_{cb} - \omega_\mu{}^c{}_b \eta_{ac} \\ &= -\omega_{\mu ab} - \omega_{\mu ba}. \end{aligned} \tag{J.33}$$

Then setting this equal to zero implies

$$\omega_{\mu ab} = -\omega_{\mu ba}. \tag{J.34}$$

Thus, metric compatibility is equivalent to the antisymmetry of the spin connection in its Latin indices. (As before, such a statement is only sensible if both indices are either upstairs or downstairs.) These two conditions together allow us to express the spin connection in terms of the vielbeins. An explicit formula expresses this solution, but in practice it is easier to simply solve the torsion-free condition

$$\omega^a{}_b \wedge e^b = -\mathrm{d}e^a, \tag{J.35}$$

using the asymmetry of the spin connection, to find the individual components.

One of the best reasons for thinking about noncoordinate bases is that they actually lead to great simplifications in certain cases, including the calculation of the curvature tensor. Let's see how this works in a simple example, a spatially flat expanding universe, with metric

$$ds^2 = -\mathrm{d}t^2 + a^2(t)\delta_{ij}\mathrm{d}x^i\mathrm{d}x^j. \tag{J.36}$$

We will use the differential-forms notation of (J.26) and (J.27); calculations such as this are good evidence that this language is practically useful as well as elegant. The metric is thus written (for any geometry)

$$ds^2 = \eta_{ab}e^a \otimes e^b. \tag{J.37}$$

We need to choose basis one-forms e^a such that this matches our metric (J.36). There are many choices (related by local Lorentz transformations), but one obvious one:

$$\begin{aligned} e^0 &= \mathrm{d}t \\ e^i &= a\mathrm{d}x^i. \end{aligned} \tag{J.38}$$

We would now like to solve for the spin connection using (J.35). The good news is that we basically can do it by guessing. First, by appropriately raising and lowering indices (with η^{ab} and η_{ab}) we derive the consequences of the antisymmetry of ω_{ab}:

$$\begin{aligned} \omega^0{}_0 &= 0 \\ \omega^0{}_j &= \omega^j{}_0 \\ \omega^i{}_j &= -\omega^j{}_i. \end{aligned} \tag{J.39}$$

We next calculate the right-hand side of (J.35),

$$\begin{aligned} \mathrm{d}e^0 &= 0 \\ \mathrm{d}e^i &= \mathrm{d}a \wedge \mathrm{d}x^i = \dot{a}\mathrm{d}t \wedge \mathrm{d}x^i, \end{aligned} \tag{J.40}$$

and then the left,

$$\begin{aligned}\omega^0{}_b \wedge e^b &= \omega^0{}_j \wedge e^j = a\omega^0{}_j \wedge \mathrm{d}x^j \\ \omega^i{}_b \wedge e^b &= \omega^i{}_0 \wedge e^0 + \omega^i{}_j \wedge e^j = \omega^i{}_0 \wedge \mathrm{d}t + \omega^i{}_j \wedge \mathrm{d}x^j.\end{aligned} \tag{J.41}$$

Plugging into (J.35) yields

$$\begin{aligned}\omega^0{}_j \wedge \mathrm{d}x^j &= 0 \\ \omega^i{}_0 \wedge \mathrm{d}t + \omega^i{}_j \wedge \mathrm{d}x^j &= -\dot{a}\mathrm{d}t \wedge \mathrm{d}x^i.\end{aligned} \tag{J.42}$$

We would like to solve these equations for $\omega^a{}_b$. It is tempting to guess $\omega^0{}_j = 0$; but then to solve the second equation we would require $\omega^i{}_j = -\dot{a}\delta^i_j \mathrm{d}t$, which is incompatible with $\omega^i{}_j = -\omega^j{}_i$ from (J.39). But we can solve the first equation by setting $\omega^0{}_j$ proportional to $\mathrm{d}x^j$ (due to the antisymmetry of the wedge product). Indeed, if we choose

$$\omega^0{}_j = \dot{a}\mathrm{d}x^j, \quad \omega^i{}_0 = \dot{a}\mathrm{d}x^i, \tag{J.43}$$

we find that both equations in (J.42) are solved by setting

$$\omega^i{}_j = 0. \tag{J.44}$$

Now that we know the spin connection, we can easily get the curvature through

$$R^a{}_b = \mathrm{d}\omega^a{}_b + \omega^a{}_c \wedge \omega^c{}_b. \tag{J.45}$$

We first calculate the exterior derivative of the spin connection forms,

$$\begin{aligned}\mathrm{d}\omega^i{}_0 &= \ddot{a}\mathrm{d}t \wedge \mathrm{d}x^i \\ \mathrm{d}\omega^0{}_j &= \ddot{a}\mathrm{d}t \wedge \mathrm{d}x^j \\ \mathrm{d}\omega^i{}_j &= 0,\end{aligned} \tag{J.46}$$

and then the wedge products,

$$\begin{aligned}\omega^0{}_c \wedge \omega^c{}_0 &= 0 \\ \omega^i{}_c \wedge \omega^c{}_0 &= 0 \\ \omega^i{}_c \wedge \omega^c{}_j &= \dot{a}^2\mathrm{d}x^i \wedge \mathrm{d}x^j.\end{aligned} \tag{J.47}$$

We therefore obtain the curvature two-form,

$$\begin{aligned}R^0{}_0 &= 0 \\ R^0{}_j &= \ddot{a}\mathrm{d}t \wedge \mathrm{d}x^j \\ R^i{}_0 &= \ddot{a}\mathrm{d}t \wedge \mathrm{d}x^i \\ R^i{}_j &= \dot{a}^2\mathrm{d}x^i \wedge \mathrm{d}x^j.\end{aligned} \tag{J.48}$$

For purposes of comparison, we can use vielbeins to convert $R^a{}_{b\mu\nu}$ to our conventional expression $R^\rho{}_{\sigma\mu\nu}$, using

$$R^\rho{}_{\sigma\mu\nu} = e^\rho{}_a e_\sigma{}^b R^a{}_{b\mu\nu}. \tag{J.49}$$

In component form the vielbeins (J.38) and their inverse are

$$e_\mu{}^a = \begin{pmatrix} 1 & & & \\ & a & & \\ & & a & \\ & & & a \end{pmatrix}, \quad e^\nu{}_b = \begin{pmatrix} 1 & & & \\ & a^{-1} & & \\ & & a^{-1} & \\ & & & a^{-1} \end{pmatrix}. \tag{J.50}$$

We will also need to evaluate the components of the wedge products of basis forms, which is straightforward enough,

$$(\mathrm{d}x^\alpha \wedge \mathrm{d}x^\beta)_{\mu\nu} = \delta^\alpha_\mu \delta^\beta_\nu - \delta^\alpha_\nu \delta^\beta_\mu. \tag{J.51}$$

Putting it all together yields the components $R^\rho{}_{\sigma\mu\nu}$,

$$\begin{aligned} R^0{}_{j0l} &= a\ddot{a}\delta_{jl} \\ R^i{}_{0k0} &= -\frac{\ddot{a}}{a}\delta^i_k \\ R^i{}_{jkl} &= \dot{a}^2(\delta^i_k\delta_{jl} - \delta^i_l\delta_{jk}), \end{aligned} \tag{J.52}$$

as well as ones obtained by antisymmetry in the last two indices. We may contract to get the components of the Ricci tensor $R_{\sigma\nu} = R^\lambda{}_{\sigma\lambda\nu}$,

$$\begin{aligned} R_{00} &= -3\frac{\ddot{a}}{a} \\ R_{i0} &= 0 \\ R_{ij} &= (a\ddot{a} + 2\dot{a}^2)\delta_{ij}. \end{aligned} \tag{J.53}$$

You can check that this agrees with our results from Chapter 8. Already in this simple example, the tetrad method was computationally simpler than the coordinate-basis method; in more complicated metrics the comparative advantage continues to grow.

In the language of noncoordinate bases, it is possible to compare the formalism of connections and curvature in Riemannian geometry to that of gauge theories in particle physics. In both situations, the fields of interest live in vector spaces that are assigned to each point in spacetime. In Riemannian geometry the vector spaces include the tangent space, the cotangent space, and the higher tensor spaces constructed from these. In gauge theories, on the other hand, we are concerned with "internal" vector spaces. The distinction is that the tangent space and its relatives are intimately associated with the manifold itself, and are naturally defined once the manifold is set up; the tangent space, for example, can be thought of as the space of directional derivatives at a point. In contrast, an internal vector space can be of any dimension we like, and has to be defined as an independent addition to the manifold. In math jargon, the union of the base manifold with the internal vector spaces (defined at each point) is a **fiber bundle**, and each copy of the vector space is called the "fiber" (in accord with our definition of the tangent bundle).

Besides the base manifold (for us, spacetime) and the fibers, the other important ingredient in the definition of a fiber bundle is the "structure group," a Lie group that acts on the fibers to describe how they are sewn together on overlapping coordinate patches. Without going into details, the structure group for the tangent bundle in a four-dimensional spacetime is generally GL(4, **R**), the group of real invertible 4×4 matrices; if we have a Lorentzian metric, this may be reduced to the Lorentz group SO(3, 1). Now imagine that we introduce an internal three-dimensional vector space, and sew the fibers together with ordinary rotations; the structure group of this new bundle is then SO(3). A field that lives in this bundle might be denoted $\phi^A(x^\mu)$, where A runs from one to three; it is a three-vector (an internal one, unrelated to spacetime) for each point on the manifold. We have freedom to choose the basis in the fibers in any way we wish; this means that "physical quantities" should be left invariant under local SO(3) transformations such as

$$\phi^A(x^\mu) \rightarrow \phi^{A'}(x^\mu) = O^{A'}{}_A(x^\mu)\phi^A(x^\mu), \tag{J.54}$$

where $O^{A'}{}_A(x^\mu)$ is a matrix in SO(3) that depends on spacetime. Such transformations are known as **gauge transformations**, and theories invariant under them are called "gauge theories."

For the most part it is not hard to arrange things such that physical quantities are invariant under gauge transformations. The one difficulty arises when we consider partial derivatives, $\partial_\mu \phi^A$. Because the matrix $O^{A'}{}_A(x^\mu)$ depends on spacetime, it will contribute an unwanted term to the transformation of the partial derivative. By now you should be able to guess the solution: introduce a connection to correct for the inhomogeneous term in the transformation law. We therefore define a connection on the fiber bundle to be an object $A_\mu{}^A{}_B$, with two "group indices" and one spacetime index. Under GCT's it transforms as a one-form, while under gauge transformations it transforms as

$$A_\mu{}^{A'}{}_{B'} = O^{A'}{}_A O^B{}_{B'} A_\mu{}^A{}_B - O^C{}_{B'} \partial_\mu O^{A'}{}_C. \tag{J.55}$$

(Beware: our conventions are different from those in the particle physics literature.) With this transformation law, the "gauge covariant derivative"

$$D_\mu \phi^A = \partial_\mu \phi^A + A_\mu{}^A{}_B \phi^B \tag{J.56}$$

transforms "tensorially" under gauge transformations, as you are welcome to check. [In ordinary electromagnetism the connection is just the conventional vector potential. No indices are necessary, because the structure group U(1) is one-dimensional.]

It is clear that this notion of a connection on an internal fiber bundle is very closely related to the connection on the tangent bundle, especially in the orthonormal-frame picture we have been discussing. The transformation law (J.55), for example, is exactly the same as the transformation law (J.23) for the spin connection. We can also define a curvature or "field strength" tensor which is a two-form

$$F^A{}_B = \mathrm{d}A^A{}_B + A^A{}_C \wedge A^C{}_B, \tag{J.57}$$

in exact correspondence with (J.29). We can parallel transport things along paths, and there is a construction analogous to the parallel propagator; the trace of the matrix obtained by parallel transporting a vector around a closed curve is called a "Wilson loop."

We could go on in the development of the relationship between the tangent bundle and internal vector bundles, but that would be another book. Let us instead finish by emphasizing the important *difference* between the two constructions. The difference stems from the fact that the tangent bundle is closely related to the base manifold, while other fiber bundles are tacked on after the fact. It makes sense to say that a vector in the tangent space at p "points along a path" through p; but this makes no sense for an internal vector bundle. There is therefore no analogue of the coordinate basis for an internal space—partial derivatives along curves have nothing to do with internal vectors. It follows in turn that there is nothing like the vielbeins, which relate orthonormal bases to coordinate bases. The torsion tensor, in particular, is only defined for a connection on the tangent bundle, not for any gauge theory connections; it can be thought of as the covariant exterior derivative of the vielbein, and no such construction is available on an internal bundle. You should appreciate the relationship between the different uses of the notion of a connection, without getting carried away.

J.1 ■ EXERCISES

1. In (J.37) we mention that the metric in an orthonormal basis can be written

$$ds^2 = \eta_{ab} e^a \otimes e^b. \tag{J.58}$$

How can this possibly be? If the components of the metric are η_{ab} everywhere, how can we know what the geometry is?

2. Calculate the connection one-forms, curvature two-forms, and hence the components of the Riemann tensor for the Mixmaster universe. The metric is given by

$$ds^2 = -\mathrm{d}t \otimes \mathrm{d}t + \alpha^2 \sigma^1 \otimes \sigma^1 + \beta^2 \sigma^2 \otimes \sigma^2 + \gamma^2 \sigma^3 \otimes \sigma^3.$$

Here α, β, γ are functions of t only and the one-forms σ^i are given by

$$\sigma^1 = \cos\psi\, \mathrm{d}\theta + \sin\psi \sin\theta\, \mathrm{d}\phi$$

$$\sigma^2 = \sin\psi\, \mathrm{d}\theta - \cos\psi \sin\theta\, \mathrm{d}\phi$$

$$\sigma^3 = \mathrm{d}\psi + \cos\theta\, \mathrm{d}\phi.$$

Bibliography

I have made no attempt to provide careful citations to the original literature in the text. In keeping with the philosophy of focusing on pedagogy, I have included references to recent review articles where appropriate. Here I list a number of books that might be useful supplements to the one you are reading; the list is not meant to be comprehensive, and focuses on books that are in print and with which I happen to be familiar.

There are two websites that are invaluable resources for keeping up with recent work in gravitational physics. The first is the ArXiv e-print server for general relativity and quantum cosmology:

`http://arxiv.org/form/gr-qc/`

This is where researchers all over the world put their most recent papers, which can then be easily downloaded. There are similar servers for other areas of physics. The other website is for Living Reviews in Relativity:

`http://www.livingreviews.org/`

Living Reviews is an on-line journal specializing in review articles in all areas of gravitational physics. It is an excellent starting point for anyone interested in exploring recent work in a topic of current interest.

Special Relativity

E. Taylor and J. Wheeler, *Spacetime Physics* (Freeman, 1992). A very nice introduction to special relativity, making a great effort to explain away the "paradoxes" this subject seems to engender.

A.P. French, *Special Relativity* (W.W. Norton, 1968). Somewhat less colorful than Taylor and Wheeler, but a straightforward introduction to special relativity.

Undergraduate General Relativity

B.F. Schutz, *A First Course in General Relativity* (Cambridge, 1985). This is a very nice introductory text, making a real effort to bridge the transition from common topics in undergraduate physics to the language and results of GR.

J.B. Hartle, *Gravity: An Introduction to Einstein's General Relativity* (Addison-Wesley, 2002). Eases the exploration of GR by concentrating on examples of curved spacetimes and the behavior of particles in them, putting physics before formalism whenever possible.

E.F. Taylor and J.A. Wheeler, *Exploring Black Holes: An Introduction to General Relativity* (Benjamin Cummings, 2002). Uses black holes as a way to introduce physical principles of GR.

Graduate General Relativity

R. Wald, *General Relativity* (Chicago, 1984). Thorough discussions of a number of advanced topics, including black holes, global structure, and spinors. An invaluable reference, this is the book to turn to if you need the right answer to a well-posed GR question.

C. Misner, K. Thorne and J. Wheeler, *Gravitation* (Freeman, 1973). The book that educated at least two generations of researchers in gravitational physics. Comprehensive and encyclopedic, the book is written in an often-idiosyncratic style that you will either like or not.

S. Weinberg, *Gravitation and Cosmology* (Wiley, 1972). A great book at what it does, especially strong on astrophysics, cosmology, and experimental tests. However, it takes an unusual non-geometric approach to the material, and doesn't discuss black holes. Weinberg is much better than most of us at cranking through impressive calculations.

R. D'Inverno, *Introducing Einstein's Relativity* (Oxford, 1992). A sensible and lucid introduction to general relativity, with solid coverage of the major topics necessary in a modern GR course.

A.P. Lightman, W.H. Press, R.H. Price, and S.A. Teukolsky, *Problem Book in Relativity and Gravitation* (Princeton, 1975). A sizeable collection of problems in all areas of GR, with fully worked solutions, making it all the more difficult for instructors to invent problems the students can't easily find the answers to.

Advanced General Relativity

S. Hawking and G. Ellis, *The Large-Scale Structure of Space-Time* (Cambridge, 1973). An advanced book that emphasizes global techniques, differential topology, and singularity theorems; a classic.

F. de Felice and C. Clarke, *Relativity on Curved Manifolds* (Cambridge, 1990). A mathematical approach, but with an excellent emphasis on physically measurable quantities.

R. Sachs and H. Wu, *General Relativity for Mathematicians* (Springer-Verlag, 1977). Just what the title says, although the typically dry mathematics prose style is here enlivened by frequent opinionated asides about both physics and mathematics (and the state of the world).

J. Stewart, *Advanced General Relativity* (Cambridge, 2003). A short but sweet introduction to some advanced topics, especially spinors, asymptotic structure, and the characteristic initial-value problem.

Mathematical Background

B. Schutz, *Geometrical Methods of Mathematical Physics* (Cambridge, 1980). Another good book by Schutz, this one covering some mathematical points that are left out of the GR book (but at a very accessible level). Included are discussions of Lie derivatives, differential forms, and applications to physics other than GR.

T. Frankel, *The Geometry of Physics: An Introduction* (Cambridge, 2001). A rich, readable book on topics in geometry that are of real use to physics, including manifolds, bundles, curvature, Lie groups, and algebraic topology.

M. Nakahara, *Geometry, Topology and Physics* (Institute of Physics, 2003). An accessible introduction to differential geometry and topology, with an emphasis on topics of interest to physicists.

F.W. Warner, *Foundations of Differentiable Manifolds and Lie Groups* (Springer-Verlag, 1983). A standard text in the field, includes basic topics such as manifolds and tensor fields as well as more advanced subjects.

Specialized Topics

J.D. Jackson, *Classical Electrodynamics* (Wiley, 1999). The classic reference for graduate-level electromagnetism. The problems have left indelible marks on generations of graduate students.

H. Goldstein et al., *Classical Mechanics* (Prentice-Hall, 2002). The classic reference for graduate-level mechanics. An updated edition adds more discussion of nonlinear dynamics.

V.I. Arnold, *Mathematical Methods of Classical Mechanics* (Springer-Verlag, 1989). A scary book for some physicists, but an inspiring treatment of classical mechanics from a mathematically sophisticated point of view. A lot of good differential geometry here.

E.W. Kolb and M.S. Turner, *The Early Universe* (Perseus, 1994). Has become a standard reference for early-universe cosmology, including dark matter, phase transitions, and inflation.

A.R. Liddle and D. Lyth, *Cosmological Inflation and Large-Scale Structure* (Cambridge, 2000). Focusing on inflation and its implications for large-scale structure, gives a careful treatment of cosmological perturbation theory.

B.S. Ryden, *Introduction to Cosmology* (Addison-Wesley, 2002). A very modern and physical introduction to topics in contemporary cosmology, aimed at advanced undergraduates or beginning graduate students.

S. Dodelson, *Modern Cosmology* (Academic Press, 2003). A graduate-level introduction to cosmology, emphasizing cosmological perturbations, large-scale structure, and the cosmic microwave background.

C.M. Will, *Theory and Experiment in Gravitational Physics* (Cambridge, 1993). A useful compendium of alternatives to GR and the experimental constraints on them, including a discussion of the parameterized post-Newtonian formalism.

S.L. Shapiro and S.A. Teukolsky, *Black Holes, White Dwarfs and Neutron Stars: The Physics of Compact Objects* (Wiley, 1983). A self-contained introduction to the physics and astrophysics of compact stars and black holes.

M.E. Peskin and D.V. Schroeder, *An Introduction to Quantum Field Theory* (Westview Press, 1995). Has quickly become the standard textbook in quantum field theory.

J. Polchinski, *String Theory* (Cambridge, 1998). The standard two-volume introduction to modern string theory, including discussions of D-branes and string duality.

C.V. Johnson, *D-Branes* (Cambridge, 2003). A detailed introduction to the extended objects called D-branes, which have become an indispensable part of string theory; prior knowledge of string theory itself is not required.

E.E. Falco, P. Schneider, and J. Ehlers, *Gravitational Lenses* (Springer Verlag, 1999). A thorough introduction to the theory and applications of gravitational lensing.

N.D. Birrell and P.C. Davies, *Quantum Fields in Curved Spacetime* (Cambridge, 1984). The standard book for those who want a practical introduction to quantum field theory in curved spacetime, including the Hawking effect.

R.M. Wald, *Quantum Field Theory in Curved Spacetime and Black Hole Thermodynamics* (Chicago, 1994). A careful and mathematically rigorous exposition of quantum fields in curved spacetimes; if you really want to know what a vacuum state is, look here.

Popular Books

K.S. Thorne, *Black Holes and Time Warps: Einstein's Outrageous Legacy* (W.W. Norton, 1994). Thorne is one of the world's leading researchers in gravitational physics of all kinds, and he offers both a history of work in GR and an introduction to very up-to-date research topics.

R. Geroch, *General Relativity from A to B* (Chicago, 1981). A truly beautiful exposition of the workings of spacetime.

B. Greene, *The Elegant Universe: Superstrings, Hidden Dimensions, and the Quest for the Ultimate Theory* (W.W. Norton, 1999). A timely and personal introduction to the physics of string theory. Not afraid to discuss quite advanced concepts, but aims at a general audience all along; very well written.

A.H. Guth, *The Inflationary Universe: The Quest for a New Theory of Cosmic Origins* (Perseus, 1998). A thorough and lucid introduction to all of modern cosmology, focusing on inflation.

L. Smolin, *Three Roads to Quantum Gravity* (Basic Books, 2002). The "three roads" are string theory, loop quantum gravity, and something more profound; Smolin is a partisan for loop quantum gravity, but the discussion should be interesting for everyone.

G. Kane, *Supersymmetry: Unveiling the Ultimate Laws of Nature* (Perseus, 2001). A nice introduction to supersymmetry, a hypothetical symmetry between bosons and fermions that may be within the reach of particle accelerators soon.

A. Einstein, H.A. Lorentz, H. Weyl, and H. Minkowski, *The Principle of Relativity* (Dover, 1924). Actually not a "popular" book at all; rather, a collection of the original research articles on special and general relativity, translated into English.

A. Pais, *Subtle Is the Lord: The Science and the Life of Albert Einstein* (Oxford, 1983). A scientific biography of Einstein, complete with equations.

Index